PRINCIPLES AND APPLICATIONS OF SPATIAL HEARING

PRINCIPLES AND APPLICATIONS OF SPATIAL HEARING

Miyagi-Zao Royal Hotel, Zao, Japan, 11 – 13 November 2009

Editors

Yôiti Suzuki

Tohoku University, Japan

Douglas Brungart

Walter Reed Army Medical Center, USA

Yukio Iwaya

Tohoku University, Japan

Kazuhiro Iida

Chiba Institute of Technology, Japan

Densil Cabrera

University of Sydney, Australia

Hiroaki Kato

NICT, Japan

World Scientific

NEW JERSEY • LONDON • SINGAPORE • BEIJING • SHANGHAI • HONG KONG • TAIPEI • CHENNAI

Published by

World Scientific Publishing Co. Pte. Ltd.
5 Toh Tuck Link, Singapore 596224
USA office: 27 Warren Street, Suite 401-402, Hackensack, NJ 07601
UK office: 57 Shelton Street, Covent Garden, London WC2H 9HE

British Library Cataloguing-in-Publication Data
A catalogue record for this book is available from the British Library.

PRINCIPLES AND APPLICATIONS OF SPATIAL HEARING

ISBN-13 978-981-4313-87-2
ISBN-10 981-4313-87-4

Printed in Singapore by Mainland Press Pte Ltd.

PREFACE

In November, 2009, an elite group of more than ninety of the world's foremost researchers in spatial hearing assembled in Zao, a secluded mountain resort near Sendai, Japan, to attend the first International Workshop on the Principles and Applications of Spatial Hearing (IWPASH 2009). Although this meeting was the first of its kind ever held in Japan, it was modeled on two earlier meetings that had shared the same goal of bringing the best researchers in the world together in one place in order to obtain a snapshot of the worldwide state-of-the-art in spatial hearing and virtual acoustics. The first of these meetings was the Conference on Binaural Hearing that was co-organized by the Air Force Research Laboratory and Wright State University in 1993. That conference resulted in a comprehensive volume entitled "Binaural and Spatial Hearing in Real and Virtual Environments" (Edited by Robert Gilkey and Tim Anderson, and often affectionately referred to as the "Purple Book") that to this day remains an essential reference for those interested in the field of spatial hearing. Ten years later, in 2003, another International Conference on Binaural and Spatial Hearing was held in Utrecht, Holland, sponsored by the Air Force Office of Scientific Research and TNO Human Factors. That conference resulted in a special issue on Binaural and Spatial Hearing, appearing in Acta Acustica united with Acustica in the spring of 2005.

It was at the conference in Utrecht where Suzuki and Brungart, who are two of the editors of this book and were the co-chairpersons of IWPASH 2009, first discussed the idea of having a third workshop in the series. Our original plan was to hold the conference five years after the second, and to have it in Japan as a way to highlight the accelerating interest in spatial hearing research that was occurring in many Asian countries. By holding the conference in Japan, we were also able to extend a series of spatial hearing conferences organized in Zao by several Japanese institutions including the Research Institute of Electrical Communication, Tohoku University. These "Workshops for Intensive and Comprehensive Discussion on HRTFs" attracted more than fifty attendees in 2002 and more than seventy attendees in 2005, reflecting widespread and fruitful research activity in this field in Japan. Although these workshops were held in Japanese, the one held in 2002 resulted in a special issue of Acoustical Science and Technology (the English language journal of the Acoustical Society of Japan) on Spatial Hearing, appearing in the fall of 2003.[i] A special issue of Applied Acoustics issued in August 2007 represented eight outstanding papers from the 2005 workshop. Based on these successes, the organizers were convinced that the third one should be organized as international conference. Thus, IWPASH 2009 held in Zao really represented the culmination of a fifteen-year trend of international spatial hearing conferences spanning three continents.

When we put the final touches on the conference program in the fall of 2009, we were already confident that we had been successful in our goal of assembling a quorum of the world's most outstanding researchers in the spatial and binaural hearing fields. Despite these high expectations, we were still surprised both by the quality and by the quantity of outstanding research presented at the conference. In total, sixty-six presentations were made, of which twenty were invited lectures. The other forty-six were contributed presentations, of which thirty-three were poster presentations and thirteen were demonstrations. The papers corresponding to these presentations were distributed to the attendees at the conference and published online as an e-proceedings shortly thereafter.[ii] However, there was general agreement among the organizers and attendees of the conference that the material presented at IWPASH warranted publication in expanded form as a book, similar to the one that resulted from the 1993 Spatial Hearing Conference in Dayton, Ohio. Hence we asked all of the invited presenters at the conference to submit an extended version of their proceedings paper, and were extremely pleased that nearly all of them agreed to contribute a chapter to this book. These nineteen chapters are marked with an asterisk in the table of contents. Furthermore, we selected around 55% of the contributed papers that excelled in terms of innovation and scientific quality and asked their authors to submit an extended version of their proceedings paper to serve as a book chapter; twenty contributing presenters provided extended manuscripts, and these chapters were reviewed by the technical program committee members of IWPASH 2009 in order to further improve quality.

Thus, the thirty-nine chapters in this collection provide a snapshot of the research on spatial hearing presented at IWPASH 2009, which we consider to be representative of the state-of-the-art in this important field as of the fall of 2009. In this volume, we have divided these papers into four distinct areas:

1) Exploring new frontiers in sound localization;
2) Measuring and modeling the head-related transfer function;

---

[i] http://www.jstage.jst.go.jp/browse/ast/24/5/
[ii] http://eproceedings.worldscinet.com/9789814299312/toc.shtml

3) Capturing and controlling the spatial sound field; and
4) Applying virtual sound techniques in the real world.

Each of these areas is described in more detail below.

1) Exploring new frontiers in sound localization

Much of the earliest research on spatial hearing was focused on measuring how well human listeners are able to identify the locations of sound sources, and on understanding the acoustic cues listeners use to make these localization judgments. After more than 100 years of research in this area, spatial audio researchers have now achieved a much clearer picture of the fundamental characteristics of human sound localization. This long history of spatial audio research is exploited in the first two chapters in this section, which describe metastudies that have drawn on numerous localization studies conducted over many years to obtain a clearer overall picture of sound localization accuracy in the free field. However, most real-world sound localization takes place in more complicated environments, so the remaining chapters in this section focus on sound localization environments wherein the localization cues are distorted by the presence of reverberation or competing maskers, or where the localization judgments themselves are influenced by context effects or by the listeners' hearing acuity.

2) Measuring and modeling the head-related transfer function

All of the basic perceptual cues that facilitate human sound localization are contained within the direction-dependent transfer function representing the sound propagation path in a free field from a sound source to a listener's ear. This head-related transfer function, or HRTF, is strongly dependent on the unique shapes of an individual listener's ears and head. Thus, highly accurate virtual audio display depends on the ability to capture the HRTF of an individual listener rapidly and accurately, or to adapt the features of a non-individualized HRTF to personalize it for an individual. In the last decade, great strides have been made in improving the techniques used to measure and model the HRTF, and the chapters in this section provide a thorough overview of recent advances in these areas.

3) Capturing and controlling the spatial sound field

Scientific spatial hearing studies must be accompanied by engineering studies to better capture and control the spatial sound field. This important area has grown rapidly in recent years as the technology used to implement spatial sound has advanced, and one of our primary goals in organizing IWPASH was to ensure that this important area was not overshadowed by perceptual or psychoacoustic studies focusing purely on the perception of idealized sound environments by human listeners. Sound field control systems play a critical role in conducting listening experiments to accumulate new scientific knowledge of spatial hearing. Moreover, the development of practical engineering systems to capture and control the spatial sound field is a necessary step that must be achieved before the benefits of spatial and binaural sound research can be accessible to end-users and consumers. All the chapters in this section introduce recent advances in this field, such as state-of-the-art high-definition systems and reasonable design methods for small size systems.

4) Applying virtual sound techniques in the real world
Generation/synthesis of three dimensional virtual sound spaces is a major application area of spatial hearing and spatial sound technologies. As the capabilities of virtual audio synthesis systems have increased and their costs have come down, greater efforts have been made to apply these systems in solving real-world problems. The papers in this section describe novel ways of applying spatial audio technologies for enhancing human interaction, extending the boundaries of musical expression, improving the welfare of the vision- and hearing-impaired, and ensuring the continued peace and security of societies.

We hope and believe that this volume will contribute to further advancement of research in spatial and binaural hearing in two ways. First, we hope it will serve as a reference that provides an insightful overview of the state of research in spatial hearing science and technology in the first decade of the 21st century. Second, and most importantly, we hope it will inspire readers with an interest in spatial and binaural technology to continue to produce even greater innovations in this field that can be presented at future meetings similar to the IWPASH.

Please allow us to express our deepest appreciation to all the chapter authors who kindly accepted our idea of publishing this volume with extended manuscripts of IWPASH 2009 e-proceedings. We also thank the technical program committee members for their efforts in reviewing the selected chapters, and the organizing committee members as well as all the attendees of IWPASH 2009 who contributed to making the conference a success.

October, 2010
Editors:

Yôiti SUZUKI, Douglas BRUNGART,
Yukio IWAYA, Kazuhiro IIDA,
Densil CABRERA and Hiroaki KATO

CONTENTS

* invited chapters

Section 4: Applying Virtual Sound Techniques in the Real World

Section 1

Exploring New Frontiers in Sound Localizations

LOCALIZATION CAPACITY OF HUMAN LISTENERS*

D. HAMMERSHØI

Section of Acoustics, Department of Electronic Systems, Aalborg University, Fredrik Bajers Vej 7 B5, 9220 Aalborg Ø, Denmark

Localization is for some scenarios and situations vital for the success of hearing, e. g. when listening out single sources in multi-source environments, or when navigating primarily by audible information. It is therefore of interest to know the limits of the human localization capacity, and its dependence on e.g. direction and distance. When addressed in laboratory experiments, the significance of other modalities are controlled in different ways, yet figures will inherently reflect properties of the test situation as well. The present paper will discuss the methodologies of localization experiments, generally and by examples.

1. Introduction

Localization is the process of linking spaces, namely that of linking the position of a given physical source, with that of the "position" of the listener's auditory event (if any). For most everyday natural situations, this is a highly meaningful part of the individual's formation of perceptual space. Localization supports navigation, facilitates communication between humans, and is believed to play an active role in attentive and possibly selective listening (e.g. the "cocktail party effect"). The localization performance in itself will rarely sufficiently describe a given human behavior in a given situation, and may often only play a supporting role.

On the other hand, successful localization may also occasionally be taken for granted. This can apply to test situations, where sound sources are spatially arranged to facilitate localization, but where the test doesn't directly address the listeners' localization performance. Other performance measures like, e. g., speech intelligibility may be determined, which may or may not depend on the subject's ability to localize. The results can be analyzed with a view to the position of the source, and may vary with position. But this can also be the case, even when the listener is unable to localize the sound, i. e., when the listener doesn't have a "positioned" auditory event. It is therefore a challenge to evaluate

* Work preliminarily reported at the ISAAR 2009, 26-28 August 2009, Elsinore, Denmark.

the significance of successful localization, when evaluating e. g. virtual acoustics as a supportive tool for e. g. communication and other tasks.

One aspect of this is the natural limits of the human localization capacity, incl. dependence on e. g. direction and distance. When addressed in laboratory experiments, the significance of other modalities are controlled in different ways, yet figures will inherently reflect properties of the test situation as well as the capacity of the individual. The present paper will discuss some of the parameters, which are only indirectly controlled in most localization experiments.

2. Methods

The experiments that give information of the human hearing localization performance naturally divide in three groups; i) explorative studies on the absolute localization performance, (e. g. Gardner [1], Oldfield and Parker [2], Makous and Middlebrooks [3], Butler and Humanski [4], Lorenzi et al. [5], and Van den Bogaert et al. [6]), ii) just noticeable differences in direction (or distance), (e. g. Mills [7], Häusler et al. [8], Morrongiello and Rocca [9], Perrott and Saberi [10], Huang and May [11], Litovsky [12], and Senn et al. [13]), and iii) direct source identification experiments (e. g. Møller et al. [14, 15, 16], Hammershøi and Sandvad [17], Minnaar et al. [18], Brimijoin et al. [19]). While many of the experiments in the first two categories have had the primary objective to explore the human hearing capacity as such, many of the experiments in the latter category serve to evaluate a given performance degradation that may inadvertently be imposed in e. g. inadequate control in recording and playback in e. g. binaural sound systems. For such experiments, it is also desirable that some of the primary parameters, e. g. the number of source locations, number of subjects, etc. represent a reasonably general range of options, for which reason the results also provide information of the localization performance more generally, although secondary to the investigation's objective.

Some of the key methodological aspects are discussed in the following, primarily based on the experience with localization experiments carried out to assess the principles of binaural recording and playback [14, 15], performance of artificial heads [16, 18], and performance of binaural synthesis under "ideal" conditions [17].

2.1. *Source representation*

The number of source positions to include and the position of these, whether physically or virtual, is always a trade-off between the primary objective, the representation required for this, and the wish to give ecologically valid surroundings for the individual during tests. If the objective is to study the capacity of distance estimation by hearing, then sources need to be placed at different distances, but this will experimentally not be possible in real life for many directions. Likewise, if the objective is accurate information on the capacity of the human directional hearing, sources needs to be represented in many different directions, with restricted options for representation of different distances, and for spatial resolution. Even with a relatively sparse representation of sources (and response options), it is possible to detect small differences in e. g. signal processing, as will be shown later in the examples.

2.2. *Visibility*

A separate and important aspect is the visibility of the sources, and what they represent. Nowadays most experiments are computer controlled, which enables the reproduction of the exact same audio stimuli over and over. Sounds are, for the latter reason, without doubt not perceived as authentic or ecologically valid. Even when speech signals are presented, it hardly presents a truly communicative situation, not even if the task challenges the intelligibility. With respect to visibility, source positions are either represented by the sound producing devices, the loudspeakers, or by purpose not made visible. In the latter case subjects have to speculate to the origin of the sound, and also the physical position of the source.

Ideally, if correctly instructed, subjects should relate to the auditory image and its position, disregarding any objects producing the sound or objectifying the possible positions for the source. Since localization is by definition about linking spaces, and to a great extent about finding the source, this ideal can probably not be mastered by the majority of typical listeners participating in given experiments.

There is also strong evidence for the significance of visual information on auditory perception, and its congruence to the task at hand. We are all familiar with the ventriloquist effect, where the spectator is easily fooled into believing that it is the puppet and not the puppeteer, which speaks. This reminds us that auditory perception is not only about the sound that enters our ears and our capacity for hearing it, but also about congruence to other modalities, in particular vision.

2.3. *Response options*

The response options are to a wide extent defined by the scope of investigation and physical setup, but yet the definition of the subject's task and his/her options for response influences the results, e. g. Perrett and Noble [20]. In identification experiments, subjects' are typically instructed to assign the position of the physical source nearest to the position of the auditory event. The task instruction may focus on the fact that the auditory event doesn't necessarily coincide with the position of the physical source. Nowadays most subjects accept this immediately, since most will have heard e. g. stereo reproduction, where the image doesn't coincide with the sound producing device. Yet it influences the subjects' expectations, and it will depend on the scope, whether this is desirable.

One aspect of the response collection method relates to the congruence between auditory and visual space, and sense of self-center. Some studies (e. g. Arthur et al. [21]) have discussed the significance of having either ego-centered response options, e. g. head turning (Makous and Middlebrooks [3], Carlile et al. [22]), using gaze direction (Hofman et al. [23], Populin [24]), calling out coordinates (Wightman and Kistler [25, 26]), versus exo-centric options, e. g. using touch screens or spheres (Gilkey et al. [27]), tablets (Møller et al. [14, 15, 16], Hammershøi and Sandvad [17], Minaar et al. [18]), computer displays (Iida et al. [28], Savel [29]), or paper drawings). It has been demonstrated, that although we interact seemingly effortless with objects in the physical world, there are sizeable misperceptions of spatial relationships, even in the nearby environment. Yet the effect of this will depend on instructions, the level of difficulty in the task presented, and to which extend the subject succeed in putting him/herself in the centre of the given response options.

3. EXAMPLE I: TEST IN STANDARD LISTENING ROOM

The test paradigm used in Møller et al. [14, 15, 16] and Minaar et al. [18] is an example of localization tests, where the test is carried out in acoustically "normal" conditions, with sound source positions at different directions and distances, with a fairly simple task for the test person, but with options for detecting even mild deteriorations of the sound reproduction. Figure 1 illustrates the test scenario.

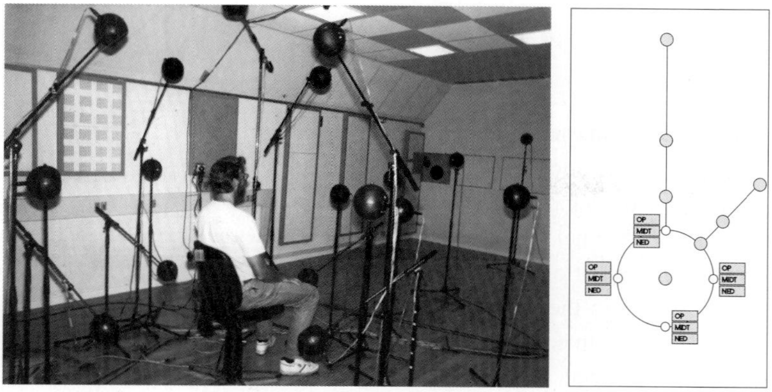

Figure 1. Left: Photo of setup for localization experiments with 19 loudspeakers. Right: Sketch appearing on tablet for response collection. Only grey zones represented valid response options. The grey boxes represented positions at different elevations, "OP" being 45° above horizontal plane, "MIDT" being in the horizontal plane, and "NED" being 45° below horizontal plane. Subjects were instructed only to look down on the tablet, when response was required, and maintain upright position during stimulus playback. This was monitored. From Møller et al. [14].

The test paradigm illustrated in Figure 1 was used for various tests of the "authenticity" of binaural reproduction: Whether individual (the person's own binaural recordings) could provide a localization performance similar to real life, whether non-individual (binaural recordings from other subjects) could, whether artificial head recordings could, whether the headphone reproduction needed individual (personal) equalization, and more.

An example of accumulated test results for the given test paradigm is given in Figure 2.

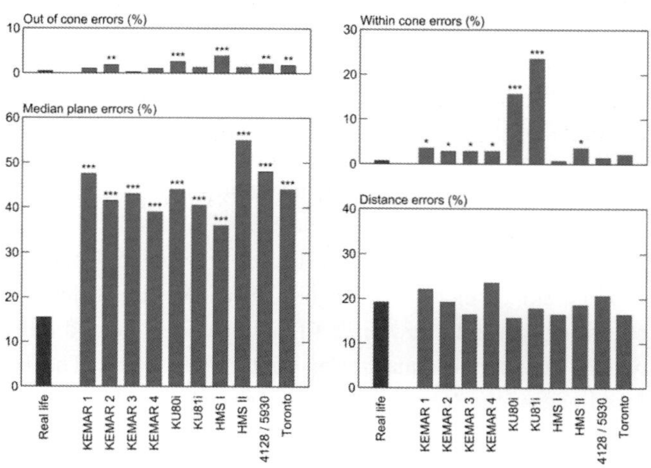

Figure 2. Accumulated responses (percentage "correct") from localization tests with various artificial recordings vs. real life listening. Data from Møller et al. [16].

Figure 2 summarizes the results for listening tests with different artificial heads, vs. the localization performance of the same listeners in the relevant real life situation (with sound played back over loudspeakers in the same setup).

From top left panel it can be seen, that a few of the artificial heads provide significantly more out of cone errors. This is a relatively severe error, since the cones represented in the setup is 45° apart (in horizontal plane), and the errors thus represent a confusion of sources relatively far apart. Such confusion would normally indicate that the arrival time of sound at left versus right ear is incorrect, which could suggest that the artificial head has an inappropriate geometry.

From left lower panel it can be seen that the artificial heads without exception give more median plane errors than the corresponding real life test. These errors represent confusions between sources in the median plane, which would indicate that spectral fingerprint of the signals do not well match what the listener normally hears. This is to a great extent controlled by the detailed geometry of the outer ear, but could also be due to imperfect headphone equalization (if not individually designed).

From the top right panel it can be seen that some artificial heads has a high number of "within-cone" errors. These errors represent confusions between sources on the cone that extends out from the listeners ears' at 45° elevation angle, e. g. the "left low" and "left high" direction.

One can again speculate to the origin of these confusions, and it is remarkable that it is the artificial heads without torso that have the most of this type of error. This suggests that the torso and the related shoulder reflections are important for sound localization for certain directions.

From the lower right panel it can be seen that all heads provided a near real distance perception. In view of the magnitude of other types of errors this would seem to indicate that distance perception is not controlled by features of the head, torso or ear, but most probably by the acoustics of the room.

In summary, the test scenario proved useful in detecting even small differences in processing, incl. the significance of individual versus non-individual recording, possible flaws in artificial head design, and individual versus non-individual headphone equalization (not shown here). The latter is normally considered one of the weaker compromises to make, but results from Møller et al. [15] showed that the difference was significant when tested.

4. EXAMPLE II: ANECHOIC TEST

Another localization test scenario is illustrated in Figure 3.

Fig. 3: Left: Photo of setup for localization experiments with binaural synthesis. From Hammershøi and Sandvad [17].

The test scenario presented in Figure 3, was used to assess the performance of "the best possible" binaural synthesis. This is theoretically obtained using the individuals' own head-related transfer functions (HRTFs) in the synthesis, and using individual headphone equalization.

The synthesis was carried out assuming only the direct sound transmission path from loudspeaker to listener, and did not include any representation of reflections. This was done to avoid the influence of possible short-comings in the room analysis and in the representation of room reflections and late reverberation. This has the consequence that the synthesis effectively simulates an anechoic environment, which is unnatural to most listeners, both from an acoustical and visual point of view.

Experiments included listening to binaural signals reproduced over headphones, and to the real life setup for two types of stimuli, noise and speech. The results of the localization tests in anechoic chamber (Figure 4) indicate that more errors are made with binaural synthesis, than in the corresponding real life situation.

Most errors are generally made between source directions that are within the cones of confusion represented in the setup. In both situations most confusion exists between directions in the upper hemisphere. This can be explained by the fact that the head-related transfer functions are quite similar in this region, thus the hearing has only few cues available for the localization process.

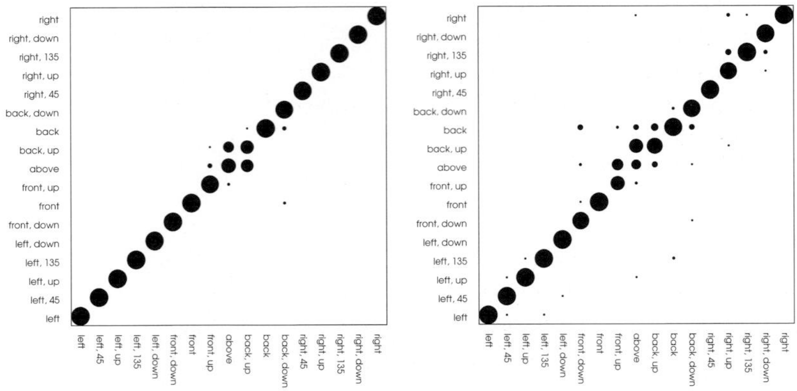

Figure 4. Results from listening tests in anechoic chamber. Left: Stimulus vs. response for the real life playback situation. Right: Stimulus vs. response for playback of individual binaural synthesis. The area of each circle is proportional to the number of responses it holds. From Hammershøi and Sandvad [17].

There is also a slight over-representation of errors going from front hemisphere positions to rear hemisphere positions, again dominated by upper hemisphere confusions. Whether this generally describes the human hearing, or whether it relates to the specific setup and task is harder to determine. If the subject doesn't see him/herself in the centre of the setup distortion can occur.

In the quest for perfection of binaural synthesis (the original motivation for the study), explanations for the difference in number of errors in the two situations are also called for.

One methodological aspect relates to the "perfectly" dry simulation. With the binaural synthesis, there are really no reflections from the room, whereas in real life, any anechoic chamber will have a minimum of reflections from the setup, floor, etc. With no room-related information at all to support that the source is positioned "out there", it is possible that some sounds were perceived within the head of the listener. This could explain the (few) responses, which shifted more than 45° horizontally.

In the design of response options, it was considered, whether the subject should have the option of indicating that he/she heard the sound within the head. One reason for not including this option anyway was that there is little ecological validity in the localization process, if the subject is left with such unnatural options for possible source positions.

This illustrates very well the most difficult challenge in the design of localization experiments: On one hand you investigate the success with which the subject successfully links the physical world with the perceptual world. On the other hand you want the uncensored report of what the subject hears (characteristics of the auditory event) in given situations. But just by asking, you bias perception.

Acknowledgments

The author would like to acknowledge the many fruitful discussions with colleagues at Aalborg University on the subject of localization experiments, this includes in particular Henrik Møller, Michael Friis Sørensen, Clemen Boje Larsen (former Jensen), and Jesper Sandvad.

References

1. M. B. Gardner, "Some monaural and binaural facets of median plane localization," *J. Acoust. Soc. Am.* **54**, 1489 (1973).
2. S. R. Oldfield and S. P. A. Parker, "Acuity of sound localization: a topography of auditory space. I. Normal hearing conditions," *Perception* **13**, 581 (1984).
3. J. C. Makous and J. C. Middlebrooks, "Two-dimensional sound localization by human listeners," *J. Acoust. Soc. Am.* **87**, 2188 (1990).
4. R. A. Butler and R. A. Humanski, "Localization of sound in the vertical plane with and without high-frequency spectral cues," *Perception & Psychophysics* **51**, 182 (1992).
5. C. Lorenzi, S. Gatehouse, and C. Lever, "Sound localization in noise in normal-hearing listeners," *J. Acoust. Soc. Am.* **105**, 1810 (1999).
6. T. Van den Bogaert, T. J. Klasen, M. Moonen, L. Van Deun, and J. Wouters, "Horizontal localization with bilateral hearing aids: without is better than with," *J. Acoust. Soc. Am.* **119**, 515 (2006).
7. A. W. Mills, "On the minimum audible angle," *J. Acoust. Soc. Am.* **30**, 237 (1958).
8. R. Häusler, S. Colburn, and E. Marr, "Sound localization in subjects with impaired hearing. Spatial-discrimination and interaural discrimination tests," *Acta Otolaryngol. Suppl.*, **400**, 1 (1983).
9. B. S. Morrongiello and P. T. Rocca, "Infants localization of sounds within hemifields— estimates of minimum audible angle," *Child. Dev.* **61**, 1258 (1990).

10. D. R. Perrott and K. Saberi, "Minimum audible angle thresholds for sources varying in both elevation and azimuth," *J. Acoust. Soc. Am.* **87**, 1728 (1990).
11. A. Y. Huang and B. J. May, "Spectral cues for sound localization in cats: effects of frequency domain on minimum audible angles in the median and horizontal planes," *J. Acoust. Soc. Am.* **100**, 2341 (1996).
12. R. Y. Litovsky, "Developmental changes in the precedence effect: estimates of minimum audible angle," *J. Acoust. Soc. Am.* **102**, 1739 (1997).
13. P. Senn, M. Kompis, M. Vischer, and R. Häusler, "Minimum audible angle, just noticeable interaural differences and speech intelligibility with bilateral cochlear implants using clinical speech processors," *Audiol. Neurootol.* **10**, 342 (2005).
14. H. Møller, M. F. Sørensen, C. B. Jensen, and D. Hammershøi, "Binaural technique: Do we need individual recordings?," *J. Audio Eng. Soc.* **44**, 451 (1996a).
15. H. Møller, C. B. Jensen, D. Hammershøi, and M. F. Sørensen, "Using a typical human subject for binaural recording," *Proc. 100th Audio Eng. Soc. Conv. Copenhagen*, May 11-14, 1996, preprint 4157 (1996b). Abstract in *J. Audio Eng. Soc.* **44**, 632.
16. H. Møller, D. Hammershøi, C. B. Jensen, and M. F. Sørensen, "Evaluation of artificial heads in listening tests," *J. Audio Eng. Soc.* **47**, 83 (1999).
17. D. Hammershøi and J. Sandvad, "Binaural auralization. Simulating free field conditions by headphones," *Proc. 96th Audio Eng. Soc. Conv. Amsterdam*, Feb. 26 – Mar. 1, 1994, preprint 3863 (1994). Abstract in *J. Audio Eng. Soc.* **42**, 395.
18. P. Minnaar, S. K. Olesen, F. Christensen, and H. Møller, "Localization with binaural recordings from artificial and human heads," *J. Audio Eng. Soc.* **49**, 323 (2001).
19. W. O. Brimijoin, D. McShefferty, and M. A. Akeroyd, "Auditory and visual orienting responses in listeners with and without hearing-impairment," *J. Acoust. Soc. Am.* **127**, 3678 (2010).
20. S. Perrett and W. Noble, "Available response choices affect localization of sound," *Perception and Psychophysics* **57**, 150 (1995).
21. J. C. Arthur, J. W. Philbeck, J. Sargent and S. Dopkins, "Misperception of exocentric directions in auditory space," *Acta Psychologica* **129**, 72 (2008).
22. S. Carlile, P. Leong, and S. Hyams, "The nature and distribution of errors in sound localization by human listeners," *Hearing Res.* **114**, 179 (1997).
23. P. M. Hofman, J. G. A. Van Riswick, and J. Van Opstal, "Relearning sound localization with new ears," *Nature Neuroscience* **1**, 417 (1998).
24. L. C. Populin, "Human sound localization: measurements in untrained, head-unrestrained subjects using gaze as a pointer," *Exp. Brain Res.* **190**, 11 (2008)

25. F. L. Wightman and D. J. Kistler, "Headphone simulation of freefield listening. I. Stimulus synthesis," *J. Acoust. Soc. Am.* **85**, 868 (1989).
26. F. L. Wightman and D. J. Kistler, "The dominant role of lowfrequency interaural time differences in sound localization," *J. Acoust. Soc. Am.* **91**, 1648 (1992).
27. R. H. Gilkey, M. D. Good, M. A. Ericson, J. Brinkman, and J. M. Stewart, "A pointing technique for rapidly collecting localization response in auditory research," *Behavior Research Methods, Instruments, and Computers* **27**, 1 (1995).
28. K. Iida, M. Itoh, A. Itagaki, and M. Morimoto, "Median plane localization using a parametric model of the head-related transfer function based on spectral cues," *Applied Acoustics* **68**, 835 (2007).
29. S. Savel, "Individual differences and left/right asymmetries in auditory space perception. I. Localization of low-frequency sounds in free field," *Hearing Research* **255**, 142 (2009).

A META-ANALYSIS OF LOCALIZATION ERRORS MADE IN THE ANECHOIC FREE FIELD*

V. BEST
University of Sydney, Sydney, NSW, Australia

D. S. BRUNGART
Walter Reed Army Medical Center, Washington, DC, USA

S. CARLILE, C. JIN
University of Sydney, Sydney, NSW, Australia

E. A. MACPHERSON
University of Western Ontario, London, ON, Canada

R. L. MARTIN, K. I. MCANALLY
Defence Science and Technology Organisation, Fishermans Bend, VIC, Australia

A. T. SABIN
Northwestern University, Evanston, IL, USA

B. D. SIMPSON
Wright-Patterson Air Force Base, Dayton, OH, USA

This chapter briefly summarizes the results of a meta-analysis that examined auditory localization accuracy for more than 80,000 trials where brief broadband stimuli were presented anechoically in one of four different laboratories. The analyses were aimed at creating a comprehensive map of localization accuracy as a function of sound source location, and characterizing the distribution of responses along the "cone of confusion". The results reveal trends in auditory localization whilst minimizing the influence of different experimental methodologies and response methods.

* This work is supported by AFOSR and a University of Sydney Postdoctoral Research Fellowship to VB.

1. Introduction

Human listeners are quite accurate at localizing broadband sounds in the free field, and do so by making use of acoustical cues that vary with the direction of the source relative to the head. Binaural differences in arrival time and intensity provide robust cues enabling sound source localization in the lateral dimension. However, these cues are inherently ambiguous with respect to locations in three-dimensional space and there exists a "cone of confusion" along which locations share the same binaural cues but differ in their vertical and front-back location [1-3]. Resolution of this ambiguity comes in part from direction-dependent filtering by the outer ear, head, and torso that results in spectral cues that can help pinpoint the location of the sound source. Spectral cues appear to be less reliable than binaural cues, and in many listening situations listeners make errors within the cone of confusion (sometimes large errors, where the sound is perceived to be in the incorrect front or back hemifield). Under ideal listening conditions, however, these large errors are quite rare (3-5% of trials [4-6]).

There have been only a few attempts to map auditory localization accuracy as a function of the location of the sound source [4-7]. Even when the methodological requirements for presenting sounds from a range of near-continuous locations are met, a limitation has been the extremely large amount of data needed to build a clear picture. The detailed characterization of front-back confusions is particularly difficult based on the data of single studies because they occur so rarely. To address this issue, a meta-analysis of localization accuracy and response patterns was conducted on the combined anechoic free-field anechoic localization data from multiple laboratories.

2. Data Sets

Sound localization data were obtained from four research laboratories located at different institutions in Australia and the USA. In total, 82568 localization responses from 161 different listeners were obtained. The data were collected under broadly similar conditions, involving the presentation of broadband stimuli in an anechoic environment and the acquisition of responses via a manual pointing method. More details can be found in previous publications [4, 8-10].

Wright-Patterson Air Force Base (Dayton, OH, USA): This data set consisted of 39930 trials from 63 different listeners. The stimuli were 250-ms noises (20-ms ramps) and were presented at 65 dB SPL. Stimuli were presented from one of 244 loudspeakers (Bose 11-cm full-range) distributed at 15° intervals on a geodesic sphere of radius 2.3 m. Only locations above -45° elevation were included. Subjects stood on a platform in the center of the sphere and responded by pointing a hand-held wand. The orientation of the wand was tracked (Intersense IS-900) and used to provide visual feedback by illuminating an LED on the loudspeaker nearest the indicated position.

Wright State University (Dayton, OH, USA): This data set consisted of 3585 trials from 3 different listeners. These trials were collected in the same facility as the Wright-Patterson data set, and the stimuli were 320-ms noises (10 ms ramps) presented at 65 dB SPL. However, responses were made using the God's-Eye Localization Pointing Technique, which required listeners to move a stylus to the position on the surface of a sphere that most closely matched the perceived location of the sound source relative to the listener's head [11].

University of Sydney (Sydney, NSW, Australia): This data set consisted of 23104 trials from 60 different listeners. The stimuli were 150-ms noises (10-ms ramps) and were presented at 70 dB SPL. Stimuli were presented in the dark from a single loudspeaker (VIFA-D26TG-35) that was moved robotically to one of 76 locations on a sphere of radius 1m. Only locations between -45° and 45° elevation were included. Subjects stood on a platform in the center of the sphere and responded by pointing their noses towards the sound and pressing a button. A head-mounted electromagnetic tracker (Polhemus Fastrak) recorded the orientation of the head upon response.

Defence Science and Technology Organisation (Melbourne, VIC, Australia): This data set consisted of 11231 trials from 29 different listeners. The stimuli were 328-ms noises (20-ms ramps) or 41-ms noises (20-ms ramps) and were presented at 60 dB SPL. Stimuli were presented from a single loudspeaker (Bose FreeSpace tweeter) that was moved robotically to one of 448 locations on a sphere of radius 1m. The loudspeaker was obscured by an acoustically-transparent cloth. Only locations between -47.6° and 80° elevation were included. Subjects were seated on a swivel chair in the center of the sphere and responded by pointing a head-mounted laser onto the cloth lining the sphere and pressing a button. A head-mounted electromagnetic tracker (Polhemus Fastrak) recorded the orientation of the head upon response.

Kresge Hearing Research Institute (Ann Arbor, MI, USA): This data set consisted of 4718 trials from 6 different listeners. The stimuli were 250-ms noises (20 ms ramps) in duration and were presented at an individual sensation level that varied from trial-to-trial between 30 and 60 dB. Stimuli were presented in the dark from one of two identical loudspeakers on a hoop that was moved robotically to one of 200 locations on a sphere of radius 1.2m. Subjects stood on a platform in the center of the sphere and responded by pointing their heads towards the sound and pressing a button. A head-mounted electromagnetic tracker (Polhemus Fastrak) recorded the orientation of the head upon response.

3. Co-ordinate System

Target and response locations on the sphere were described using the interaural polar co-ordinate system. This system comprises the lateral angle θ, which ranges from -90° to +90° and the polar angle ϕ, which ranges from -180° to +180° (Fig. 1). The interaural polar co-ordinate system is particularly convenient for describing sound localization data because it captures these two dimensions for which binaural (lateral angle) and spectral cues (polar angle) are most informative.

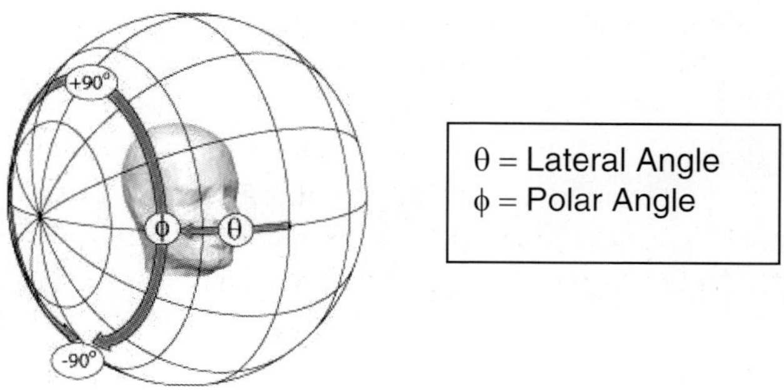

Figure 1. The interaural polar co-ordinate system.

4. Distribution of Target Locations

By combining data from different laboratories using different stimulus positions, we ultimately obtained relatively good coverage of the sphere. The contour plot in Fig. 2 shows the number of trials collected at each point across the sphere. The plot was constructed by first dividing the surface of the unit sphere into a grid with 10° resolution in polar and lateral angle, and then counting all trials in the combined data set that fell within within 10° of great-circle arc from each grid point. The results show that the highest density of stimulus locations occurred directly in front and behind the listener. However, coverage elsewhere on the unit sphere was fairly uniform, with a minimum of roughly 400-600 trials occurring at virtually all grid locations in the analyzed region above the -45° in elevation. This kind of continuous distribution is something that individual studies have generally not been able to provide.

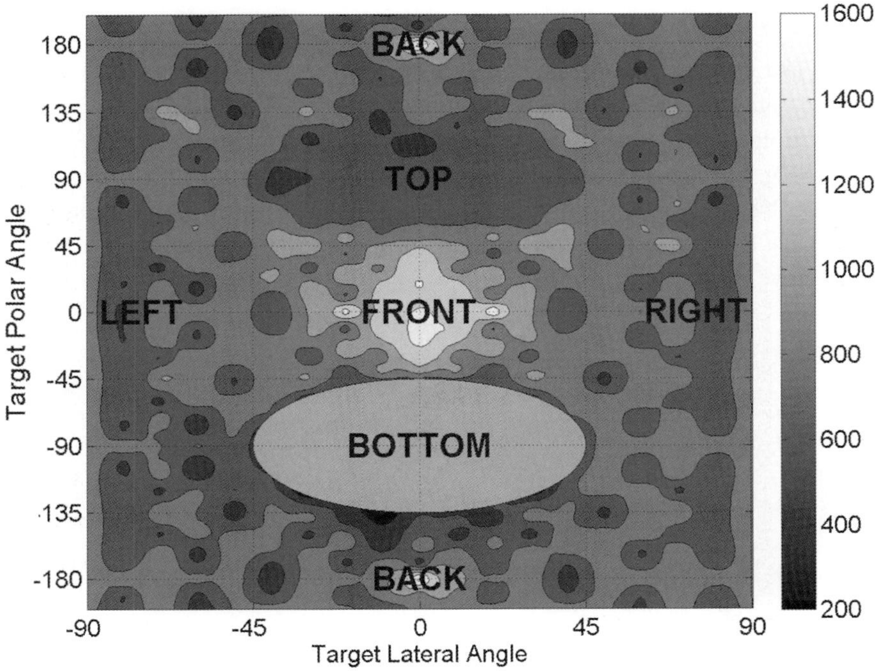

Figure 2. Contour plot showing the distribution of target locations in the combined data set (shade bar shows absolute numbers of trials).

5. Example Analyses

5.1. *Distribution of Errors and Response Biases*

One type of analysis that is made possible by the large number of data points contained in the combined data set is a high-resolution analysis of the relationship between the location of a sound source and the accuracy with which a listener is able to localize that sound. The contour plot shown in Fig. 3 shows how the great-circle localization error in degrees varied as a function of sound source location in the interaural polar co-ordinate system. The plot was constructed by first dividing the surface of the unit sphere into a 10° by 10° grid (as for Fig. 1), and then finding all trials in the combined data set that fell within 10° of great-circle arc of each grid point. In cases where no significant difference was found between the overall angular errors for grid points located at mirror image locations across the median plane (using a Bonferonni-corrected two-tailed t-test), left-right symmetry was assumed and the data were pooled across the two mirror image locations in the calculation of the mean error metric.

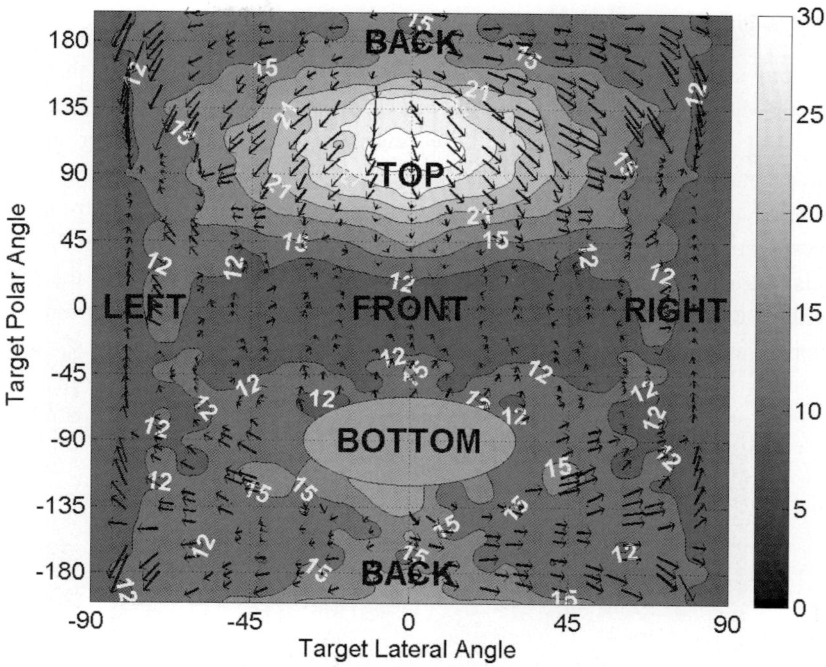

Figure 3. Contour plot showing the mean great-circle error (shade bar in degrees) as a function of source position. Superimposed arrows indicate the size and direction of local population response biases for different source positions.

The arrows in the figure provide a rough estimate of the directional bias in the population responses for each source location. For each point on the 10° by 10° grid, an arrow has been drawn from the mean target vector across all trials within that bin to the mean response vector for all trials within that bin.

From the figure, a number of important trends are apparent:

- Localization accuracy is best for sound sources in front of the listener
- In the front hemisphere, optimal performance extends over polar angles from -45° to +30°
- Localization accuracy is worst, by a wide margin, for sounds above and slightly behind the listener
- In the front hemisphere, responses are biased towards locations just above the horizontal plane
- In the rear hemisphere, responses are biased forward and toward the left and right poles

5.2. *Hemisphere Reversals*

Previous studies have noted the occurrence of particular large errors that have the correct lateral angle but fall in the wrong front-back (or up-down) hemifield [4-6, 12]. As discussed above, these reversals are infrequent under optimal experimental conditions. Using the strictest definition of these errors, which includes any response that crosses the frontal or horizontal plane, front-back reversals occurred on 6.6% of trials and up-down reversals occurred on 3.7% of trials in this large data set. When small errors occurring within 10° of the plane dividing the hemifields were excluded, as is commonly done, the front-back reversal rate came to 5.9% and the up-down reversal rate to2%.

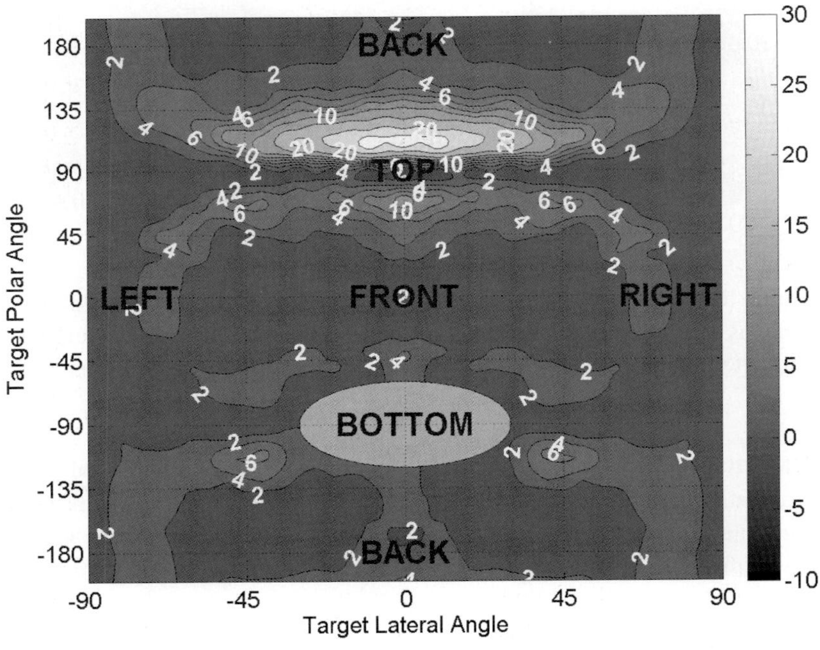

Figure 4. Contour plot showing the frequency of front-back reversals (shade bar in percent of trials) as a function of source position. Target positions within 10° of the frontal plane have been excluded.

Front-back reversals were tallied and plotted as a percentage of trials for each point in the same 10° by 10° grid used to generate Figs. 2 and 3. Fig. 4 indicates that front-back reversals are very rare across most possible source locations. However, the number of these errors is quite high for sound sources located above and behind the listener (this is consistent with the findings of [6]). With the exception of these high locations in the back, there is not a very strong bias in favor of either front to back or back to front confusions.

5.3. *Polar Angle Response Patterns*

Although front-back reversals are commonly described, this may be simply because they are large and obvious, or because many early studies were restricted to the horizontal plane. It is possible that front-back reversals are just one example of a more general kind of error in which listeners respond randomly on the cone of confusion. However, there is certainly a prevailing notion amongst researchers in the field that front-back reversals are a special class and moreover that they tend to be "perfect" reversals, in which the hemifield is confused but the vertical component of the polar angle is preserved (*i.e.*, they are mirror reversed). A similar observation has been made about up-down reversals.

To shed light on this matter, the data were collapsed across lateral angle, and histograms of response polar angle were generated for each target polar angle (in 10° bins). Histograms were normalized by the number of contributing trials and log scaled. Fig. 5 shows six example histograms for target polar angles in the frontal hemifield (top row) and the rear hemifield (bottom row). The vertical lines in each panel represent the target polar angle (solid line), the perfect front-back reversal angle (dashed line), and the perfect up-down reversal angle (dotted line). All histograms showed a primary peak at the target polar angle, and in some cases there was evidence of a secondary peak at the front-back reversal location. Very rarely did a secondary peak occur at the up-down reversal angle.

When the histograms for the 36 different target polar angles were shifted circularly so that they were centred on the target polar angle and pooled, a strong composite peak was visible (Fig. 6, left panel). When the histograms were shifted so that they were centred on the perfect front-back angle and pooled, a smaller composite peak emerged (Fig. 6, middle panel). No peak was visible when the histograms were shifted so that they were centred on the perfect up-down angle (Fig. 6, right panel). In fact, because the perfect up-down reversal location differs from the perfect front-back reversal location by 180°, then by definition the up-down figure is a shifted version of the front-back figure. This analysis confirms our observations that perfect front-back reversals occurred at least slightly more frequently than would be predicted simply from errors distributed randomly around the target location, and dominated perfect up-down reversals.

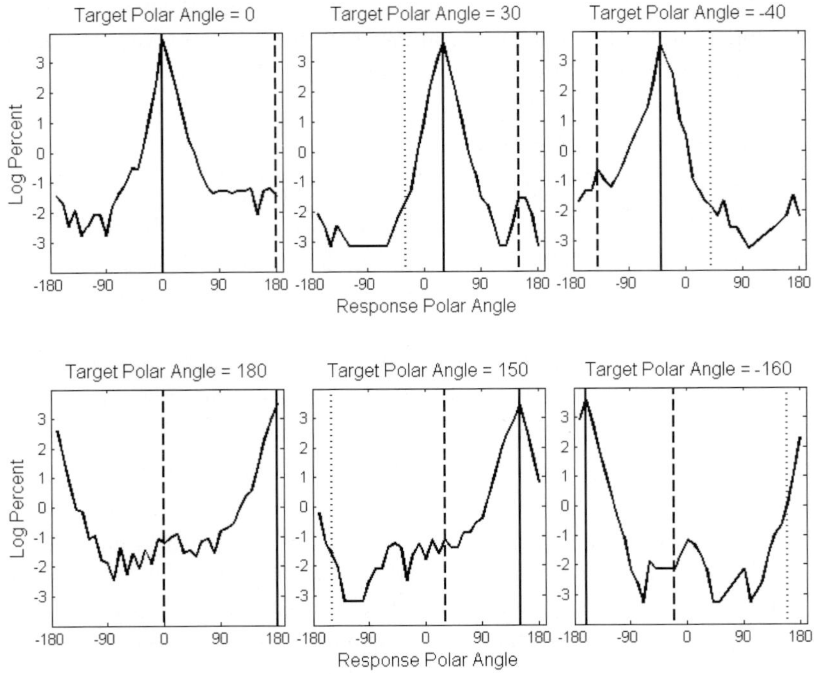

Figure 5. Example histograms of polar angle responses, expressed in log percent. Vertical lines show the polar angle of the target (solid), the perfect front-back reversal angle (dashed), and the perfect up-down reversal angle (dotted).

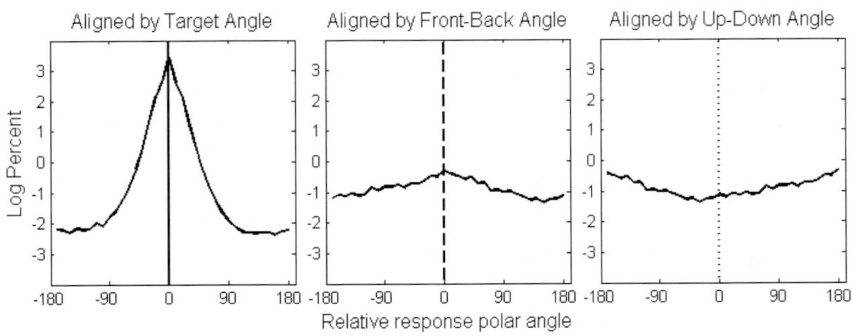

Figure 6. Composite histograms of polar angle responses, expressed in log percent. Histograms for each polar angle were pooled after shifting to align them by the target polar angle (left), the perfect front-back reversal location (middle), or the perfect up-down reversal location (right). Plotted are means across the 36 polar angles.

We conclude that while errors along the cone of confusion are dominated by local errors centered on the target location, the large polar angle errors that are occasionally observed show a slight tendency towards being "perfect" front-back reversals. This result has several possible interpretations. For one, it may suggest simply that these pairs of locations give rise to spectral cues that are highly similar (by some measure). Alternatively, the result may be taken to imply that there are independent decisions made about lateral angle, front/back hemifield and vertical position as has been suggested previously [13, 14].

References

1. A.W. Mills, Auditory localization, in *Foundations of Modern Auditory Theory*, J.V. Tobias, Editor, Academic Press, New York, 303 (1972).
2. H. Wallach, On sound localization, *J. Acoust. Soc. Am.* **10**, 270 (1939).
3. J. Blauert, *Spatial Hearing: The Psychophysics of Human Sound Localization*, MIT Press, Cambridge (1997).
4. S. Carlile, P. Leong, and S. Hyams, The nature and distribution of errors in sound localization by human listeners, *Hearing Res.* **114**, 179 (1997).
5. F.L. Wightman and D.J. Kistler, Headphone simulation of free field listening II: Psychophysical validation, *J. Acoust. Soc. Am.* **85**, 868 (1989).
6. J. Makous and J.C. Middlebrooks, Two-dimensional sound localization by human listeners, *J. Acoust. Soc. Am.* **87**, 2188 (1990).
7. S.R. Oldfield and S.P. Parker, Acuity of sound localisation: a topography of auditory space. I. Normal hearing conditions, *Perception* **13**, 581 (1984).
8. R. Bolia, W. D'Angelo, P. Mishler, and L. Morris, Effects of hearing protectors on auditory localization in azimuth and elevation, *Human Factors* **43**, 122 (2001).
9. A. Sabin, E. Macpherson, and J. Middlebrooks, Human sound localization at near-threshold levels, *Hearing Res.* **199**, 125 (2005).
10. R.L. Martin, D.B. Watson, S.E. Smith, K.I. McAnally, and D.L. Emonson, Effect of normobaric hypoxia on sound localization, *Aviat. Space Environ. Med.* **71**, 991 (2000).
11. R.H. Gilkey, M.D. Good, M.A. Ericson, J. Brinkman, and J.M. Stewart, A pointing technique for rapidly collecting localization responses in auditory research, *Behav. Res. Meth. Instr. Comp.* **27**, 1 (1995).
12. F.L. Wightman and D.J. Kistler, The dominant role of low-frequency interaural time differences in sound localization, *J. Acoust. Soc. Am.* **91**, 1648 (1992).
13. F. Asano, Y. Suzuki, and T. Sone, Role of spectral cues in median plane localization, *J. Acoust. Soc. Am.* **88**, 159 (1990).
14. E.H.A. Langendijk and A.W. Bronkhorst, Contribution of spectral cues to human sound localization, *J. Acoust. Soc. Am.* **112**, 1583 (2002).

AUDITORY PERCEPTION IN REVERBERANT SOUND FIELDS AND EFFECTS OF PRIOR LISTENING EXPOSURE

P. ZAHORIK*, E. BRANDEWIE AND V. P. SIVONEN†

Department of Psychological and Brain Sciences, University of Louisville
Louisville, KY 40205, USA

Although sound reflective surfaces are ubiquitous in everyday listening environments and their acoustical effects (e.g. echoes and reverberation) are easily identifiable through physical measurement, they are often perceptually unnoticed in normal listening situations. Previous research suggests that this perceptual suppression of reflected sound may be due to a type of adaptation or calibration that results through prior listening exposure to the particular spatial configuration of the source and reflection(s). Because this suppression effect has been studied almost exclusively using only a single simulated reflection, it is important to determine the extent to which the effect generalizes to more natural listening conditions, such as rooms with multiple reflections and reverberation. Here we summarize recent and ongoing research in our laboratory that addresses this issue and demonstrates that the effect has important implications for complex listening tasks in reverberant rooms such as sound localization, speech intelligibility, and loudness comparisons.

1. Introduction

In everyday life, the sound that reaches our ears is a complex combination of acoustic waves propagating through the environment, often from many sources. The problem of segregating these combined waves into meaningful auditory perceptual events is one of the central issues in psychological acoustics, and spatial hearing plays an important role in solving this problem. One specific aspect of this scene segregation problem results from acoustic reflections caused by objects and surfaces in the environment. How are we able to segregate a source of sound from its reflection?

It is clear that the auditory system and brain effectively solve this problem, since we are seldom aware of any competing sound information from acoustic reflections, even though such competition is easily identifiable through physical measurement. A phenomenon known as the *precedence effect* is often used to explain the dominance of the first arriving wavefront in specifying the spatial

* Corresponding author. Email: pavel.zahorik@louisville.edu

† Current affiliation: Dept. of Signal Processing & Acoustics, Helsinki Univ. of Technology, Finland

position of the sound source and the apparent suppression of information from the reflected sound. This effect has received considerable scientific study throughout the past half century and has now come to be associated with a suite of phenomena all related to auditory perception in acoustic environments with echoes. While some of the original observations of this effect were concerned exclusively with the localization of the sound source [1], other research clearly demonstrates the significance of this effect for an important segregation application: understanding speech in reverberant environments [2]. Brainstem-level neural correlates of the precedence effect have also been identified for a variety of species (see [3] for review).

More recent results [4,5] challenge the view that precedence is subserved solely by low-level neural processes, suggesting instead that portions of the precedence effect may be products of more central brain mechanisms that implement a form of "smart" echo suppression in which models of the acoustic environment are dynamically constructed and interpreted [5]. Aspects of this adaptive type of echo suppression are evident in the results from a number of experiments that show an increase, or "buildup," of echo suppression when prior listening exposure (duration of exposure typically on the order of tens of seconds) to the acoustic environment is provided [6-8], and also dramatic decreases, or "breakdown," in echo suppression when abrupt changes are made to the spatial configurations [4], relative delays [5], spectral characteristics [9], or temporal patterns [10] of source and reflection that would be implausible in natural listening situations [5,10-12]. Precedence effect buildup has also been shown to be somewhat greater when the reflection (simulated) occurs on the listener's left rather than right side, which suggests a role of cortical level processing in the effect [13], and unilateral ablation of the auditory cortex in cats has been shown to impair the echo suppression observed in the precedence effect for echo locations ipsilateral to the lesion [14]. It is important to note that buildup effects do not appear to be mediated by cognition, however, since they have been shown to be resistant to practice and learning [11]. These results all suggest that high-level unconscious perceptual processing plays an important role in perceptual echo suppression by representing and evaluating physically plausible models of the listening situation.

Because this model-building process has been evaluated only under restricted stimulus conditions typically containing a single reflection, it is important to determine whether these results generalize to more complex and more realistic environments such as rooms with multiple echoes and reverberation, and how aspects of the environment's acoustical properties may impact the effect. It is additionally important to determine how this suppression may affect other listening tasks that can potentially depend on the processing of indirect sound, such as in understanding speech or determining the loudness of sounds in reverberant environments.

This chapter summarizes recent and ongoing work in our laboratory that addresses these issues and demonstrates that room acoustics and dynamic echo suppression have important implications for complex listening tasks in reverberant rooms such as sound localization, speech intelligibility, and loudness comparisons. The work makes extensive use of binaural technology for effectively simulating and rapidly switching between various room acoustic environments. The simulation techniques have been described in detail elsewhere [15], and although significantly simplified relative to state-of-the-art methods used primarily for simulating concert hall acoustics (see [16] for a comprehensive review), these techniques allow complete control over the simulation and have been shown to yield results that are perceptually similar to those derived from measurements in a real room [15].

2. Echo Suppression Buildup

Previous studies demonstrating dynamic buildup in the precedence effect with repeated exposure to a source and a single reflection provide important insights into the nature of the neural mechanisms subserving echo suppression in humans [4,7,13]. An important next step is to probe the generality of this result. Does precedence effect buildup also occur in more natural listening situations, such as in rooms that produce complicated patterns of reflected and reverberant sound? The following study [17] was designed to address exactly this issue, using a paradigm in which listeners were asked to judge whether a set of reflections was shifted spatially either to the left or to the right of midline. If buildup also occurs in room environments, then spatial information from the room reflections should be similarly suppressed, resulting in degraded directional discrimination performance. This directional discrimination paradigm is an extension of a single-interval paradigm used to assess precedence effects with single reflections [18], except that here more than one reflection is spatially shifted. Virtual auditory space techniques [15] were used to simulate a reverberant room with dimensions of 5.7 × 4.3 × 2.6 and a broadband (125 – 4000 Hz) reverberation time (T_{60}) of approximately 0.4 s. Using these techniques, the spatial distribution of the early reflections was manipulated by imposing either a leftward or rightward lateral shift to the natural reflection locations. The number of shifted reflections was varied, ranging from the first 2 reflections following the floor and ceiling reflections to more than 500 following reflections, as shown in Fig. 1a. Examples of the spatial distributions of natural and left-shifted reflections 3 – 511 are shown in Figs. 1b and 1c. The simulated source direction (direct-path) was always directly in front of the listener at ear level at a distance of 1.4 m, in the center of the simulated room. The source signal was a 140-μs pulse. All stimuli were presented over equalized headphones (Beyerdynamic DT-990-Pro) at moderate level (approximately 65 dB SPL) within a double-walled sound isolation chamber (Acoustic Systems).

Seventeen normal-hearing listeners participated in each of two listening conditions: an Exposure condition and a No Exposure condition. In the Exposure condition, listeners were presented with 12 repetitions of the source signal (2 Hz rate) in the naturally simulated room on every trial (see Fig. 1b), followed by the test stimulus: a single presentation of the source signal with lateral reflection shift imposed (e.g. Fig. 1c). In the No Exposure condition, only the test stimulus was presented. Within a block of trials 20 left and 20 right shifts were presented for each of 8 reflection ensembles in randomized order. Exposure condition was held constant within a block of trials, but alternated between blocks. Each subject completed an initial practice block (excluded from data analysis) and 4 additional blocks per condition. Feedback was provided after all responses.

Results from a single representative listener are displayed in Fig. 1d, showing estimates of the psychometric functions that relate task performance (proportion of correct responses adjusted for potential biases resulting from the single-interval task) to the maximum reflection delay of the laterally shifted reflection ensemble (see Fig. 1a for delay values). Each point represents the result of 80 responses. Psychometric functions were estimated with logistic function fits using a maximum-likelihood criterion. From Fig. 1d, it is apparent that for this listener, prior listening exposure to the reverberant listening room resulted in both an elevated threshold for discriminating reflection shift direction, as well as a flattening in the slope of the psychometric function. These trends are generally representative of that observed in the sample of listeners tested in this task. A statistically significant increase in threshold of approximately 20 ms on average was observed between the No Exposure and Exposure conditions, $t(16) = 2.13$, $p < 0.05$, which suggests that prior listening exposure results in a suppression of reflection location information. To our knowledge, this is the first demonstration of precedence effect buildup in realistic room listening conditions with multiple reflections and reverberation.

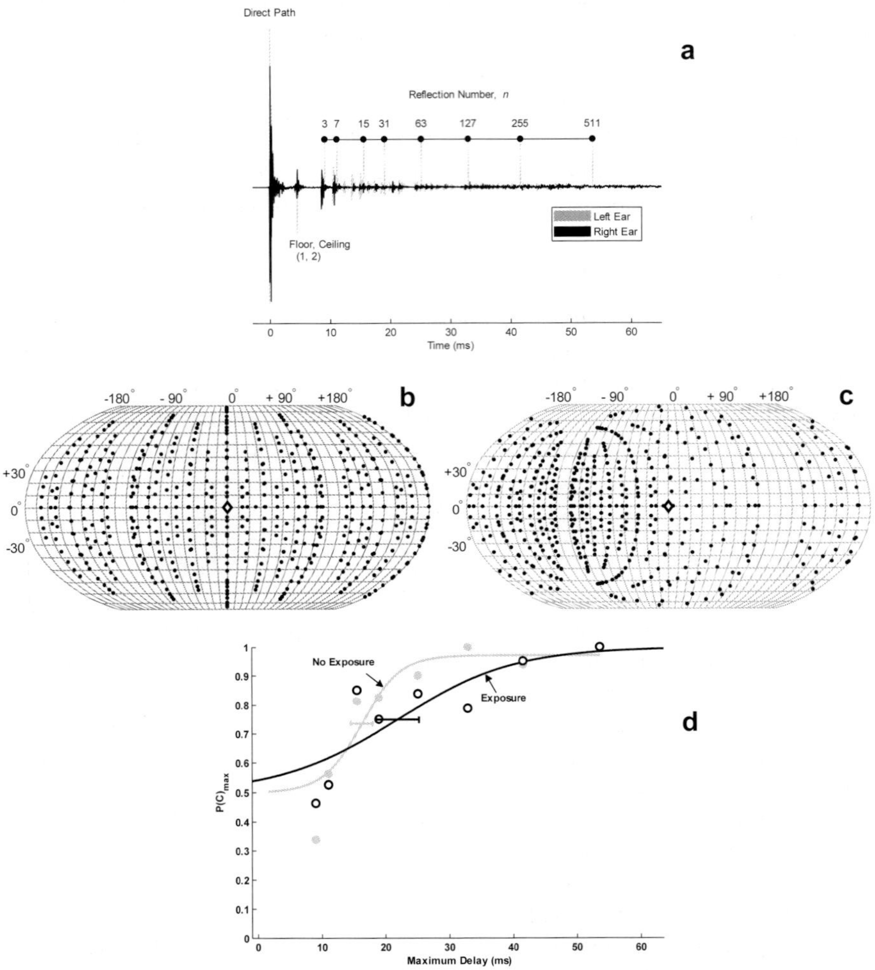

Figure 1. **a.** Example of a laterally shifted binaural room impulse-response (left-shift, n = 511). Shifted reflection ensembles included reflections 3 through n, where n is the cumulative number of reflections following the direct-path. The eight tested reflection ensembles sizes, n, and their corresponding maximum delays are also indicated. **b.** Directional distribution of naturally occurring reflections 3 - 511 in the simulated room environment relative to the listener. The diamond symbol indicates the direct-path direction. **c.** Directional distribution of reflections 3 - 511 with leftward lateral shift imposed. **d.** Proportion of correct directional shift responses (adjusted for single-interval response biases) as a function of maximum delay in the reflection ensemble for a single listener (LDC) both with and without prior listening exposure to the simulated reverberant room. Threshold values and 95% confidence limits (estimated via a bootstrapping procedure [19]) are indicated on each fitted function.

A statistically significant decrease in slope of the fitted logistic functions was also observed, $t(16) = 4.15$, $p < 0.01$. Interpretation of this latter result is less clear, but may be indicative of change in the processes underlying the spatial position decisions between the two experimental conditions. Follow-up testing in our lab is now examining important next questions such as the extent to which the effect may depend on specific acoustical characteristics of the listening room environment and how rapidly the suppression activates.

3. Speech Intelligibility Improvement

Although there is a well-known debilitating effect of reverberation on speech intelligibility [20,21], little research has been conducted to determine how prior exposure to a reverberant environment can affect intelligibility. Experiments by Watkins [22,23] indicate that a carrier phrase presented in a room acoustic context that is congruent with the target word can facilitate word identification performance at the phonemic level. This suggests that given the right echoic context for a room, perceptual compensation can attenuate the effects of reverberant energy. Work by Djelani and Blauert [24] also suggests that precedence effect buildup occurs with speech signals, at least in simplified acoustical situations with a single echo. Here we describe a study [25] that extends this previous work by testing whether word-level speech intelligibility can be modulated by prior listening exposure to a speech source in a reverberant room environment. We measured closed-set speech intelligibility for 19 listeners in a simulated reverberant room with a spatially separated noise masker using the Coordinate Response Measure (CRM) speech corpus [26]. Two room exposure conditions were tested: one designed to provide prior listening exposure to a reverberant room environment, and one designed to minimize such exposure. If suppression of competing reflections and/or reverberant sound energy is enhanced with prior listening exposure to the room, then intelligibility should improve with room exposure.

Two extremes of listening exposure were tested by varying the length of the carrier sentence phrase in the CRM. In a No Exposure condition, listeners were presented only with the color-number targets from the CRM, and the simulated room was varied randomly from trial-to-trial (from a set of 3 potential rooms) in order to limit any across-trial room exposure. In an Exposure condition, a two-sentence carrier phrase (approximately 10 s duration) preceded the color-number target. In both conditions, a competing broad-band noise masker was presented at one of nine signal-to-noise (SNRs) ratios ranging from -28 to +4 dB. All speech and masker signals were presented in a reverberant room simulated using virtual auditory space techniques [15]. The dimensions of the room were $5.7 \times 4.3 \times 2.6$ m, with a broadband (125 – 4000 Hz) T_{60} of 0.4 s. The two additional rooms used to limit any across-trial carry-over effects in the

No Exposure condition had the same dimensions, but different surface absorption properties, resulting in broadband T_{60} values of 0.3 and 3 s. In all cases, the target speech was simulated at a spatial location directly in front of the listener at a distance of 1.4 m, and a broadband noise masker was presented at a simulated location opposite the listener's right ear, also at 1.4 m. The two exposure conditions were run in separate blocks of 54 trials (6 repetitions at each of 9 SNRs). Listeners completed 5 blocks for each condition. All sounds were presented over equalized headphones (Beyerdynamic DT-990-Pro) at moderate level (approximately 65 dB SPL) within a double-walled sound isolation chamber (Acoustic Systems). Listeners entered their responses on a GUI and received feedback as to the correctness of the response after every trial.

Figure 2a displays the proportion of correct color-number identifications in the CRM corpus as a function of SNR for a single listener (LNN) in both exposure conditions. Logistic functions were fit to the data (maximum-likelihood criterion) and 95% confidence limits were obtained for each fitted function's threshold value, P(C) = 0.516, using a bootstrapping procedure [19]. For this listener, it is clear that prior listening exposure significantly decreased speech reception threshold (SRT) by approximately 3 dB, which corresponds to an improvement in speech intelligibility of approximately 17%. Figure 2b provides a summary of SRT data from all listeners in the experiment. Nearly all listeners demonstrated significantly improved intelligibility with prior listening exposure to the room, as indicated by points that lie below the diagonal line in Fig. 2b. The median improvement in SRT was 2.7 dB, which corresponded to a median improvement in intelligibility of 16.7%. To our knowledge, this is the first study to demonstrate a facilitating effect of prior listening exposure for speech intelligibility in reverberant rooms. Subsequent testing in our lab has demonstrated that this facilitation effect is absent in anechoic space and under monaural listening conditions. These results suggest that the facilitation is specific to reverberant rooms and requires binaural input. Additional testing has also revealed that the effect is influenced by the specific acoustical properties of the listening room environment, and appears to be strongest for moderate levels of reverberation ($0.4 \leq T_{60} \leq 1$ s) [27]. Results from a related study [28] that measured open-set intelligibility demonstrated what appears to be a similar adaptation effect, but only for natural binaural room simulations. Unnatural simulations, such as when the simulated binaural room impulse response was reversed in time, or when listening to diotic room simulations, failed to produce an adaptation effect.

These results of speech intelligibility improvement with prior listening exposure to a reverberant room are similar to precedence effect buildup results in at least two respects. First, both phenomena do not occur immediately, but instead require some time to become fully active. Presumably this adaptation time reflects a form of perceptual calibration to the specifics of the acoustically reflective listening situation. A second similarity is that both effects appear to depend critically on binaural input. In the case of precedence effect buildup, the detection or localization of single reflections depends critically on information contained in binaural input signals. Speech intelligibility, though primarily supported by the monaural auditory system, also benefits from binaural input in many situations, such as when speech targets are imbedded in backgrounds of one or more competing, but spatially separated sources. Given the spatial configuration of target and masker in experiments described here, it is perhaps not surprising that binaural input appears critical for the perceptual suppression of room acoustic effects. Such a result does not imply that other monaural aspects of speech de-reverberation do not also exist, however. The perceptual compensation for room reverberation described by Watkins [22,23] appears to operate with nearly equal strength in both monaural and binaural conditions, and perhaps serves to remove spectral colorations caused by room acoustics. This result appears fundamentally similar to observations described by Toole [29] regarding loudspeaker reproduction in rooms, where insensitivity to measureable room coloration was observed after listening exposure to the room. These seemingly contradictory results regarding the necessity of binaural input could be rooted in differences between objective speech intelligibility tasks versus subjective speech perception tasks [22,23]. It is also possible that they

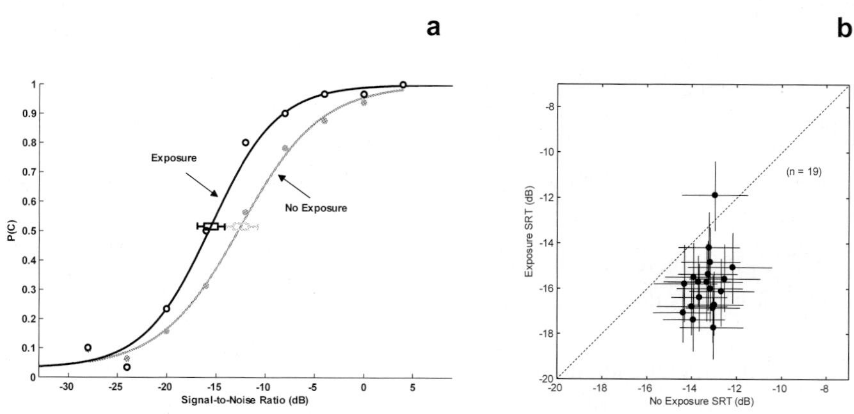

Figure 2. **a.** Psychometric functions for both exposure conditions for a single representative listener (LNN). Threshold values and their 95% confidence limits are indicated on each function. **b.** Scatterplot of speech reception thresholds (SRTs) and their 95% confidence limits for all listeners in the experiment (*n* = 19).

result from separate yet complementary aspects of room effect suppression: one that relates to spatial configurations within the room and therefore is facilitated by binaural input, and one that is concerned primarily with removing monaural coloration caused by room acoustics. Further study will be needed to test this two-system hypothesis.

4. Loudness

The perceived intensity of sound, *loudness*, has typically been investigated by playing back monaural, diotic, or dichotic stimuli over headphones to listeners. The loudness of sound fields, on the other hand, has often been investigated in anechoic environments free from reflections, or in the presence of only moderate amounts of reverberation. Hence, relatively little is known about the perception of loudness in reverberant, everyday environments.

Although loudness differences between sound fields have been accurately predicted by differences in magnitude spectra at the listeners' ears in a variety of listening situations [30,31], recent data obtained under headphone playback suggest that, in addition to at-the-ear spectra, interaural correlation also has an effect on binaural loudness [32]. Interaurally uncorrelated signals (interaural correlation coefficient, IACC = 0) are generally perceived as louder than interaurally correlated signals (IACC = 1), with the effect being largest at low frequencies. This finding was corroborated in a recent study from our laboratory using reverberant stimuli [33], where at-the-ear signals were decorrelated using a computational room model [15]. Interestingly, IACC had the smallest effect on loudness in the most reverberant condition, the effect of decreasing IACC on loudness running somewhat counter to the amount of reverberation in the stimuli. This may be related to differential weighing of direct and reverberant sound in loudness comparisons, as proposed by Stecker and Hafter [34] for stimuli with equal energy, but temporally asymmetric amplitude envelopes. Further support for such perceptual parsing comes from a study on loudness constancy [35], where subjects were able to accurately judge the sound power of a source at various distances in a reverberant room, despite profound changes in at-the-ear exposures. The extent to which constancy plays a role, and how direct and reverberant sounds are weighted for loudness in everyday listening of sounds from various distances and directions remain to be uncovered. At present, there is no evidence to suggest that loudness is in any way affected by prior listening exposure to a reverberant room.

5. Summary and Conclusions

Prior listening exposure to a reverberant room impairs listeners' abilities to discriminate changes in the spatial locations of the reflections and enhances speech intelligibility. These effects appear to require binaural input and are not observable in the absence of acoustic reflections. Although prior listening exposure does not appear to affect sound loudness per se, binaural decorrelation caused by room reverberation results in an increase in loudness.

Research on room adaptation effects is in its infancy. While we believe that there is now accumulating evidence for the existence of such effects in a variety of different listening task domains, there are still many unanswered questions related to these adaptation effects. How long does it take to become adapted? Does the adaptation depend on the specifics of the source signal and/or spatial configuration of the source and listener within the room? What are the neural mechanisms that underlie the effect? Further study will be needed to answer these and other important questions. Results from this research, in addition to contributing to our scientific understanding of the auditory system and its perceptual processing abilities, will be important initial steps to better understanding the often profound effects of reflected and reverberant sound on individuals with hearing impairment.

Acknowledgments

Thanks to Devan Haulk, Noah Jacobs, Laricia Longworth-Reed, Joanna Ohlendorf, and Jeremy Schepers their contributions and assistance in conducting the listening experiments described in this chapter. Financial support provided by NIH-NIDCD R01 DC008168.

References

1. Wallach, H., Newman, E. B., and Rosenzweig, M. R., *Am. J. Psychol.* 62, 315 (1949).
2. Haas, H., *J. Aud. Eng. Soc.* 20 (2), 146 (1972).
3. Litovsky, R. Y., Colburn, H. S., Yost, W. A., and Guzman, S. J., *J. Acoust. Soc. Am.* 106 (4 Pt 1), 1633 (1999).
4. Clifton, R. K., *J. Acoust. Soc. Am.* 82 (5), 1834 (1987).
5. Clifton, R. K., Freyman, R. L., Litovsky, R. Y., and McCall, D., *J. Acoust. Soc. Am.* 95 (3), 1525 (1994).
6. Clifton, R. K. and Freyman, F. L., *Percept. Psychophys.* 46 (2), 139 (1989).
7. Freyman, R. L., Clifton, R. K., and Litovsky, R. Y., *J. Acoust. Soc. Am.* 90 (2), 874 (1991).

8. Thurlow, W. R. and Parks, T. E., *Percept. Mot. Skills* 13, 7 (1961).
9. McCall, D. D., Freyman, R. L., and Clifton, R. K., *Percept. Psychophys.* 60 (4), 593 (1998).
10. Freyman, R. L. and Keen, R., *J. Acoust. Soc. Am.* 120 (6), 3957 (2006).
11. Clifton, R. K. and Freyman, R. L., in *Binaural and Spatial Hearing in Real and Virtual Environments*, edited by R. H. Gilkey and T. R. Anderson (Erlbaum, Mahwah, New Jersey, 1997), pp. 233.
12. Clifton, R. K., Freyman, R. L., and Meo, J., *Percept. Psychophys.* 64 (2), 180 (2002).
13. Grantham, D. W., *J. Acoust. Soc. Am.* 99 (2), 1118 (1996).
14. Cranford, J. L., Ravizza, R., Diamond, I. T., and Whitfield, I. C., *Science* 172, 286 (1971).
15. Zahorik, P., *J. Acoust. Soc. Am.* 126 (2), 776 (2009).
16. Vorländer, M., *Auralization.* (Springer-Verlag, Berlin, 2008).
17. Zahorik, P. and Haulk, D., *Abstr. Midwinter Res. Meet. Assoc. Res. Otolaryngol.* 30, 311 (2007).
18. Yang, X. and Grantham, D. W., *Percept. Psychophys.* 59 (7), 1108 (1997).
19. Wichmann, F. A. and Hill, N. J., *Percept. Psychophys.* 63 (8), 1314 (2001).
20. Bolt, R. H. and MacDonald, A. D., *J. Acoust. Soc. Am.* 21 (6), 577 (1949).
21. Knudsen, V. O., *J. Acoust. Soc. Am.* 1 (1), 56 (1929).
22. Watkins, A. J., *J. Acoust. Soc. Am.* 118 (1), 249 (2005).
23. Watkins, A. J., *Acta Acust. united Ac.* 91 (5), 892 (2005).
24. Djelani, T. and Blauert, J., *Acta Acust. united Ac.* 87 (2), 253 (2001).
25. Brandewie, E. and Zahorik, P., *Abstr. Midwinter Res. Meet. Assoc. Res. Otolaryngol.* 31, 296 (2008).
26. Bolia, R. S., Nelson, W. T., Ericson, M. A., and Simpson, B. D., *J. Acoust. Soc. Am.* 107 (2), 1065 (2000).
27. Zahorik, P. and Brandewie, E., *Abstr. Midwinter Res. Meet. Assoc. Res. Otolaryngol.* 32, 145 (2009).
28. Longworth-Reed, L., Brandewie, E., and Zahorik, P., *J. Acoust. Soc. Am.* 125 (1), EL13 (2009).
29. Toole, F. E., *J. Aud. Eng. Soc.* 54 (6), 451 (2006).
30. Sivonen, V. P., *J. Acoust. Soc. Am.* 121 (5 Pt1), 2852 (2007).
31. Sivonen, V. P. and Ellermeier, W., *J. Acoust. Soc. Am.* 119 (5 Pt 1), 2965 (2006).
32. Edmonds, B. A. and Culling, J. F., *J. Acoust. Soc. Am.* 125 (6), 3865 (2009).
33. Sivonen, V. P. and Zahorik, P., *Abstr. Midwinter Res. Meet. Assoc. Res. Otolaryngol.* 32, 146 (2009).
34. Stecker, G. C. and Hafter, E. R., *J. Acoust. Soc. Am.* 107 (6), 3358 (2000).
35. Zahorik, P. and Wightman, F. L., *Nat. Neurosci.* 4 (1), 78 (2001).

THE IMPACT OF MASKER FRINGE AND MASKER SPATIAL UNCERTAINTY ON SOUND LOCALIZATION *

B. D. SIMPSON†, R. H. GILKEY

Air Force Research Laboratory and Wright State University,
Wright-Patterson Air Force Base, Ohio 45433, USA
†E-mail: brian.simpson@wpafb.af.mil

D. S. BRUNGART

Army Audiology and Speech Center, Walter Reed Medical Center
Washington, DC 20307, USA

N. IYER, J. D. HAMIL

Air Force Research Laboratory
Wright-Patterson Air Force Base, Ohio 45433, USA

Tone-in-noise detection improves when the masker duration is greater than the signal ("masker fringe") relative to the case in which the signal and masker are pulsed on/off simultaneously. This has been attributed to the fact that the fringe provides a baseline set of stimulus parameters that serves as a context against which the signal may be detected. Conversely, when the fringe parameters are inconsistent with those of the masker, signal detectability can be reduced. In this chapter, the impact of masker fringe on sound localization is examined in four experiments. The results demonstrate the importance of stimulus parameters prior to, and subsequent to, the portion of the stimulus containing the signal for sound localization.

1. Introduction

The threshold for a tonal signal presented in a masking noise may be reduced by 5 dB or more when the masker is turned on prior to the signal onset (forward masker fringe) and/or the masker is turned off subsequent to the signal offset (backward masker fringe) relative to the case in which the signal and masker are pulsed on and off simultaneously [1,2]. Moreover, detectability has been shown to improve as the duration of this fringe increases. It has been argued that this masker fringe provides a baseline set of stimulus parameters against which the signal may be more easily detected as a change in those parameters [3,4].

* This work was supported by a grant from the Air Force Office of Scientific Research

A slightly different, but related, explanation would suggest that the fringe provides the listener with information about the parameters of the masker stimulus present during the signal interval. As such, the fringe reduces masker uncertainty. This interpretation may be related to studies on informational masking, where it has been shown that a 'preview' (e.g., through cuing) of the spectral components of a masker improves performance on a signal detection task, presumably as a result of reduced masker uncertainty (e.g., [5]). It is possible that forward masker fringe similarly provides a preview of the masker characteristics and it is the reduction in masker uncertainty afforded by this preview that leads to a release from masking. Recent results have also demonstrated the impact of spatial uncertainty on signal detection [6] and speech intelligibility [7]. The goal of the four experiments described in this chapter was to determine the impact of masker fringe and masker spatial uncertainty on sound localization in noise and to examine how such effects might be related to binaural detection and informational masking.

2. Methods

2.1. *Participants*

In Experiments 1 and 2, five listeners (three male, two female) participated. In Experiments 3 and 4, the same five listeners plus an additional female participated. Listeners were 20-25 years of age, had normal hearing (i.e., thresholds $\leq$ 15 dB HL from .125-8.0 kHz), and all had previously participated in studies on sound localization. All listeners were paid for their participation.

2.2. *Apparatus*

The study was conducted in the Auditory Localization Facility at the Air Force Research Laboratory at Wright-Patterson Air Force Base (Figure 1). This facility consists of a geodesic sphere (4.3 m in diameter) with 277 full-range loudspeakers mounted on its surface. The sphere is housed within an anechoic chamber, the walls, floor, and ceiling of which are covered in 1.1-m thick fiberglass wedges. For this study, only those loudspeakers above -45° in elevation were utilized. Mounted on the front of each loudspeaker is a cluster of four light-emitting diodes (LEDs).

Fig. 1. The Auditory Localization Facility at Wright-Patterson Air Force Base.

2.3. *Procedure*

Listeners stood on a platform in the Auditory Localization Facility with their head positioned in the center of the sphere. At the beginning of each trial, the listener was required to orient toward the loudspeaker at 0° azimuth, 0° elevation, and remain in a fixed position during the stimulus presentation. After the stimulus presentation, listeners were required to point a hand-held tracking device at the perceived location of the target signal and depress a button on the device to record the localization response. Trial-by-trial feedback was provided by activating the LED cluster at the actual target location.

3. Experiment 1

The target signal was a 100-Hz, random-phase click train, with a bandwidth of 0.2 to 14.5 kHz and a duration of 250 ms with 25-ms $\cos^2$ on/off ramps. The masker was a Gaussian noise with the same bandwidth and duration as the target, and was preceded by, and followed by, a masker fringe of either 10 ms or 500 ms. The masker and fringe were presented at 60 dB SPL from the loudspeaker directly in front of the listener (0° azimuth, 0° elevation). The target level was 50, 55, or 60 dB SPL and its location was randomly varied from trial to trial across 142 loudspeaker locations distributed throughout the sphere.

The results from Experiment 1 are shown in Figure 2. Here, the overall angular localization errors, averaged across all listeners, are plotted as a function of signal-to-noise ratio (SNR) in the two fringe conditions (10-ms and 500-ms). Localization performance for the target presented in isolation was also measured as a baseline condition, and the average angular error was found to be approximately 15° (diamond symbol). As can be seen, when the target was presented in noise, angular errors were larger and increased in both fringe conditions as the SNR decreased. Moreover, errors increased more rapidly in the 10-ms condition than the 500-ms condition. The horizontal separation of the curves indicates a 5-6 dB benefit of a longer masker fringe. These effects are similar in magnitude to those observed in binaural detection [1].

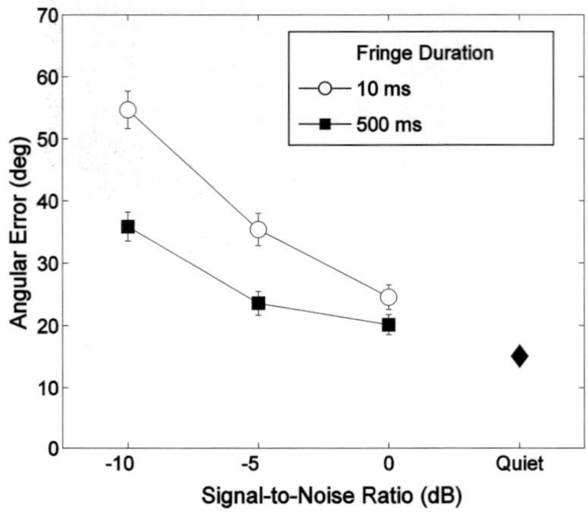

Fig. 2. Angular localization errors, averaged across trials, plotted as a function of SNR for the 10-ms (open circles) and 500-ms (filled squares) masker fringe conditions. The filled diamond depicts mean localization error for the target in quiet. Error bars indicate ±1 standard error across trials.

4. Experiment 2

Cuing a listener to characteristics of the masker (e.g., spectral or spatial parameters) can improve performance by reducing masker uncertainty [5,6,7]. It is conceivable that masker fringe similarly serves to reduce masker spatial uncertainty in a localization task by cuing the location of the masker. This notion was directly examined in Experiment 2 by comparing performance with a fixed, known-location masker (low uncertainty) to performance when the masker spatial location was randomly varied from trial to trial across 13 locations (high uncertainty). The target and masker were similar to those employed in Experiment 1, but the duration of each was reduced to 80 ms, and the 10-ms fringe condition was replaced with a no-fringe or '0-ms' condition.

The results from Experiment 2 are shown in Figure 3. Here, overall angular localization errors, averaged across all listeners, are plotted for the low and high spatial uncertainty cases for the 0-ms and 500-ms fringe conditions. Performance was found to be better in the 'low uncertainty' condition overall. In the 0-ms condition, masker spatial uncertainty resulted in a 20° increase in localization errors relative to the case in which the masker location was known. However, when a 500-ms fringe was provided, the difference between the low and high uncertainty conditions had diminished to 5°, indicating that masker fringe may help to reduce masker spatial uncertainty.

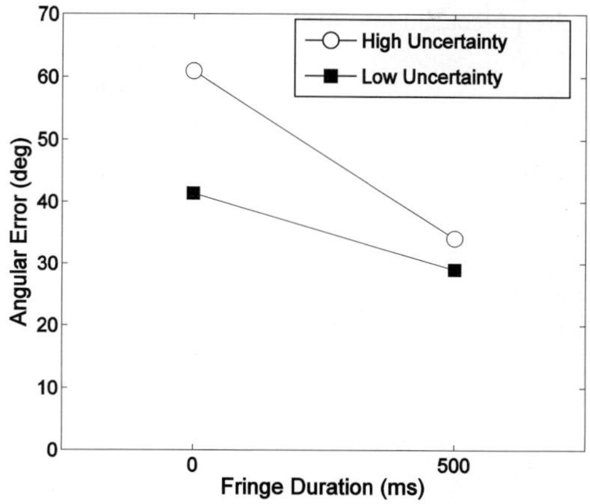

Fig. 3. Angular localization errors, averaged across trials, plotted as a function of the duration masker of forward and backward masker fringe. Error bars indicate ±1 standard error across trials.

5. Experiment 3

For Experiment 3, the effect of forward and backward masker fringe duration on sound localization was examined. The target was similar to the target employed in Experiments 1 and 2 but was 60-ms in duration with 5-ms $\cos^2$ ramps and was presented at 63 dB SPL. The masker was a Gaussian noise of the same duration and bandwidth as the target, and was presented simultaneously with the target. The fringe was constructed from multiple pulses with the same parameters as the masker. The pulses were presented sequentially such that there were no temporal gaps between the offset of one pulse and the onset of the next. All masker and fringe pulses were presented at 60 dB SPL and came from the same location. The duration of the fringe was varied by changing the number of pulses.

It was hypothesized that performance would improve as a function of the duration of the masker fringe, in agreement with the binaural detection literature [5]. In Figure 4, overall angular localization errors, averaged across all listeners, are plotted as a function of the duration of the masker fringe. Note that each value on the abscissa refers to the duration of the forward *and* backward fringe (e.g., a value of 120 ms indicates 120 ms of forward fringe and 120 ms of backward fringe). As can be seen, localization errors decreased as the duration of the masker fringe increased from 0 ms to 240 ms, consistent with the results from the binaural detection literature. Of note is the fact that even 60 ms of fringe leads to a substantial reduction in errors relative to the case in which the masker and target are pulsed on and off simultaneously (0 ms of masker fringe).

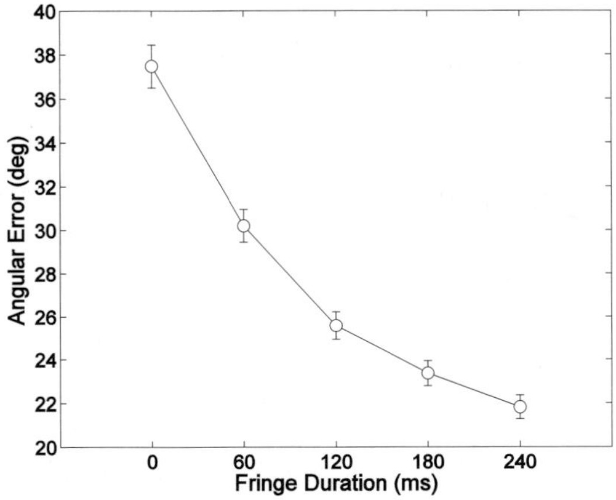

Fig. 4. Angular localization errors, averaged across trials, plotted as a function of the duration of forward and backward masker fringe. Error bars indicate ±1 standard error across trials.

6. Experiment 4

In Experiment 4, spatial uncertainty was manipulated within a trial by fixing or varying the spatial location of the individual pulses that comprise each portion of masker fringe (forward and backward). Specifically, in some cases, all fringe pulses came from the same location as the masker ('fixed' masker fringe); in other cases, each fringe pulse came from a different, randomly-selected location ('random' masker fringe). In addition, in some cases, the forward or backward masker fringe, or the masker itself, was absent (a quiet portion of the stimulus). A graphical depiction of the stimuli employed is shown in Figure 5. In each panel, a 3-letter designation indicates the parameters of the condition being depicted. The format is [*forward fringe*][*masker*][*backward fringe*]. [M] refers to the *masker;* [F] refers to a fringe in which all pulses are presented from a *fixed* location that is the same as the masker; [R] refers to a fringe in which individual pulses are presented from *randomly*-selected locations. [Q] refers to a *quiet* portion of the stimulus in which no masker or fringe is present. Only one condition was tested within each block.

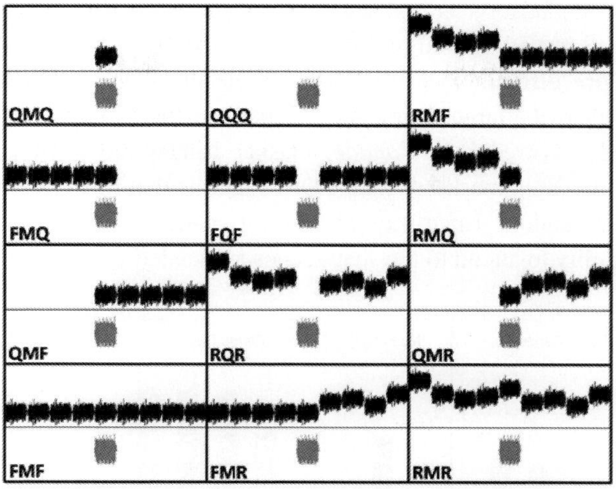

Fig. 5. Graphical representation of the stimuli employed in the study. The black pulses are independent broadband noise pulses that comprise the forward/backward fringes and the masker. The target is depicted in gray. The noise pulse simultaneous with the target is the masker; the pulses preceding and following the masker comprise the fringe. Variation in the vertical position of pulses on the ordinate of each panel indicates variations in spatial locations of the individual pulses.

The results from Experiment 4 are shown in Figure 6. The black bar in the second panel represents the condition in which the stimulus contained no masker fringe (the 0-ms condition replotted from Figure 4). The bars in the first panel indicate conditions in which masker fringe enhances performance relative to the 'No Fringe' condition. As can be seen, localization errors are reduced by up to 18° when a forward fringe is presented from the same location as the masker (FMQ). Moreover, localization errors in this forward masker fringe condition are roughly 13° lower than errors in the backward masker fringe condition (QMF). This result is consistent with results from the binaural detection literature, which demonstrate that forward masker fringe provides a greater benefit than backward masker fringe [2]. The bars depicted in the third panel of Figure 6 indicate conditions in which the masker fringe degrades performance. As can be seen, the addition of a random masker fringe nearly always leads to greater errors than those found in the 'No-Fringe' condition. Moreover, the negative effects of adding a random backward fringe appear to be more severe than the effects of adding a random forward fringe.

The results in the rightmost panel of Figure 6 ('Quiet') are conditions in which there is no masker during the target interval. The increase in errors seen when the fringe is presented suggests that fringe itself may act like a masker.

One possible explanation is that temporal integration could lead to the fringe being averaged in with the target interval, thus reducing the effective SNR and making it more difficult to recover the localization cues associated with the target. However, the large degradation in performance seen when the target is preceded and followed by a *random* masker fringe (the RQR condition) as compared to the *fixed* masker fringe (FQF) suggests that this effect is at least in part due to the random fringe acting as a informational masker, interfering with a listener's ability to attend to the spatial cues associated with the target.

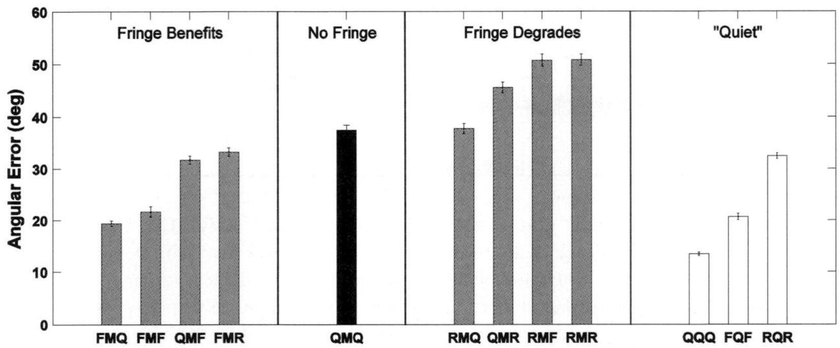

Fig. 6. Angular localization errors, averaged across trials, plotted for each condition examined. Error bars indicate ±1 standard error across trials.

The effect of adding various forward and backward masker fringes is depicted in Figure 7. In each panel, one type of fringe (fixed or random) is added to each of three conditions. In the first (leftmost) panel, a fixed-location forward fringe is added. As can be seen, adding this type of fringe to any stimulus configuration always led to a substantial reduction in localization errors, suggesting that this type of fringe effectively facilitates localization. In contrast, the results in the fourth panel show that the addition of a random-location backward fringe always resulted in a substantial *increase* in localization errors, indicating that this type of fringe somehow interferes with a listener's ability to recover localization cues. The effects of adding a random forward fringe (second panel) or a fixed backward fringe (third panel) are less consistent across stimulus conditions. Adding a random forward fringe to a stimulus that contains a backward fringe leads to an increase in localization errors, but when the stimulus contains no backward fringe, no change in performance is seen (e.g., difference between QMQ and RMQ). This might suggest that listeners can effectively ignore the random forward fringe and attend to the last pulse in the sequence, a strategy that fails in cases where a backward fringe of any type follows the target. This is consistent with results indicating a recency effect in conditions involving spatial uncertainty and informational masking [8].

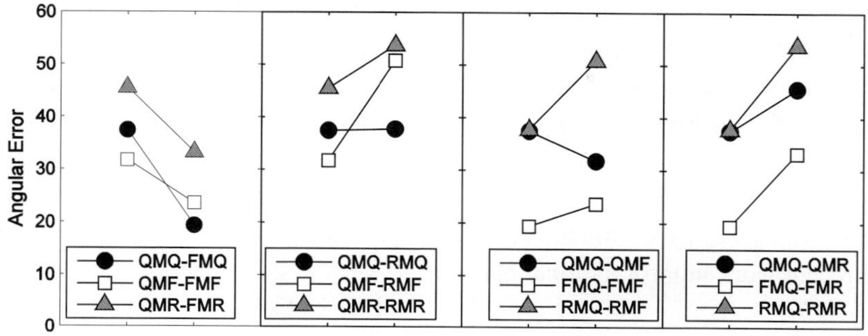

Fig. 7. In each panel, a comparison is made between average angular localization errors for a given stimulus condition and for that condition with a specific forward or backward masker fringe added.

7. Discussion and Conclusions

The results from Experiment 1 suggest that a masker fringe with spatial parameters consistent with those of the masker leads to improved performance relative to the case in which a target and masker are pulsed on and off nearly simultaneously. The results from Experiment 2 suggest that these effects might result from the fact that masker fringe can reduce masker spatial uncertainty. Data from Experiment 3 indicate that the benefit of masker fringe increases as the duration of the fringe increases, consistent with results from the binaural detection literature. Finally, the results from Experiment 4 indicate that both forward and backward masker fringe lead to improved localization performance when the fringe parameters are consistent with those of the masker (the fixed-fringe conditions). Furthermore, in these cases, a forward masker fringe more effectively facilitates performance than does a backward fringe. Conversely, when masker fringe is presented from a location (or locations) that differ from the location of the masker (the random fringe conditions), the fringe provides no benefit for localization, and in most cases actually leads to a degradation in localization performance. In this context, the random fringe seems to function as an informational masker. The fact that this random fringe causes more informational masking when it occurs after the target (backward fringe) is consistent with the suggestion that a backward informational masker is more effective than a forward informational masker [9].

In general, the results of this study suggest that having information about the spatial parameters of the masker can improve performance in a localization-in-noise task. These results appear to be consistent with results demonstrating that providing a preview of the masker was the most effective means of cuing in a signal detection task [5]. The suggestion that cuing the masker allows a listener to establish a "rejection filter" to minimize the interference caused by the masker [10] could be relevant in this situation, where a listener would apply such a filter to a region of space associated with the masker. A random masker fringe does not afford the listener an obvious region to which such a filter should be applied. The overall results suggest that the effects of masker fringe are complicated, but it appears that some insights may be gained by considering these results within the context of the binaural detection literature as well as more recent work in informational masking.

References

1. D.M. McFadden, Masking-level differences with continuous and with burst masking noise, *J. Acoust. Soc. Am.* **40**, 1414-1419 (1966).
2. C. Trahiotis, T.R. Dolan, T.E. Miller, Effect of 'backward' masker fringe on the detectability of pulsed diotic and dichotic tonal signals, *Percept & Psychophys.* **12**, 335-338 (1972).
3. W.A. Yost, Prior stimulation and the masking-level difference, *J. Acoust. Soc. Am.* **78**, 901-907 (1985).
4. R.H. Gilkey, B.D. Simpson, J.M. Weisenberger, Masker fringe and binaural detection, *J. Acoust. Soc. Am.* **88**, 1323-1332 (1990).
5. V.M. Richards, D.L. Neff, Cuing effects for informational masking, *J. Acoust. Soc. Am.* **115**, 289-300 (2004).
6. W.L. Fan, T.M. Streeter, N.I. Durlach, Effect of spatial uncertainty of masker on masked detection for nonspeech stimuli, *J. Acoust. Soc. Am.* **124**, 36-39 (2008)
7. N. Kopco, V. Best, S. Carlile, Localizing a speech target in a multitalker mixture, *J. Acoust. Soc. Am.* **125**, 2691 (2009)
8. D.W. Chandler, D.W. Grantham, M.R. Leek, Effects of uncertainty on auditory spatial resolution in the horizontal plane, *Acta Acust.* **91**, 513-525 (2005)
9. I. Pollack, Auditory informational masking, *J. Acoust. Soc. Am.* **57**, S5 (1975).
10. N.I. Durlach, C.R. Mason, G. Kidd, T.L. Arbogast, H.S. Colburn, B.G. Shinn-Cunningham, Note on informational masking (L), *J. Acoust. Soc. Am.* **113**, 2984-2987 (2003)

BINAURAL INTERFERENCE: THE EFFECTS OF LISTENING ENVIRONMENT AND STIMULUS TIMING*

D. W. GRANTHAM

Vanderbilt Bill Wilkerson Center, Vanderbilt University School of Medicine, Nashville, Tennessee 37232, USA

N. B. H. CROGHAN

Speech, Language, & Hearing Sciences, University of Colorado, Boulder, Colorado 80309, USA

C. R. CAMALIER

Department of Psychology, Vanderbilt University, Nashville, Tennessee 37203, USA

L. R. BERNSTEIN

Departments of Neuroscience and Surgery (Otolaryngology), University of Connecticut Health Center, Farmington, Connecticut 06030, USA

Interaural time difference (ITD) thresholds for a high-frequency stimulus are elevated in the presence of a low-frequency interferer [McFadden and Pasanen, *J. Acoust. Soc. Am.* **59**, 634 (1976)]. This phenomenon, known as "binaural interference," was further explored in the present series of experiments. In experiment 1 binaural interference was measured in the free field. The minimum audible angle for a high-frequency target was considerably elevated when a low-frequency interferer was pulsed on and off with the target from a position directly in front of the subject. However, if the low-frequency interferer was presented continuously, or if its spatial position was different from that of the target, the interference was practically eliminated. Experiment 2 showed that the degree of binaural interference in an ITD task decreased monotonically with the duration of interferer fringe that came before, after, or both before and after the target. These results are consistent with the notion that interference from a spectrally remote low-frequency interferer may be influenced by the extent to which the target and interferer are fused into a single perceptual object. Experiment 3 showed that binaural interference increased as the interstimulus interval between the two target presentations decreasesd This unexpected finding indicates that the weights assigned to target and interferer in an interference paradigm are governed by temporal windows that may be even longer than those associated with binaural sluggishness.

* This work was supported in part by a T35 training grant to Vanderbilt University (DC008763) which funded the second author's visit to Vanderbilt University during the summer of 2008.

1. Introduction

In 1976 McFadden and Pasanen showed that interaural time difference (ITD) thresholds for a high-frequency noise were elevated in the presence of a low-frequency diotic noise [1]. Specifically, ITD threshold for a target narrow band of noise centered at 4000 Hz was 71 μs when presented alone (averaged across four subjects). However, when a second narrow band of noise centered at 500 Hz (the "interferer") was presented diotically and pulsed on and off along with the target, ITD threshold for the target noise increased by more than a factor of two (to 152 μs).

This phenomenon has become known as "binaural interference" and has been widely studied since McFadden and Pasanen's seminal paper (see review [2]). The low-frequency interferer has been shown to disrupt not only ITD processing, but also interaural level difference processing [3, 4], binaural release from masking [4], and laterality judgments of high-frequency stimuli [2, 5].

Because of the wide spectral separation of target and interferer, binaural interference cannot be attributed to energetic masking effects. Rather, the presence of the low-frequency interferer has some detrimental effect on the ability of the binaural system to process ITDs in the high-frequency target. The generally accepted explanation for the interference effect is that subjects form an "obligatory combination" of the target and interferer, and that the ITD cue in the target is thus "diluted" by the diotic interferer. Quantitative "weighted combination" models have been reasonably successful in describing the interference effect [5, 6].

There are two especially interesting aspects of binaural interference that have been widely reported. First, binaural interference is generally asymmetric. That is, although a low-frequency diotic interferer results in elevated ITD thresholds in a high-frequency target, a high-frequency interferer does not generally result in elevated ITD thresholds for a low-frequency target (*e.g.*, [1]). This asymmetry is predicted by the weighted-combination models because they give more weight to low than to high frequencies when they are combined in an ITD task. Specifically, the respective weights can be estimated from measuring ITD thresholds for the low-frequency and high-frequency targets when they are presented in isolation. Because ITD threshold is lower for the low-frequency target than for the high-frequency target, the low-frequency information gets more weight when the two stimuli are combined.

A second widely reported finding concerning binaural interference is that it depends on the relative timing of target and interferer. Specifically, interference is greatest when target and interferer are pulsed on and off together. When the interferer is continuously present, the interference is markedly reduced [4, 7].

Best *et al.* suggested that this effect of timing can be explained on the basis of object formation [2]. They suggested that when the target and interferer are pulsed on and off together, subjects are likely to perceive the two stimuli as a single object, and consequently there would be a greater "obligatory combination" of the two stimuli. However, if the target is presented along with a continuous interferer, subjects are likely to perceive the two stimuli as separate objects, resulting in there no longer being an "obligatory combination."

The purpose of our first experiment was to measure binaural interference in the free field using a task analogous to the ITD detection task employed in headphone studies. It is of interest to investigate this phenomenon in the free field, where different cues dominate spatial resolution for low-frequency (ITD) and high-frequency (ILD) stimuli [8]. Accordingly, we measured the minimum audible angle (MAA) for narrow bands of noise (centered at 500 and 4000 Hz), in the absence and presence of interfering bands of noise. In parallel with the headphone studies, we investigated the asymmetry of interference across frequency and the effect of relative timing of target and interferer. In addition, we investigated the effect of the spatial separation of target and interferer.

2. Experiment 1: Binaural Interference in the Free Field

2.1. *Method*

2.1.1. *Subjects*

Subjects were seven normal-hearing young adults (6 female, 1 male), aged 25-31.

2.1.2. *Environment and stimuli*

The experiment was conducted in the Bill Wilkerson Center Anechoic Chamber (4.6 m x 6.4 m x 6.7 m). Subjects were tested individually while seated in the center of a 64-loudspeaker array that spanned a full 360° in the horizontal plane (distance 1.95 m).

Two stimuli were employed: a 400-Hz wide noise, centered at 4000 Hz (HIGH), and a 400-Hz wide noise, centered at 500 Hz (LOW). Each was presented at a nominal level of 70 dB SPL, as measured with a microphone at the position of the subject's head. Each stimulus, HIGH and LOW, served in turn as Target or Interferer, as described below.

2.1.3. *Procedure*

To estimate the MAA, an adaptive 2-interval forced-choice task was employed that tracked the 79% correct level [9]. On each trial, the subject heard two presentations of the Target presented from slightly different azimuths, centered in front. The subject's task was to indicate whether the second presentation was to the left or right of the first presentation. Angular separation between the two presentations was varied from trial to trial, and the final MAA threshold estimate was taken as the average of the last six of eight total reversals in the tracking procedure. Stimuli could be presented from any arbitrary azimuth by employing a stereophonic balancing algorithm applied to adjacent loudspeakers in the array [10]. Four to eight threshold runs were completed per condition, and the final reported MAA was the mean of all 4-8 estimates.

Figure 1 (upper portion) illustrates the structure for a single trial in the experiment in which the angular separation between the two presentations is 4°. Duration of each presentation was 150 ms (including 20-ms raised cosine rise-decay times), and the interstimulus interval was 700 ms.

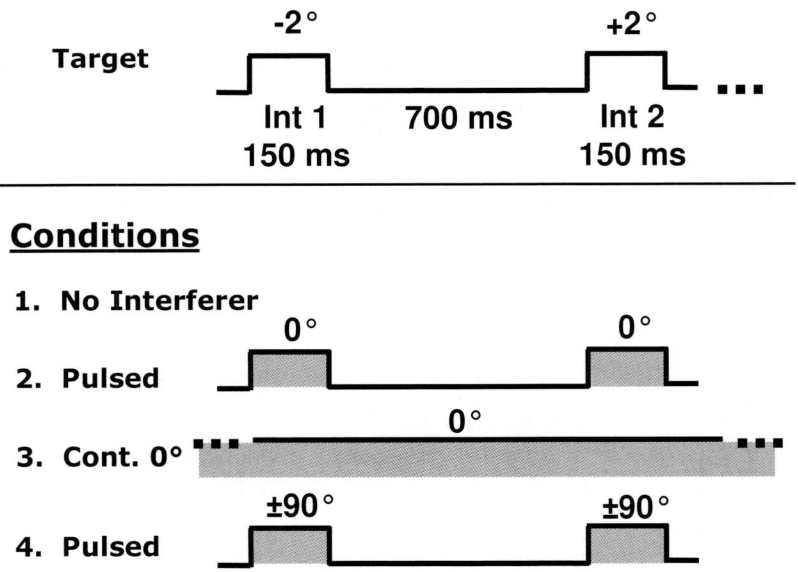

Figure 1. Upper portion: trial structure in the MAA task. In this example the stimulus in the second interval is presented 4° to the right of the stimulus in the first interval. Lower portion: illustration of the interferer conditions.

2.1.4. *Conditions*

Four basic conditions were employed, as illustrated in the lower portion of Figure 1.

1. No interferer;
2. Pulsed 0° - the interferer was pulsed on and off with the target, presented from 0° azimuth;
3. Continuous 0° - the interferer was on continuously during the threshold run, presented from 0° azimuth;
4. Pulsed ±90° - the interferer was pulsed on and off with the target, and was composed of independent noises presented from -90° and +90°.

For the HIGH target, all four conditions were run. For the LOW target, only conditions (1) and (2) were run. In the latter case, pilot data indicated that there was no interference in the condition most likely to produce interference (condition 2), so it was deemed unnecessary to run conditions (3) and (4).

2.2. *Results and Discussion*

For the HIGH target, MAAs for the four conditions are shown in the left panel of Figure 2. In the No-Interferer (NONE) condition, average threshold across the 7 subjects was 2.1°, which is in agreement with previous studies that have measured MAAs with high frequency stimuli [11, 12]. For each of the three interferer conditions, the MAA was elevated (2.4° for Continuous 0°; 13.4° for Pulsed 0°; 3.1° for Pulsed ±90°). Each of these thresholds was significantly different from the NONE threshold, as tested by two-tailed paired t-tests ($p < 0.05$). However, as can be seen from the figure, only the Pulsed 0° condition resulted in a large threshold increase relative to the NONE condition.

For the LOW target, MAAs for the two conditions are shown in the right panel of Figure 2. Mean threshold across subjects was 1.5° in the NONE condition and 1.6° in the Pulsed 0° condition. This difference was not statistically significant ($p = 0.17$).

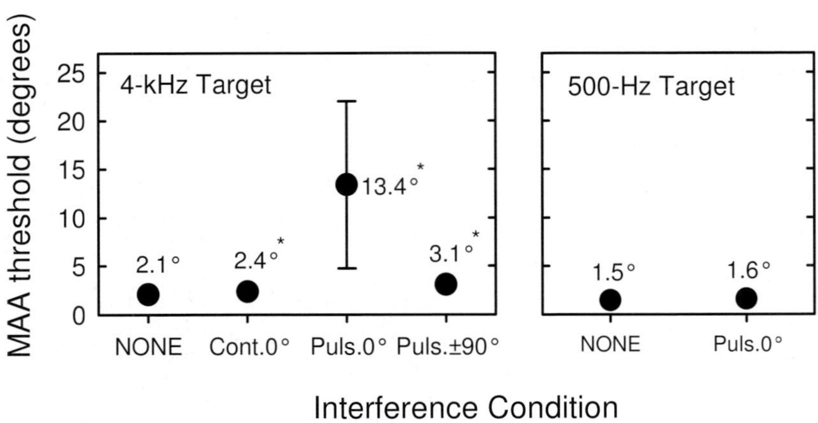

Figure. 2. Mean MAA across subjects for the different interferer conditions (± 1 standard deviation). Error bars not shown when standard deviations were smaller than the symbol. Left panel: 4.0 kHz target; right panel: 500-Hz.

2.2.1. *Asymmetry of binaural interference*

As with the headphone results [1], we found that binaural interference occurred with the HIGH target, but not with the LOW target. In discussing this asymmetry it should be borne in mind that the cues for spatial discrimination in the free field are different for low- and high-frequency stimuli. Spatial resolution for low-frequency targets is based on discrimination of interaural temporal differences (ITDs), while spatial resolution for high-frequency targets is based primarily on discrimination of interaural level differences (ILDs) [8, 13].

Thus, our results show that a low-frequency interferer can disrupt ILD processing for high-frequency targets, but that a high-frequency interferer does not disrupt ITD processing in low-frequency targets. Leaving aside for a moment which cues underlie performance in the two frequency regions, the direction of the asymmetry observed can be predicted by the weighted-combination models proposed to predict headphone presented stimuli. Specifically, because the MAA is significantly lower for the 500 Hz target in isolation than for the 4000-Hz target in isolation (1.5° vs 2.1°), a model based on the weighted combination of information across frequency would place more weight on the low-frequency target. For both ITD and MAA tasks, low-frequency information dominates in some sense.

2.2.2. *Relative timing of target and interferer*

For the HIGH target, we found that there is greater interference when the interferer is pulsed on and off with the target than when it is continuously present (Pulsed 0° MAA is significantly greater than Continuous 0° MAA; $p < 0.02$). This result has been reported frequently with headphone-presented stimuli (e.g., [4, 7, 14]). The results both in free field and under headphones are consistent with the notion that auditory grouping plays a strong role in binaural interference [2]. According to this notion, when the interferer is continuously present, subjects are likely to hear the target and interferer as separate objects and thus be able to listen "analytically." Under such an assumption the target ITD would be "diluted" less, or not at all, by the interaural information in the interferer. On the other hand, when the target and interferer are pulsed on and off together, subjects are likely to perceive the combined signal as a single object, and thus would become more susceptible to the obligatory combination of the signals (i.e., they will be less able to listen analytically).

2.2.3. *Effect of spatial position of the interferer*

For the HIGH target, there was significantly greater interference in the Pulsed 0° condition than in the Pulsed ±90° condition ($p < 0.03$). This result is also consistent with an auditory grouping notion, as described above. Specifically, for an interferer that is spatially separated from the target, subjects are likely to hear the two signals as separate events and thus be able to listen more analytically to the stimulus combination than when the interferer and target are spatially near each other.

Interestingly, this has not been the finding generally reported in the headphone studies. Introducing different interaural manipulations to the target and the interferer (such as employing an interaurally uncorrelated interferer or an interferer with a non-zero ITD) has led to greater interference [4], lesser interference [7], or no change in interference [6] compared to the use of a diotic interferer. Possibly the effect of spatial separation of target and interferer on the ability to segregate auditory objects is greater in the free field, in which multiple cues can contribute to spatial perception, than under headphones, in which only a single cue is manipulated.

3. Experiment 2: The Effect of Interferer Fringe on Binaural Interference

The results of experiment 1 are consistent with the notion that interference from a spectrally remote low-frequency interferer occurs to the extent that the target and interferer are fused into a single perceptual object [2]. Specifically, we showed that a pulsed interferer results in large interference effects, while a continuous interferer results in minimal interference. This result replicates previous binaural interference effects observed with headphone-presented stimuli (e.g., [4, 7]).

It is of interest to measure binaural interference in *intermediate* conditions, in which the interferer precedes and/or follows the target stimulus by specified finite durations. One might hypothesize that some degree of onset (or offset) interferer fringe would enable a subject to hear two objects. If interference persists for measurable degrees of interferer fringe despite the ability to segregate the stimulus into two auditory objects, one might conclude that binaural interference is not simply related to auditory object formation.

In fact, two previous studies did find that binaural interference persisted, relatively undiminished, for interferer fringes up to 320 ms [15, 16]. Accordingly, the authors of these studies came to the conclusion noted above: binaural interference is not based on simple object formation. In other words, segregation of auditory objects may not be sufficient to prevent interference. However, it should be noted that both studies employed only low-frequency targets and interferers (800 Hz and below). The effect of interferer fringe on binaural interference for high-frequency targets is still unknown.

The purpose of experiment 2 was to measure binaural interference as a function of interferer fringe duration for the condition when the target was a high-frequency stimulus and the interferer a low-frequency stimulus. Interaural time difference (ITD) thresholds were measured under headphones.

3.1. *Method*

3.1.1. *Subjects*

Subjects were four normal-hearing females, aged 25-31. None of the four had participated in experiment 1.

3.1.2. *Environment and stimuli*

Subjects were tested individually in a sound-insulated test booth. The same two stimuli as described in experiment 1 were employed: a 400-Hz wide noise, centered at 4000 Hz (HIGH), and a 400-Hz wide noise, centered at 500 Hz (LOW). Stimuli were presented over TDH-49 headphones. The level of each stimulus, when presented continuously in isolation, was 70 dB SPL.

3.1.3. *Procedure*

To estimate the ITD threshold, an adaptive 2-interval forced-choice task was employed, which tracked the 70.7% correct level [9]. On each trial, the subject heard two presentations of the Target. One of the presentations was presented diotically, and the other had a specified ITD favoring the left ear. Stimulus duration was 150 ms (20-ms rise decay time), and the interstimulus interval was 700 ms. The subject's task was to indicate whether the second presentation was to the left or right of the first presentation. Trial by trial feedback was provided. ITD was varied from trial to trial, and the final ITD threshold estimate was taken as the average of the last six of eight total reversals in the tracking procedure. Four to eight threshold runs were completed per condition, and the final reported ITD threshold was the mean of all 4-8 estimates.

3.1.4. *Conditions*

Threshold ITD was measured for the HIGH target in isolation and in the presence of the LOW interferer. The LOW interferer was always presented diotically in both intervals of every trial. In addition to continuous and pulsed interferers (as illustrated in the bottom portion of Figure 1), interferers were presented with fringe durations of 0-400 ms. The fringe duration was either all at the Onset, at the Offset, or divided evenly between Onset and Offset (see Figure 3).

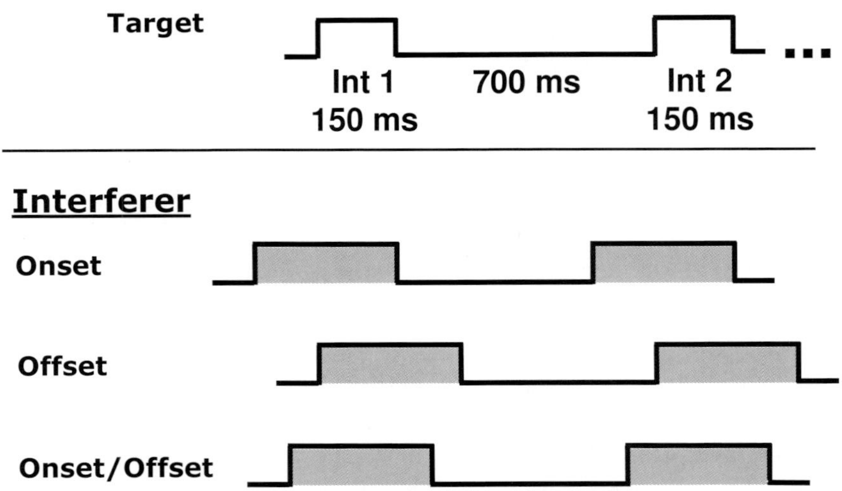

Figure 3. Illustration of the trial structure in the ITD task. Upper portion: HIGH target. Lower portion: LOW interferer. In the case of Onset fringe, the interferer was turned off with the target. In the case of Offset fringe, the interferer was turned on with the target. For Onset/Offset fringe, the fringe was divided evenly between onset and offset.

3.2. *Results and Discussion*

Average data for the four subjects are displayed in Figure 4. ITD thresholds were normalized to the threshold measured for the HIGH target in isolation, as indicated by the arrow to the right of the panel.[a] The data point on the left (Fr 0) represents the case when target and interferer are pulsed on and off together (0 Fringe). The data point on the right (Cont.) is the average normalized threshold in the case of a continuous interferer. As shown in experiment 1 and in earlier studies, there is much greater interference in the former than in the latter condition, which indeed formed the rationale for this experiment.

The data points plotted at Fr200 and Fr400 represent thresholds obtained when the total fringe duration was 200 and 400 ms, respectively. The parameter indicates the type of fringe (Onset, Offset, or Onset/Offset). As can be seen, there is no consistent effect of type of fringe. ITD thresholds decrease monotonically as fringe duration increases for each type of fringe. Collapsing across type of fringe, there was a significant linear trend effect of fringe duration on the log transformed data from our four subjects ($p = 0.03$).

[a] Thresholds for the 4000-Hz (HIGH) target presented in isolation for the four subjects were 62 μs, 159 μs, 65 μs, and 83 μs.

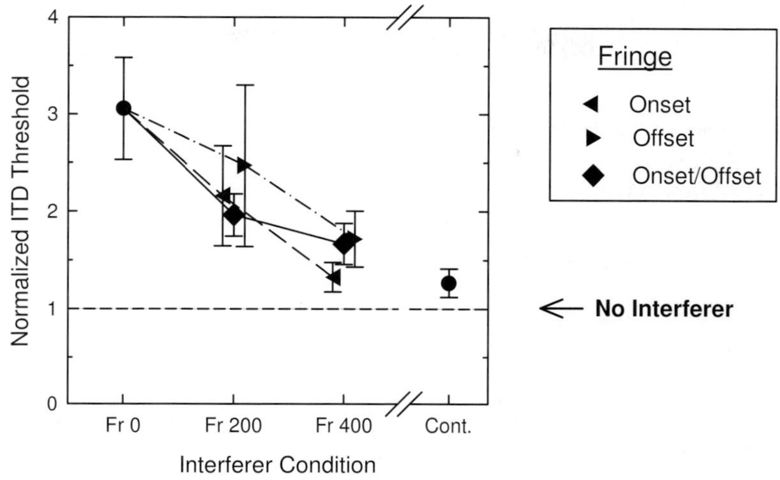

Figure 4. Mean normalized ITD thresholds for four subjects in experiment 2. Error bars indicate ± 1 standard deviation. Thresholds are plotted as a function of fringe duration (0, 200, 400 ms), with threshold for continuous fringe shown at the right. Parameter shows type of fringe.

Unlike the results from experiments in which targets and interferers were both low-frequency stimuli, the present data show that interference decreases monotonically as fringe duration increases over the range tested. Now, to the extent that subjects are more likely to perceive a pair of segregated auditory images as finge duration increases, these results are consistent with the notion that, for high-frequency targets, interference is strongly influenced by the degree to which the target and interferer are heard as separate objects. While the notion that the probability of object segregation increases as fringe duration increases is intuitively appealing, firm conclusions regarding the influence of object segregation on binaural interference will have to await empirical data that address directly how object segregation varies with fringe duration in conditons like those tested here.

4. Experiment 3. The Effect of Interstimulus Interval on Binaural Interference

During pilot data collection leading up to experiment 2, we noticed an interesting and unexpected trend in the data. Experiment 2 required us to use relatively long interstimulus intervals (ISIs) of 700 ms between the two target presentations on every trial. This was necessary in order to have sufficient time to place the interferer fringe before and/or after the target. The unexpected finding was that interference (in a pulsed interferer condition) was consistently less than we had observed in earlier pilot data obtained with short ISIs. This trend suggested that binaural interference may interact in a complicated way with binaural temporal processing.

The purpose of experiment 3 was to systematically investigate the effect of ISI on the degree of binaural interference, using the same target and interferer stimuli as employed in experiments 1 and 2.

4.1. *Subjects, Environment, and Stimuli*

Subjects were four normal-hearing females, aged 25-31. One (RA) had participated in experiment 1, and the other three were new. The same two stimuli as described in experiments 1 and 2 were employed: a 400-Hz wide noise, centered at 4000 Hz (HIGH), and a 400-Hz wide noise, centered at 500 Hz (LOW). As in experiment 2, stimuli were presented over TDH-49 headphones. The level of each stimulus, when presented continuously in isolation, was 70 dB SPL.

4.1.1. *Procedure*

The procedure was the same as described for experiment 2. Threshold ITD was measured for the HIGH target in isolation and in the presence of the LOW interferer. The LOW interferer was always presented diotically in both intervals of every trial. In this experiment, the interferer, when present, was always pulsed on and off with the target. The target (and interferer) duration was 150 ms (e.g., see Figures 1 and 3). Interstimulus interval was 50, 100, 300, or 700 ms. The order of conditions was counterbalanced across subjects.

4.2. *Results and Discussion*

Thresholds for the four subjects are plotted in Figure 5 as a function of ISI. Open symbols show the results when the target was presented in isolation, and filled symbols show thresholds in the presence of the low-frequency diotic interferer.

Considering first the targets presented in isolation (open symbols), thresholds are relatively independent of ISI (except for subject AH). This result is similar to that reported for ITD thresholds for low-frequency [17] and broadband [18] signals, although it is known from these earlier studies that ITD threshold does increase considerably as ISI further decreases below 50 ms. It is believed that this worsening of ITD discrimination as the ISI becomes increasingly small is related to binaural sluggishness (e.g., [17]. Here, with ISI greater than or equal to 50 ms (for these 150-ms markers), binaural sluggishness is apparently not a factor.

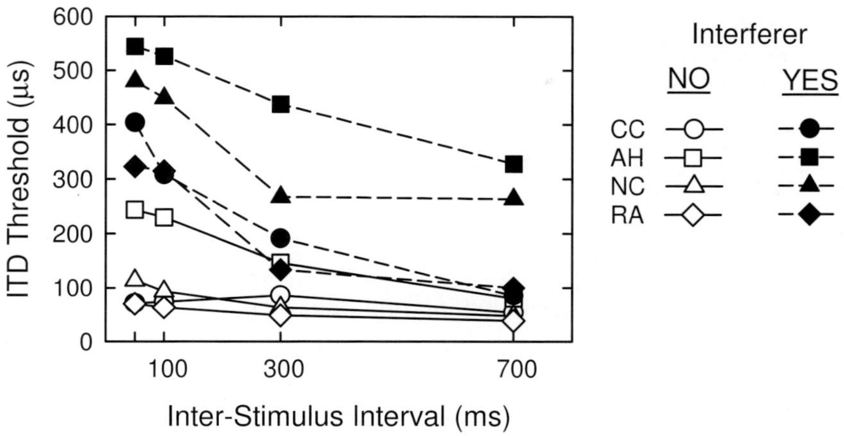

Figure 5. ITD thresholds for the high-frequency target for four subjects obtained with (filled symbols) and without (open symbols) the low-frequency interferer, plotted as a function of interstimulus interval.

The thresholds for the target in the presence of the interferer (solid symbols) show a consistent decline as a function of ISI. For all subjects except AH (who had the highest thresholds in both conditions), this decrease in threshold is much greater than the slight changes seen in the thresholds obtained in isolation, confirming the finding of our pilot data that binaural interference decreases with increasing ISI. These effects are supported by the results of a two-factor repeated measures ANOVA (4 ISIs x 2 interference conditions). Both main effects were statistically significant, but more importantly for the question of whether ISI affects the amount of interference, the interaction between the two factors was also significant ($p = 0.035$). This interaction is illustrated more directly in Figure 6 which plots the amount of interference (the difference between thresholds obtained with and without the interferer) as a function of ISI.

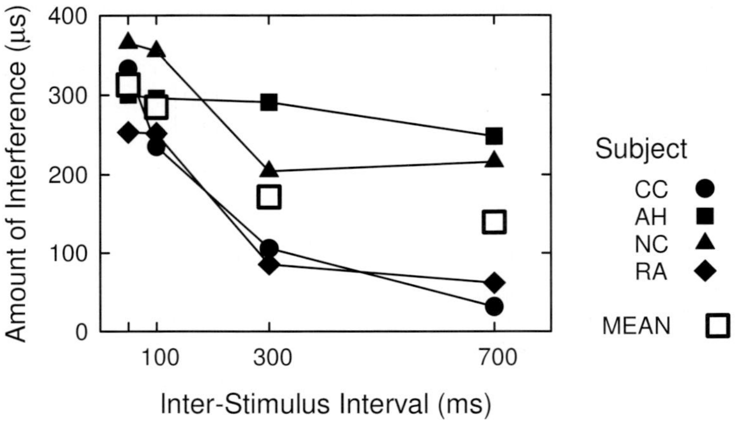

Figure 6. Amount of interference (difference between thresholds obtained with and without the interferer) as a function of interstimulus interval for four subjects. The large open squares show the means across subjects.

As noted above, the apparent decrease in binaural interference as a function of ISI was a surprising finding, and it is as yet unclear the source of this effect. The temporal course of the effect appears to be longer than that associated with binaural sluggishness, where it has been observed that the "minimum integration time" (that time required to perform optimally in a spatial resolution task) is on the order of 150-300 ms [19]. Here, it appears that the minimum time required to reach optimum performance is somewhat greater than 300 ms. These data indicate that the basic models proposed to describe binaural interference [5, 6] will need to be augmented to account for such long-term temporal effects. For example, the processing variances associated with target and interferer may be governed by different temporal window functions that cause their relative contributions to vary over the course of a trial presentation. Further experiments will be required to elucidate the operation of such temporal windows.

5. Summary and Conclusions

1. Experiment 1 showed that the MAA for a high-frequency target is considerably elevated when a low-frequency interferer is pulsed on and off with the target from a position directly in front of the subject. However, if the low-frequency interferer is presented continuously, or if its spatial position is different from that of the target, the interference is practically eliminated. A high-frequency interferer does not elevate the MAA for a low-frequency target. These results suggest that binaural interference as

traditionally measured under headphones may play a significant role in human auditory spatial resolution in complex environments.

2. Experiment 2 showed that the degree of binaural interference in an ITD task decreases monotonically with the duration of interferer fringe that comes before, after, or both before and after the target.
3. The results of both experiments 1 and 2 are consistent with the notion that interference from a spectrally remote low-frequency interferer is influenced by the extent to which the target and interferer are fused into a single perceptual object [2]. If cues are provided that would be expected to promote perceptual segregation (such as temporal onset differences or spatial location differences), the interference is reduced or eliminated.
4. Experiment 3 showed that binaural interference increases as the interstimulus interval between the two target presentations decreases. This unexpected finding indicates that the weights assigned to target and interferer in an interference paradigm are governed by temporal windows that may be even longer than those associated with binaural sluggishness.

References

1. D. McFadden and E. G. Pasanen, Lateralization at high frequencies based on interaural time differences, *J. Acoust. Soc. Am.* **59,** 634 (1976).
2. V. Best, F. J. Gallun, S. Carlile and B. G. Shinn-Cunningham, Binaural interference and auditory grouping, *J. Acoust. Soc. Am.* **121,** 1070 (2007).
3. L. M. Heller and V. M. Richards, Binaural interference in lateralization thresholds for interaural time and level differences, *J. Acoust. Soc. Am.* **In press** (2010).
4. L. R. Bernstein and C. Trahiotis, Binaural interference effects measured with masking-level difference and with ITD- and IID-discrimination paradigms, *J. Acoust. Soc. Am.* **98,** 155 (1995).
5. L. M. Heller and C. Trahiotis, Extents of laterality and binaural interference effects, *J. Acoust. Soc. Am.* **99,** 3632 (1996).
6. T. N. Buell and E. R. Hafter, Combination of binaural information across frequency bands, *J. Acoust. Soc. Am.* **90,** 1894 (1991).
7. C. Trahiotis and L. R. Bernstein, Detectability of interaural delays over select spectral regions: Effects of flanking noise, *J. Acoust. Soc. Am.* **87,** 810 (1990).
8. F. L. Wightman and D. J. Kistler, The dominant role of low-frequency interaural time differences in sound localization, *J. Acoust. Soc. Am.* **91,** 1648 (1992).
9. H. Levitt, Transformed up-down methods in psychoacoustics, *J. Acoust. Soc. Am.* **49,** 467. (1971).

10. V. Pulkki and M. Karjalainen, Localization of amplitude-panned virtual sources. I. Stereophonic panning, *J. Audio Eng. Soc.* **49,** 739 (2001).
11. E. R. Hafter, T. N. Buell, D. A. Basiji, and E. E. Shriberg, Discrimination of direction for complex sounds presented in the free-field, in *Basic Issues in Hearing: Proceedings of the 8th International Symposium on Hearing*, edited by H. Duifhuis, J. W. Horst, and H. P. Wit (Academic, London), pp. 394-401.(1988).
12. E. R. Hafter, K. Saberi, E. R. Jensen, and F. Briolle, Localization in an echoic environment, in *Auditory Physiology and Perception*, edited by Y. Cazals, K. Horner, and L. Demany (Pergamon, Oxford), pp. 555-561.(1992).
13. A. W. Mills, Lateralization of high-frequency tones, *J. Acoust. Soc. Am.* **32 ,** 132 (1960).
14. L. M. Heller and C. Trahiotis, Interference in detection of interaural delay in a sinusoidally amplitude-modulated tone produced by a second, spectrally remote sinusoidally amplitude-modulated tone, *J. Acoust. Soc. Am.* **97,** 1808 (1995).
15. L. R. Bernstein and C. Trahiotis, Spectral interference in a binaural detection task: Effects of masker bandwidth and temporal fringe, *J. Acoust. Soc. Am.* **94,** 735 (1993).
16. W. S. Woods and H. S. Colburn, Test of a model of auditory object formation using intensity and interaural time difference discrimination, *J. Acoust. Soc. Am.* **91,** 2894 (1992).
17. D. W. Grantham, Auditory motion perception: Snapshots revisited, in *Binaural and Spatial Hearing in Real and Virtual Environments*, edited by R. H. Gilkey and T. R. Anderson (Lawrence Erlbaum, Mahwah, NJ), pp. 295-313.(1997).
18. D. R. Perrott and S. Pacheco, Minimum audible angle thresholds for broadband noise as a function of the delay between the onset of the lead and lag signals , *J. Acoust. Soc. Am.* **85,** 2669 (1989).
19. D. W. Grantham, Detection and discrimination of simulated motion of auditory targets in the horizontal plane, *J. Acoust. Soc. Am.* **79,** 1939 (1986).

EFFECTS OF TIMBRE ON LEARNING TO REMEDIATE SOUND LOCALIZATION IN THE HORIZONTAL PLANE

D. YAMAGISHI and K. OZAWA†

Interdisciplinary Graduate School of Medicine and Engineering,
University of Yamanashi, Kofu 400-8511, Japan
†E-mail:ozawa@yamanashi.ac.jp
www.ccn.yamanashi.ac.jp/~ozawa/lab.htm

Previous studies have shown the efficacy of sound localization training with non-individualized head-related transfer functions (HRTFs) in a virtual auditory display. However, because the training and test phases used the same types of sound, the efficacy in those studies might have been based on timbre perception. This study re-examined the efficacy using phases with different types of sound; noise and music were used in the training and test sessions, respectively. The experimental results show that the training was effective, indicating that learning was based on the HRTFs rather than the timbre.

1. Introduction

At least three cues are included for sound localization in Head-Related Transfer Functions (HRTFs) for both ears: interaural time difference (ITD), interaural level difference (ILD), and spectral cues [1]. Previous studies [2, 3] have demonstrated adaptation to degraded spectral cues over a few weeks.

Recently, a study by Zahorik *et al.* [4] demonstrated the efficacy of a sound localization training procedure with non-individualized HRTFs in a Virtual Auditory Display (VAD), using short periods of training (two 30-min sessions) to significantly reduce the rate of front-back reversal. These effects were observed at untrained locations, and persisted at least four months after the training. Additionally, Ueno *et al.* [5] demonstrated a learning effect in the perception of sound location. These two studies indicate that training is effective even when HRTFs are different from those of the individual listener. However, the efficacy might have been based on timbre perception because the training and test phases used the same types of sound. Herein, we constructed a sound localization training system and reexamined the efficacy using phases with different types of sound; noise and music were used in the training and test sessions, respectively.

Because it has been suggested that auditory spatial perception is affected by other sensory modalities, e.g. vision and proprioception [6, 7], Zahorik *et al.* [4] used auditory, visual, and proprioceptive/vestibular feedback in their training. However, the contribution of such multimodal feedback to the fast adaptation is unclear. Hence to evaluate the effectiveness of multimodal feedback, we provided only auditory feedback.

2. Method

2.1. *Apparatus*

We constructed a training and test system for sound localization with a VAD. The system consisted of a personal computer (HP, f2096b), high-quality stereo headphones (Sennheiser, HD 600), and a touch display (I-O data, LCD-AD171F-T). The system, including the graphical user interface (GUI), was developed using the Visual Studio 2005. Figure 1 shows an example of the GUI used in both the training and test phases from a bird's eye view where the head of the subject is located at the center.

Subjects were presented with a sound stimulus via headphones, and then used their finger to point to the location of the perceived sound image on the GUI. In Fig. 1, the filled circle is the perceived sound image of the stimulus. In a training session, if the subject pushed the "feedback" button, the correct

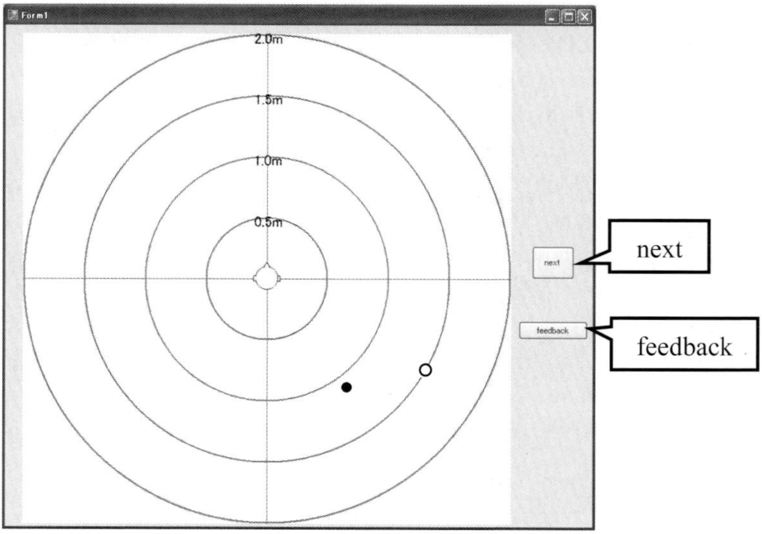

Fig. 1. Picture of a GUI. Filled circle: Position of the perceived sound image of a stimulus. Hollow circle: Simulated position of the stimulus.

position of the simulated sound source was displayed by a hollow circle and the stimulus was presented again. The subject could receive the feedback for as long as he or she desired. By pressing the "next" button, the next stimulus was presented.

2.2. *Stimuli*

In a VAD, a stimulus is generated by convolution of a source sound signal and a HRTF. As source signal, a sample of white noise with a duration of 1 s was used in both the test and training phases, whereas a piece of music with a duration of 6 s (H's Band, "The Ramp," RWC-MDB-G-2001 [8], No. 33) was used only in the test phase.

The HRTFs used in this study were measured in an anechoic room at the Research Institute of Electrical Communication, Tohoku University using a dummy head (Koken, SAMRAI) at the eardrums of both ears with C-type couplers and microphones (B&K, 4134). In the horizontal plane, 72 HRTFs were measured at 5-degree increments.

Prior to reproducing a stimulus via the headphones, the inverse characteristics of the headphone transfer function (HpTF) [9, 10] were convolved to the signal for calibration. This calibration dissolved frequency characteristic duplication of the outer ear during recording and reproduction as well as compensating for the frequency characteristics of the headphones.

2.3. *Procedure*

Figure 2 shows the three phases of the experiment. The first phase was a pre-training test session, which is abbreviated as "Pre" hereafter. The second phase consisted of two training sessions called "Train1" and "Train2", while the third incorporated the three post-training test sessions referred to as "Post1," "Post2," and "Post3". Sessions from Pre to Post1 were carried out on two consecutive days. On the first day, a subject engaged in the Pre test then Train1, whereas the subject participated in Train2 and Post1 on the second day. Post2 and Post3 tests were conducted one week and one month after Post1, respectively.

The 72 HRTFs were divided into three sets: each set consisted of 24 HRTFs at 15-degree increments. To avoid subjects from remembering the timbre

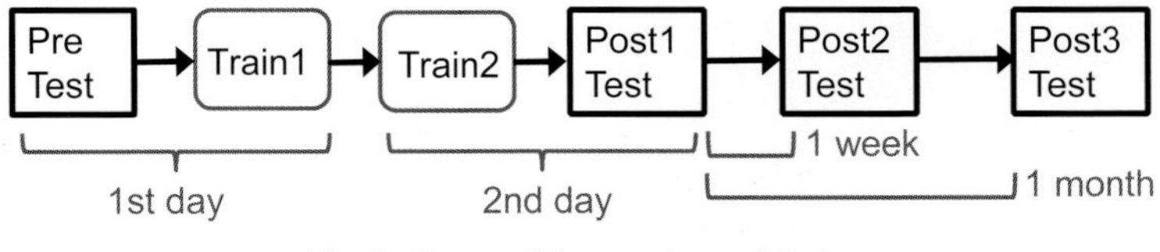

Fig. 2. Phases of the experimental design.

associated with a sound position, test sessions Pre and Post1–Post3, used the set in which the sample direction of 0 degrees was included, whereas Train1 and Train 2 used one of the other two sets. In each test session of Pre or Post1–Post3, a subject listened to a stimulus and answered the perceived position of its sound image. Each test session of Pre or Post1–Post3 consisted of 144 stimuli, because two types of sound (noise and music) were each presented three times, from 24 directions, in random order. However, only 72 stimuli were presented in the training sessions, Train1 or Train2, because the type of stimulus was noise, which was randomly presented three times from 24 directions. Additionally, in the training sessions the subject received feedback on the correct position after every sound stimulus.

Eleven subjects (six male and five female, age range: 21–24 years) with normal hearing participated in the experiment.

3. Results

3.1. *Localization results*

As examples, Figs. 3 and 4 represent the experimental results of a subject (Subject 2) for noise and music stimuli, respectively. In the Pre condition, numerous front-back confusions occurred, where stimuli presented in front were perceived as from the rear for both types of signal. This type of front-back reversal occurs frequently when non-individualized HRTFs are used

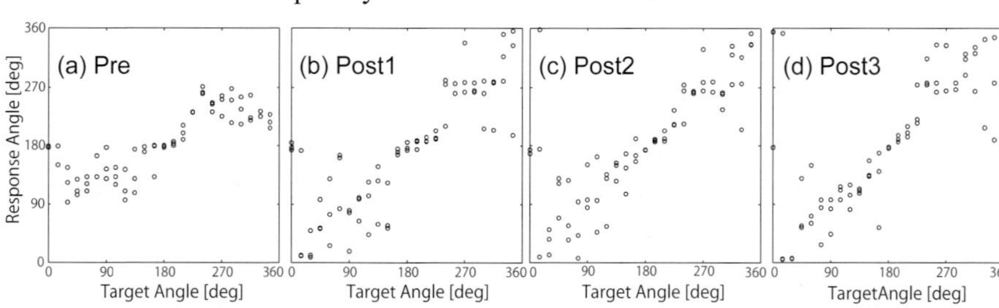

Fig. 3. Scatterplots of the target sound source angle versus response angle for noise (Subject 2).

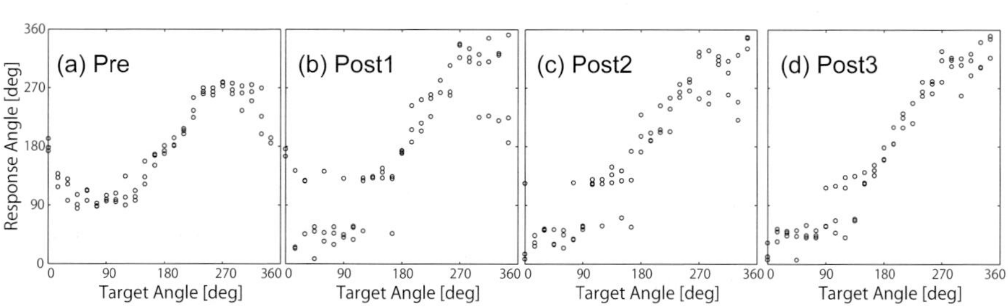

Fig. 4. Scatterplots of the target sound source angle versus response angle for music (Subject 2).

in a VAD. However, two training sessions using noise stimuli resulted in a marked improvement for both the noise and music stimuli (Post1). This finding indicates that the subjects' judgments are not based on the timbre of sound stimuli, but on the characteristics of HRTFs, i.e. spectral cues included in the sound stimuli. Additionally, this study demonstrates that fast learning to remediate sound localization is possible without visual and proprioceptive/vestibular feedback, although the previous study [4] used auditory, visual, and proprioceptive/vestibular feedback. Moreover, the results of Post2 and Post3 demonstrated that the efficacy of the training persisted for at least one month, which is consistent with the previous study [4].

3.2. *Summary of localization results*

Figure 5 shows the mean unsigned error and the standard deviation (SD) across subjects, which were calculated according to the previous study [4]. The mean unsigned error is defined as the mean unsigned deviation of the response angle from the target angle in units of degrees. Initially, the mean unsigned error in the right-left dimension was calculated and used to estimate the ambiguity in localization, ε_{est}. Then reversals in the front-back dimension where the response angle occurred in the incorrect hemi-field were identified if the deviation was larger than ε_{est}. Finally, after the front-back reversals were resolved, the mean unsigned error in the front-back dimension was calculated. As shown in Fig. 5, the two panels exhibited nearly identical results. This strongly suggests that localization errors are consistent in both the right-left and front-back dimensions when the front-back reversals are resolved.

The data in Fig. 5 were subjected to two-way analysis of variance (ANOVA) in which the factors were the test session (Pre and Post1–Post3) and the type of signal (noise and music). The results show that the two factors are

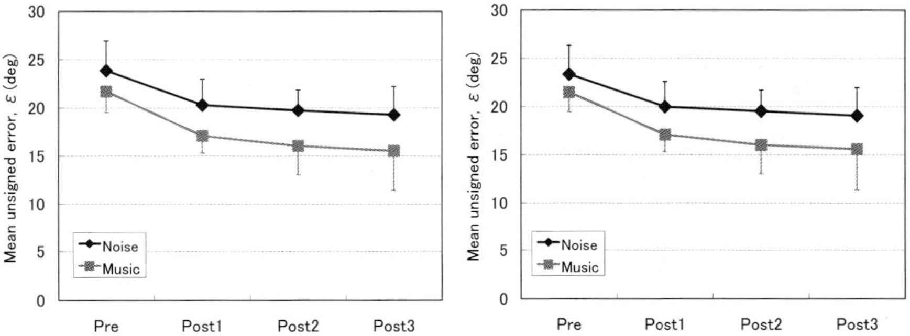

(a) Right-left Dimension (b) Front-back Dimension

Fig. 5. Summary of the mean unsigned error and SD across subjects.

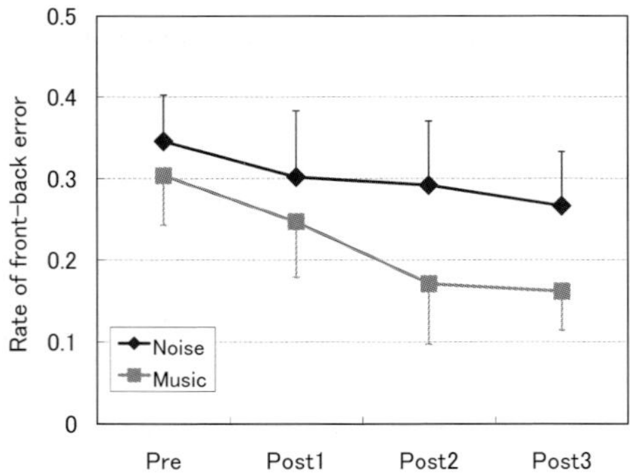

Fig. 6. Summary of the rate of front-back error and SD across subjects.

statistically significant, but the interactions between the two factors are not significant for either direction (Right-left direction: $F(3, 80) = 16.19$, $p < 0.001$ for the test session; $F(1, 80) = 28.78$, $p < 0.001$ for the type of signal; $F(3, 80) = 0.42$, $p > 0.1$ for the interaction. Front-back direction: $F(3, 80) = 16.05$, $p < 0.001$ for the test session; $F(1, 80) = 28.44$, $p < 0.001$ for the type of signal; $F(3, 80) = 0.38$, $p > 0.1$ for the interaction.). The results of the least square difference (LSD) tests were Pre > Post1 = Post2 = Post3 for both directions (Front-back direction: LSD = 1.72; Right-left direction: LSD = 1.20). These results confirm that the training is effective, even if the type of signal differs between the test and training sessions.

Figure 6 shows the rate of front-back reversals and SD across subjects. Two-way ANOVA revealed that the main effects of the two factors are significant, but the interaction is insignificant ($F(3, 80) = 11.75$, $p < 0.001$ for the test session; $F(1, 80) = 31.26$, $p < 0.001$ for the type of signal; $F(3, 80) = 1.76$, $p > 0.1$ for the interaction). The results of the LSD test were Pre > Post1 > Post2 = Post3 (LSD = 0.04), indicating that the training effectively reduces front-back reversals.

4. Discussion

4.1. *Difference between music and noise*

As shown in Figs. 5 and 6, localization errors for music stimuli are significantly lower than those for noise. One reason for this phenomenon is that the duration of a music stimulus (6 s) is longer than that of a noise stimulus (1 s). A longer duration may make it easier to obtain directional information such as the spectral cues. Another possibility is that the source signal of music was fixed, whereas a fresh noise was used for every presentation. Hence, it may be easier to gather directional information from a fixed source signal.

4.2. *Proposal of a new definition of front-back errors*

Figure 7 shows another example of the results for the Pre and Post1 tests with respect to Subject 3. The training appears to be effective for this subject because the responses are distributed along the diagonal for Post1 test, whereas most of the responses are allocated to the rear for Pre test. However, the rate of front-back reversals calculated according to the literature [4] indicates that the rate of Post1 is higher than that of Pre test. This contradiction seems to arise from the definition of the rate of front-back reversals.

Figure 8(a) schematically represents the definition of a front-back error in the literature [4] where ε denotes the ambiguity in localization as described in Sec. 3.2. For example, if the target angle is 45 degrees, a response angle of 90 degrees is not regarded as a front-back error, but a response angle of 91 degrees is regarded as a front-back error. The same phenomenon occurs near a response angle of 270 degrees. As shown in Fig. 7(a), numerous responses are on the 90-

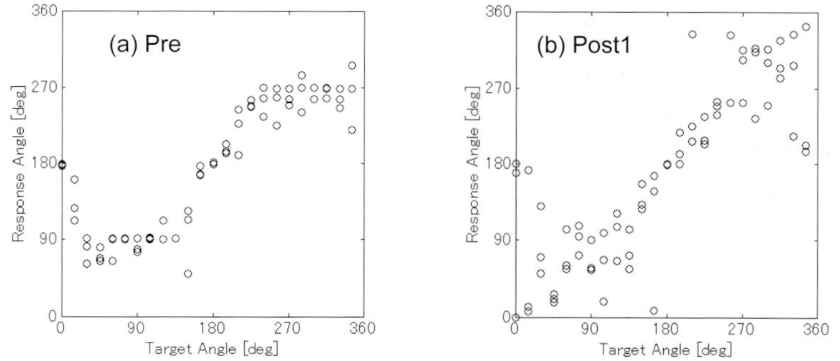

Fig. 7. Scatterplots of the target sound source angle versus response angle for music (Subject 3).

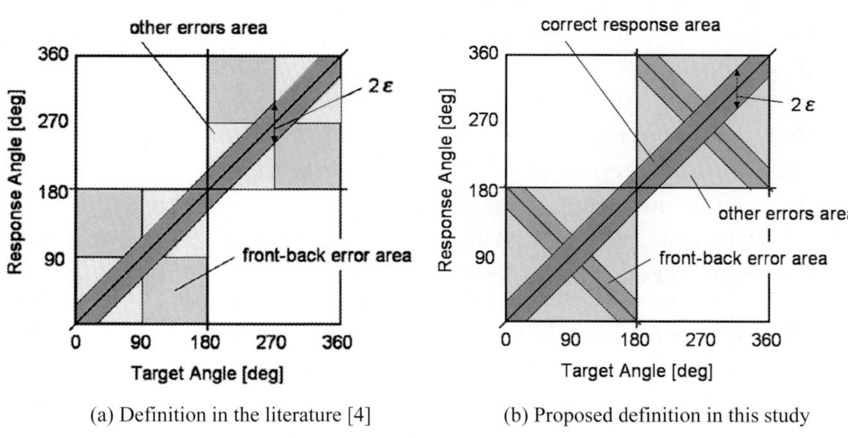

Fig. 8. Schematic representation of the front-back error on a scatterplot diagram.

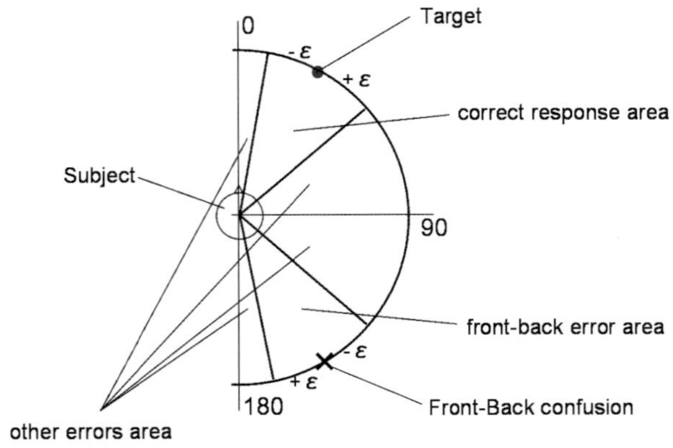

Fig. 9. Proposed definition of a front-back error area.

and 270-degree lines. This is because the rate of front-back reversals is smaller for the Pre test according to this definition.

Herein, we propose a new definition of a front-back error as shown in Fig. 8(b). Similar to the literature [4], a response within $\pm\varepsilon$ of its target angle is regarded as a correct one. As shown in Fig. 9, we defined the front-back error area as the reversal of the correct response area. The area where the correct response and the front-back error areas overlap near 90 or 270 degrees is regarded as the correct response area. Consequently, incorrect responses are divided into two measures: "front-back errors" and "other errors." According to this definition, the summary of the two measures is calculated and displayed in Fig. 10. The results of two-way ANOVA show that the factor of test session is significant for the two measures, and the signal type is significant for the front-back error (Other errors: $F(3, 80) = 12.83$, $p < 0.001$ for the test session; $F(1, 80) = 2.09$, $p > 0.1$ for the type of signal; $F(3, 80) = 0.53$, $p > 0.1$ for the interaction. Front-back error: $F(3, 80) = 3.11$, $p < 0.05$ for the test session; $F(1, 80) = 13.64$, $p < 0.001$ for the type of signal; $F(3, 80) = 1.52$, $p > 0.1$ for the interaction). The results of the LSD test are Pre > Post1 = Post2 = Post3 for the measure of other errors (LSD = 0.04), whereas those for the measure of front-back errors are complex: Pre = Post1 > Post2 = Post3, but Pre = Post2 (LSD = 0.03). Thus, the proposed measures of error indicate that the training is effective mostly for other errors corrected in our experiment.

4.3. *Discrepancy between the previous and present studies*

The results of this study conflict with the previous study [4] where the efficacy of training was seen primarily as front-back error remediation. We hypothesize that this conflict is due to the difference in the HRTFs used in VADs. In the

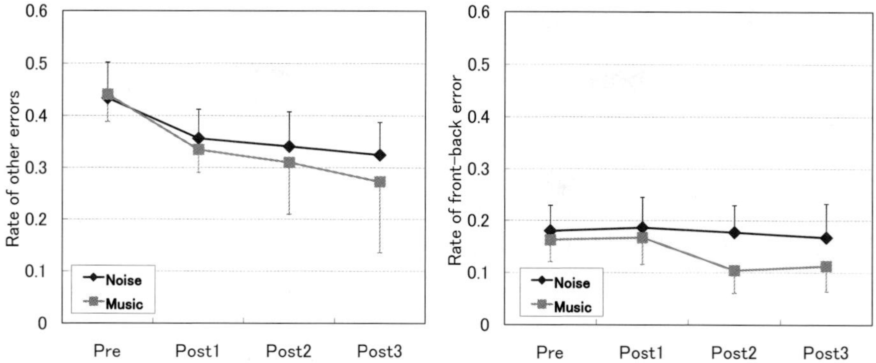

(a) Rate of other errors

(b) Rate of front-back error

Fig. 10. Summary of the error rates and SD across subjects.

results of our Pre test, the tendency for responses to cluster around 90 or 270 degrees is clearly observed for five subjects, including Subjects 2 and 3 as seen in Figs. 4(a) and 7(a), respectively. This suggests that the HRTFs used in our experiment have larger ITD and/or ILD cues than the subjects' own HRTFs; thus, the sound image tends to be localized laterally. If this is true, learning of ITD and/or ILD should be available in short periods of training. Additionally, difference in the spectral cues from the listener's own HRTF is obvious because front-back confusion is apparent near 0 degrees, but as the figures show, training remediates this issue. Unfortunately, this remediation is not represented clearly in our proposed measure.

Although we proposed a new definition for the errors, further consideration is necessary. The measure defined in the literature [4] is too sensitive for errors around 90 and 270 degrees, while the proposed measure is too strict for evaluating front-back errors.

5. Summary

In this study, we examined the efficacy of a sound localization training procedure with non-individualized HRTFs in a VAD. The results indicate that short periods of training with noise stimuli are effective for localization of music stimuli. Hence, learning to remediate sound localization is not based on timbre perception of a stimulus, but is based on the directional information included in the stimulus, such as spectral cues in HRTFs. Furthermore, learning occurs without proprioceptive/vestibular feedback.

To elucidate the relative importance of the spectral cues on learning, further research is required because the ITD and ILD cues affect sound localization in the horizontal plane [11].

Acknowledgments

The authors wish to thank Dr. K. Watanabe and Mr. M. Itokazu of University of Yamanashi for their help in measuring HRTFs. We are also grateful to Ms. H. Ishii and Ms. Y. Nakazawa of University of Yamanashi for their assistance in constructing the experimental setup. Part of this study was carried out under the Cooperative Research Project of the Research Institute of Electrical Communication, Tohoku University (H20/A10).

References

1. J. Blauert, *Spatial Hearing. The psychophysics of human sound localization (Revised Ed.)*, (MIT Press, Cambridge, 1997).
2. P. V. Hofman, J. G. Van Riswick and A. J. Van Opstal, Relearning sound localization with new ears, *Nat. Neurosci.* **1**, 417 (1998).
3. M. M. Van Wanrooij and A. J. Van Opstal, Relearning sound localization with a new ear, *J. Neurosci.* **25**, 5413 (2005).
4. P. Zahorik, P. Bangayan, V. Sundareswaran, K. Wang and C. Tam, Perceptual recalibration in human sound localization: Learning to remediate front-back reversals, *J. Acoust. Soc. Am.* **120**, 343 (2006).
5. K. Ueno, H. Tsuchiya, S. Ise and M. Otani, Study on learning in localization and searching of sound source, *Tans. Tech. Comm. Psychol. Physiol. Acoust. of Acoust. Soc. Jpn.* **36**, 645 (2006).
6. J. Lewald, Opposing effects of head position on sound localization in blind and sighted human subjects, *Eur. J. Neurosci.* **15**, 1219 (2002).
7. M. P. Zweirs, A. J. Van Opstal and G. D. Paige, Two-dimensional sound localization behavior of early-blind humans, *Exp. Brain Res.* **140**, 206 (2001).
8. M. Goto, H. Hashiguchi, T. Nishimura and R. Oka, RWC Music Database: Popular, Classical, and Jazz Music Databases, in *Proc. 3rd Int. Conf. on Music Information Retrieval (ISMIR 2002)*, (Paris, France, 2002).
9. H. Møller, Fundamentals of binaural technology, *Appl. Acoust.* **36**, 171 (1992).
10. K. Fukudome, Equalization for the dummy-head-headphone system capable of reproducing true directional information, *J. Acoust. Soc. Jpn. (E)* **1**, 59 (1980).
11. K. Watanabe, K. Ozawa, Y. Iwaya, Y. Suzuki and K. Aso, Estimation of interaural level difference based on anthropometry and its effect on sound localization, *J. Acoust. Soc. Am.* **122**, 2832 (2007).

EFFECT OF SUBJECTS' HEARING THRESHOLD ON SIGNAL BANDWIDTH NECESSARY FOR HORIZONTAL SOUND LOCALIZATION

D. MORIKAWA AND T. HIRAHARA

Faculty of Engineering, Toyama Prefectural University, 5180 Imizu, Toyama 939-0398, Japan

This paper presents examination of the signal bandwidth necessary to localize real sound sources, and the relationship between bandwidth and the listener's audible frequency range. Horizontal sound localization experiments were conducted with 10 listeners using white noise, high-pass noise with cut off frequencies (F_c) of 2, 4, 8, 12, or 16 kHz, or low-pass noise with F_c of 0.5, 1, 2, 4, or 8 kHz. Hearing threshold levels from 0.125 to 18 kHz were also measured for each subject. Sound localization was very difficult for high-F_c high-pass noises. Sound localization performance was 74% for 12-kHz high-pass noise, although it was 34% for 16-kHz high-pass noise. By contrast, sound localization was possible even for 500-Hz low-pass noise. Sound localization performances were, respectively 78% and 67% for 1-kHz and 500-Hz low-pass noise. A signal bandwidth from 2 kHz to 12 kHz was necessary to perform good horizontal sound localization. Furthermore, listeners' hearing threshold levels at high frequencies and sound localization performances for high-pass noise were correlated.

1. Introduction

Three-dimensional (3-D) sound can be reproduced either with binaural, transaural, wave-field synthesis, or multi-channel surround-sound technology. Perceptual characteristics of these 3-D sounds have been investigated extensively. Interaural time difference (ITD), interaural level difference (ILD) and spectral cues are well known to contribute greatly to 3-D sound localization [1]. Nevertheless, few detailed reports in the literature describe how broad a signal bandwidth must be to reproduce the signal as 3-D sound.

Regarding to real sound sources, Nakabayashi reported that horizontal localization is incomplete with one octave bandwidth noise [2]. Blauert described on directional bands, certain frequency band signals are subject to localization at particular directions [1]. Kondou *et al.* reexamined this directional band phenomenon [3]. Morimoto *et al.* investigated the role of low-frequency components in median plane localization. They found that the higher frequency components are dominant in median plane localization, although the lower frequency components do not contribute significantly to localization [4]. Recently, Nojima *et al.* showed that front-back judgment is difficult for 500-Hz low-pass noise when head movement is not allowed [5]. Brungart and Simpson reported that low-pass filtering has a profound effect on localization accuracy in the vertical dimension [6]. With regard to virtual sound sources, Arrabito and Mendelson reported that 14-kHz high-pass noise contributes little to vertical and horizontal sound localization [7].

Furthermore, no reported study has elucidated how a listener's audible frequency range affects the required signal bandwidth. In most studies, subjects are said to be young healthy adults or students who have normal hearing. Even if subjects' audiograms were measured, the highest tone frequency tested must have been 8 kHz because this is the highest frequency tone used in a standard audiometer. However, signals used in sound localization experiments usually include frequency components higher than 8 kHz. It is also known that the spectral cues for sound localization are located in the 6-16 kHz band [8]. Moreover, humans gradually lose high-frequency hearing with age, so potential users of 3-D sound reproduction systems are not necessarily limited to young adults: they can include older people with high-frequency hearing loss.

Therefore, this article presents examination of the signal bandwidth necessary for sound localization and elucidates the relationship between the bandwidth and subjects' audible frequency range.

2. Subject Hearing Thresholds

2.1. *Procedure*

Hearing thresholds were measured for pure tones at 0.125, 0.25, 0.5, 1, 2, 4, 8, 12, 16 and 18 kHz. Tones were generated in a PC with a sampling frequency of 48 kHz with 16 bit accuracy. A USB audio interface (UA-101; Roland) was used as a digital-to-analog converter (DAC). The tone was presented monaurally via closed-type circumaural headphones (HDA200; Sennheiser). The sound pressure levels of the tones were calibrated using an IEC-60711 coupler installed in a head and torso simulator (4281D; Brüel & Kjær) with an audio analyzer (PULSE Analyzer; Brüel & Kjær). A high pass filter was inserted only for the 16 and 18 kHz tones to eliminate spurious audible low-frequency signals generated by the DAC. An attenuator (PA5; TDT) was used to adjust the maximum sound pressure level of each tone. Hearing thresholds were determined using a transformed up-down method using PSYLAB; a collection of MATLAB scripts for controlling interactive psychoacoustical listening experiments [9]. The tone amplitude of tones was manipulated digitally.

All measurements were conducted in a sound proof room within which the A-weighted level of the background noise was 16 dB.

2.2. *Subjects*

Ten subjects participated in the experiment: six males and two females were in their 20s; one male was in his 30s; and one male was in his 50s. None reported hearing loss and no specific problem was found in audiograms taken during their annual health checkup.

2.3. *Results*

Figure 2(a) shows the mean hearing threshold of 10 subjects. Compared to the absolute hearing threshold curve of ISO 226:2003 [10], only five of the subjects in their 20s showed small hearing losses—less than 25 dB—from 125 Hz to 18 kHz. The hearing loss at 18 kHz was over 25 dB for two subjects, hearing losses at 16 and 18 kHz were over 25 dB for the other three subjects. Very-high-frequency hearing sensitivity was severely reduced in one subject in his 20s and in one subject in his 30s. A subject in his 50s did not hear 16 or 18 kHz tones presented at 100 dB.

Figure 2(b) presents the mean difference in the hearing threshold between the right and left ears of all subjects. Sensitivities of the two ears were unequal. The right–left mean difference is a few decibels at less than 1 kHz, but it is greater than 6 dB at frequencies higher than 8 kHz.

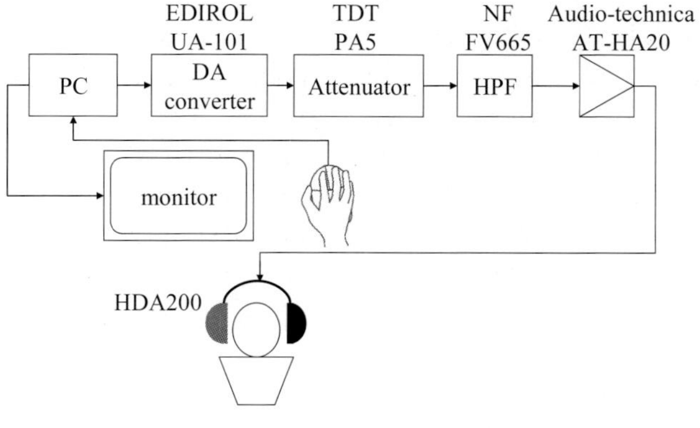

Figure 1. Block diagram of the system used for measuring the hearing threshold.

3. Horizontal sound localization experiments

3.1. *Experimental system*

The experimental system consisted of a Windows PC, two 8-channel DACs (UA-101; Roland), 12 power amplifiers (1705II; BOSE.), and 12 4-inch full-range loudspeakers (MG10SD-09-08; Vifa). The sampling frequency of the DACs was 48 kHz. Loudspeakers were placed around a chair in a horizontal circumference in a 1-m radius at 30 degree intervals. The loudspeaker height was 1.1 m. The sound localization experiment was conducted in an experimental room, of which the walls and ceiling were covered with sound-absorbing materials. The A-weighted level of the background noise of the room was 53 dB. Figure 3 portrays the experimental system and setup.

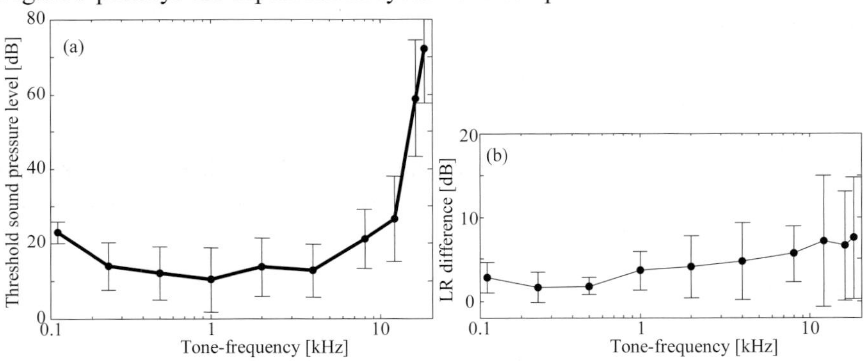

Figure 2. (a) Mean hearing threshold of 10 subjects. (b) Mean hearing threshold difference between right and left ears of 10 subjects. Vertical bars represent standard deviations.

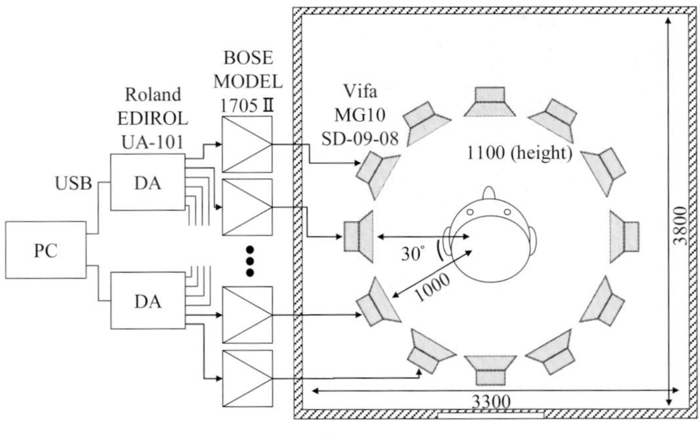

Figure 3. Experimental system and setup.

3.2. *Stimuli*

White noise, low-pass (F_c = 0.5, 1, 2, 4, 8 kHz) and high-pass (F_c = 2, 4, 8, 12, 16 kHz) filtered noises were used as stimuli. The low-pass and high-pass filters were 512 tap finite impulse response (FIR) filters with a -60 dB stop-band attenuation. The FIR filters were designed using a window method. Frequency responses of the loudspeakers were not corrected; they are portrayed in Fig. 4.

The stimulus duration and inter-stimulus interval were both 3 s. A 30 ms linear ramp was used at the beginning and end of the stimuli. The sound pressure level of the white noise stimuli was 80 dB at the head center position. The sound pressure level of the low-pass and high-pass filtered noise stimuli decreased depending on the filtering.

3.3. *Subjects and procedure*

Ten subjects whose hearing thresholds had been measured participated in the sound localization experiments. The experimental procedure was as follows: Subjects sat on a chair placed in the center of the speaker array. The subjects listened to stimuli reproduced from one loudspeaker. They were asked to localize horizontal sound image positions of the real sound sources and to mark the located position on an answer sheet. The method was forced choice method of one of 12 directions in the horizontal plane. Subjects had to close their eyes and keep their head still when a stimulus was reproduced.

Each session consisted of 60 trials; the stimuli were presented in random order from the 12 loudspeakers. One experiment consisted of four sessions, resulting in 20 trials from each of the 12 directions. Experiments were conducted separately for each low-pass, high-pass and white noise stimulus.

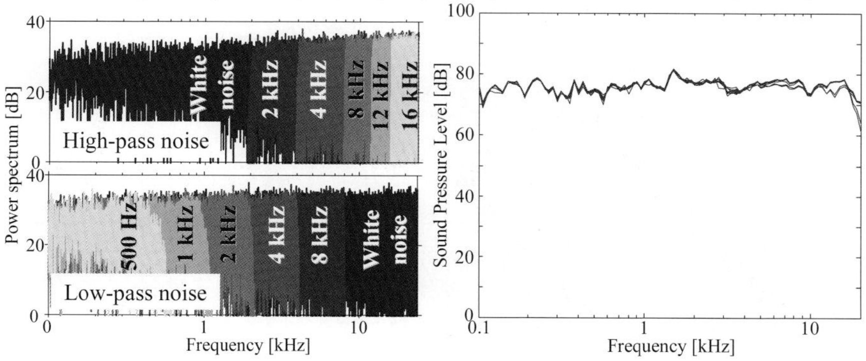

Figure 4. Low-pass and high-pass filtered noise stimuli and frequency responses of 12 loudspeakers used in the experiments at 1 m distance.

3.4. *Results*

Sound images were reported to be externalized from the head for all stimuli. Figure 5 shows the mean sound localization performance and standard deviations of the 10 subjects against the filter cut off frequency for all stimuli. Sound localization was almost perfect (96%) for the white noise. The sound localization performance was better than 86.6% for 8-kHz and 4-kHz low-pass noise. The performance dropped gradually with the lowering of the cut-off frequency of the low-pass filter. By contrast, the sound localization performance was greater than 86.6% with the 2-, 4- and 8-kHz high-pass noise, but the performance dropped rapidly for the 12- and 16-kHz high-pass noise. Figure 6 presents sound localization results for some stimuli. Figure 6(a), white noise, and (b), 500-Hz low-pass noise, are the results for 10 subjects of all groups. Figure 6(c), 16-kHz high-pass noise, shows the result for eight subjects who can hear the stimulus. Figure 6(d), 16-kHz high-pass noise, is the result for two subjects who were unable to hear the stimulus. Front–back confusion increased for 500-Hz low-pass noise. The sound localization performance decreased for 16-kHz high-pass noise (*i.e.,* most stimuli were just roughly localized, even for the subjects who could hear the stimuli). Sound localization was impossible for the high-pass noise with the subjects who could not hear the stimuli because of severe hearing loss at frequencies higher than 16 kHz.

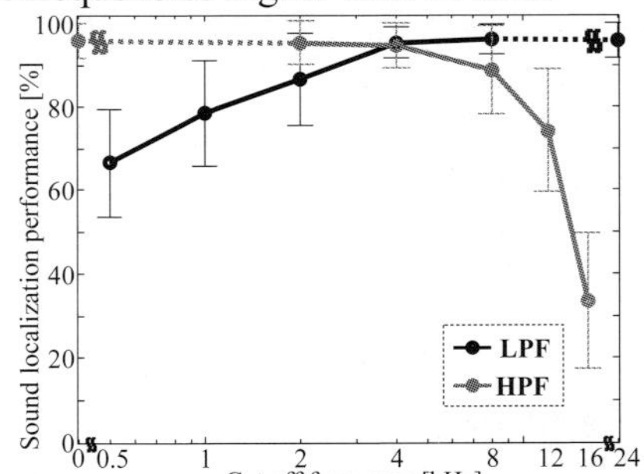

Figure 5. Mean sound localization performance and standard deviations of the 10 subjects.

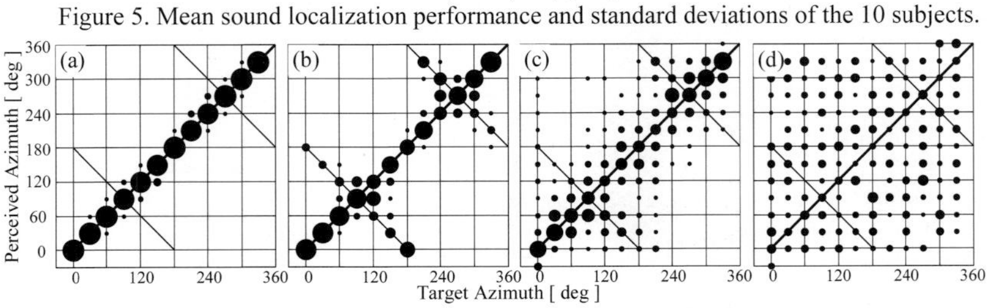

Figure 6. Horizontal sound localization results: (a) white noise for 10 subjects, (b) 500-Hz low-pass noise for 10 subjects, (c) 16-kHz high-pass noise for 8 subjects who can hear the noise, (d)16-kHz high-pass noise for 2 subjects who cannot hear the noise. The ordinate of each panel represents the perceived azimuth, and the abscissa shows the target azimuth. The size of each circle is proportional to the number of answers.

Figure 7 shows the relationship between sound localization performance for 500-Hz low-pass noise and 16-kHz high-pass noise and hearing threshold levels at 0.5, 1, 12, and 16 kHz for each subject. The hearing threshold at 16 kHz is highly correlated with sound localization performance for the 16-kHz high-pass noise. The correlation coefficient is -0.91, which is statistically significant ($p<0.001$). The hearing threshold at 12 kHz is also highly correlated with sound localization performance for the 16-kHz high-pass noise. The correlation coefficient is -0.79, which is statistically significant ($p<0.01$). The correlation coefficients between sound localization performance for the 500-Hz low-pass noise and the hearing threshold at 500 Hz and 1 kHz are -0.39 and -0.35, which are not statistically significant ($p>0.05$).

4. Discussion

The signal bandwidth necessary to perform horizontal sound localization appears to be from 2 kHz to 12 kHz: as presented in Fig. 5, sound localization is difficult for stimuli that consist of only low-frequency components of less than 2 kHz or high-frequency components higher than 12 kHz. Particularly, the sound localization performance decreases greatly for the 16-kHz high-pass noise (*i.e.,* sound localization is extremely difficult for stimuli which have only very high spectral components).

One reason for the poor sound localization performance with 16-kHz high-pass noise is that no low spectral components exist. For this stimulus, sound image locations must be calculated only from the ILD of high spectral components. The ITD cannot be used as a cue because it is calculated from low spectral components below a few kilohertz at the medial superior olive (MSO)

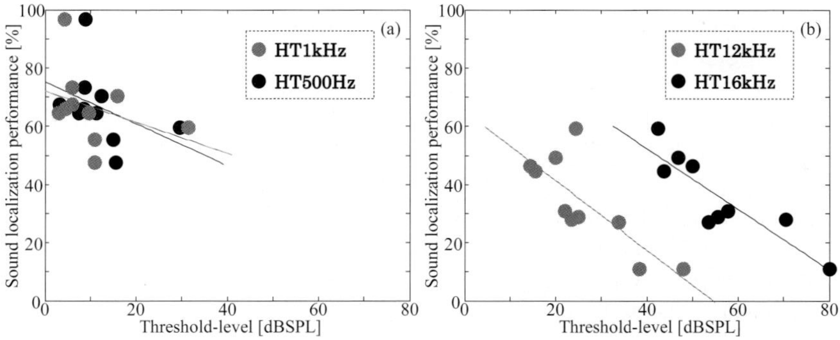

Figure 7. Relationship between sound localization performance and hearing threshold of each subject.
(a) Filled circles (●) are for 500-Hz low-pass noise vs. hearing threshold at 500 Hz. Gray circles (●) are for 500-Hz low-pass noise vs. the hearing threshold at 1 kHz.
(b) Filled circles (●) are for 16-kHz high-pass noise vs. hearing threshold at 16 kHz. Gray circles (●) are for 16-kHz high-pass noise vs. the hearing threshold at 12 kHz.

in mammals [11-13]. The head-related transfer function (HRTF) spectrum around 3–14 kHz, in which several prominent spectral peaks and notches appear [7], can not be used as a cue for sound localization in the stimulus.

Another possible reason for the poor sound localization performance with 16-kHz high-pass noise is the low sensation level (SL) of the stimuli. Hebrank and Wright showed that localization performance was independent when white noise stimuli was presented over 40 dB SL, but the performance decreased for the stimuli less than 40 dB SL [14]. Inoue reported that the sound localization performance for white noise stimuli was low at 0–20 dB SL [15]. In fact, the sound pressure level of the 16-kHz high-pass noise used in our experiment is 72 dB; its SL is 0–30 dB depending on the subjects. The low SL of the stimuli is therefore a possible cause for the low sound localization performance of the stimulus.

Sound localization is also difficult for stimuli that consist only of low-frequency components less than 2 kHz. However, the sound localization performance for the 500-Hz low-pass noise is much better than that for 16-kHz high-pass noise. For the 500-Hz low-pass noise, the major cue for sound localization is the ITD calculated from the low spectral components. The ILD of the low spectral component is useful as another cue, but the ILD is calculated mainly from high spectral components at the lateral superior olive (LSO) [11]. The HRTF spectral peaks and notches around 3–14 kHz cannot be used as cues for sound localization for the stimuli. The ITD and possibly some ILD from low spectral components provide a better estimation of the sound location than the ILD from high spectral components alone.

Figure 7 shows that sound localization performances for the high-pass noise stimuli drop with the lowering of the hearing thresholds at high frequencies. Meanwhile, hearing thresholds at low frequencies do not necessarily contribute to sound localization performances for the low-pass noise stimuli. One reason for this difference is that the hearing loss at high frequencies is greater than that at low frequencies. High-frequency sounds exist physically, but they are not coded neurally well because the sensitivity for high-frequency is much lower. Another reason is that the temporal coding does not require a high signal-to-noise ratio of the stimulus: the ITDs can be coded well with low SL stimuli.

Some visually impaired people who cannot see anything can nevertheless predict visual stimulus location—a phenomenon known as blind sight [16]. Our experiment showed that subjects who cannot hear the 16-kHz high-pass noise cannot localize the sound at all, as depicted in Fig. 6(d). The two subjects' hearing loss at high frequency is most likely the result of malfunctions at the sensor stage by aging. Consequently, it is not surprising that they cannot hear the stimuli and that they cannot localize the stimuli.

As presented in Fig. 2, the hearing threshold of right and left ears were not the same in every subject. However, sound localization performance was almost perfect for the white noise and wide band limited noises. Front sound image location does not shift to the ear for which the hearing threshold is low. Instead, the hearing threshold difference of the right and left ears does not affect formation of the internal representation of the ILD and calculation of the sound image location in the brain. For the loud sound, possessing a level far above hearing threshold level, the internal representation of the ILD is apparently the same as its physical ILD.

5. Conclusion

Sound localization experiments for low-pass and high-pass noise were conducted with 10 subjects whose hearing thresholds were measured precisely. The obtained results are as follows:

(1) Sound localization for wide-band noises was almost perfect even when the hearing thresholds of the right and left ear were not the same.
(2) Sound localization was extremely difficult for high-pass noises. Sound localization performance was 74% for 12-kHz high-pass noise and 34% for 16-kHz high-pass noise.
(3) Sound localization was impossible for high-pass noises with subjects who could not hear the stimuli because of severe hearing loss for frequencies higher than 16 kHz.
(4) Sound localization was difficult for low-pass noise with 500-Hz cut off frequency. Sound localization performance was 78% for 1-kHz low-pass noise and 67% for 500-Hz low-pass noise.
(5) Hearing thresholds at high frequencies and sound localization performance for high-pass noise are highly correlated; the correlation was statistically significant.
(6) Signal bandwidth from 2 kHz to 12 kHz was necessary to perform horizontal sound localization.

Acknowledgments

The authors thank the subjects who participated in the experiments. They also thank Ms. Nozomi Shimakura for her great help in building the sound localization experimental system and measuring 12 speaker responses.

References

1. J. Blauert, *Spatial Hearing*, The MIT Press, 36-176 (1997).
2. K. Nakabayashi, "Sound Localization on the Horizontal Plane," *J. Acoust. Soc. Jpn.* **30(3)**, 151-160 (1974).

3. T. Kondou, Y. Kumamoto and J. Michiyama, "A study of Influence of limited frequency Band on Sound Image Localization Processing," *Proceedings of the IEICE General Conference* **1996**, 284 (1996).
4. M. Morimoto, M. Yairi, K. Iida and M. Itoh, "The role of low frequency components in median plane localization," *Acoustical Science and Technology* **24(2),** 76-82 (2003).
5. R. Nojima, M. Morimoto and H. Sato, "Head movements during sound localization," *Proc. Autumn Meeting, Acoust. Soc. Jpn.* **2009,** 515-516 (2009).
6. D. S. Brungart and B. D. Simpson, "Effects of bandwidth on auditory localization with a noise masker," *J. Acoust. Soc. Am.* **126(6)**, 3199-3208 (2009).
7. G. R. Arrabito and J. R. Mendelson, "The relative impact of generic head-related transfer functions and signal bandwidth on auditory localization: Implications for the design of three-dimensional audio displays," *Defence and Civil Inst of Environmental Medicine* **DCIEM-TR-2000-67**, 1-31 (2000).
8. E. H. A. Langedijk and A. W. Bronkhorst, "Contribution of spectral cues to human sound localization." *J Acoust Soc Am.* **112(4)**, 1583-96 (2002).
9. http://www.hoertechnik-audiologie.de/web/file/Links/psylab.php
10. H. Takeshima, Y. Suzuki, M. Kumagai, T. Fujimori and H. Miura, "Threshold of hearing for pure tone under free-field listening conditions," *J. Acoust. Soc. Jpn. (E)* **15(3)**, 159-169 (1994).
11. J. O. Pickles, *An Introduction to the Physiology of Hearing (Second Edition)*, Academic Press, 87, 179-188 (1988).
12. B. Grothe and G. Neuweiler, "The function of the medial superior olive in small mammals: temporal receptive fields in auditory analysis." *Journal of Comparative Physiology A* **186(5)**, 413-423 (2000).
13. D. McAlpine and B. Grothe, "Sound localization and delay lines – do mammals fit the model ?" *TRENDS in Neurosciences* **26(7)**, 347-350 (2003).
14. J. Hebrank and D. Wright, "The effect of stimulus intensity upon the localization of sound sources on the median plane." *Journal of Sound and Vibration* **38(4)**, 498-500 (1975).
15. J. Inoue, "Effects of Stimulus Intensity on Sound Localization in the Horizontal and Upper-hemispheric Median Plane." *Journal of UOEH* **23(2)**, 127-138 (2001).
16. R. Fendrich, C.M. Wessinger and M.S. Gazzaniga, "Residual Vision in a Scotoma: Implications for Blindsight." *Science* **258(5087)**, 1489-1491 (1992).

THE 'PHANTOM WALKER' ILLUSION: EVIDENCE FOR THE DOMINANCE OF DYNAMIC INTERAURAL OVER SPECTRAL DIRECTIONAL CUES DURING WALKING

W. L. MARTENS and D. CABRERA
Faculty of Architecture, Design and Planning,
University of Sydney, NSW 2006 Australia
E-mail: william.martens@sydney.edu.au, densil.cabrera@sydney.edu.au
www.sydney.edu.au

S. KIM
Yamaha Corporation, 203 Matsunokishima
Iwata, Shizuoka 438-0192, Japan
E-mail: sungyoung_kim@gmx.yamaha.com

It is well established that the changes in interaural cues that occur during active localization normally aid listeners in directionally resolving environmental sound sources, for example, in terms of their position in front or in back of a listener. However, the relative importance of such dynamic interaural directional cues, which are naturally available to a walking listener, have not been examined under conditions in which these cues are placed in conflict with natural spectral cues to direction. In order to investigate the relative strength of head-motion-coupled directional cues as compared to spectral cues associated with the filtering effects of the listener's pinnae, an experiment was executed in which directional judgments were made by listeners who were fitted with a binaural hearing instrument that allowed for the signals reaching their ears to be interchanged. This binaural hearing instrument was configured in such a way as to preserve the spectral cues associated with each listener's own pinnae, and at the same time allowed the sound arriving at the blocked entrance of the left ear canal to be captured so that it could be reproduced via an acoustically isolated earphone inserted into the right ear, and *vice versa*. Testing under various combinations of conditions (walking/motionless; normal/interchanged ears) has established that interchanging ear signals does not cause a reversal of elevation judgments when listeners are asked to stand still; however, when listeners with interchanged ear signals were asked to walk past a continuously presented speech sound source emanating from a fixed spatial position, the sound source was heard to be located in a spatial region that was reversed with respect to all three spatial axes: left for right, front for back, and above for below. Furthermore, despite having the stationary physical source of the sound in clear view as listeners walked toward it, the sound was invariably heard to be approaching them from behind, and the voice of this illusory 'Phantom Walker' overtook listeners as they passed by the physically stationary source.

1. Introduction

1.1. *An Auditory Illusion of Source Motion*

The topic of this chapter is the auditory perception of sound sources that seem to move through space for a walking listener, but are perceived as correctly stationary when the listener holds still. Such auditory illusions of source motion, which depend upon listener movement, have been known for many years (e.g., [1][2]), and recent experimental results (e.g., [3][4]) provide evidence for the dominance over spectral directional cues of the dynamic interaural cues available to moving listeners. Unlike the research described in many of the chapters in this book, the research reported here is concerned with the perception of a familiar sound source presented continuously over seconds (rather than milliseconds), under conditions in which listeners are instructed to walk through a fairly normal reverberant space and encouraged to attempt actively to localize that sound source (cf., [5]).

The observed auditory illusion of source motion occurs under such conditions for listeners wearing a binaural hearing instrument that allows the sound arriving at the blocked entrance of the left ear canal to be captured and reproduced via an acoustically isolated earphone inserted into the right ear, and *vice versa*. In this 'Interchanged' listening mode, sound sources that are actually located on the listener's left side are naturally perceived to be located on the right. To enable controlled comparisons, the ear signals can be quickly switched back into a 'Normal' listening mode, so that dynamic interaural and spectral directional cues give consistent information regarding source locations. Because a binaural hearing instrument can be configured with miniature microphones positioned at the blocked entrance to the listener's ear canals, spectral cues to source azimuth and elevation that are associated with the filtering effects of the head and pinnae can be adequately maintained during its use. Therefore, when the ear signals are switched (i.e., to an Interchanged listening mode), a relatively normal listening experience results when the listener hold still, though there is a left-right reversal of perceived source locations. What is most interesting about the comparison between this and the Normal listening mode is what happens when listeners are instructed to walk toward a talker producing continuous speech. Although the stationary physical source of a speech sound may be in clear view in front of the listeners, the auditory image of the source will most often be heard to be located behind them, and it will approach them from behind as they approach the talker. Of course, if listeners walk straight along a path that keeps the talker on their left hand side, they will eventually reach a point at which the auditory image would be directly on their right, opposite the physical source on their left. Figure 1 shows a diagram of what happens as listeners walk forward, away from a stationary talker located directly on their left. When listening in an ear-interchanged mode, the auditory image of the speech sound source is initially located directly to the listener's right, but as the listener walks forward, the perceived location of the talker moves forward rather than staying behind, and moves at twice the speed at which the listener is walking. Since no physical source of the speech sound is visible, the talker's voice seems to belong to an invisible 'Phantom Walker' moving away from the listener so quickly that it can never be caught!

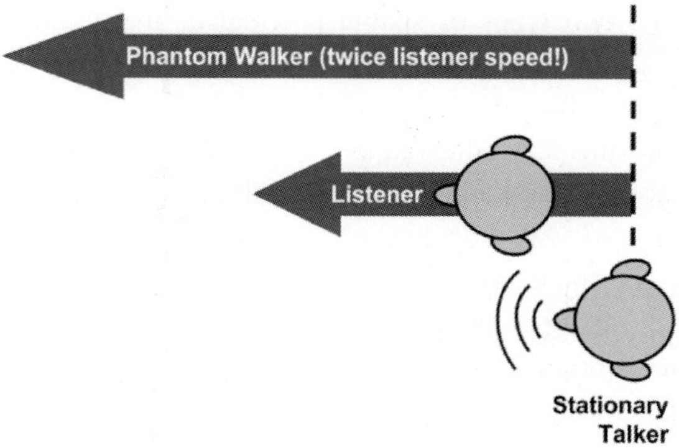

Fig. 1. Diagram showing the spatial path taken by the Phantom Walker when a listener with interchanged ear signals walks away from a stationary talker initially positioned on the listener's left.

It is easy to understand the left-right reversal of perceived source location that occurs in the Interchanged listening mode, but it takes a bit of thinking to figure out why a stationary source located in front of the listener is heard reflected to a position behind the listener. The observed front-back reversal is just what would be expected given a reversal of the dynamic interaural cues that normally aid listeners in correct front-back discrimination, as will be explained below.

1.2. *Auditory Reversals of Source Location*

Listeners with interchanged ear signals who are presented with a sound source emanating from a fixed spatial location may experience the sound source as positioned in a spatial region that is reversed from its original spatial region in a number of ways. To simplify subsequent discussion, some of the types of reversals that might be expected are presented in this subsection, in reference to the two-dimensional head-related coordinate system illustrated in Figure 2. The figure shows a listener with head located at the intersection of the two axes indexing points on the listener's horizontal plane, which is a plane defined by all points at the listener's ear level.

If a sound source were to be located directly to the listener's left, at a point on the listener's interaural axis (shown as the black double-headed arrow drawn along the left-right line intersecting both ears), then a simple left-right reversal along that axis might be expected for listeners with interchanged ear signals. If, however, a source were shifted forward in space as well, to a location in the front-left quadrant of the horizontal plane (as illustrated by the radiant sphere in Figure 2), then two types of spatial reversal of the auditory image might be anticipated. The first could be described as a reflection with respect to the median plane dividing the listener's space into left and right hemifields, in which case the image would remain in the frontal hemifield, but show up at equal but opposite lateral angle (at the end of the darker grey arrow in Figure 2). The second type of spatial reversal could be described as a reflection through the origin of the coordinate system (located at the center of the listener's head), in which case the auditory image of the source actually located in the front-left quadrant would show up at a position in the rear-right quadrant of the listener's horizontal plane (at the end of the lighter grey arrow in Figure 2).

Fig. 2. Diagram showing two possible types of reversals that might be experienced in reference to a two-dimensional head-related coordinate system, with the listener's interaural axis shown as the black double-headed arrow. The darker arrow depicts a left to right reversal while the lighter arrow depicts a reversal from the listener's front-left quadrant to rear-right quadrant.

Of course in everyday experience, sound sources are rarely contained within a listener's horizontal plane; rather, they also can be given a third coordinate with location indexed along a third axis that extends from below the listener to above the listener. Hence for a source displaced below the listener's ear level (toward which the listener's finger is pointing in Figure 2), a third type of spatial reversal can be described as a reflection through the origin of a three-dimensional head-related coordinate system: In this case the auditory image of a source actually located in the lower-front-left region would show up at a position in the upper-rear-right region. To foreshadow the experimental results to be presented in this chapter, it is pointed out that this third type of spatial reversal is the most common one to be experienced by listeners with interchanged ears when they walk past a stationary sound source. This sort of 'diametrical' reversal was observed in early studies of auditory reversals of source location, as described in the next section.

1.3. *Early Studies of Auditory Reversals of Source Location*

An early observation of location reversal in auditory perception by a listener wearing a binaural hearing instrument was reported in 1928 by Young [1]. He called the binaural hearing instrument that he employed in his listening experiments 'The Reversing Pseudophone,' since with it he could present to a listener's right ear the sound arriving at a receiving trumpet placed just in front of the listener's left ear, and *vice versa*. Young went to some trouble to try to match the angle of the ear trumpet to the forward facing angle of the listener's natural pinna, but recognized that the character of the sound was not matched to that heard in unaided listening. In fact, Young's Reversing Pseudophone largely eliminated spectral cues to source direction while maintaining interaural cues to source lateral angle that were nonetheless reversed as a result of the left-right ear interchange (as confirmed in his follow-up study [7]). What is most interesting in Young's personal experience wearing his Reversing Pseudophone is that a visible source of sound was occasionally heard at its correct location, despite the interchange of the ears. In contrast, an unseen source, arriving from an unknown direction and localizaed purely through audition, was almost always heard as reversed.

So, in the case of "purely auditory localization," Young found that he trusted his ears, but when a sound source was in plain sight, through ventriloquism or through some behavioral adaptation, he often was able to counteract the reversal (i.e., to localize sources correctly). There was one exception to this rule of visual influence that occurred when Young walked while wearing the Reversing Pseudophone. It is worth reviewing this experience that was described by Young in his 1928 report as follows:

> *"While walking along the sidewalk I heard the voice of two ladies and their steps approaching and overtaking me from behind on the right. Quite automatically I stepped to the left making room for them to pass. I looked back and found that I had stepped directly in front of them. My automatic reaction as well as my localization was reversed."*
>
> (Young [1], p. 409)

Even though he could clearly see the two ladies approaching him from in front and to his left, Young experienced their speech as approaching him from in back and to his right (a phenomenon that has been described in the current chapter as the Phantom Walker Illusion). Under these conditions, however, Young did not report any reversals in elevation. On the other hand, in a replication of his study by Willey, Inglis, and Pearce [2], one of their subjects who wore a Reversing Pseudophone for an extended period of time did in fact experience a reversal in elevation, which was reported as follows:

> *"One subject, however, reported an instance of a diametrical reversal of the sound of an airplane; on this occasion the characteristic sound seemed to be coming out of the ground."*
>
> (Willey, et al. [2], p. 120)

Despite the listener's knowledge that the sound of the airplane must have been arriving from above ear level, the auditory image experienced was reversed to a spatial region below ear level, as well as being reversed left for right and front for back. It is no surprise that these reversals occur when wearing ear trumpets that largely eliminated spectral cues to source location, since the remaining dynamic interaural cues are quite effective in supporting directional distinctions when listeners are allowed to move their heads freely. As will be explained in greater detail below (in the discussion section), dynamic interaural cues from horizontal head rotations resolve front-back ambiguities, and those from head rolling (dropping an ear towards one's shoulder) resolve above-below ambiguities. The question of central concern in this chapter concerns what may be found when the spectral cues to source direction are not eliminated but preserved by a Reversing Pseudophone, in which case dynamic interaural cues may be placed in conflict with those spectral cues. This question was probably first addressed in the early studies of Wallach [6], which are described next.

Just 20 years after Young published the results he found wearing ear trumpets, Wallach [6] provided strong experimental evidence that dynamic cues to source direction could dominate localization judgments even when spectral cues were unmodified. His experimental apparatus allowed listeners to use their own pinnae while he presented dynamic interaural cues that were synthetically reversed by switching a sound source between multiple loudspeakers. The switching mechanism was coupled to the listener's own head rotation, and the dynamic interaural cues were reversed by directing an audio signal to a series of loudspeakers so as to produce lateral shifts along the listener's interaural axis consistent with sources in back of the listener, though the loudspeakers were actually in front of the listener. The dynamic interaural cues, so synthesized, were strong enough to produce front to back reversal in apparent source direction when spectral cues associated with the listener's pinnae were in conflict with those dynamic cues that were coupled with the rotation of the listener's head. Not only did Wallach [6] produce these important experimental results, but in a subsequent paper [8] he also proposed principles of spatial hearing that might explain these results, and that is the topic of the next section.

1.4. *Principles Explaining Auditory Reversals of Source Location*

The auditory reversals of source location reported by Young [1] and by Willey, Inglis, and Pearce [2] for listeners wearing a Reversing Pseudophone are well explained by principles of spatial hearing proposed by Wallach [8]. And although the movement-dependent Phantom Walker Illusion that is the focus of the current chapter was not discussed by Wallach [8], the phenomenon is closely related to the results he obtained under conditions in which listener movements were constrained to head turning alone. These principles will be summarized briefly here, and restated only slightly in order to due justice to the Phantom Walker Illusion that had also been reported by Young [1] (though the phenomenon had not be named as such). The first so-called "selective" principle Wallach [8] identified was based upon the recognition of a fundamental problem for moving listeners who attempt to localize sound sources, as is described next.

When walking through an environment populated by sound sources of uncertain spatial location, the fundamental problem for human listeners is to determine if the changes in binaural signals at their ears are due only to changes in head orientation, or whether there are also at the same time changes in source locations relative to head position in space. For a continuous sinusoidal sound signal presented from an otherwise unknown location, the problem is virtually impossible to solve due to the ambiguity of interaural time and intensity differences that change in a way that depends upon the sequence of lateral angles through which the source varies. These lateral angles are those that the source makes with the listener's median sagittal plane (the imaginary plane splitting the human body symmetrically into left and right halves, traveling vertically from the top to the bottom of the body, and thereby dividing space into left and right hemifields). For Wallach [6], how listeners solve this fundamental problem is the basis for a fundamental principle of spatial hearing:

> *"We know that almost every lateral angle can be represented by a number of different directions, and that a given sequence of lateral angles can thus be represented by a nearly endless variety of patterns of subsequent directions. But in a given case, this sequence of lateral angles, no matter how it is produced, gives rise to one percept only, that of a stationary direction which is compatible with the sequence."*
>
> (Wallach [6], p. 359)

Wallach claimed that this first general "selective" principle is implied in all sound localization, but he proceeded to make a more succinct statement of the above, which he termed the "principle of rest," as follows: *"Of all the directions which realize the given sequence of lateral angles, that one is perceived which is covariant with the general content of surrounding space."* Of course, this principle as stated does not address the complexity of the situation faced by the listener when both head and sound source are moving through the listener's environment, and the perception of a stationary direction for the source is not supported by the sensory data (especially with reference to laboratory experiments that synthetically or otherwise modify patterns of sound stimulation). In recognition of these prospects, a broader principle was proposed:

> *"This broader principle was demonstrated in synthetic production of sound directions when sequences of lateral angles were presented for which no stationary direction existed. It was named the principle of least displacement. If a sequence of lateral angles is presented to which no stationary direction corresponds, the sound is perceived in the region where it has to undergo the smallest displacement in space while realizing the given sequence of lateral angles."*
>
> (Wallach [8], p. 360)

While these principles go a long way toward explaining the results of Wallach's studies, a further extension is required to explain the Phantom Walker Illusion that is described in this paper. Since the experience of the Phantom Walker Illusion depends not only upon changes in the orientation of the listener's head, but also upon changes in the listener's position, the circumstances do not match those in Wallach's stimulus presentation. Indeed, listeners can only experience the linear displacement of the Phantom Walker when they are also linearly displaced. So, the principle of least displacement must be understood in this context to mean that the sound will undergo the smallest **angular** displacement in space consistent with the given sequence of lateral angles, in contrast to a **linear** displacement in its spatial position, which may indeed be quite large due to its rapid movement in the direction that the listener is walking. This proposed restatement of the principles that Wallach laid out is not at all inconsistent with those principles. In fact, it may be that Wallach's original intention, though not specified as such, was that it is the **angular** displacement that will be minimized in human auditory image formation, since it was shifts in lateral angle (i.e., along the interaural axis) that were the focus of his research.

Before proceeding to describe the current experimental study of the Phantom Walker Illusion, a further explanation should be given to make it clear how these principles explain the results of Wallach's earlier work [6]. These studies clearly showed that head-turning has the potential to create distinct percepts of sources located in frontward versus rearward spatial regions, despite sources being located in the opposite regions. Under normal listening conditions, the auditory image of a sound source initially located on the horizontal plane directly in front of the listener will shift by a lateral angle of 10° to the right when the listener's head is turned 10° to the left (the angle of such a leftward turn of the head typically being specified as a change in yaw angle, to distinguish head-turning angles from source angle variation). Using his experimental apparatus, Wallach was able to switch sound reproduction between loudspeakers in response to head turning such that a sound source initially located directly in front of the listener would shift to the left by a lateral angle of 20° when the listener's head was turned to the left by a yaw angle of 10° (i.e., in the same leftward direction, but at twice the rate of head turning). The auditory image would thus be heard to shift to the left by a lateral angle of 10°, which is the amount of shift expected for a source located toward the rear of the listener when the head is turned to the left by an equivalent yaw angle. Even when broadband signals such as orchestral or piano music were presented via the array of selectable loudspeakers positioned in clear view, right in front of the listeners, the misperception of frontward incidence for rearward incidence seems to have been the only result possible. This is quite in contrast to what Wallach's listeners experienced when those same sources were reproduced over the same loudspeakers, but without "pseudophonic" changes coupled with the listener's head turning. The ordinary listening experience allowed frontward and rearward sources to be distinguished easily even without head movement, in which case listeners could rely upon the spectral transformation of the respective sources by the individual's pinnae, a factor which Wallach termed the 'pinna factor.' This led Wallach to conclude his 1939 paper as follows:

> *"Under ordinary circumstances, discrimination between front and back on the basis of the pinna factor alone, i.e., without head movement, is quite reliable. The fact that this factor is invariably overcome in the synthetic production experiment indicates clearly its subordinate role."*
>
> (Wallach [8], p. 274)

The study summarized in the following section of this chapter was designed to examine the relative salience of this pinna factor when it is put in conflict with other head-motion-coupled directional cues during active localization by listeners using a binaural hearing instrument. When the instrument effected an interchange of their ear signals through a 'pseudophonic hookup' (a term used to describe a system designed to deliver false information to the ears), it was of course surprising to listeners that auditory imagery was reversed front for back as well as left for right. However, because the experiment to be reported here also enabled rapid switching between normal localization and that with interchanged ear signals, the contribution of the pinna factor to elevation judgments could also be placed in conflict with dynamic interaural cues to elevation. As will be seen, the cues to sound source location due to the pinna factor are overcome not only in making front-back distinctions for the location of the Phantom Walker, but also in making above-below distinctions for the Phantom Walker's voice. This elevation reversal is a second surprise that was hidden from listeners until after their participation in the Phantom Walker experiments was concluded: It will be shown that listeners wearing an ear-interchanging binaural hearing instrument experience such reversals in a blind test of elevation discrimination ability of listeners who are walking past a vertical array of loudspeakers.

2. A Study of the Phantom Walker Illusion

An experimental study of the perceived elevation of the voice of the Phantom Walker was undertaken in order to document under which testing conditions the different types of auditory reversals are to be expected. In the study to be summarized here, five listeners unfamiliar with the Phantom Walker illusion heard the illusion for the first time, and completed the experiment in a single session before they were given time to think about the experience.

2.1. *A Pseudophonic Binaural Hearing Instrument*

The five volunteers (ages ranging from 21 to 51) participated in a brief listening session that required them to wear a custom binaural hearing instrument that combined two Etymotic ER-2 insert earphones with two Sennheiser KE-2 electret microphone capsules that were fixed to the occluding foam pad of the ER-2 earphones at the ear canal entrance of each of the listener's ears. The two Sennheiser microphone capsules were attached to independent preamplifier housings (from Sound Devices) via a long audio cable, making it easy for the experimenter to switch the connection from each to either of the two ER-2 inset earphones that were powered by a Crown power amplifier.

2.2. *Stimuli and Procedure*

A recording of continuous speech (taken from Diana Deutsch's 2003 "Speech Illusions" CD) was reproduced via one of five loudspeakers arranged in a vertical array such that one of the loudspeakers was located at the same height as the subject's ears. Two of the loudspeakers were located above ear level, and two below, with spacing at roughly 20-degree intervals, so that the overall vertical extent of the array was roughly from -40° to +40° in elevation (this was the extent when the array was positioned directly to the listener's side, 90° azimuth). The task for the listeners was to indicate whether a continuously presented speech sound stimulus was arriving from above or below their ear level. In the 'Standing Still' condition, listeners were asked to keep their heads as still as possible, and to make an elevation choice as soon after the onset of the speech sound source as possible. In the 'Walking' condition, listeners were asked to wait until the playback of the speech had begun before beginning to walk toward the vertical array of loudspeakers, and to wait until after they had walked past the vertical array of loudspeakers before making an elevation choice. The task required a two-alternative forced choice (2AFC) response as to whether the source sounded as if it above or below the listener's ear level, with no option for expressing uncertainty or an intermediate 'ear level' direction. Listeners completed 10 forced choice trials in each of the four conditions.

2.3. *Results*

The obtained 2AFC discrimination data are shown in Figure 3 for the four treatments defined by the factorial combination of listening modes and listener movement conditions. The combined proportions of 'Above' responses from all five listeners are plotted in the left panel of Figure 3 for both listening modes in the Standing Still listening condition. These results illustrate that it was not perfectly simple for listeners to judge if the presented speech sound sources were located above or below ear level for listeners who were standing still. These results also show that the pattern of response proportions did not depend upon whether or not the signals delivered to their insert earphones were interchanged left for right with the signals captured at their ears. Plotted in the right panel of Figure 3 are combined response proportions for the same five listeners in the Walking condition, showing a clearly different pattern of results for listeners with ear signals interchanged left for right as compared to the Normal condition.

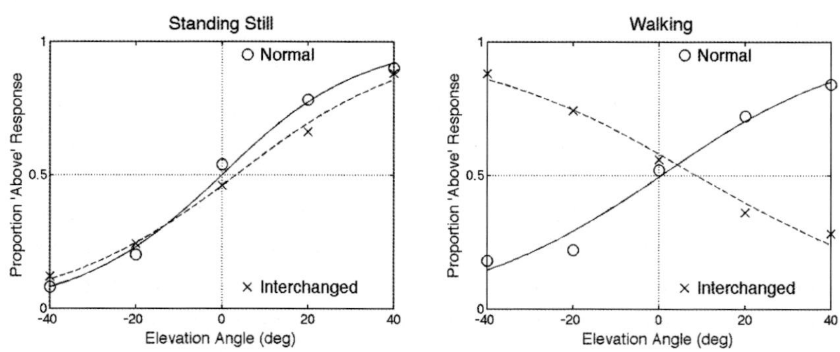

Fig. 3. Proportion of 'Above' responses resulting from summing the elevation discrimination data obtained from five listeners in two different listening modes and under two different listener movement conditions. The panel on the left shows their forced choice discrimination performance in the 'Standing Still' movement condition, and the panel on the right shows performance in the 'Walking' movement condition. In both panels, circular symbols are used to plot response proportions as a function of the actual elevation of the sound source when listeners were in a Normal listening mode (in which the left earphone emitted the left earmic signal, and the right earphone the right); the 'x' plotting symbol in both panels was used to display performance when listeners were in an Interchanged listening mode (in which the left earphone emitted the right earmic signal, and *vice versa*). The solid curves show the logit functions fit to the data obtained in the 'Normal' listening mode, while the dashed curves show the fit to the 'Interchanged' listening mode data.

Whereas the proportions of 'Above' responses increase with elevation angle of the sound source in the Normal listening mode, as they did in the Standing Still condition, for walking listeners in the Interchanged listening mode, the 'Above' response proportions decrease with elevation angle.

2.4. *Analysis*

While listeners wearing the experimental binaural hearing instrument were standing still, their forced choice performance in discriminating source elevation in the 'Normal' and the 'Interchanged' listening modes was quite similar. A chi-square analysis was performed on the two-way contingency table formed by summing all forced choice data across five listeners in this 'Standing Still' movement condition, and the results provided no evidence for rejecting the Null Hypothesis (i.e., the results supported the retention of hypothesis that performance was the same here in the 'Normal' and the 'Interchanged' listening modes). In contrast, a chi-square analysis of the contingency table data summarizing performance in the 'Walking' movement condition provides ample evidence for rejecting the Null Hypothesis ($X^2 = 112.7$; $p < .001$). Of course, the failure to reject the Null in the former case does not in any sense prove that discrimination ability is truly the same under the 'Normal' and the 'Interchanged' listening modes; and indeed, the statistical significance of the chi-square result in the latter 'Walking' movement condition is no surprise at all when viewing the panel on the right, given the obvious reversal in the trend of the data when comparing between the 'Normal' and the 'Interchanged' listening mode.

What could possibly explain the reversal in elevation judgments that occurs between Normal and the Interchanged listening modes for listeners who walk past the target sound source whose elevation is in question? It cannot be the head turning that is implicated in explaining front-back reversals, for the shifts in interaural differences that result from horizontal rotation of the head provide ambiguous cues to whether a sound source is located above or below ear level. Rather, it must be the rolling of the listeners' heads during walking that is providing the disambiguating cues to elevation, which would be consistent with observed dependencies in vertical plane sound localization results found for seated listeners (cf. [9]). Confidence in the correctness of this explanation would be increased if it could be shown that listeners do in fact exhibit greater amount of head rolling while walking rather than standing still, and so a set of measurements were made to confirm this hypothesis.

Now it is easy to imagine that the relative amount of head rolling generally increases when a listener begins to walk; yet empirical measurements should support the argument made here. The relevant series of measurements were made using a Polhemus FASTRAK head-tracking system to document the extent to which head rolling increases for walking listeners. Measurements were completed for five subjects over consecutive 3-second sampling periods while they were engaged either in talking while standing still or in walking towards or away from the source. After combining the head-tracking data from all five subjects, and using the standard deviation of roll values observed during talking as a baseline measure, it was found that walking produced more than four times as much modulation in roll angle over time (the respective angular standard deviations were 12.8° vs. 2.7°). In contrast, the standard deviation of head turning values during walking was only twice that observed during the 3-second samples of talking (with respective standard deviations of 13.5° vs. 6.3°). While either of these two amounts of head turning was enough to produce a strong dynamic front/back cue that contradicted the 'pinna factor' in the Interchanged listening mode, in the case of elevation judgment, it may be that the dynamic above/below cue was not as strong. Therefore, in contrast to head turning, it is suggested that overriding the 'pinna factor' here may have required the greater amount of modulation in roll angle that was associated with walking rather than standing still.

To support this conclusion, histograms of typical head turning and head rolling behavior of five listeners, made while they were walking or standing still, are presented in Figure 4. The plot in the left panel of Figure 4 shows that there is very little difference in the distribution of head turning (yaw angle variation) exhibited by listeners when walking vs. talking (i.e., standing still as possible, but moving as they will during talking). The plot in the right panel of Figure 4 shows a particularly large increase in head rolling during walking, supporting the contention that dynamic interaural cues associated with head rolling must be produced to a much greater extent during walking. This difference between head rolling during walking and talking could well be responsible for producing the observed reversal in elevation judgments between these two listener movement conditions.

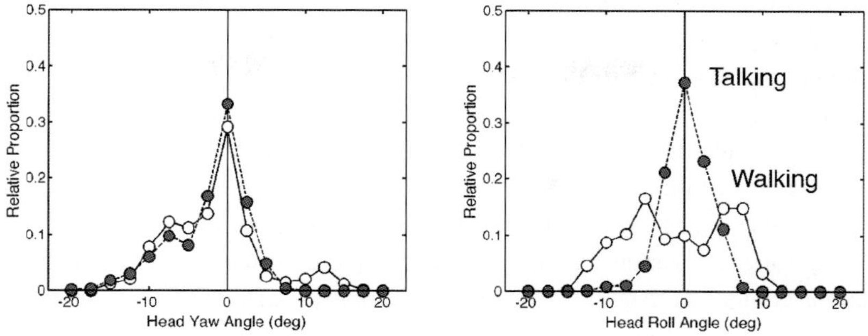

Figure 4: Histograms constructed for 2 types of head rotation measurements (yaw, roll) made using the Polhemus FASTRAK system under 2 different behavioral conditions (Talking, Walking). The left panel shows the relative proportion of head yaw angles in histogram bins with a width of 2.5° observed within a 40° range of possible yaw angles, with filled symbols for the talking condition and open symbols for the walking condition. The right panel shows the relative proportion of head roll angles for the talking (filled symbols) and the walking (open symbols) conditions, again with histogram bin width of 2.5° and a 40° range of possible roll angles (see text).

3. Discussion

The strong influence of dynamic interaural cues on directional hearing has long been known, i.e., that changes in the signals reaching a listener's ears that are directly caused by a listener's head movements provide useful directional information in addition to the spectral cues provided by the filtering effect of the pinna (via measureable head-related transfer functions) [10, 11]. However, it has not been so well established which cues will dominate directional judgment when dynamic interaural cues are placed in conflict with the spectral cues (that is, the direction cues available without listener motion, which comprise what Wallach [8] termed the pinna factor). The experimental results reported above show that dynamic interaural cues can dominate elevation judgments of walking listeners under conflict conditions, although the so-called pinna factor may have been somewhat reduced in its influence by the experimental binaural hearing instrument that was employed here. That is to say, unaided localization performance under natural conditions might rely more upon pinna-based spectral cues than localization using the electroacoustic mediation of sound signals provided by combined earmic and earphone.

Although the observed proportion of 'Above' responses plotted in the left panel of Figure 3 seem to suggest that elevation discrimination performance of the listeners is reasonably good, it is important to note that no results are reported here for unaided localization given the same stimuli and task. It may be that superior performance could be observed when listening without the binaural hearing instrument, and that the pinna factor would exert a stronger influence on performance were there no electroacoustic mediation. But with no means for interchanging ear signals without this or some similar mediation, the only experimental control that was included in the design of the current study was that of walking versus standing still.

Data on elevation perception itself might be useful here, since only the simple choice of 'Above' vs. 'Below' responses was required in the current experiment. Previous results [3] obtained under similar conditions using interchanged ear signals, but using a different task that required identification of the source elevation angle, do give some indication that perceived elevation was modulated by dynamic interaural cues. However, that study also did not include a control condition in which unaided elevation percepts could be compared with though using the binaural hearing instrument. A more comprehensive test of the relative salience of dynamic interaural vs. spectral directional cues would include such a control. Also, there is another factor that must be considered in future studies of the Phantom Walker Illusion, and that is the sound source's distribution of energy over frequency.

There is reason to suppose that dynamic interaural directional cues might not dominate spectral cues if sources contained no low-frequency content. For example, in a related study of illusory sound source motion produced by conflicting spectral and dynamic interaural cues, Macpherson [12] found that signals containing predominantly or exclusively high-frequency content had directional percepts that were apparently more influenced by spectral directional cues. In further discussion of the role of head motion and spectral cues reported in this book, Macpherson [13] concludes that Wallach's 'principle of least displacement' may be violated under these conditions, and sources that might otherwise be heard to be located at fixed position behind the listener during head turning, are heard to be moving quickly left and right in front of listeners as they turn their heads. However there are other important differences between his studies and that reported herein, perhaps most important being the linear displacement of the walking listener that was included in the current studies, and was absent from the studies of Macpherson [13]. Furthermore, the speech sound source that was presented in the current study was presented for a much longer duration than the noise sources used in those he reported. Indeed, he found that the moving stimulus needed to be presented for only 50 to 100 ms to be useful to the listener in distinguishing source incidence angles. He also found that a shift of only 5 to 10 deg in source azimuth angle was required for dynamic interaural cues to effectively support front-back discrimination between virtual source locations. It will be interesting to learn more from envisioned related studies in which the spectral and temporal parameters of the stimuli are investigated. It is also interesting to compare the current results with those found in previous studies of head movement and sound localization.

3.1. *Related Work on the Role of Head Movement in Sound Localization*

In what is believed to be the earliest systematic study on the role of various types of head movements in sound localization, Thurlow and Runge [14] focused upon the influence involuntary head movements: In effect, they studied localization performance by manually inducing the head movements of their listeners rather than allowing them to generate their head movements voluntarily. They examined errors in both azimuth and elevation judgments for a number of types of head movement. Without delineating the specifics of their experiments, their general results may be summarized as follows: Relative to a condition in which no head movement was allowed, rotation of the head reduced errors in azimuth judgment as would be expected. However, head rotation did not significantly reduce errors in elevation judgments. If, on the other hand, the listener's head was rolled from side to side while listening, so that first one ear was dropped closer to the shoulder on the same side, and then the other was dropped towards its shoulder, elevation error was reduced and azimuth error was not. When the head was pitched forward and back (looking up then looking down, as when indicating a "yes" response), neither error rate was reduced significantly. In order to determine how frequently human listeners use different types of head movements as an aid in determining the location of a sound source, Thurlow, Mangels, and Runge [15] observed over 50 subjects during a free-field localization task. First, they noted that most subjects always included some amount of horizontal head rotation (i.e., yaw variation) in their exploration of the sound stimulus. Interestingly, they observed that their listeners rotated their heads 42° on the average when the sound source was located at a high elevation, but only rotated their heads an average of 29° when the source was at a lower elevation. In fact, it was most common to find listeners rotating their heads while also pitching their heads (which they termed tipping). The way in which they classified observed head movements used a criterion of 3° for each of the three types of variation in head orientation. Their results can be summarized as follows:

The type of head movement observed most often was a combination of rotation and pitching, without substantial rolling of the head (which they termed pivoting). The second most common type of head movement observed was rotation without pitching or rolling, while the third most common was a combination of rotation, pitching, and rolling. Fourth was rotation with rolling. The relatively high frequency of this fourth type of movement was reported as somewhat surprising by Thurlow, et al. [15] because they observed that head rotation and roll produced confounding lateral shifts of the auditory image that should not disambiguate direction. That is, because rotation-induced binaural changes most strongly cue front/rear distinctions, and roll-induced binaural changes most strongly cue above/below distinctions, the interpretation of the binaural changes accompanying combined rotation and rolling movements were regarded by Thurlow, et al. [15] as providing only uncertain information. In contrast, the experience of listeners in the current study supports the contention that the horizontal rotation and rolling of the head during walking yields unambiguous angular impressions, perhaps due to the familiarity of the albeit complex variation in head orientation that occurs.

3.2. *Demonstration of the Phantom Walker Illusion*

An effective means for demonstrating the Phantom Walker Illusion is presented in Figure 5. In addition to the curious fact that the speech sound source of the Phantom Walker will be perceived to be moving away from a walking listener, what is to be demonstrated is the following: When the voice of the Phantom Walker is presented from below a listener's ear level, the voice is judged to be arriving from above ear level, and a voice arriving from above will be judged to be lower. It is better at this point to use second person perspective, and so the text in the following paragraph is switched to this perspective to describe the situation depicted in Figure 5, as an indication of the explanation a participant might receive.

Consider what you might experience given the presented situation in which two manikins that have speakers at their mouth positions and these are used to reproduce samples of recorded continuous speech. Since these two manikins are positioned on either side of the hall down which you will walk, one with mouth positioned above your ear level, and the other with mouth positioned below your ear level, consider what you might experience were your ears to be reversed. What about their apparent height relative to your own? While it is certain that the voice of the taller manikin positioned on your left will be heard on your right, and will be heard to be approaching from the rear as you walk toward this taller manikin, it is also almost certain that the voice of this taller manikin will be heard as arriving from the mouth of the shorter manikin. Fortuitously, this illusion is visually reinforced here, since the voice of the shorter manikin positioned on your right will be heard on your left, where the taller manikin is positioned. To summarize, the voice coming from the mouth of the short manikin on the right will be heard as arriving from a tall manikin on the left, but this manikin's voice will be sneaking up behind you as you walk forward between the two manikins.

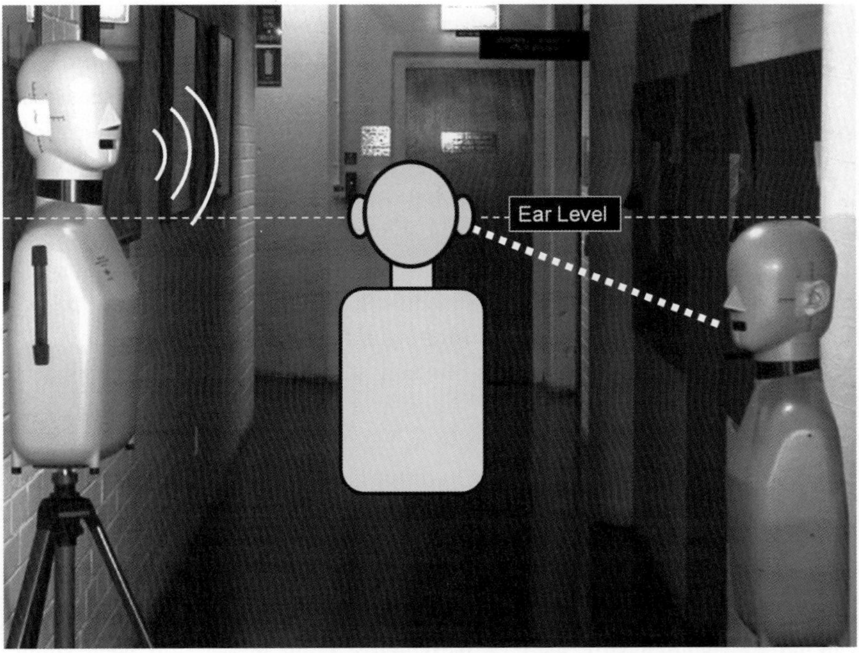

Figure 5: Photograph of the hallway in which the Phantom Walker illusion has been demonstrated. Superimposed upon the photograph is a graphic depiction of a listener walking between the two manikins of differing height, one manikin with mouth above the listener's ear level (with white curved lines representing the emitted speech sound), and the other manikin with mouth below (from which the speech sound seems to emanate when a listener wears an ear-interchanging binaural hearing instrument).

4. Conclusion

An illusion of auditory motion termed the Phantom Walker motivated this study of auditory reversals associated with the interchange of a listener's ears using a binaural hearing instrument. While left-right and front-back reversal of a target sound source occurred virtually always under all conditions tested, it was shown that a reversal in elevation of the sound source depended upon whether the listener was walking or standing still. The fact that the speech sound source of the Phantom Walker is perceived to be moving away from a walking listener was shown to be consistent with Wallach's [6] *principle of least displacement* that was described in this chapter. The results of the experiment reported herein provide evidence (as promised in the chapter title) for the dominance of dynamic interaural directional cues that are available to a walking listener under conditions in which these cues are placed in conflict with natural spectral cues to source location.

In the most effective form of the Phantom Walker Illusion, the listener walks toward a source that is actually located both above ear level and in front of the listener, and that source seems to be approaching the walking listener from the rear and at an elevation angle below ear level. The results of controlled experiments reported herein reveal that dynamic interaural lateralization cues naturally available to walking listeners determine the perceived elevation of a sound source when listeners are presented with conflicting spectral cues to elevation cues associated with the filtering effects of the pinna. In order to confirm that the functioning of the binaural hearing instrument is adequate to support correct elevation discrimination on the basis of static cues, the same five listeners were also asked to participate in the discrimination experiment while not moving. By testing under various combinations of conditions (walking/motionless; normal/interchanged ears) it was established that interchanging ear signals does not cause a reversal of elevation judgments when listeners are asked to stand still. Furthermore, measurements of typical head turning and head rolling behavior of five listeners, made while they were walking or standing still, show a particularly large increase in head rolling during walking, supporting the contention that dynamic interaural cues associated with head rolling are responsible for producing the observed reversal in elevation judgments. Of course, such large head movements are not required for spectral cues to be dominated by dynamic interaural cues in resolving front/back ambiguity, as was well documented in early studies of virtual source localization [16] that placed these cues directly in conflict. Nonetheless, it is suggested that the study of sound localization during walking might provide a particularly rich area of inquiry, as is becoming apparent in research on changes during walking observed in other sensory modalities such as vision [17].

References

1. P.T. Young, Auditory localization with acoustical transposition of the ears. J. Exp. Psychol. **11**, 399-429 (1928).
2. C.F. Willey, E. Inglis, and C.H. Pearce. Reversal of auditory localization. J. Exp. Psychol. **20**, 114-130 (1937).
3. W.L. Martens and S. Kim, "Dominance of head-motion-coupled directional cues over other cues during active localization using a binaural hearing instrument," Proc. WESPAC X, 10th Western Pacific Acoustics Conference, Beijing, China (September, 2009).
4. W.L. Martens, D. Cabrera, and S. Kim, "Dynamic auditory directional cues during walking: An experimental investigation using a binaural hearing instrument. In: International Workshop on the Principles and Applications of Spatial Hearing, Zao, Miyagi, Japan (November, 2009).
5. J.M. Loomis, C. Hebert and J.G. Cicinelli, Active localization of virtual sounds. J. Acoust. Soc. Am. **88**,1757-1764 (1990).
6. H. Wallach, The role of head movements and vestibular and visual cues in sound localization. J. Exp. Psych., 27, 339-368 (1940).
7. P.T. Young, The role of head movements in auditory localization. J. Exp. Psychol. **14**, 95-124 (1931).
8. H. Wallach, On sound localization. J. Acoust. Soc. Am. **10**, 270-274 (1939)
9. S. Perrett, W. Noble. The effect of head rotations on vertical plane sound localization. J. Acoust. Soc. Am., **102**, 2325–2332 (1997).
10. Y. Iwaya, Y. Suzuki and D. Kimura, Effects of head movement on front-back error in sound localization. Acoust. Sci. & Tech., **24**, 322–324 (2003).
11. F.L. Wightman, D. Kistler, Resolution of front-back ambiguity in spatial hearing by listener and source movement. J. Acoust. Soc. Am., **105**, 2841–2853 (1999).
12. E.A. Macpherson, Illusory sound source motion produced by conflicting spectral and dynamic interaural cues. In: International Workshop on the Principles and Applications of Spatial Hearing, Zao, Miyagi, Japan (November, 2009).
13. E.A. Macpherson, Head motion, spectral cues, and Wallach's 'principle of least displacement' in sound localization. in IWPASH 2009 Book, eds. Y. Suzuki and D. S. Brungart (World Scientific, 2010).
14. W. Thurlow and P.S. Runge Effect of induced head movements on localization of direction of sounds. J. Acoust. Soc. Am., **42**, 480-488 (1967a).

15. W. Thurlow, J. W. Mangels and P.S. Runge Head movements during sound localization. J. Acoust. Soc. Am., **42**, 489-493 (1967b).
16. J. Kawaura, Y. Suzuki, F. Asano, T. Sone. Sound localization in headphone reproduction by simulating transfer functions from the sound source to the external ear. J. Acoust. Soc. Jpn.(E), 12, 203-216 (1991).
17. F.H. Durgin, When walking makes perception better. Current Directions in Psychological Science, **18**, 43-47 (2009).

HEAD MOTION, SPECTRAL CUES, AND WALLACH'S 'PRINCIPLE OF LEAST DISPLACEMENT' IN SOUND LOCALIZATION

E. A. MACPHERSON

National Centre for Audiology, University of Western Ontario,
London, Ontario N6G 1H1, Canada
E-mail: ewan.macpherson@nca.uwo.ca
www.uwo.ca/nca

The dynamic interaural cues produced by listener head rotation provide unambiguous information about the front/rear location of a source only if the source is assumed to be stationary. Moving-source trajectories can be devised that produce dynamic cues similar to those for a stationary source, but Wallach (1940) reported that the stationary interpretation is preferred — the 'principle of least displacement'. Front/rear information derived from spectral cues can conflict with a stationary-source interpretation of dynamic cues when sources follow such trajectories during head motion or when high-frequency, narrowband source spectra create illusory location percepts. Recent results from both freefield and virtual auditory space testing suggest that salient high-frequency spectral cues tend to dominate in cases of spectral/dynamic-cue conflict, leading to veridical or illusory source-motion percepts. Thus, head movements cannot resolve front/rear errors or ambiguity for all stimuli, and the applicability of Wallach's principle of least displacement is stimulus-dependent.

Keywords: head movement; interaural cues; spectral cues; dynamic cues; head-related transfer functions

1. Introduction

1.1. *Dynamic sound localization cues*

Under natural conditions, a listener can take an active role in sound localization by means of moving the head. In that case, dynamic information about the location of a sound source is provided by the relationship between the motion of the head and the resulting changes in the interaural time- and level-difference cues (ITD and ILD). In particular, for head rotation about the vertical axis, the direction of change of ITD and ILD for a stationary source in the front hemisphere is opposite to that for a source in the rear. The benefits of head movements in improving localization accuracy are confirmed by consistent reports of reductions in front/rear reversals for subjects localizing in the freefield with restricted-bandwidth stimuli[1,2] or distorted pinna cues,[3,4] and in virtual auditory space generated with individualized[5] or non-individualized[6,7] head-related transfer functions.

The geometrical theory of localization based on dynamic cues that was worked out in detail some 70 years ago by Wallach[8,9] predicts that information derived from head motion can also contribute to localization in the vertical dimension. For example, in the present volume, Martens *et al.*[10] report that dynamic cues produced by head roll (side-to-side rotation about the front/rear axis) contribute significantly to the perception of source elevation. Additionally, the *rate* of change of interaural difference cues relative to head rotation rate is a cue to the vertical displacement of a source, but this cue has been shown to be effective only for low-frequency noise stimuli,[11,12] which suggests that low-frequency ITD information may be required for significant influence of the dynamic cue.

Some studies have failed to show substantial effects of head movement on localization accuracy. Pollack and Rose[13] conducted a series of experiments in which listeners localized horizontal-plane, broadband click and noise-burst sources presented during head motion. A beneficial effect of head movement was found only for long-duration sources initially situated off the median plane of the head. Head movement thus served to bring these sources onto the median plane where localization on the basis of interaural difference cues is most acute. Such weak effects of head movement prompted Middlebrooks and Green[14] to propose that the main benefit of head movement is to allow the listener to orient toward the source, thus bringing it into the region of greatest azimuthal acuity. For long-duration stimuli, spontaneous head movements do indeed typically involve a large-angle rotation in the direction of the source.[2,5,15]

Unequivocal support for the salience of the dynamic "Wallach" cue is, however, provided by the results reported in this chapter, by Martens *et al.* in this volume, and by two studies conducted by Perrett and Noble.[1,12] Recognizing that head movements that allow the listener to zero-in on the source location confound dynamic cues with changes in static binaural acuity, Perrett and Noble required their listeners to make stereotyped head movements over restricted ranges. Listeners either oscillated their head position rapidly between -30 and +30 degrees azimuth during a 3-s stimulus duration,[12] or made a single rotation from 0 to -45 degrees azimuth, initiated at the onset of a 0.5- or 3-s signal.[1] For sources in the horizontal plane, the oscillatory motion was sufficient to virtually eliminate front/rear reversals for low-pass, broadband, and high-pass noise stimuli. For low-pass sources in the vertical median plane, that motion permitted listeners to report source elevations strongly correlated with the physical locations, although the apparent displacements from the horizontal plane were only ~2/3 of the physical values. For 3-s low-pass stimuli, the 45-degree rotation movement was found to be as effective as free head motion for the perception of elevation, and smaller portions of that rotation occurring during 0.5-sec low-pass stimuli were sufficient to eliminate front/rear reversals.

1.2. *Static sound localization cues*

In the absence of head motion, the interaural difference cues provide information only about the lateral location of a sound source. Psychophysical studies have established that low-frequency ITD carried by waveform fine structure is the dominant lateral angle cue in stimuli containing low-frequency components (below ~1.5 kHz), whereas in high-pass or high-frequency stimuli, ILD is the dominant lateral cue.[16–18] Any specific value of ITD or ILD is unique not to a single direction, but to a locus of directions encircling the interaural axis at approximately the same lateral displacement from the median plane (a "cone of confusion"[19]).

Under static-head conditions, then, information about the front/rear and vertical dimensions that serves to disambiguate location on the cone of confusion must be provided by the spectral cues created by the direction-dependent filtering imposed on incident sound by the pinnae, head and torso. That filtering can be characterized by the magnitude spectrum of the head-related transfer function (HRTF). The most salient cues in HRTFs appear to be the spectral features lying above ~4 kHz.[20–22] Stimuli lacking energy in the high-frequency region or with bandwidth insufficient to reveal the high-frequency spectral features cannot be localized accurately in the vertical plane. Instead, such stimuli appear to originate from phantom locations that depend on their frequency content rather than their physical location. The apparent position for low-pass stimuli is on or slightly below the front horizontal plane for most listeners.[20,23] Narrow-band sources can appear to originate in front, overhead or behind a listener depending on center frequency,[24,25] and this effect varies between listeners.[26,27]

1.3. *Dynamic cue ambiguity and conflicting spectral cues*

Dynamic interaural cues generated by head movement provide unambiguous information about front/rear location only if the source is assumed to be motionless. In the absence of other front/rear information, there are two possible sets of horizontal-plane source trajectories consistent with the changes in ITD and ILD produced by a given head rotation. These are illustrated in the lower panel of Fig. 1; the first trajectory consists of a stationary, rear-hemisphere source location (filled circle), and the second consists of a source in the front hemisphere that rotates at twice the angular velocity of the head (open circles). The moving trajectory corresponds to the path traversed by the stationary source location after moment-by-moment front/rear reversal across the interaural axis (Fig. 1, upper panels).

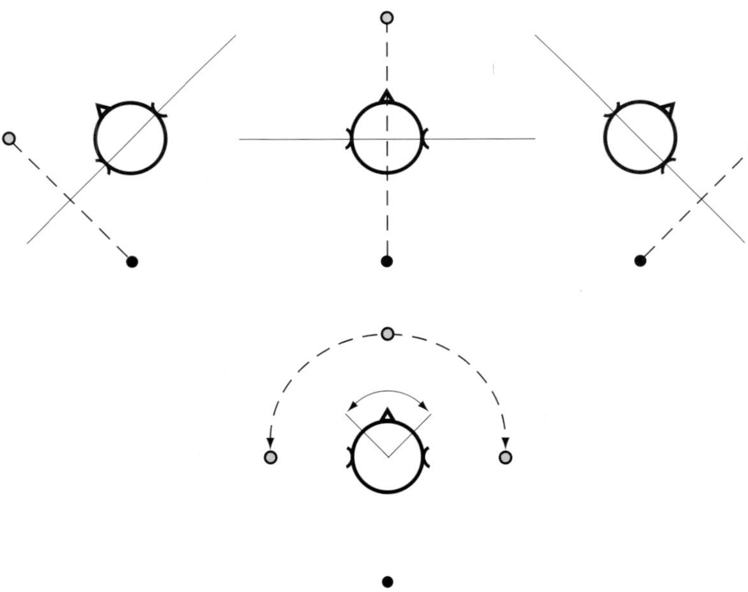

Fig. 1. Front/rear ambiguity for stationary or moving sources. Upper panels: Rear- and front-hemisphere source locations producing similar interaural cues at three different head angles. Lower panel: Stationary (filled circle) and moving (shaded circles) source trajectories compatible with the changing interaural cues generated by a head rotation.

In addition to the possibility of veridically perceiving either type of trajectory, should they physically occur, a listener might also make one of the two forms of perceptual error illustrated in Fig. 1, lower panel. These are: 1) *reversal-induced motion*, in which the spectrum of a stationary source produces a front/rear reversal and consequently an illusory percept of motion; and 2) *motion-induced reversal*, in which a source that executed the proper rotation in one hemisphere during a head movement might be perceived as stationary in the opposite hemisphere.

An indication as to whether stationary or moving trajectories are perceived more frequently was provided by Wallach,[8,9] who reported the results of several localization experiments employing a horizontal, circular array of loudspeakers and a signal commutator mechanically coupled to a listener's head. When the commutator was arranged to produce the front-hemisphere, double-speed rotation illustrated in Fig. 1, listeners perceived the source as stationary in the rear. When the commutator was arranged to keep the active speaker in front of the listener as the head was rotated, listeners perceived the source as stationary overhead, consistent with the unchanging, near-zero interaural difference cues. An overhead percept was also generated for a fixed frontal speaker when listeners experienced illusory rotation induced by a moving visual background. From these results, Wallach derived the 'principle of least displacement' — that for such ambiguous stimuli, the preferred auditory percept corresponds to the trajectory involving the smallest movement of the source in space. A similar 'assumption of stationarity' principle has been proposed for the visual system.[28] Wallach referred to the illusory stationary location corresponding to an appropriately moving source as the *synthetic direction.*

When the physical trajectory of a moving source differs from the synthetic direction, the spectral cues consistent with the true motion are placed in conflict with the stationary-source interpretation of the dynamic cues. The issue then arises of the relative weighting of those two cues by the auditory system. Wallach was aware that listeners could localize in the vertical and front/rear dimensions in the absence of head motion, and correctly attributed this to "the pinna factor",[8] about which he commented that ...

> In every case of a successful synthetic production the pinna factor is overcome by the cues procured by the head movement, for here the perceived direction is quite different from the direction from which the sound actually arrives at the head.
>
> ...
>
> The fact that this [pinna] factor is invariably overcome in the synthetic production experiment [in the horizontal plane] indicates quite clearly its subordinate role.

In addition to the work of Wallach,[8,9] only one previous study has explicitly addressed the weighting of conflicting spectral and dynamic cues. Kawaura *et al.*[6] presented music filtered with static HRTFs corresponding to front or rear locations with an additional, head-coupled, dynamic ITD consistent with the opposite hemisphere. In those cases of conflict, two of three listeners reported apparent locations corresponding to the dynamic cue, while the third experienced front-to-rear reversals.

Because narrow-band sounds generate apparent locations determined by frequency and not by physical location, the handful of studies[1,11,12] in which such sounds have been presented in the free field with head movement might also be considered as examples of cue conflict. For example, a 2-kHz low-pass noise signal presented in the median plane is typically localized to the front horizontal plane regardless of its physical location. Rotation of the head generates dynamic cues that allow listeners to correctly identify, for example, overhead or rear-hemisphere locations for such a source that are in conflict with the erroneous spectral cues. For a 3-sec low-pass signal, Perrett and Noble[1] found that for some listeners, 45 degrees head rotation was sufficient for perception of such a source as elevated, suggesting dominance of the dynamic cues. Thurlow and Runge[11] found improvements in elevation accuracy with 45-degree head rotation for low-frequency (0.5–1 kHz), but not for high-frequency (7.5–8 kHz), narrow-band noise stimuli.

1.4. *Aims and methodological issues*

The aims of the present study were twofold: a) to explore the relative weighting of spectral and dynamic localization cues and the principle of least displacement for stimuli with varying degrees of spectral-cue reliability; and b) to explore the possibility that reversal-induced motion explains the failure of head movements to resolve front/rear reversals for high-frequency narrowband noise (described below in Experiment 1).

The experiments described here employed a novel method of stimulus presentation. Listeners made single large-amplitude head rotations ("head sweeps") similar to those used by Perrett and Noble,[1] but rather than stimulus onset cuing head movement, stimuli were gated on and off once the head was in motion based on real-time tracking of head position. Head movement initiated by the onset of the stimulus is problematic because head-movement latency is rather large (100's of ms[19]) and rather variable,[1] making it difficult to control precisely how much access listeners have to the dynamic cues. Vliegen *et al.*[29] have demonstrated accurate localization of sources presented during head saccades, which provided confidence that listeners would be able to localize accurately under such conditions. Distortions of auditory space associated with head saccades have also been reported,[30,31] but only for very brief (10-ms) stimuli presented in the period between the presentation of a visual saccade target and the onset of head motion.

2. Experiment 1: Salience of dynamic cues as a function of stimulus spectrum

More complete results of Experiment 1 have been previously reported,[32,33] but selected relevant findings are described here briefly to provide motivation and the methodological background for Experiment 2.

2.1. *Methods*

Experiments were conducted in a darkened, anechoic room containing an apparatus that could position a loudspeaker at any desired azimuthal position in the horizontal plane. Blocks of trials were run either with the head fixed (stimulus duration 200 ms) or with the head in motion at approximately 50 deg/s, as described below. Stimuli were bands of wideband (WB, 0.5–16 kHz), low-frequency narrow-band (LNB, 0.5–1 kHz), or high-frequency narrowband (HNB, 6.0–6.5 kHz) noise, and were presented from locations spaced by approximately 30 degrees around the horizontal plane. Four normally hearing young adults participated.

The paradigm for head-motion trials is illustrated in Fig. 2. Each motion trial began with the listener's head turned 45 degrees to one side. The listener then began a head rotation at the practised velocity (50 deg/s) while head orientation was tracked continuously using an electromagnetic sensor (Polhemus Fastrak). When the orientation entered a selected spatial window centered straight ahead (widths, Θ, between 2.6 and 20 degrees), the stimulus was gated on. When the head's orientation exited the window, the stimulus was gated off, while the listener continued the head rotation to 45 degrees on the other side. Following each head-fixed or head-moving stimulus presentation, the listener indicated the apparent direction of the stimulus by turning the body and orienting with the head. A final reading of head orientation constituted the listener's response. It is important to note that the positions of the spatial window and of the the targets were entirely independent.

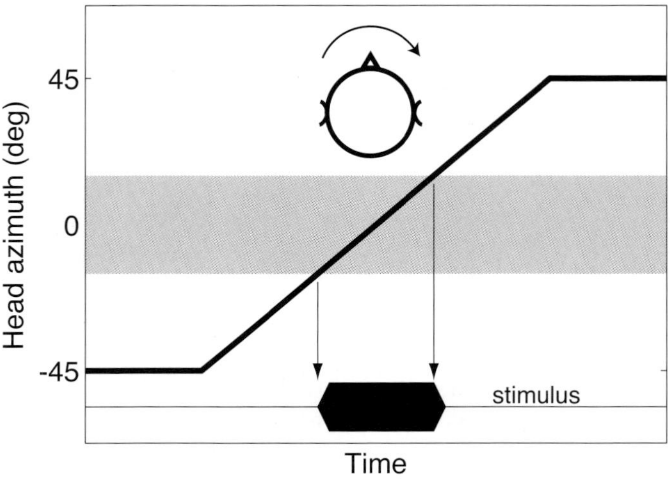

Fig. 2. Head-sweep stimulus presentation paradigm. The target stimulus (located at an arbitrary azimuth) was gated on and off as the head orientation entered and exited a definable spatial window.

2.2. *Results*

For each listener and stimulus condition (a particular combination of noise spectrum and spatial window), response azimuth was plotted versus target azimuth. In such a plot, veridical responses lie on the positive diagonal and front/rear reversals lie near the negative diagonal.

Target-response plots for one typical listener (S203) in selected conditions (head-fixed and 50-deg/s motion with spatial windows of 5 and 20 degrees) are shown in Fig. 3. Each column corresponds to one spatial window width. With the head fixed (leftmost column), WB stimuli were localized accurately, but all of this listener's responses to the LNB and HNB noises fell in the front hemisphere, producing many back-to-front reversals. In head-motion conditions: responses to wideband noise were somewhat more scattered, but generally accurate; performance for LNB noise improved dramatically with increasing spatial window width; similar improvement was not observed for HNB noise.

The result for the HNB noise stimuli is remarkable; the fact that responses lie near either the positive or negative diagonal shows that the listener could certainly use the available interaural cues to report accurately the left/right component of the target location, but the listener apparently could not use the head-motion-related *change* in those cues to disambiguate

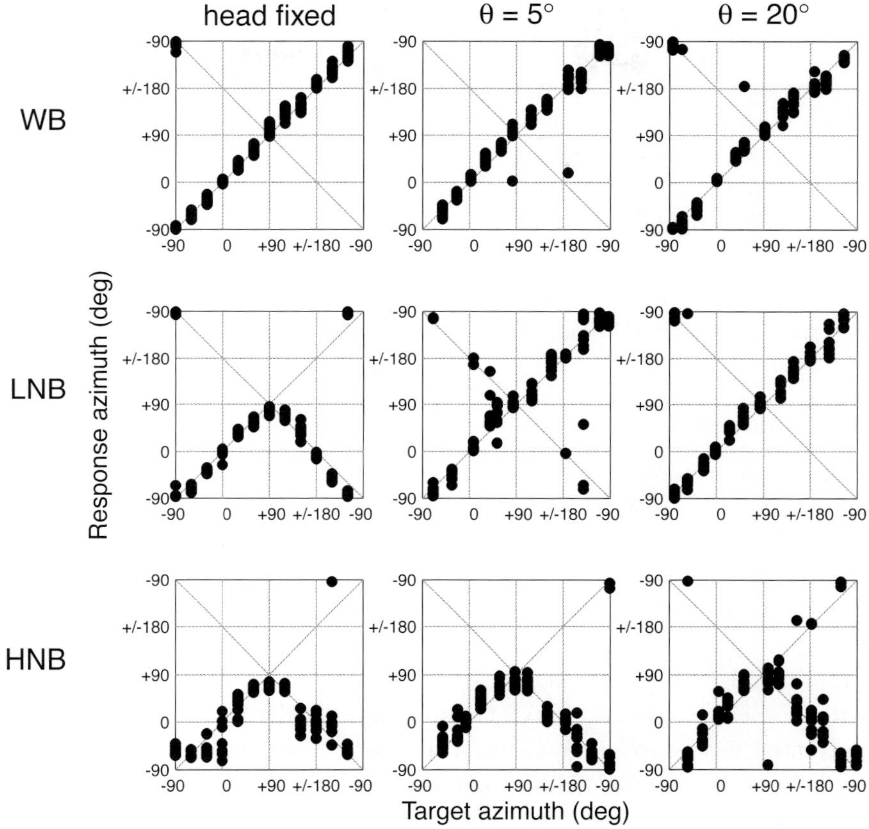

Fig. 3. Azimuthal target-response plots for one typical listener (S203) in head-fixed and selected 50-deg/s head-motion conditions of Experiment 1. Top, middle, and bottom rows: WB, LNB, and HNB noises respectively.

front from rear for the high-frequency targets.

One possible interpretation of these data is that the low-frequency ITD information carried by the WB and LNB noises is the critical dynamic cue, and that this source of information is simply lacking in the HNB stimulus, which must be lateralized primarily on the basis of ILD. A second interpretation is that the spectrum of the LNB noise provided only ambiguous front/rear information that was easily dominated by the dynamic cues, whereas the HNB noise provided misleading, but unambiguous, spectral cues resulting in consistent rear-to-front reversal and possibly the perception of reversal-induced motion for sources located in the rear hemisphere.

3. Experiment 2: Relative influence of spectral and dynamic cues for front/rear location — virtual stimuli

Information about the front/rear location of a sound source can be provided both by dynamic interaural cues and by spectral cues. The aim of this experiment was to explore the applicability of the principle of least displacement and the relative influence of those two sources of information as a function of the reliability of the spectral cues provided by the stimulus spectrum. A second point of interest was exploration of the hypothesis that the failure in Experiment 1 of head rotation to resolve rear-to-front reversals for the HNB noise was due to illusory reversal-induced motion.

3.1. *Methods*

The spatial-window/head-movement paradigm described above was used with a head velocity of 50 deg/s and spatial window widths of 0 (head fixed, stimulus duration 200 ms), 5, 10, 20, and 40 degrees. The 40-degree width was added to aid listeners in detecting perceived source motion, if any. Stimuli consisted of the WB, LNB, and HNB bands of noise used in Experiment 1 plus high-pass (HP, 4–16 kHz) noise. The WB and HP noises were intended to provide access to accurate high-frequency spectral cues with or without low-frequency ITD information. The LNB and HNB noises prevented access to accurate spectral cues. Stimuli were presented over headphones in virtual auditory space (VAS) using individually measured directional transfer functions (DTFs)[34] for each listener. The real-time, head-tracked VAS synthesis was performed by a Tucker-Davis Technologies RX6 Multifunction Processor, with head-orientation data provided at 120 Hz from a Polhemus Fastrak interfaced with a TDT HTI3 head tracker interface module. Four normally hearing listeners participated.

Stimulus locations were spaced by 30 degrees around the horizontal plane, as in Experiment 1, but were presented in both *natural* and *synthetic*[8] modes. In the natural mode, the real-time VAS synthesis applied the left- and right-ear DTF filter pair corresponding to the position of a stationary source relative to the instantaneous head position — a faithful rendering of the freefield conditions of Experiment 1. In the synthetic mode, the DTFs corresponded to the double-speed, opposite-hemisphere source trajectory necessary to produce a motion-induced reversal — a VAS re-creation of Wallach's commutator and speaker array apparatus. In synthetic-mode presentation, there is conflict between the direction indicated by spectral cues and the synthetic direction derived from a static-source interpretation of the dynamic interaural cues.

The natural and synthetic modes were implemented by the same program on the TDT RX6 processor by using, respectively, either *normal* or *front/rear reversed* DTF filter sets. The normal DTF filter sets provided a faithful VAS synthesis. In the reversed DTF filter sets, the filter pairs for symmetrical front and rear locations were swapped. Thus the filter pair labeled 0-azimuth, 0-elevation (straight ahead) actually consisted of the filters measured at 180-azimuth, 0-elevation (straight behind) and vice versa. Similarly, filters measured at 30 degrees azimuth were swapped with those for 150 degrees, those at 60 degrees with those at 120 degrees, and so on, as illustrated in Fig. 4a. Figure 1 can be interpreted to illustrate how front/rear DTF swapping produces a moving virtual source when a head rotation is performed.

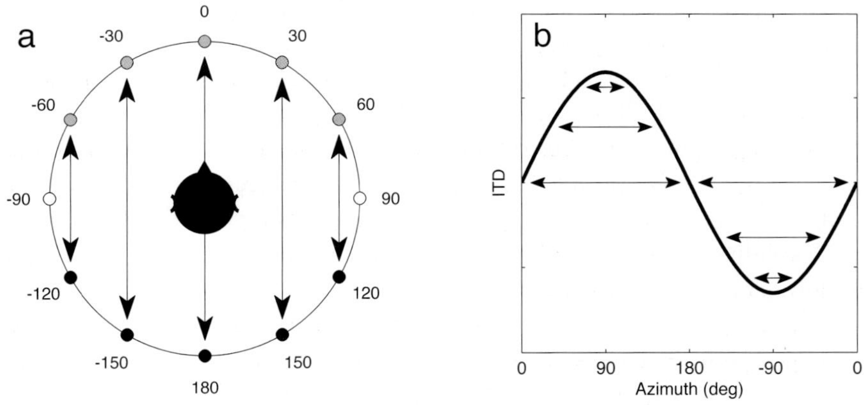

Fig. 4. Front/rear reversal of DTF filter set to produce synthetic-mode stimuli. a) DTF filter pairs for front/rear symmetrical locations were swapped. b) Cartoon illustrating that the manipulation had little impact on the interaural difference cues for lateral angle.

This reversed-DTF method also provided a straightforward extension of the synthetic direction concept to the static head situation; in the absence of head motion, the swapped spectral cues were intended to prompt the listener to make responses which appeared to be consistent front/rear reversals with respect to the filter labels (i.e., the synthetic directions). This manipulation should have little effect on the lateral component of perceived location because interaural difference cues (and in particular ITD[35]) are approximately front/rear symmetrical as shown in cartoon form in Fig. 4b.

In separate blocks of trials, three of the four listeners re-performed the head-movement task for a subset of the stimulus conditions (spatial window widths of 10 and 40 degrees). In those blocks, they simply reported whether each stimulus appeared to be "stationary" or "moving" in space during the head rotation, and did not provide localization responses.

3.2. *Results*

We present here preliminary results from Experiment 2. Target-response azimuth data are plotted for one typical listener in Fig. 5 for head-fixed and 20- and 40-degree spatial window conditions and all combinations of stimulus spectrum (WB, HP, LNB, HNB) and presentation mode (natural, synthetic). In synthetic-mode cases, the plotted target azimuth corresponds to the synthetic direction. Percentage values to the right of each row give the mean proportion of trials (across three listeners) in which source movement was reported for each stimulus type with the 40-degree spatial window width. The corresponding values (not shown) for the narrower 10-degree window were very low, ranging from 2–11% with a median of 3%.

In all normal-DTF conditions for both WB and HP stimuli (rows 1 and 3) there were few front/rear reversals (found near the negative diagonal in these plots), and the proportion of trials reported as "moving" was low (10% and 9% for WB and HP noise, respectively). In the head-fixed, reversed-DTF conditions for both WB and HP stimuli (column 1, rows 2 and 4), the majority of responses were front/rear reversals induced by the DTF filter swapping manipulation. Head movement with spatial window widths of up to 40 degrees did not effectively eliminate these reversals either with (WB) or without (HP) low-frequency stimulus components. A much higher proportion of trials were reported as "moving" in the reversed-DTF conditions for WB and HP noise (43 and 58% , respectively).

For the LNB noise, front/rear reversals were observed in the head-fixed conditions for both normal and reversed DTFs (column 1, rows 5 and 6), but were completely resolved by head movements for both DTF types. The rates of "moving" reports were low (12 and 14%). Thus, the stationary-source percept was favored for these stimuli regardless of DTF type.

For the HNB noise, results were also similar between normal- and reversed-DTF conditions. Many rear-to-front reversals were observed in the head-fixed condition and for spatial window widths up to 20 degrees. At the 40-degree window width, the reversal rate was reduced for both DTF types, indicating that the dynamic cues were beginning to be effective in resolving rear-to-front reversals. The reported rates of "moving" stimuli were low

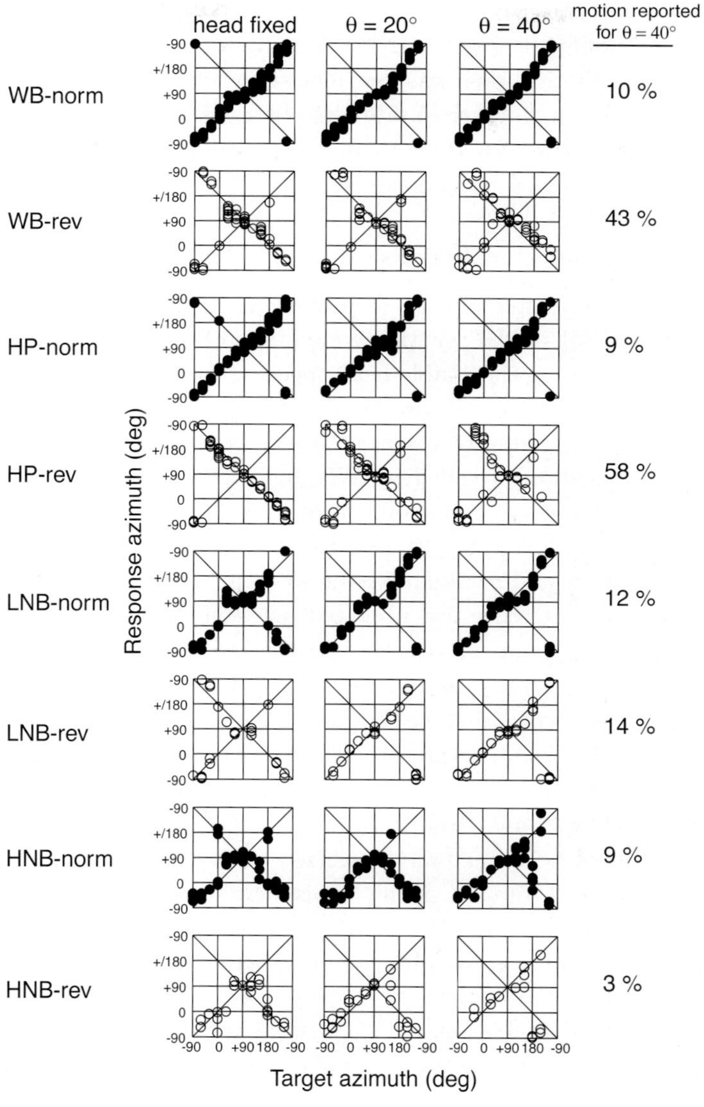

Fig. 5. Azimuthal target-response plots for one typical listener (S220) in head-fixed and selected 50-deg/s head-motion conditions of Experiment 2. Data for WB, HP, LNB, and HNB noise are presented in pairs of rows from top-to-bottom, with odd-numbered rows (filled symbols) corresponding to natural-mode presentation (normal DTFs) and even-numbered rows (open symbols) corresponding to synthetic-mode presentation (front/rear-reversed DTFs).

(9 and 3% for normal and reversed DTFs). The stationary-source percept was therefore dominant for the HNB noise stimuli at the 40-degree spatial window width. Motion rate was not measured for the 20-degree window width, for which a high reversal rate was observed. Thus, the present data are unfortunately uninformative about the prevalence of perceived motion of rear-to-front reversed HNB noise stimuli.

4. Discussion and Conclusions

Contrary to the principle of least displacement, in Experiment 2, listeners often perceived WB and HP, synthetic-mode sources as reversed and moving (presumably executing the double-head-speed rotation illustrated in Fig. 1), even though this motion was correlated with the listeners' head motion in a manner highly unlikely to occur naturally. The accurate spectral cues for front/rear location provided by WB and HP noise stimuli apparently dominated over the more parsimonious, stationary-source interpretation of dynamic cues even for large (40 degrees) spatial window widths. Because no window width tested resulted in reduced dominance of the spectral cues, and therefore no difference was observed between WB and HP stimuli, it is not possible to comment on the hypothesized dominance of dynamic low-frequency ITD relative to dynamic high-frequency ILD.

Although the reversal-induced motion hypothesized to account for the failure of small head movements to resolve front/rear confusions for HNB noise would be illusory motion, the motion perceived for WB and HP stimuli in Experiment 2 was *not* illusory. In all reversed-DTF, head-moving conditions, the set of DTFs actually traversed on each trial corresponded to a double-head-speed rotation in the hemisphere opposite the nominal synthetic location of the source. A failure to resolve the reversals and thereby to perceive the source motion simply corresponds to veridical perception of the stimulus trajectory.

In the case of the LNB noise, which provided impoverished spectral cue information, listeners resolved reversals under both normal- and reversed-DTF-set conditions at relatively small spatial window widths and reported low rates of source motion. These observations indicate that the preferred interpretation of the cues was of a stationary source even when the source was moving, as it was with the reversed DTFs. Thus, LNB stimuli were susceptible to motion induced reversal, and provide a case for which the principle of least displacement does seem to apply. An explanation for the demonstrated reduced salience of dynamic cues for the HNB noise is still not clear, however the present results do suggest that potent spectral cues (either accurate or misleading) can dominate over a least-displacement interpretation of dynamic cues, and thus that head movements cannot resolve front/rear reversals for all stimuli.

The present results are somewhat at odds with those reported by Martens *et al.*[10] in this volume. In that study, listeners were fitted with a 'binaural hearing instrument' that captured the acoustical signal at the entrance to each ear canal and presented it to the opposite ear via insert headphones. Target stimuli were recordings of speech played from one of five loudspeakers mounted in a vertical array to the listener's side. The listener's task was to indicate the whether the speech appeared to originate from a loudspeaker above or below head level. When standing stationary next to the array, listeners' responses were consistent with accurate identification of the active speaker (although the left-right position was reversed) suggesting that the hearing instrument preserved the necessary spectral cues for vertical-plane localization. When walking past the loudspeaker array, however, the amplitude of spontaneous head rotation and roll motions was sufficient to induce front/rear and up/down reversal of the apparent loudspeaker location, giving rise to the "phantom walker" illusion.[10] That result suggests that the dynamic interaural cues dominated over the available accurate spectral cues, the opposite of the findings in the present study for WB and HP noise, and for HNB noise for spatial window widths of up to 20 degrees.

The differences in the results of these two studies are likely attributable to differences in methodology and stimuli. The experimental environment used by Martens *et al.*[10] was a naturally reverberant hallway, whereas the data described in this chapter were obtained in an anechoic chamber. Listeners in the former study stood or walked naturally and verbally classified the perceived elevation of the active loudspeaker, whereas listeners in the present study listeners stood still while performing a stereotyped unidirectional head-sweep.

A final difference, and one which might be most easily addressed by parametric manipulation, was the chosen acoustic stimulus. Whereas Martens *et al.*[10] used recordings of speech, the present study used flat-spectrum noise of various bandwidths. Although the speech stimuli clearly had sufficient energy in the region of the high-frequency spectral cues to permit accurate vertical localization, the majority of the energy in the signal lay at lower frequencies. This might explain the observed dominance of dynamic cues over spectral cues, particularly if low-frequency ITD is indeed the most salient dynamic cue. Consistent with this, Kawaura *et al.*[6] demonstrated dominance of dynamic ITD cues over static spectral cues in virtual auditory space using recorded jazz as the stimulus, and Wallach[9] employed "orchestra or piano music from victrola records", which was likely of limited bandwidth. Contrary to this trend, however, McAnally and Martin[36] failed to replicate Wallach's findings involving illusory listener rotation with both lowpass noise and musical stimuli.

In the present study, the WB and HP noise stimuli carried most or all of their energy between 4 and 16 kHz, perhaps accounting for the observed spectral-cue dominance and failure of the principle of least displacement described in this chapter. An obvious means of experimentally addressing the spectral dependence of these results would be to repeat Experiment 2 of the present study with speech or speech-shaped noise stimuli, or with broadband noise with variable high-frequency roll-off in order to identify the "tipping point" between spectral- and dynamic cue dominance.

Acknowledgements

The author is very grateful to Devin Kerr, Zekiye Onsan, Chris Ellinger and David Grainger for technical assistance, to Yôiti Suzuki for drawing his attention to Ref. 6, and to William Martens for enlightening discussions. This work was supported by funds provided by the US National Institutes for Health (R01 DC00420 and P30 DC05188, NIDCD), the US National Science Foundation (0717272, Perception, Action and Cognition Program), and the University of Western Ontario.

References

1. S. Perrett and W. Noble, The contribution of head motion cues to localization of low-pass noise, *Percept. Psychophys.* **59**, 1018 (Oct 1997).
2. Y. Iwaya, Y. Suzuki and D. Kimura, Effects of head movement on front-back error in sound localization, *Acoust. Sci. & Tech.* **24**, 322 (2003).
3. H. G. Fisher and S. J. Freedman, The role of the pinna in auditory localization, *J. Aud. Res.* **8**, 15 (1968).
4. M. Kato, H. Uematsu, M. Kashino and T. Hirahara, The effect of head motion on the accuracy of sound localization, *Acoust. Sci. & Tech.* **24**, 315 (2003).
5. F. L. Wightman and D. J. Kistler, Resolution of front-back ambiguity in spacial hearing by listener and source movement, *J. Acoust. Soc. Am.* **105**, 2841 (1999).
6. J. Kawaura, Y. Suzuki, F. Asano and T. Sone, Sound localization in headphone reproduction by simulating transfer functions from the sound source to the external ear, *J. Acoust. Soc. Japan (E)* **12**, 203 (1991).
7. J.-R. Wu, C.-D. Duh and M. Ouhyoung, Head motion and latency compen-

sation on localization of 3D sound in virtual reality, in *Proc. ACM VRST*, 1997.
8. H. Wallach, On sound localization, *J. Acoust. Soc. Am.* **10**, 270 (1939).
9. H. Wallach, The role of head movements and vestibular and visual cues in sound localization, *J. Exp. Psychol.* **27**, 339 (1940).
10. W. L. Martens, D. Cabrera and S. Kim, The Phantom Walker Illusion: XXXXX NEED FINAL TITLE AND PAGE NUMBERS XXXXX, in *XXXXX NEED FINAL TITLE XXXX - IWPASH 2009 Book*, eds. Y. Suzuki and D. S. Brungart (World Scientific, 2010) pp. ?????–?????
11. W. R. Thurlow and P. S. Runge, Effect of induced head movements on localization of direction of sounds, *J. Acoust. Soc. Am.* **42**, 480 (1967).
12. S. Perrett and W. Noble, The effect of head rotations on vertical plane sound localization, *J. Acoust. Soc. Am.* **102**, 2325 (1997).
13. I. Pollack and M. Rose, Effect of head movement on the localization of sounds in the equatorial plane, *Percept. Psychophys.* **2**, 591 (1967).
14. J. C. Middlebrooks and D. M. Green, Sound localization by human listeners, *Ann. Rev. Psychol* **42**, 135 (1991).
15. W. R. Thurlow, J. W. Mangels and P. S. Runge, Head movements during sound localization, *J. Acoust. Soc. Am.* **42**, 489 (1967).
16. J. W. Strutt, On our perception of sound direction, *Philos. Mag.* **13**, 214 (1907).
17. F. L. Wightman and D. J. Kistler, The dominant role of low-frequency interaural time differences in sound localization, *J. Acoust. Soc. Am.* **91**, 1648 (1992).
18. E. A. Macpherson and J. C. Middlebrooks, Listener weighting of cues for lateral angle: the duplex theory of sound localization revisited, *J. Acoust. Soc. Am.* **111**, 2219 (2002).
19. R. S. Woodworth and H. Schlosberg, *Experimental Psychology* (Holt, Rinehart and Winston, New York, 1954).
20. J. H. Hebrank and D. Wright, Spectral cues used in the localization of sound sources on the median plane, *J. Acoust. Soc. Am.* **56**, 1829 (1974).
21. F. Asano, Y. Suzuki and T. Sone, Role of spectral cues in median plane localization, *J. Acoust. Soc. Am.* **88**, 159 (1990).
22. M. Morimoto and H. Aokata, Localization cues of sound sources in the upper hemisphere, *J. Acoust. Soc. Japan* **5**, 165 (1984).
23. S. Carlile, S. Delaney and A. Corderoy, The localisation of spectrally restricted sounds by human listeners, *Hear. Res.* **128**, 175 (Feb 1999).
24. J. Blauert, Sound localization in the median plane, *Acustica* **22**, 205 (1969/70).
25. R. A. Butler and C. C. Helwig, The spatial attributes of stimulus frequency in the median sagittal plane and their role in sound localization, *Am. J. Oto.* **4**, 165 (1983).
26. J. C. Middlebrooks, Narrow-band sound localization related to external ear acoustics, *J. Acoust. Soc. Am.* **92**, 2607 (1992).
27. M. Itoh, K. Iida and M. Morimoto, Individual differences in directional bands in median plane localization, *Applied Acoustics* **68**, 909 (2007), special

issue:Head-Related Transfer Function and its Applications.
28. M. Wexler and J. J. A. van Boxtel, Depth perception by the active observer, *Trends Cog. Sci.* **9**, 431 (Sep 2005).
29. J. Vliegen, T. J. V. Grootel and A. J. V. Opstal, Dynamic sound localization during rapid eye-head gaze shifts, *J. Neurosci.* **24**, 9291 (Oct 2004).
30. J. Cooper, S. Carlile and D. Alais, Distortions of auditory space during rapid head turns, *Exp. Brain Res.* **191**, 209 (Nov 2008).
31. J. Leung, D. Alais and S. Carlile, Compression of auditory space during rapid head turns, *Proc. Natl. Acad. Sci. USA* **105**, 6492 (Apr 2008).
32. E. A. Macpherson and D. M. Kerr, Minimum head movements required to localize narrowband sounds, in *Am. Aud. Soc. 2008 Annual Meeting, Scottsdale, AZ*, 2008.
33. E. A. Macpherson and D. M. Kerr, The salience of dynamic sound localization cues as a function of head velocity and stimulus frequency, in *2008 Auditory Perception, Cognition, and Action Meeting, Chicago, IL*, 2008.
34. J. C. Middlebrooks, Virtual localization improved by scaling nonindividualized external-ear transfer functions in frequency, *J. Acoust. Soc. Am.* **106**, 1493 (1999).
35. F. L. Wightman and D. J. Kistler, Factors affecting the relative salience of sound localization cues, in *Binaural and spatial hearing in real and virtual environments*, eds. R. H. Gilkey and T. R. Anderson (Laurence Erlbaum Associates, Mahwah, New Jersey, 1997) pp. 1–23.
36. K. I. McAnally and R. L. Martin, Sound localisation during illusory self-rotation, *Exp. Brain Res.* **185**, 337 (Feb 2008).

DEVELOPMENT OF VIRTUAL AUDITORY DISPLAY SOFTWARE RESPONSIVE TO HEAD MOVEMENT AND A CONSIDERATION ON DERATION OF SPATIALIZED AMBIENT SOUND TO IMPROVE REALISM OF PERCEIVED SOUND SPACE[†]

Y. IWAYA[1)], M. OTANI[2)], Y. SUZUKI[1)]

1) Research Institute of Electrical Communication, Tohoku University,
2-1-1 Katahira, Aoba-ku,
Sendai, Miyagi, 980-8577, Japan
E-mail: {iwaya, yoh}@riec.tohoku.ac.jp

2) Faculty of Engineering, Shinshu University,
4-17-1 Wakasato, Nagano, 380-8553, Japan
E-mail: otani@cs.shinshu-u.ac.jp

Recently, three-dimensional sound rendering methods for immersive sound spaces have been developed. Such systems are called virtual auditory displays (VADs). In a VAD system, it is common to render a sound associated with a specific sound object by convolving it with particular head-related transfer functions (HRTFs) to render positional information. Furthermore, sound images can be made stable in their absolute positions by continuously switching HRTFs in response to a listener's head movements. When this is done, the positions of sound images are stable in the world coordinate, which is similar to the real world, and the sound localization accuracy is much better than when HRTFs are not switched. The authors have put much effort into developing high-performance VAD software on personal computers. This VAD works with a native CPU of a personal computer on Windows (Microsoft Corp.) or Linux and outputs audio signals for headphones with a three-dimensional position sensor. This chapter overviews the VAD middleware we developed. Furthermore, to improve the realism of the virtual sound space, subjective evaluations were performed to clarify the relation between the

[†]This work was partly supported by Grant-in-Aid for Scientific Research (C) No.20500110 from JSPS, a Grant-in-Aid for Specially Promoted Research No.19001004 from MEXT Japan, Consortium R&D Projects for Regional Revitalization (15G2025), Ministry of Economy, Trade and Industry, and the Cooperative Research Project Program from the Research Institute of Electrical Communication, Tohoku University (H19A05).

perceived reality of virtual sound space with ambient sound and the listener's head movements.

1. Overview of virtual auditory display software

We can perceive the direction of a direct sound clearly using binaural cues and spectral cues involved in head-related transfer functions (HRTFs) [1], which represent free field transfer characteristics from a sound source to listeners' ears. Recently, three-dimensional sound rendering methods for immersive sound spaces have been developed for telecommunications systems to realize tele-existence or virtual reality. A technique often used in such systems is that of a virtual auditory display (VAD) system. By the late 1980s, simple real-time VAD systems became possible with a DSP chip. Recently, signal processing for VAD has become quite easy with a native CPU in a personal computer. For that reason, several VAD systems based on PCs have been developed, for example, SLAB developed at NASA [2] and DIVA developed at the Helsinki University of Technology [3]. Moreover, some commercial and GNU software systems [4] are available.

We also have devoted much effort to developing a high-performance VAD software engine on personal computers. This VAD works with a native CPU of a personal computer on Windows (Microsoft Corp.) or Linux operating system and outputs audio signals for headphones installed with a three-dimensional position sensor. With the position sensor, the VAD can respond to a head and body movement which enables highly precise auditory rendering [5]. Our VAD software is designated as Simulation environment for three-dimensional Acoustic Software (SifASo). In the next two subsections, a SifASo of each type is introduced.

1.1. *SifASo on Linux operation system*

A software VAD system has been developed on the Linux operating system [6]. The system consists of a set of headphones, a magnetic position sensor, and a personal computer (3.06 GHz Pentium 4 CPU, 2 Gbyte memory) running a Linux (kernel 2.6) operating system. Electrostatic open-back type headphones are used (ear-speaker, SRS-2020; STAX Ltd.) as the output device. The magnetic position sensor used is FASTRAK (Polhemus), which can provide six degrees of freedom sensing. The sensing capability of translation with FASTRAK enables this SifASo to respond not only to the head but also body movement. The FASTRAK receiver is mounted on the top of the headband of STAX headphones to acquire position and direction data at a rate of

120 samples/s. System latency (SL) is a very short value of approximately 12 ms, including the latency of the position sensor. The SifASo on Linux is simple application software intended mainly for localization tests with high accuracy. Therefore, it is not easy for programmers to use, or to extend, it to develop various applications based on the SifASo on Linux. The authors' group revealed the detective threshold of SL of VAD as being around 75 ms with SifaSo of this type [7].

1.2. *SifASo on Windows (Microsoft Corp.) operating system*

A middleware engine of VAD was developed as SifaSo on Windows (Microsoft Corp.). It has high performance in rendering a three-dimensional auditory space, including presentation of multiple sound sources by convolving individualized [8] or non-individualized HRTFs, Doppler effects [9], first-order reflections, and reverberations. The HRTFs are smoothly interpolated according to head and sound source movements. Another strong point of SifASo is its total system latency of only around 30 ms, including a latency of tacking listener's head movements, which is shorter than that of most existing engines and much shorter than the detection threshold of the delay [7]. Because of these advantages, SifASo realizes stable, precise and natural positioning of rendered sound images, especially moving sounds. SifASo is developed as middleware in a form of a set of dynamic link libraries (DLLs) so that it can be called easily from application software systems. Therefore, it facilitates the development of programs including advanced signal processing techniques. The class diagram of the main part of SifASo is presented in Fig. 1. We intended to apply SifASo on Windows (Microsoft Corp.) to develop a system to train space cognition for visually impaired people. Visually impaired people must recognize spaces without visual information, but especially with auditory three-dimensional information. For that reason, training of the spatial cognition ability is regarded as an extremely important subject to learn for pupils and those who have lost their sight. Based on the middleware engine, we developed several application software systems that are apparently intended as games for entertainment (edutainment). They are expected to be useful not only for such training purposes, but also for improving quality of life (QoL) of visually impaired people.

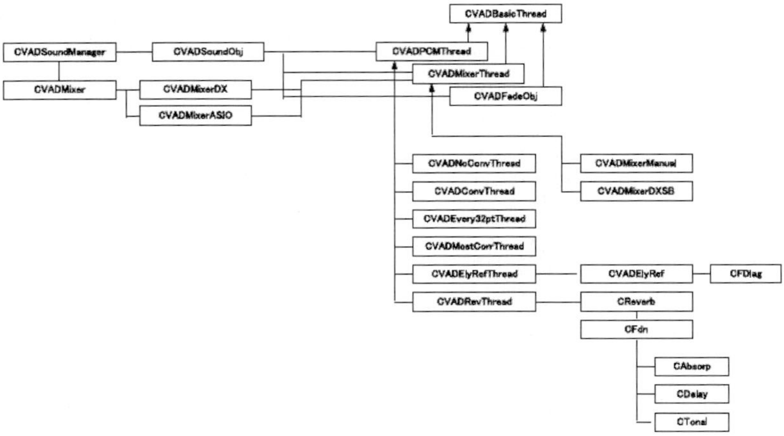

Fig. 1: Structure of a class diagram of SifASo on Windows (Microsoft Corp.).

'Bee Bee Beat' is an action-type game that trains sound localization ability (Fig. 2a). This game resembles a popular 'whack-a-mole game.' In this game, sounds of honeybees appear instead of pesky moles. A player has a plastic hammer to hit honeybees. A three-dimensional position sensor and a vibration unit are installed in the hammer. The player should localize the sound of a honeybee quickly and hit it with the hammer. When a position of the hammer and that of the bee coincide within a certain range, the hammer vibrates and some points are given. Experimental results revealed that playing the game engenders several positive transfer effects for practical life [10], including the ability of avoidance from sound coming, and face-contact ability, which is known to be a very important human communication skill.

'Sound formula' (Fig. 2b) is a car-race game that can be played using only sounds provided by the auditory display. In this game, the Doppler effects are implemented for rendering sounds of race machines.

'Mental Mapper' (Fig. 2c) is a maze game of a kind and aims at directly training an acuity of drawing mental/cognition maps. The player's task is to reach a goal, walking in a maze built in a virtual sound space. Walking in the maze is virtually controlled using a joy pad. In this game, first-order reflections from walls are rendered. If a player reaches one of some specified preset places in a maze, then one specified animal cry begins as a sound mark, which indicates that the player has passed the place. A maze editor is also prepared to allow trainers and teachers to create arbitrary mazes by themselves. We are continuously developing software VAD system with high accuracy and a greater number of rendering methods, such as high-order reflections, ambient sounds, and expression of distance. Individualization and compensation of HRTFs are extremely important for improving the reality of a virtual sound space.

(a) Bee Bee Beat

(b) Sound formula

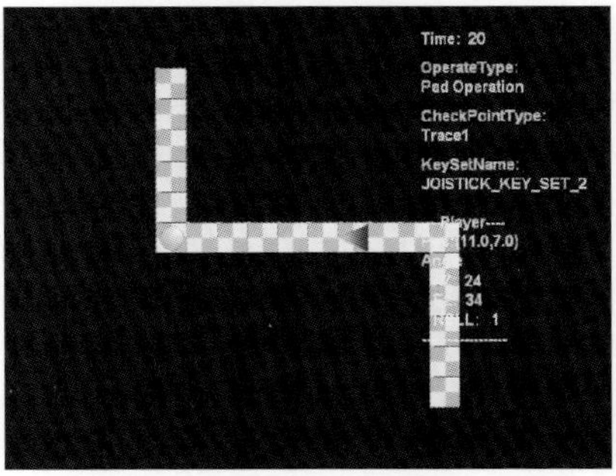

(c) Mental Mapper

Fig. 2: Examples of three-dimensional sound games for visually impaired people.

2. Effects of spatialized ambient sound on virtual auditory space

We are surrounded by various sounds in our daily life. These sounds arrive at our ears after being emitted from sound sources and interacting with an environment through various physical phenomena, such as reflection, reverberation, diffraction, Doppler shift. SifASo on Windows (Microsoft Corp.) can render these physical phenomena to some degree. In conventional VADs, only a sound source position is rendered, and other physical phenomena tend not to be rendered. In contrast to this situation, in a real sound space, we usually hear not only a target direct sound, but also ambient sounds. A lack of ambient sounds often engenders an unnatural perception of virtual auditory space presented by VADs. Therefore, ambient sounds should be included in the VAD system rendering. In some of modern VADs, certain methods to reproduce ambient sound have been proposed. Lokki *et al.* used an actual background sound, which was recorded using dummy-head microphones [11]. Seki *et al.* developed an auditory training system for visually impaired persons, which can render background sound sources by locating sound sources very far from a target sound source [12]. However, these methods necessitate the recording of real background sounds. In addition to the difficulty in obtaining recordings, it is almost impossible to record these sounds for every direction (in azimuth and elevation) to allow the system to be responsive to listeners' head movements. Therefore, we investigated a relation between the reality of sound space with ambient sounds and a listener's head movement using subjective evaluations to develop a rendering method of more realistic virtual sound spaces.

2.1. *Experiment I: Effects of spatialized ambient sound on reality of sound space perception with head movement*

2.1.1. *Sound stimuli*

A subjective evaluation was conducted to investigate the effects of spatialized ambient sound on a reality of perceived sound space. In this experiment, we used auditory stimuli consisting of two sounds: target and ambient sounds. For the target sound, musical instrument sounds of four kinds (cello, flute, oboe, and violin) from the SMILE library [13] were used. The target sound was generated by convolving Head-Related Impulse Responses (HRIRs, the time domain representation of the HRTF) of a dummy head (SAMRAI; Koken Co. Ltd.) for a given location of the target sound. To fix the position of the target sound in sound space, HRIRs were switched according to the head movement. As the ambient sound, we used red noise based on results of a preliminary investigation of frequency spectral features of ambient sound in the real world.

The ambient sound source (red noise) was convolved with HRIRs of various positions to place the sound source outside the listener's head. The HRIR positions were distributed from 0° to 350° with 10° steps in azimuth, and from –30° to 30° with 10° steps in elevation. Results show that HRIRs from 252 (=36×7) directions were used. The HRIRs were measured using a spherical loudspeaker array [14] installed in an anechoic room at the authors' institute. The HRIRs for sound sources located 1.5 m from the center of the spherical array were measured. Because the torso, the chair, and apparatus located below the chair are influential for elevations under -30°, HRIRs under -30° were not used for this study. To maintain symmetry, HRIRs for elevations over 30° were also excluded. The SifASo was used on a Linux system in the experiments. In this system, a magnetic position-sensing device (FASTRAK; Polhemus) was used to acquire position data at a 120-Hz sampling rate. A receiver was mounted on the top of the headband of the headphones. The length of the main response of each HRIR was set at 256 points, with a sampling frequency of 48 kHz. Consequently, both the target sound and ambient sound were spatialized.

2.1.2. *Method*

Eight listeners (ages 21–24) participated in the experiment. All had normal hearing. The experiment was conducted in a soundproof room. The participants were asked to judge the reality of sound stimuli with a paired comparison by choosing which of the sound stimuli presented in a pair contained sounds that were more likely to be in the real world. Listeners' head movements were not restricted. Three experimental conditions were prepared: In the *Simple* condition, the target sound and simple monaural red noise were presented. In the *Real* condition, the target sound and a binaurally recorded ambient noise were presented. The ambient noise was recorded using dummy-head microphones placed in a seminar room with only background noise from a central air-conditioning system. In the *Surround* condition, we presented the target sound and an ambient sound generated by convolutions of a red noise with HRIRs of a dummy head, as described above. However, the ambient sound was not responsive to head movement but the target sound was. The sound pressure of the target sound was set to L_{Aeq} = 60 dB. The SNR between the target and the ambient sound was selected from four values (-5, 0, 5, 10 dB). The duration of each auditory stimulus was about 10 s. We paired stimuli of these three conditions for each SNR and for each target sound type. Thereby, 48 pairs ($_3C_2 \times 4 \times 4$) were constructed and presented to a listener in a randomly determined order.

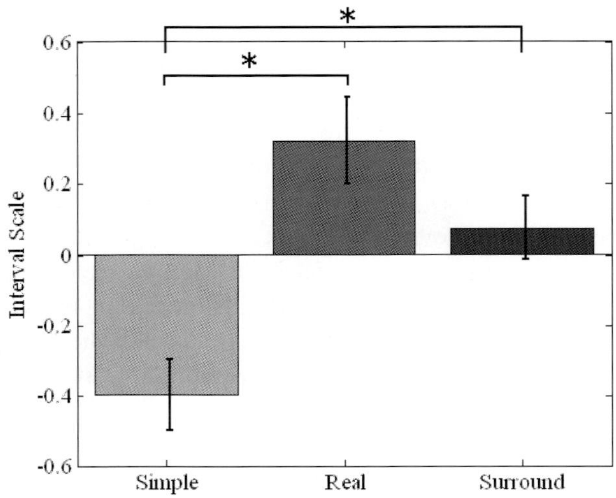

Fig. 3: Mean interval scales and standard errors for all participants in Experiment I.

2.1.3. *Results and discussion*

Figure 3 shows mean interval scale values and standard errors among all participants by pooling the scale values for SNRs and the target sound types. The mean interval scale was the highest in the *Real* condition. A one-way repeated-measures ANOVA on the mean interval scales reveals a significant difference effect for the ambient noise condition (F (2, 23) = 12.06, $p < 0.01$). All pair-wise comparisons, using t-tests with Bonferroni's correction, reveal statistically significant differences between *Simple* and *Real* conditions ($p < 0.05$) and *Simple* and *Surround* conditions ($p < 0.05$). These results indicate that participants perceived greater reality in the *Real* condition than in the *Simple* condition. In contrast, no significant difference was found between *Real* and *Surround* conditions. These results suggest that a "surround" ambient noise affects the realism of virtual sound space, irrespective of the type of the ambient noise, virtually synthesized or binaurally recorded in an actual environment. The experiment results show that presentation of spatialized ambient sounds is effective for improving the realism of the perceived sound space. In other words, it would be possible to produce more realistic virtual sound spaces by adding artificial ambient sounds. However, some listeners reported that the presented virtual sound space was unnatural, perhaps because the ambient sound was not responsive to the listener's head movements.

2.2. *Experiment II: Effects of head movement in spatialized ambient noise*

The results of experiment I showed that spatialized ambient sound is effective for improving the reality of VAD. However, in experiment I, the ambient sound was not responsive to a listener's head movement and whether or not ambient sound is responsive to listener's head movement might affect the reality of virtual sound space. Therefore, we performed an experiment (experiment II) to confirm the mechanism by which responsiveness of ambient noise to head movement affects the realism.

2.2.1. *Sound stimuli and method*

In experiment II, because the computation power of an ordinary PC was insufficient to spatialize all 252 red noises in real time, 72 sets of spatialized red noise were previously prepared: each set corresponds to one of 72 horizontal angles of five-degree steps. The sets were switched according to the listener's horizontal head movement. The experimental setup is portrayed in Fig. 4.

In the real world around us, sound pressure levels of ambient sounds are not spatially uniform. Therefore, a specific sound pressure distribution was given to ambient sounds. A two-dimensional Hanning window was used as a spatial distribution function to generate the sound pressure distributions, as illustrated in Fig. 5.

Listeners' tasks were the same as those for experiment I. In the experiment, two conditions were prepared: In the *Fix* condition, a listener was not allowed to make any head movement, but in the *Move* condition, the listener was allowed free head movement. The four target sound sources and four SNRs used in experiment I were also used here. Each listener made 16 judgments in all ($_2C_2 \times 4 \times 4$).

2.2.2. *Method*

Results show that the mean interval scale in the *Move* condition was significantly higher than that in the *Fix* condition ($t(8) = 2.48$, $p < 0.05$), which indicates that the spatialized ambient sounds responsive to head movement are effective to improve the perceived reality of sound space. Most listeners reported that they did not notice the relative change of ambient sounds responsive to the head movement, which implies that an unconscious cognition process of ambient sounds related to head movement improves the reality of the perceived sound space.

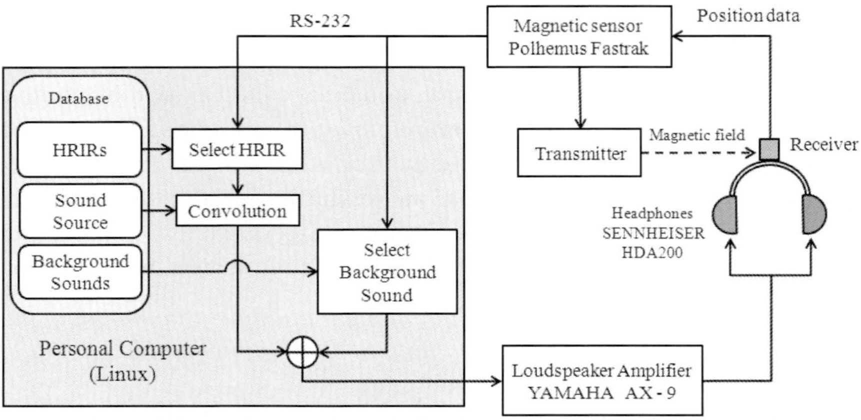

Fig. 4: Schematic overview of experimental setup in Experiment II.

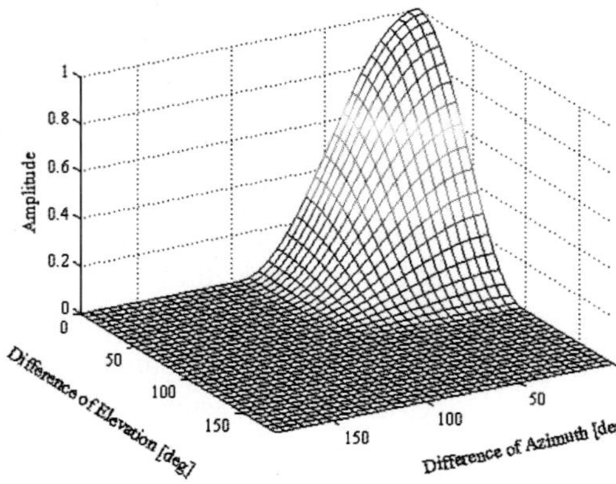

Fig. 5: Overview of the two-dimensional Hanning window in Experiment II.

2.3. *Experiment III: Effects of sound pressure distribution on reality of the perceived sound space*

As described in subsection 2.2, listeners perceive a higher degree of realism with an ambient noise that is responsive to their head movement. The ambient noise used in the previous experiment consists of multiple virtual sound sources surrounding a listener. The virtual sound sources have variation in their sound pressure levels so that the sound pressures from specific directions were larger than those from other directions. Therefore, to clarify how spatial distribution of an ambient noise affects the realism perception, a subjective experiment was performed with systematically varied spatial sound pressure distributions.

2.3.1. *Sound pressure distributions of ambient noise*

To generate several sound pressure distributions, a two-dimensional Gaussian window, written as

$$w(\theta,\phi) = \exp(-\frac{\theta^2 + \phi^2}{\sigma^2}),$$

was used. Here, w signifies weight of sound pressure at position (θ, φ), θ and φ respectively denote the azimuthal and elevation angle. The spatial distribution was varied systematically by changing the value of σ. In the experiment, five windows were used, as generated with σ^2= 0, 8, 64, 512 and ∞, as portrayed in Fig. 6. Hereinafter, conditions with σ^2= 0, 8, 64, 512 and ∞ are labeled respectively as the A_0 condition, A_8 condition, A_{64} condition, A_{512} condition, and A_∞ condition. In A_0 condition, ambient sound was identical to a point source located in front of a listener in the initial position. In this condition, two point sources—a target sound and a red noise—were responsive to head movement. However, in the A_∞ condition, the sound pressure level was spatially uniform (flat). This condition was similar to the *Surround* condition in experiment I. The sound pressure level at the center of listener's head was adjusted so that the total energy of ambient noise was constant among conditions and was set as L_{Aeq} = 60 dB. The sound pressure of the target sound was set to L_{Aeq} = 70 dB; therefore, the SNR was 10 dB.

2.3.2. *Procedure*

Eleven listeners (ages 21–24) participated in the experiment. All had normal hearing. The experiment was conducted in a soundproof room. The participants' task was the same as that of experiment I. Each listener made 80 judgments in all ($_5P_2$ × 4 target sounds).

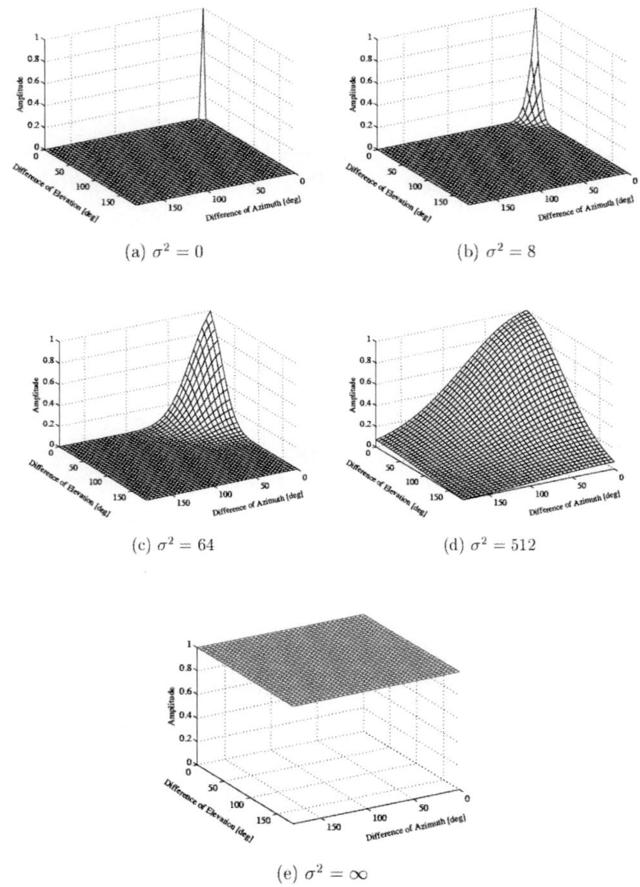

(a) $\sigma^2 = 0$

(b) $\sigma^2 = 8$

(c) $\sigma^2 = 64$

(d) $\sigma^2 = 512$

(e) $\sigma^2 = \infty$

Fig. 6: Sound pressure distributions controlled with σ in Experiment III.

2.3.3. *Results and discussion*

Figure 7 shows the mean interval scale values among all participants pooled for the target sound types. The mean interval scale value was the highest in the A_{64} condition. A one-way repeated-measures ANOVA on the mean interval scales revealed a main effect of the sound distribution (F (4, 54) = 22.335, $p < 0.01$).

Consequently, a pair-wise comparison using a *t*-test with Bonferroni's correction, revealed that the value in the A_{64} condition is significantly higher than in the A_0 condition, the A_8 condition, or the A_∞ condition (A_{64}-A_0, $p < 0.05$; A_{64}-A_8, $p < 0.05$; A_{64}-A_∞, $p < 0.05$). The value in the A_{512} condition is significantly higher than that in either the A_0 condition or the A_8 condition (A_{512}-A_0, $p < 0.05$; A_{512}-A_8, $p < 0.05$), and the value in the A_∞ condition is significantly higher than in the A_0 condition (A_∞-A_0, $p < 0.05$). The value was smallest in the A_0 condition.

In the A_{64} condition and A_{512} condition, the values are significantly higher than that in the A_∞ condition, which indicates that the ambient sounds should have a somewhat uneven distribution, rather than uniform. However, the results show that the distribution should not be too sharp; the value in the A_0 condition was the smallest among the conditions and the values in the A_0 and A_8 conditions are significantly lower than those in the A_{64} and A_{512} conditions. Namely, reality cannot be improved effectively by adding a sound with a sharp sound image other than a target sound. In summary, to improve the realism of perceived sound space, ambient sounds should be rendered so that they come from a broad spatial range, *i.e.*, they should be surrounding a listener but should have certain images of an appropriate size. It can be speculated that ambient noises with such spatial sound pressure distributions can act as a spatial reference as a positional pivot that would facilitate sound localization, thereby providing higher reality.

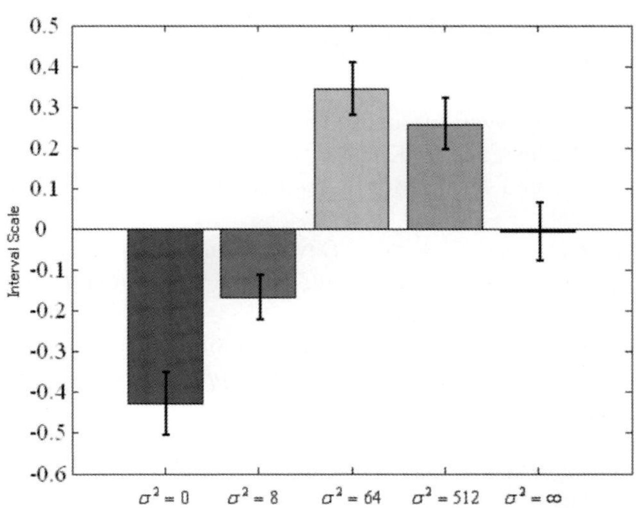

Fig. 7: Mean interval scales and standard errors for all participants in Experiment III.

3. Summary

This paper first introduces virtual auditory displays (VADs) that we have been developing. Then, we discussed the effects of ambient sounds—which surrounds a listener and which are responsive to listener's head movement—on the realism of virtual sound space presented by the VAD. The experimental results support that rendering of surrounding sounds responding to head movement is more effective than head-locked rendering. Furthermore, a spatial sound pressure distribution of ambient sounds can be optimized to improve the realism of the perceived sound space.

Acknowledgements

The authors thank Prof. S. Yairi for his cooperation in developing SifASo on Linux and Prof. A. Honda, and Prof. J. Gyoba for their cooperation in the evaluation of the edutainment application software. Moreover, the authors deeply appreciate Mr. T. Chiba and Dr. M. Kobayashi for their cooperation in conducting experiments I–III.

References

1. J. Blauert, Spatial Hearing. The MIT Press, Cambridge, Massachusetts, 1997.
2. http://human-factors.arc.nasa.gov/SLAB/
3. L. Savioja, J. Huopaniemi, T. Lokki, and R. Väänänen: "Creating interactive virtual acoustic environments", *J. Audio Eng. Soc.* 47(9), 675–705, 1999.
4. For example, http://www.3dsoundsurge.com, http://www.ircam.fr.
5. Y. Iwaya, Y. Suzuki, and D. Kimura, "Effects of head movement on front-back error in sound localization," *Acoust. Sci. & Tech.* **24**(5), 322–324, 2003.
6. S. Yairi, Y. Iwaya, and Y. Suzuki, "Development of virtual auditory display software responsive to head movement," *Trans. Virtual Reality Soc. Jpn.* (in Japanese) **11**(3), 437–446, 2006.
7. S. Yairi, Y. Iwaya, and Y. Suzuki, "Estimation of detection threshold of system latency of virtual auditory display," *Applied Acoustics* **68**(8), 851–863, 2007.
8. Y. Iwaya, "Individualization of head-related transfer functions with tournament-style listening test: Listening with other's ears," *Acoust. Sci. & Tech.* **27**(6) 340–343, 2006.
9. Y. Iwaya and Y. Suzuki, "Rendering moving sound with the Doppler effect in sound space," *Applied Acoustics* **68**(8), 916–922, 2007.

10. A. Honda, H. Shibata, J. Gyoba, Y. Iwaya, and Y. Suzuki, "Transfer effects on communication and collision avoidance behavior from playing a three-dimensional auditory game based on a virtual auditory display," *Applied Acoustics* **70**, 868-874, 2008.
11. T. Lokki, and H. Järveläinen, "Subjective evaluation of auralization of physics-based room acoustics modeling," *Proc. 2001 International Conference on Auditory Display*, Espoo July 29 – August 1, 2001.
12. Y. Seki and T. Sato, "Development of auditory obstacle perception training system for the blind using 3-D sound," *Proc. Conference and Workshop on Assistive Technologies for Vision and Hearing Impairment 2006*, Kufstein, 2006.
13. K. Kawai, K. Fujimoto, T. Iwase, T. Sakuma, Y. Hidaka, and H. Yasuoka, "Introduction of sound material in living environment 2004 (SMILE 2004): A sound source database for educational and practical purposes," *4th Joint Meet. ASA/ASJ* (Honolulu), *J. Acoust. Soc. Am.* **120**, 3070–3071, 2006.
14. S. Yairi, Y. Iwaya, and Y. Suzuki, "Individualization feature of head-related transfer functions based on subjective evaluation," *Proc. International Conference on Auditory Display (ICAD2008)*, Paris, June 24–27, 2008.

Section 2

Measuring and Modeling the Head-Related Transfer Function

RAPID COLLECTION OF HEAD RELATED TRANSFER FUNCTIONS AND COMPARISON TO FREE-FIELD LISTENING

D.S. BRUNGART*

Army Audiology and Speech Center, Walter Reed Army Medical Center
Washington, DC, 20307, USA
**E-mail: douglas.brungart@us.army.mil*

G. ROMIGH, B.D. SIMPSON

Air Force Research Laboratory,
Wright-Patterson AFB, OH 45433, USA

Although virtual audio display systems have now been in existence for more than 20 years, the procedures used to collect head-related transfer functions (HRTFs) on human listeners and implement them in head-tracked virtual audio display systems have not been standardized. In this chapter, we describe a procedure for virtual audio synthesis that allows a full set of individualized HRTFs to be collected in less than ten minutes and implemented immediately in a real-time, head-tracked virtual audio display. We also present the results of a validation experiment that show that this procedure produces virtual auditory localization accuracy that is comparable to that obtained with real sound sources in the free field.

Keywords: Head-Related Transfer Functions, Virtual Audio Displays, Auditory Localization

1. Introduction

Virtual audio synthesis is a relatively mature technology with a history of successful laboratory implementations that now stretches back more than two decades.[1] Yet, despite this long track record, virtual audio display technology has not yet achieved widespread acceptance in real-world applications. In part, at least, this lack of widespread success probably stems from the fact that high-quality virtual sound continues to require the collection of individualized Head-Related Transfer Functions (HRTFs) on the specific ears of the intended user of the virtual audio system. Although many researchers have attempted to overcome this problem through the development of techniques for customizing HRTFs that do not rely on physical acoustic measurements of the HRTF,[2,3] these attempts have met with only limited success. At the present time, accurate sound synthesis in a virtual audio display simply cannot be achieved without acoustically measuring individualized HRTFs on a human listener.

Unfortunately, the collection of individualized HRTFs for virtual audio synthesis has traditionally been a tedious process requiring a great deal of patience by both the experimenter and the subject. HRTF collection has generally required a listener to sit in a fixed orientation relative to a movable speaker system and remain essentially motionless for a period ranging from tens of minutes to up to two hours while individual measurements are taken from several hundred locations relative to the listener [e.g.[4]]. Although some faster alternative HRTF measurement procedures have been proposed, including one based on the principle of reciprocity that places sound sources into the ears of a listener surrounded by an array of spatially-separated microphones,[5] none of these alternative methods has been shown to produce the level of fidelity that can be obtained in the best HRTF measurements, where listeners are unable to distinguish between the real and virtual sources.[6,7] In this paper, we present a technique that has been developed at the Air Force Research Laboratory that allows the collection of an entire set of HRTFs in roughly 6-10 minutes and produces virtual sounds that are very nearly equivalent to those produced by loudspeakers in the corresponding locations in the free field.

2. HRTF Measurement Procedure

2.1. *Microphones*

Careful placement of the measurement microphones inside the ear canals is a critically important factor in the measurement of HRTFs. At AFRL, HRTF measurements are made with Knowles FG 23329 subminiature microphones that have been encapsulated into a small rubber fittings and embedded into Westone Oto-dam earmold dams (Figure 1). The assemblies are positioned into the subject's ears by hand and carefully guided into the ear canal with the use of a lighted oto-probe. This procedure ensures that the microphone opening faces out of the canal and the microphone is inserted at least a few millimeters inside the canal opening.[8]

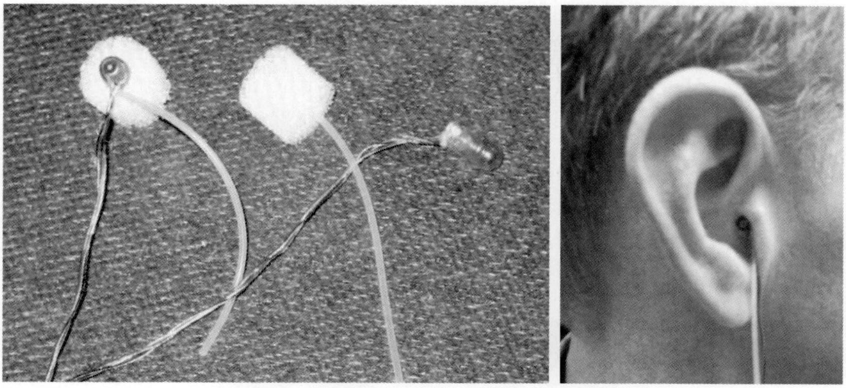

Fig. 1. Microphones used for HRTF Collection

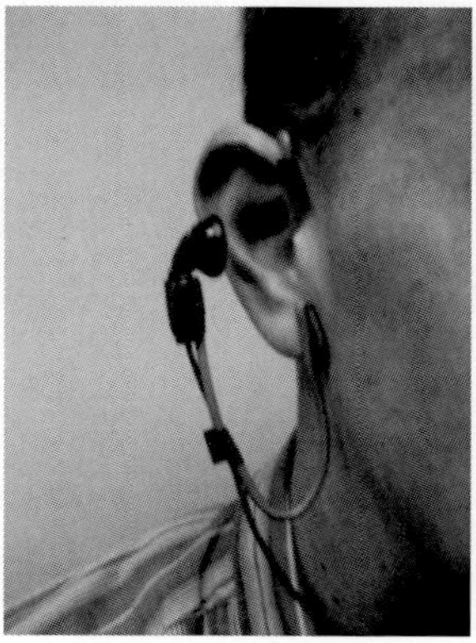

Fig. 2. Headphones used for HRTF Collection and Resynthesis

2.2. *Microphones and Headphones*

The small changes in the transfer function from the headphone to the eardrum that occur when a headphone is removed and replaced on a listener's head can seriously impair the accurate reproduction of virtual sounds.[8–10] Because of this limitation, most previous experiments that have been able to show perceptual equivalence between real and virtual sound sources in auditory localization have synthesized the virtual sound using headphones that did not need to be removed between the measurement of the HRTF and the presentation of the virtual sound.[6,7] The HRTF measurement procedure described in this chapter used custom-made earphones consisting of Sennheiser MX760 earbuds that were held in place with large-gauge wire (left panel of Figure 3). This setup allowed the headphones to be placed on the listener's head and removed without disturbing the microphones, and the microphones to be removed from the listener's ears without disturbing the placement of the headphones.

2.3. *Facility*

Fig. 3. Auditory Localization Facility used for HRTF collection

The HRTF measurements were made in the Auditory Localization Facility (ALF) (Figure 3), a geodesic sphere 4.3 m in diameter that is equipped with 277 full-range loudspeakers spaced roughly every 15° along its inside surface. The ALF facility is connected to a high-powered signal switching system that allows up to 16 different speakers to be simultaneously connected through high-power Crown amplifiers to a multichannel digital soundcard (RME).

2.4. *Procedure*

The HRTF measurements were made with listeners standing in the middle of the ALF while wearing a head-mounted position tracking device (Intersense IS-900). First, a bank of 16 speakers was selected and switched on in the ALF facility. Then a sequence of seven 2048-point periodic chirp signals (sweeping from 100 Hz to 15 kHz at a 44.1 kHz sampling rate, as shown in Figure 4) was generated for each of the 16 speakers, with roughly 250

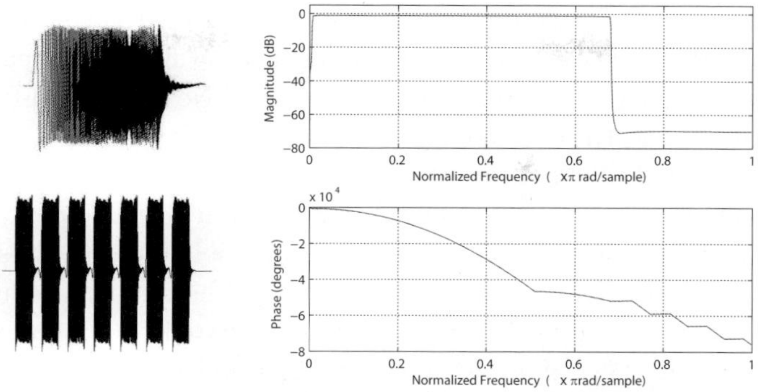

Fig. 4. Periodic Chirp used to measure HRTF

ms of silence between each 325-ms periodic chirp sequence. These periodic chirp stimuli were pre-filtered to eliminate differences in the frequency responses of the individual speakers, and the middle 5 chirps from each 7-chirp sequence were used to calculate the HRTF for each ear for that speaker location. The listener's head position was measured during each 16-speaker HRTF measurement block, and if the head moved more than 3 degrees during that measurement, that 16-speaker block was repeated. Otherwise, the head position from the tracker was used to calculate the location of each measured speaker relative to the listener's head, and these relative positions were saved along with the left and right ear impulse responses from each speaker location. The full measurement procedure required approximately 5-6 minutes to complete.

After the HRTF measurements were complete, an additional set of transfer function measurements was made for the headphones in the experiment using the same procedure as in the speaker measurements. The 2048-point left and right ear impulse responses from these headphone measurements were used to create inverse headphone filters by windowing them with a 512-point Hanning window centered on the peak of the impulse and then taking the inverse FFT of the resulting windowed waveforms.

2.5. *HRTF Processing*

Once the HRTF measurements were collected, further processing was needed to make them implementable in a real-time head-tracked virtual audio display. First, the 2048-point Head-Related Impulse Responses (HRIRs) were windowed with 401-point Hanning windows centered at the peak of the impulses to remove any residual reverberation that might have been present in the ALF facility. Then the HRIRs were inverse filtered to remove the measured frequency response of the headphones. Next, the low-frequency components of the HRTFs were reconstructed by setting the DC gain at all locations equal to the mean frequency response across all the measurements at 300 Hz and linearly interpolating between this DC value and the measured value at 300 Hz. The interaural time delay was then estimated by finding the best-fitting linear time delay over the frequency range from 430 Hz to 1500 Hz. The HRTFs were then converted to minimum phase, and truncated to 256 points. Finally, the minimum-phase HRTFs and the measured time delays were interpolated onto a grid with 5° resolution in azimuth and elevation using a nearest-neighbor interpolation procedure. The resulting 256-point HRTF filters were then formatted to work with the NASA-developed Sound Lab (SLAB) software package[11] to allow rendering of the virtual sound in real time.

3. Validation Procedure

A validation experiment was conducted to determine how accurately the rapidly-collected HRTFs could reproduce the spatial cues associated with real sound sources. Six listeners (3 male, 3 female) participated in this experiment. Each data collection session of this experiment started with a full HRTF measurement using the procedure outlined above. Once the HRTF collection process was complete, the external wirebud headphones were carefully placed on the listener's ears by the experimenter, and the headphone transfer function was measured. Then, the microphones were carefully removed from the ears without disturbing the headphones. Once the microphones were removed, the listener participated in a series of 40-trial blocks where they were asked to face the front of the ALF sphere and were presented with one of three different types of stimuli from a random direction: **Speaker:** a free-field stimulus generated from the loudspeakers in the ALF facility; **Impulse Response:** a non-headtracked virtual stimulus produced from the entire 2048-point left and right ear impulse responses measured in the HRTF (corrected only for the measured headphone response); or **SLAB:** a head-tracked virtual stimulus that was generated from HRTFs that were converted into the SLAB format using the procedure outlined in the previous section.

In the speaker and SLAB conditions, the stimuli were presented with two durations: a 250-ms noise burst, or a continuous noise that remained on until the listener pressed the button to make a localization response. In the impulse-response condition, the virtual sound was not head-tracked, so only the 250-ms burst condition was tested.

Once the stimulus was presented, the listener was asked to use a hand-held wand to point to the perceived location of the sound source. The orientation of the wand was tracked and used to illuminate an LED at the location of the speaker most closely aligned with the current direction of the wand, so the listeners always had a visual cue indicating the speaker to which they were pointing. Once the correct speaker location was selected with the LED cursor, the listener pressed a button and the LED cursor moved to the location of the actual target stimulus. Then the listener reoriented to the front of the sphere and the next stimulus presentation began.

Each data collection session lasted roughly 30 minutes, and the listeners participated in several sessions across several days until they had collected a total of 64 trials for each of the three 250-ms burst stimulus conditions and 32 trials for each of the continuous stimulus conditions. Thus a total of 1536 trials were collected in the experiment.

4. Results

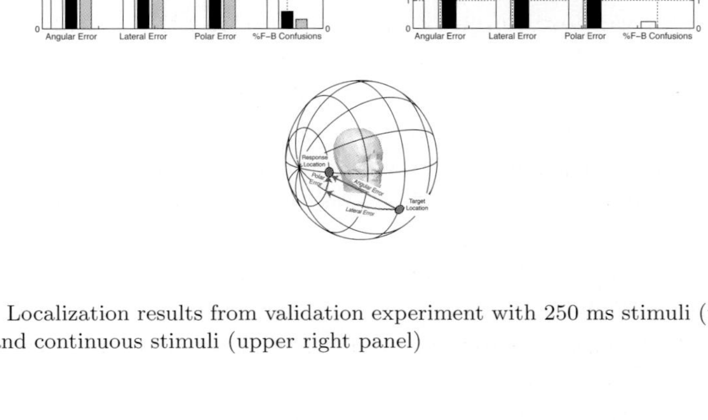

Fig. 5. Localization results from validation experiment with 250 ms stimuli (upper left panel) and continuous stimuli (upper right panel)

The results of the validation experiment are shown in Figure 5. When the stimulus was 250 ms in length (left panel), performance was essentially identical in the speaker and impulse response conditions. In the SLAB condition, performance was similar to the other two conditions in terms of angular error and lateral error, but the mean polar angle error was slightly worse. These results show that the impulse response collected in the HRTF measurement process accurately captured the localization cues present in the free-field HRTF, and that processing the full HRTF into a minimum phase filter to allow interpolation between virtual source locations in a head-tracked virtual audio display has only a minor impact on localization accuracy.

When the stimulus was on continuously (right panel), performance was consistently slightly worse in the SLAB condition, but the overall errors in both conditions were extremely small (less than 5°). These continuous stimulus results provide further evidence that the rapid HRTF collection procedure is capable of producing very high quality HRTFs that result in localization errors approaching those achieved with free-field sources. They also indicate that the HRTF processing and the SLAB-based virtual synthesis system did remarkably well not only at producing the correct localization cues, but also at maintaining alignment of the virtual image with the simulated free-field source by accurately compensating for the listener's rotational and translational head movements.

5. Discussion and Conclusions

In this chapter, we outlined a measurement procedure that allows rapid collection of HRTFs on human listeners. We have also shown that these rapidly-measured HRTFs can produce head-tracked virtual sound sources that listeners can localize nearly as accurately as real sound sources presented in the free-field. While the goal of perceptual equivalence between real sound sources and headtracked virtual sources was not quite achieved in this study, we believe the results compare favorably with those of earlier studies that have used similar techniques to examine auditory localization accuracy both with real free-field sound sources and with individualized HRTFs implemented in head-tracked virtual auditory displays. In one such study, Wightman and Kistler[12] reported that the mean correlation coefficient between stimulus elevation angle and response elevation angle fell from 0.90 to 0.81 when a real sound source was replaced with an individualized virtual sound source and no head motion was permitted. In another study, Bronkhorst[13] found that localization errors in a head-pointing task with unlimited head motion increased from roughly 9° to roughly 12° when a broadband free-field source was replaced with a virtual source, and that the total number of quadrant confusions more than doubled when a real source was replaced with a virtual source and no head movements were permitted. Thus, although limited data are available, we believe that HRTFs that have been collected and processed using the procedures outlined in this chapter yield results that are comparable, and possibly superior, to those that have been used in previous virtual auditory localization studies reported in the literature.

One of the key advantages of this HRTF measurement procedure is that it is fast enough to allow a new set of individualized HRTFs to be collected for each experimental session. This eliminates the need to remove and replace the headphones between the actual HRTF measurement procedure and the subsequent use of the measured HRTFs for data collection in psychophysical experiments. We hope that by eliminating the small errors in virtual sound synthesis that are inevitably caused by headphone placement and replacement errors,[14] we will be able to measure the impact that small variations in the virtual synthesis process have on human sound localization. This will allow a more precise definition of the minimum technical requirements that must be met to achieve optimal localization performance in a virtual audio display.

References

1. E. Wenzel, *Presence* **1**, 80 (1991).
2. J. C. Middlebrooks, E. A. Macpherson and Z. A. Onsan, *The Journal of the Acoustical Society of America* **108**, 3088 (2000).
3. J. C. Middlebrooks, *The Journal of the Acoustical Society of America* **106**, 1480 (1999).
4. F. Wightman and D. Kistler, *Journal of the Acoustical Society of America* **85**, 858 (1989).
5. D. N. Zotkin, R. Duraiswami, E. Grassi and N. A. Gumerov, *The Journal of the Acoustical Society of America* **120**, 2202 (2006).
6. A. Kulkarni, S. Isabelle and H. Colburn, *Journal of the Acoustical Society of America* **105**, 2821 (1999).
7. E. H. A. Langendijk and A. W. Bronkhorst, *The Journal of the Acoustical Society of America* **107**, 528 (2000).
8. H. Moller, M. Sorensen, D. Hammershoi and C. B. Jensen, *Journal of the Audio Engineering Society* **43**, 300 (1995).
9. A. Kulkarni and H. S. Colburn, *The Journal of the Acoustical Society of America* **107**, 1071 (2000).
10. K. I. McAnally and R. L. Martin, *Journal of the Audio Engineering Society* **50**, 263 (2002).
11. J. Miller, Slab: a software-based real-time virtual acoustic environment ren-

dering system, in *Proceedings of the International Conference on Auditory Display (ICAD 2001), Espoo, Finland, July 29-August 1*, 2001.
12. F. Wightman and D. Kistler, *Journal of the Acoustical Society of America* **105**, 2841 (1999).
13. A. Bronkhorst, *Journal of the Acoustical Society of America* **98**, 2553 (1995).
14. D. Pralong and S. Carlile, *The Journal of the Acoustical Society of America* **100**, 3785 (1996).

EFFECTS OF HEAD MOVEMENT IN HEAD-RELATED TRANSFER FUNCTION MEASUREMENT

T. HIRAHARA

Department of Intelligent Systems Design Engineering, Toyama Prefectural University, 5180 Kurokawa, Imizu, Toyama 939-0398, Japan

D. MORIKAWA

Department of Intelligent Systems Design Engineering, Toyama Prefectural University 5180 Kurokawa, Imizu, Toyama 939-0398, Japan

M. OTANI

Department of Information Engineering, Shishu University 4-17-1 Wakasato, Nagano, Nagano 380-8553, Japan

The effects of head movement during head-related transfer function (HRTF) measurements are evaluated. Head movements are measured simultaneously with HRTF measurements and spectral differences of the HRTFs are compared among repeated measurements. Without a head support aid, the human subjects' heads move considerably in all directions. HRTFs for the front position, measured at the beginning of each measurement session, differed 4 to 6.1 dB from those at the end of the measurement session. Even with a head support aid, HRTFs for the front position, measured at the beginning and the end of each measurement session, also differed by up to 6 dB, suggesting that the human subjects' heads moved during the measurements

1. Introduction

Virtual three-dimensional sounds are reproducible using head-related transfer function (HRTF)-based signal processing technologies. By convolving a left ear and right ear HRTF associated with a certain sound source location with a sound-source signal, a virtual sound image of the source can be located anywhere in three-dimensional space [1]. HRTFs are unique to a listener because HRTFs depend on head, pinnae, and body shapes. Therefore, a listener perceives a somewhat distorted virtual three-dimensional sound space when his/her own HRTFs are not used. Many HRTF sets have been measured with various human as well as dummy heads [2, 3] to study variability of HRTFs.

HRTFs, however, are not identical even when measured repeatedly with the same subject. Known factors that affect HRTFs are room temperature and humidity, instability of the HRTF measurement system, and head movement during HRTF measurement. Among these factors, subject head movement appears to have the greatest impact. Møller *et al.* analyzed three repeated HRTF measurements of several subjects and judged the measured HRTFs to be sufficiently similar across measurements [4]. Their subjects stood on a rotatable platform and the subjects were able to correct their head azimuth by monitoring a paper marker attached on the top of their head via a video system. The variation in the measured HRTF, however, was not small for high-frequency regions nor for HRTFs measured at the side opposite to the sound source. Riederer investigated the effect of various static head poses and various head movements on measured HRTFs [5]. His subjects, seated on a chair attached to a rotating turntable, were monitored via video cameras and the seat direction was corrected when necessary during the HRTF measurements. He concluded that a head position's tilt and movements change HRTFs by approximately 1 dB/1°. Hirahara *et al.* measured HRTFs repeatedly for three subjects and three dummy heads [6]. The HRTFs of each subject differed by 2.6–4.4 dB in averaged spectral distance (SD) for repeated measurements, whereas those of the dummy heads differed by 1.5–2.5 dB, even when the dummy head was removed and replaced at each HRTF measurement session.

As just described, head movement when measuring HRTFs is inevitable. However, little attention has been paid to this issue. In fact, no data have yet been reported on head movement during long HRTF measurement sessions and on the assessment of its effects on these measurements. We report the degree to which a human subject moves his/her head during a 95-min. HRTF measurement session and the manner in which head movement affects these measurements.

2. Head movement during HRTF measurement session

2.1. *Method*

The head movements of two adult males were recorded during a 95-min. HRTF measurement session at the HRTF measuring site in NTT Communication Science Research Labs. At this site, HRTFs were measured at 143 positions by moving a loudspeaker on a rotating traverse arm. The distance from the center of the head to the loudspeaker was 1.2 m. The range of the HRTF measurement was $0° \leq \theta \leq 350°$ in the azimuth angle θ and $-40° \leq \varphi \leq 90°$ in the elevation angle φ. In the median plane and horizontal planes, θ or φ was set at an interval of 10°. Other measured positions, (θ, φ) were set at intervals of less than 20° between adjacent positions in the azimuth and elevation angles. The HRTF of the front position (θ, φ) = (0°, 0°) was measured three times in one HRTF measurement session, *i.e.*, at the beginning, in the middle, and the end of a session.

The head movements were measured using a head tracker (Fastrak; Polhemus Inc.) placed on top of a subject's head with strings (Fig. 1). The head tracker detects yaw, pitch and role angles of the head at a 120-Hz sampling rate with an angular accuracy of 0.5°. The subjects sat on a chair in an anechoic room, and the initial head positions were carefully calibrated using three laser pointers. The subjects were then asked to keep their heads as still as possible during the session without the aid of any head-fixing tools.

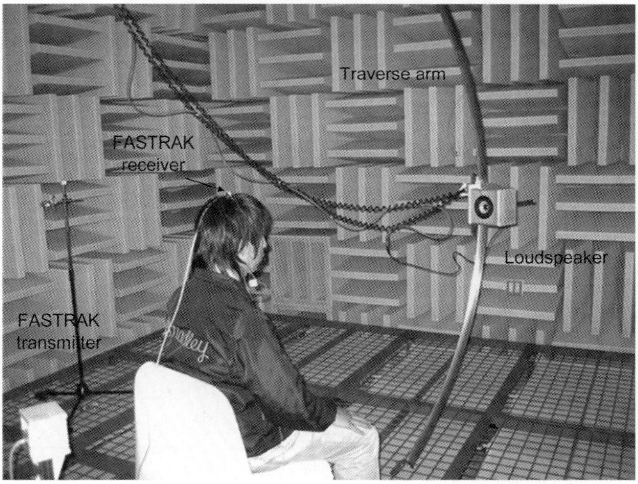

Figure 1. View of experimental setup for measuring head movement and HRTFs at NTT CS-Lab

2.2. *Results*

We failed to measure the head movements of subject 1. As the head tracker cable was not fixed to the chair, it pulled his head. His head moved gradually to the right and upward during measurement.

Figure 2 shows roll, pitch, and yaw head-movement angles (left column) and a normalized histogram of the movement angles (right column) of subject 2 during HRTF measurement sessions. This subject kept his head almost still for the first few minutes but the head began to move thereafter. The head moved between -4.5° and +1.8° in roll, ±14° in pitch, and -8.0° and +1.7° in yaw. The head position was the same in roll but differed by 7.1° in pitch and 5.0° in yaw between the beginning and end of the HRTF measurement session. Standard deviations of the head movements were, respectively, 0.97°, 3.9°, and 1.7° for roll, pitch, and yaw.

His head moved rapidly and greatly in the pitch direction. However, this excessive head movement did not always occur. Figure 3 shows the detailed head-movement trajectory of the subject, with a magnified time scale between 47-54 min. in the pitch trajectory of Fig. 2. As shown in Fig. 3, pitch movement occurred intermittently. The subject reported that he tried to keep his head still when time-stretched pulse (TSP) signals were emitted, and that he relaxed when the loudspeaker was moving from one HRTF measurement position to the next.

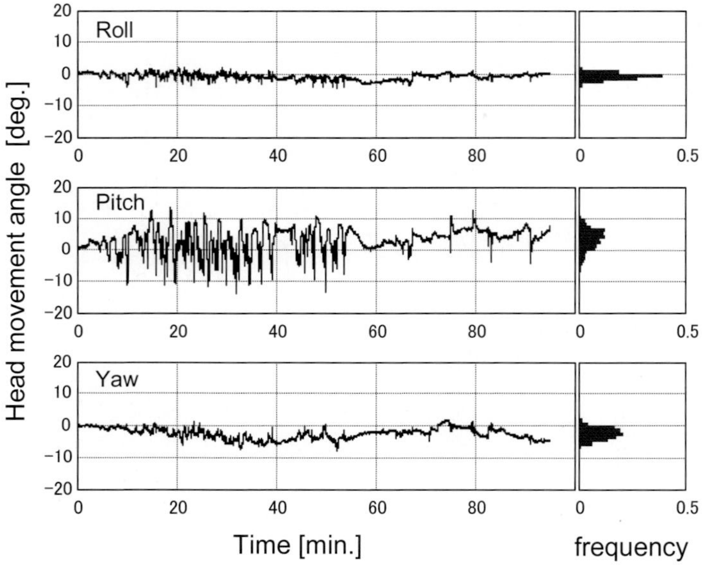

Figure 2. Roll, pitch, and yaw head-movement angles (left column) and normalized histogram of movement angle (right column) of subject 2

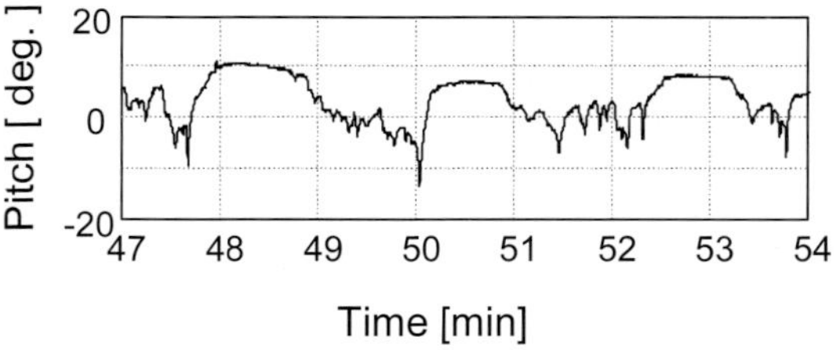

Figure 3. Detailed pitch trajectory of subject 2 with magnified time scale between 47-54 min. of Fig. 2

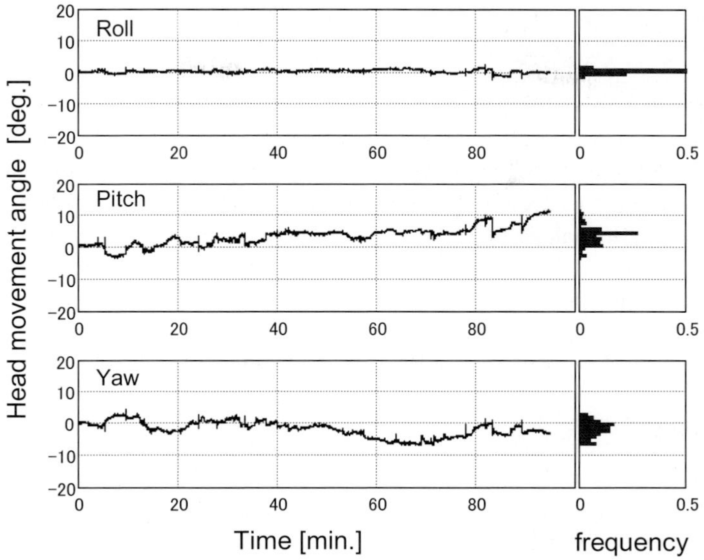

Figure 4. Roll, pitch, and yaw head-movement angles (left column) and normalized histogram of movement angle (right column) of subject 3

Figure 4 shows head-movement data of subject 3. Subject 3 also kept his head almost still for the first 5 minutes, but his head began to move thereafter. His head-movements were smaller than those of subject 2. The head, however, moved slowly ±2.3° in roll, between -3.6 to +12° in pitch, and -7.2 to +4.3° in yaw. The head position was the same in roll but differed by 11° in pitch, and -3.2° in yaw between the beginning and the end of the HRTF measurement session. Standard deviations of the head movement were, respectively, 0.5°, 2.9°, and 2.4° for roll, pitch, and yaw.

2.3. *Discussion*

The results illustrate that the head moves greatly when measuring HRTFs. Head positions at the beginning and end of the 95-min. HRTF measurement session differed by less than 1° in roll, but by as much as 10° in the pitch and yaw dimensions. Demonstrably, measured HRTFs include variations attributable to such head movements. However, it is only natural that humans cannot maintain a fixed head position for a long period and that HRTFs therefore vary according to the differences in head position differences. The more strictly the acoustical conditions are controlled, the smaller the expected variation in HRTFs. However, strict acoustical conditions often result in neck pain in subjects because an extraordinary effort is needed or more time is necessary to complete the HRTF measurement. For our subjects, the maximum limit for keeping their heads still was about 5 minutes.

Yairi measured head movements of four subjects listening to music for 5 minutes. [7]. He reported that, for each subject, the head always moved slightly, even when the subject was relaxed and seemed to keep the head still. His data show that head movements were large in the roll and pitch direction, and those in the yaw direction were small. Our results, obtained from a much longer measurement period, show that head movements were large in the pitch and yaw direction while those in the roll direction were small, which does not necessarily coincide with Yairi's results.

3. Effect of head movements on measured HRTF, ITD and ILD

3.1. *Method*

At site 1, five adult male subjects participated in HRTF measurements for 143 positions, which took 95 minutes. Head movements were measured for two of these subjects. As noted in section 2.1, the HRTF of the front position was measured three times in one HRTF measurement session, *i.e.* at the beginning (t = 0 min.), middle (t = 12 min.), and end (t = 95 min.). The difference between the HRTFs was evaluated using the SD of two HRTF amplitude spectra. Gross acoustical cues for sound localization, inter-aural time difference (ITD) and inter-aural level difference (ILD), were also calculated from the right- and left-ear HRIRs and evaluated.

3.2. *Results*

Figure 4 shows the right-ear HRTFs associated with the location directly in front for subjects 2 and 3. For both subjects, the three HRTFs were not identical. With regard to the HRTFs of subject 2, rough spectral shapes of the three HRTFs were comparable; however the detailed spectral shapes were different. The frequency, depth, and bandwidth of the first spectral notch around 6.5 kHz were different among the three HRTFs. The frequency of the second notch, seen around 11 kHz, also differed. With regard to the HRTFs of subject 3, the first spectral notch at 6.5 kHz was a single deep valley in the HRTF measured at the beginning, but it was formed by multiple shallow dips in those measured at the middle and end. In contrast, the second notch around 10 kHz was prominent in the HRTF measured at the mid and end, but it is formed by multiple shallow dips in the HRTF measured at the beginning. For subject 2, the SD between the HRTFs measured at the beginning and those at the middle was 1.3 dB, and that measured at the beginning and end was 5.4 dB. Those of subject 3 were 2.6 and 3.3 dB, respectively. The mean SD between the HRTFs measured at the beginning and middle for all five subjects was 1.4 dB with a standard deviation of 1.7 dB, and that measured at the beginning and end was 2.9 dB with standard deviation of 0.65 dB.

The ITD calculated from the HRIR measured at the beginning was zero for all subjects. It remained zero at the middle and end of measurements in 3 out of 5 subjects, but it increased to a maximum of 40 μs for the other two subjects. The ILD calculated from the HRIR measured at the beginning was not always 0 dB for all subjects. It also increased to a maximum of 2 dB at the middle and end of the session.

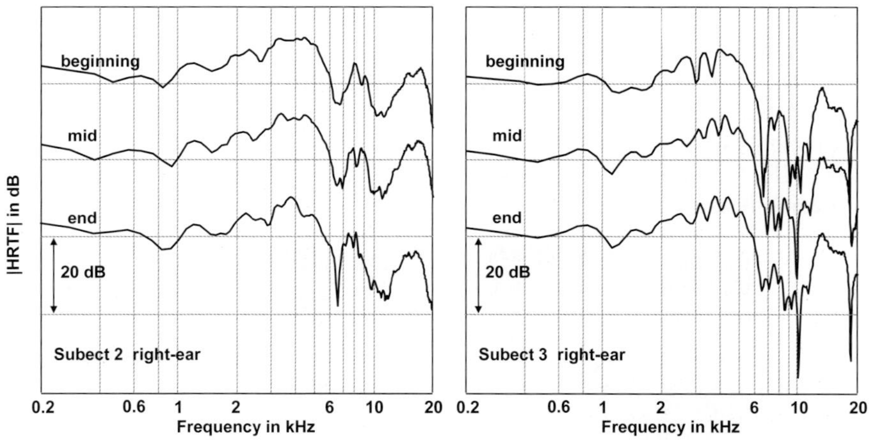

Figure 4. Right-ear HRTFs of front position measured at beginning (t = 0 min.), middle (t = 12 min.), and end (t = 95 min.) of session with subjects 2 and 3. Subjects' heads were not supported with any head fixing devices.

3.3. *Discussion*

As anticipated from the head movement data, the HRTFs, ITDs and ILDs measured at the beginning, middle and end of the 95-min. HRTF measurement session were not the same. The differences in the HRTFs, ITDs and ILDs among the three measurements depend on subjects. Subjects 2 and 3 were relatively restless in terms of head movement. The differences were almost negligible in some subjects, who must have kept their heads still like a mannequin during the session. It is, however, very difficult for a subject to keep his/her head still for a long time without the aid of any head fixing devices. A head rest, a neck fixing device or a belt to fix the head is sometimes used in measuring HRTFs. In the next section, we discuss whether the use of a simple headrest is effective for suppressing the head movement.

4. Efficacy of a headrest to prevent the head movement

4.1. *Method*

At RIEC, Tohoku Univ., HRTFs were measured with six subjects and three dummy-heads. The subjects sat on a chair with a head rest, which was a semi-circular U-shaped metal rod. The subjects just pressed his/her head against the headrest without any device to fix the head to it. HRTFs were measured at 613 positions by moving a loudspeaker array on a rotating motor. The distance from the center of the head to the loudspeakers was 1.5 m. The range of the HRTF measurement was $0° \leq \theta \leq 350°$ in the azimuth angle θ and $-80° \leq \varphi \leq 90°$ in the elevation angle φ at an interval of 10°. The HRTF of the front position $(\theta, \varphi) = (0°, 0°)$ was measured two times in one HRTF measurement session, *i.e.* at the beginning (t = 0 min) and end (t = 43 min). The difference between the two front HRTF amplitude spectra was evaluated using the SD. Head movements were not measured at this site.

Figure 5. View of experimental setup for measuring HRTF at the RIEC. A semi-circular U-shaped metal rod is used to support a head

4.2. *Results*

Figure 6 shows a comparison of the two left-ear HRTFs of the front position measured at the beginning and end of a session for subjects 12 and 14. The SD was 2.1 dB for subject 12 and 1.8 dB for the subject 14. It should be noted that the first and/or second notch frequency were obviously different even though the SD value was not so large. Even when the head was supported with a headrest, two HRTFs measured 43 minutes apart were not the same. As for the dummy-heads, the HRTFs were identical. Therefore, the discrepancies are not due to the HRTF measurement system but to human heads movements.

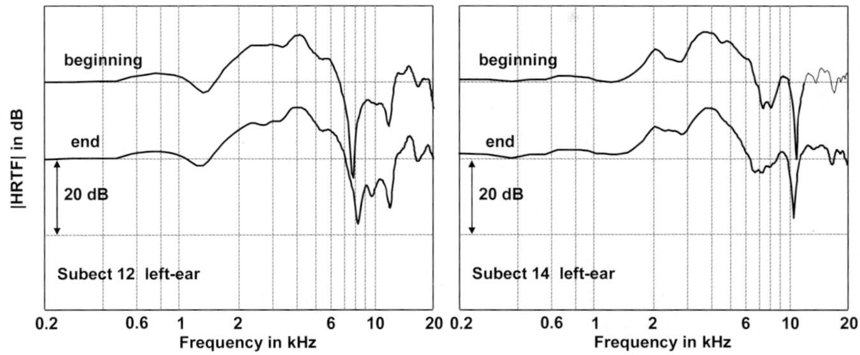

Figure 5. Left-ear HRTFs of front position measured at beginning (t = 0 min.) and end (t = 43 min.) of session with subjects 12 and 14. Subjects' heads were supported with a headrest.

5. Conclusion

Without aids for head support, subjects' heads moved considerably during the 95-min HRTF measurement and the measured HRTFs involve a high margin of error due to the uncertainty of the head position. Even when a headrest was used to fix the head position, HRTFs measured 43-minutes apart were not the same, suggesting that head movement occurs during the measurement. Consequently, it should be noted that HRTFs measured without the use of head-support aids or those with a simple headrest are likely to have a high margin of error because of head movements.

One way to lessen the head-movement issue is to use a large geodesic sphere equipped with many loudspeakers. Brungart *et al.* succeeded in collecting HRTFs of 277 positions in roughly 6 to 10 minutes [8]. Another way to obviate the head-movement issue is to simultaneously measure HRTFs at multiple positions based on Helmholtz' acoustic reciprocity principle. The fast HRTF measurement method based on the reciprocity principle was developed by Zotkin *et al.* [8, 9], and recently re-examined by Matsunaga *et al.* [10].

Acknowledgments

Part of this work was carried out under the Cooperative Research Project Program of the RIEC, Tohoku University. Authors are grateful to Profs. Y.

Suzuki, and Y. Iwaya for supporting us to use the HRTF measuring system of the RIEC. They specially thank Mr. T. Kurokawa and Ms. N. Shimakura for assisting the HRTF measurements at RIEC. Authors are also grateful to Dr. I. Toshima, Mr. H. Sagara and Mr. N. Endo for helping us to measure head movements and HRTFs simultaneously at NTT Communication Science Labs. in Atsugi.

References

1. E. M. Wenzel, "Localization in virtual acoustic displays," *Presence*, **1**, 80-107 (1992).
2. V. R. Algazi, R. O. Duda, D. M. Thompson, and C. Avendano, "The CIPIC HRTF Database," *Proc. IEEE Workshop on Applications of Signal Processing to Audio and Electroacoustics*, 99-102, (2001).
3. S. Takane, D. Arai, T. Miyajima, K. Watanabe, Y. Suzuki, and T. Sone, "A database of Head-Related Transfer Functions in whole directions on upper hemisphere," *Acoust. Sci. and Tech.* **23(3)**, 160-162 (2002).
4. H. Møller, M. F. Sørensen, D. Hammershøi, and C. B. Jensen, "Head-Related Transfer Functions of Human Subjects," *J. Audio Eng. Soc.* **43**, 300-321 (1995).
5. K. A. J. Riederer, "Effect of Head Movements on Measured Head-Related Transfer Functions," *Proc. 18th International Congress on Acoustics, Kyoto*, 795-798 (2004).
6. T. Hirahara, M. Otani, S. Yairi, Y. Iwaya, and I. Toshima, "Discussions on head-related transfer functions," (in Japanese), *Trans. Technical Committee of Psychological and Physiological Acoustics, The Acoustical Society of Japan*, **37(11)**, 867-872 (2008).
7. S. Yairi, "A Study on System Latency of Virtual Auditory Display Responsive to Head Movement," (in Japanese), Ph.D. Dissertation, Tohoku University, (2006).
8. D. S. Brungart and G. Romigh, and B. D. Simpson, "Rapid Collection and Enhancement of Individualized HRTFs," *Proc. International Workshop on the Principles and Applications of Spatial Hearing (IWPASH)*, I-13, (2009).
9. D. N. Zotkin, R. Duraiswami, E. Grassi, and N. A. Gumerov, "Fast head-related transfer function measurement via reciprocity," *J.Acoust.Soc.Am.* **120(4)**, 2202-2215 (2006).
10. R. Duraiswami, D. N. Zotkin, and A. E. O'Donovan, "Capturing and Recreating Auditory Virtual Reality," *Proc. International Workshop on the Principles and Applications of Spatial Hearing (IWPASH)*, I-17, (2009).
11. N. Matsunaga and T. Hirahara, "Re-examination of an HRTF measurement system via reciprocity," *Proc. International Workshop on the Principles and Applications of Spatial Hearing (IWPASH)*, P-16, (2009).

INDIVIDUALIZATION OF THE HEAD-RELATED TRANSFER FUNCTIONS ON THE BASIS OF THE SPECTRAL CUES FOR SOUND LOCALIZATION*

K. IIDA† and Y. ISHII

Department of Electrical, Electronics, and Computer Engineering ,
Chiba Institute of Technology,
Tsudanuma, Narashino, Chiba 275-0016, Japan
†E-mail: kazuhiro.iida@it-chiba.ac.jp
http://www.binaural-lab.com/

To provide the appropriate Head-Related Transfer Functions (HRTFs) to a listener by extracting HRTFs from an HRTF database, the authors propose to utilize the difference in spectral cues to describe the individual differences of HRTFs and to create a database of a minimal number of HRTFs. The following three issues are discussed in this study: (1) the essence of spectral cues for vertical and front-back localization; (2) an appropriate physical measure for the individual differences in HRTFs; (3) a method to provide individualized HRTFs by utilizing physical measures and the database of a minimal number of HRTFs. Systematic localization tests and observation of measured HRTFs revealed the following: (1) the lowest first and second spectral notches (N1 and N2) above 4 kHz can be regarded as spectral cues; (2) Notch Frequency Distance (*NFD*), which is the difference in the frequencies of N1 and N2, is a proper physical measure for individual differences of HRTFs; (3) the acceptance range of *NFD* for front localization is 0.1-0.2 octave; (4) a database of minimal HRTFs, consisting of 38 parametric HRTFs for the front direction is obtained by dividing the distribution range of N1 and N2 frequencies by the acceptance range of *NFD*; (5) the appropriate parametric HRTFs for the front direction for each listener can be selected from the minimal database by a brief localization test; (6) the individualized parametric HRTFs for various directions can be generated by a regression equation on the N1 and N2 frequencies.

1. Introduction

It is well known that accurate sound image localization is accomplished when the listener's own Head-Related Transfer Functions (HRTFs) are reproduced [1].

* A part of this work is supported by the "Academic Frontier" Project for Private Universities: matching fund subsidy from MEXT (Ministry of Education, Culture, Sports, Science and Technology).

A localization error and reduction in reality often occur with other people's HRTFs, which differ from those of the listener due to the individual differences in the shape and size of the head and pinnae. One of the methods to solve this problem is to provide each listener the appropriate HRTFs extracted from the HRTF database.

Middlebrooks reported that appropriate scaling of HRTFs in the frequency domain reduces the localization error even when the listener uses others' HRTFs [2,3]. He also proposed a psychophysical procedure by which a listener identifies appropriate scale factors [4]. Iwaya proposed a tournament-style listening test to select the appropriate HRTFs for a listener from the HRTF database [5]. However, as the number of HRTFs in the database increases, so does the time and effort to find the appropriate HRTFs required by these methods.

A proper physical measure of individual differences in HRTFs and a database of a minimal number of HRTFs are necessary to provide the appropriate HRTFs to a listener quickly and easily. The authors propose to utilize the differences in spectral cues to describe the individual differences of HRTFs and to create database of a minimal number of HRTFs. In this study, the following three issues are discussed:

(1) the essence of spectral cues for vertical and front-back localization

(2) an appropriate physical measure for individual difference in HRTFs

(3) a method to provide individualized HRTFs by utilizing physical measures and the minimal HRTF database.

2. What are the spectral cues for vertical and front-back localization?

It is generally known that spectral information is a cue for median plane localization. Most previous studies showed that spectral distortions caused by pinnae in the high-frequency range above approximately 5 kHz act as cues for median plane localization [6-16]. Mehrgardt and Mellert [12] showed that the spectrum changes systematically in the frequency range above 5 kHz as the elevation of a sound source changes. Shaw and Teranishi [7] reported that a spectral notch changes from 6 kHz to 10 kHz when the elevation of a sound source changes from -45° to +45°. Iida *et al.* [16] carried out localization tests and measurements of HRTFs after occlusion of the three cavities of the pinnae, scapha, fossa, and concha. They concluded that spectral cues in median plane localization exist in the high-frequency components above 5 kHz of the transfer function of the concha.

Hebrank and Wright [10] carried out experiments with filtered noise and reported the following. The spectral cues of median plane localization exist between 4 and 16 kHz; front cues are a 1-octave notch having a lower cut-off frequency between 4 and 8 kHz and increased energy above 13 kHz; an above cue is a 1/4-octave peak between 7 and 9 kHz; a behind cue is a small peak between 10 and 12 kHz with a decrease in energy above and below the peak. Moore *et al.* [17] measured the thresholds of various spectral peaks and notches. They showed that the spectral peaks and notches that Hebrank and Wright regarded as the cues of median plane localization are detectable by listeners, and the thresholds for detecting changes in the position of sound sources in the frontal part of the median plane can be considered thresholds for the detection of differences in the center frequency of spectral notches.

Butler and Belendiuk [11] showed that the prominent notch in the frequency response curve moves toward the lower frequencies as the sound source moves from above to below the aural axis in the frontal half of the median plane. Raykar *et al.* [18] noted that one of the prominent features observed in the Head-Related Impulse Response (HRIR) and another feature that has been shown to be important for elevation perception are the deep spectral notches attributed to the pinnae. They proposed a method of extracting the frequencies of pinna spectral notches from the measured HRIR, distinguishing them from other features. The extracted notch frequencies are related to the physical dimensions and shape of the pinnae.

The results of these previous studies imply that spectral peaks and notches due to the transfer function of the concha in the frequency range above 5 kHz prominently contribute to the perception of the elevation of a sound source. However, it has been unclear which component of HRTF plays an important role as a spectral cue.

This section clarifies the spectral cues for vertical localization by systematic localization tests and careful observations of the characteristics of HRTFs [19].

2.1. *Parametric HRTFs*

Iida *et al.* [19] proposed a parametric HRTF model to clarify the contribution of each spectral peak and notch as a spectral cue for vertical localization. The parametric HRTF is recomposed only of the spectral peaks and notches extracted from the measured HRTF, and these spectral peaks and notches are expressed parametrically by the frequency, level, and sharpness. Localization tests were carried out in the upper median plane by using the subjects' own measured HRTFs and the parametric HRTFs with various combinations of spectral peaks and notches. As mentioned above, the spectral peaks and notches in the frequency range above 5 kHz prominently contribute to the perception of sound source elevation. Therefore, the spectral peaks and notches are extracted from the measured HRTFs regarding the peaks around 4 kHz, which are independent of sound source elevation [12], as a lower frequency limit. Then, labels are put on the peaks and notches in order of frequency, e.g., P1, P2, N1, N2 and so on (Fig. 1). The peaks and notches are expressed parametrically with frequency, level, and sharpness. The amplitude of the parametric HRTF is recomposed of all or some of these spectral peaks and notches.

In order to extract the essential spectral peaks and notches, the microscopic fluctuations of the amplitude spectrum of HRTF were eliminated by Eq. (1),

$$HRTF_w(k) = \sum_{n=-n_1}^{n_1} HRTF\,(k+n)W(n)\ , \qquad (1)$$

where $W(n)$ is a Gaussian filter defined by Eq. (2). k and n denote discrete frequency. The sampling frequency was 48 kHz, and the duration of HRTFs was 512 samples. In this study, n and σ were set to be 4 and 1.3, respectively.

$$W(n) = \frac{1}{\sqrt{2\pi}\sigma} e^{\frac{-n^2}{2\sigma^2}}\ . \qquad (2)$$

The spectral peak and notch are defined as the maximal and minimal levels of $HRTF_w$, respectively. Thus, the frequencies and the levels of the spectral peaks and notches are obtained. The sharpness of the peak and notch is set to be their envelopment fit with that of $HRTF_w$. Fig. 2 shows examples of the parametric HRTFs recomposed of N1 and N2. As shown in the figure, the parametric HRTF reproduces all or some of the spectral peaks and notches accurately and has flat spectrum characteristics in other frequency ranges.

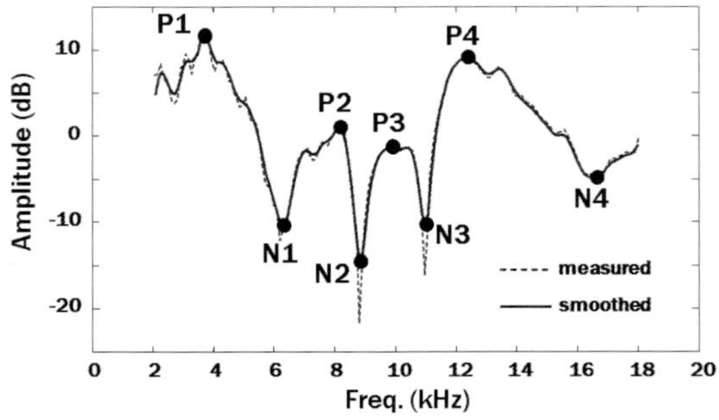

Fig. 1. Examples of extracted spectral peaks and notches from a measured HRTF.

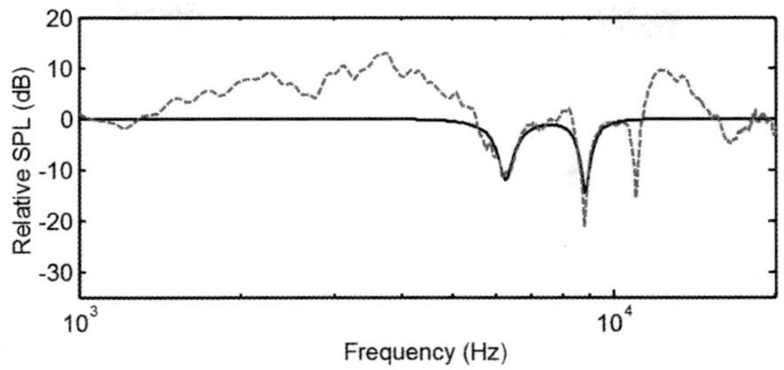

Fig. 2. An example of a parametric HRTF. Dashed line: measured HRTF, solid line: parametric HRTF recomposed of N1 and N2.

2.2. *Method of localization tests*

Localization tests in the upper median plane were carried out using the subjects' own measured HRTFs and the parametric HRTFs. A notebook computer (Panasonic CF-R3), an audio interface (RME Hammerfall DSP), open-air headphones (AKG K1000), and the ear microphones [19] were used for the localization tests.

The ear microphones were fabricated using the subject's ear molds (Fig.3). Miniature electret condenser microphones of 5 mm diameter (Panasonic WM64AT102) and silicon resin were put into the ear canals of the ear molds and consolidated (Fig.4). The diaphragms of the microphones were located at the entrances of the ear canals. Therefore, this is so called the "meatus-blocked condition" [7], in other words, the "blocked entrances condition" [20].

The subjects sat at the center of the listening room. The ear microphones were put into the ear canals of the subject. Then, the subjects wore the open-air headphones (Fig.5), and the stretched-pulse signals were emitted through them. The signals were received by the ear microphones, and the transfer functions between the open-air headphones and the ear microphones were obtained. Then, the ear microphones were removed, and stimuli were delivered through the open-air headphones. Stimuli $P_{l,r}(\omega)$ were created by Eq. (3):

$$P_{l,r}(\omega) = S(\omega) \times H_{l,r}(\omega) / C_{l,r}(\omega)\,, \tag{3}$$

where $S(\omega)$ and $H_{l,r}(\omega)$ denote the source signal and HRTF, respectively. $C_{l,r}(\omega)$ is the transfer function between the open-air headphones and the ear microphones.

The source signal was a wide-band white noise from 280 Hz to 17 kHz. The measured subjects' own HRTFs and the parametric HRTFs, which were recomposed of all or a part of the spectral peaks and notches, in the upper median plane in 30-degree steps were used. For comparison, stimuli without an HRTF convolution, that is, stimuli with $H_{l,r}(\omega)$=1, were included in the tests.

A stimulus was delivered at 60 dB SPL, triggered by hitting a key of the notebook computer. The duration of the stimulus was 1.2 s, including the rise and fall times of 0.1 s, respectively. A circle and an arrow, which indicated the median and horizontal planes, respectively, were shown on the display of the notebook computer. The subject's task was to plot the perceived elevation on the circle, by clicking a mouse, on the computer display. The subject could hear each stimulus over and over again. However, after he plotted the perceived elevation and moved on to the next stimulus, the subject could not return to the previous stimulus. The order of presentation of stimuli was randomized. The subjects responded ten times for each stimulus.

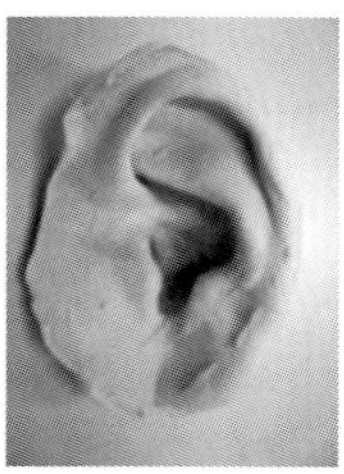

Fig. 3 An ear mold of a subject.

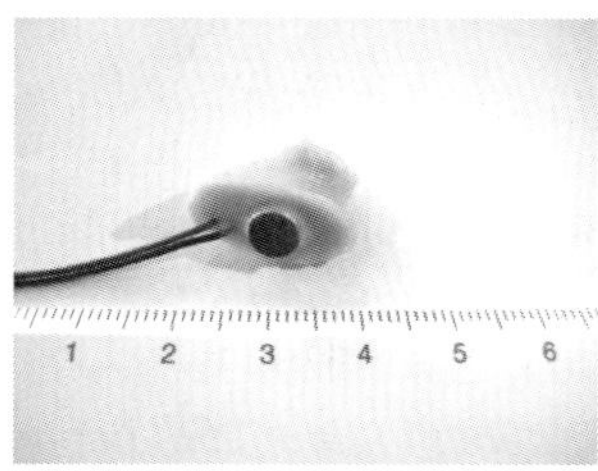

Fig. 4 An ear microphone.

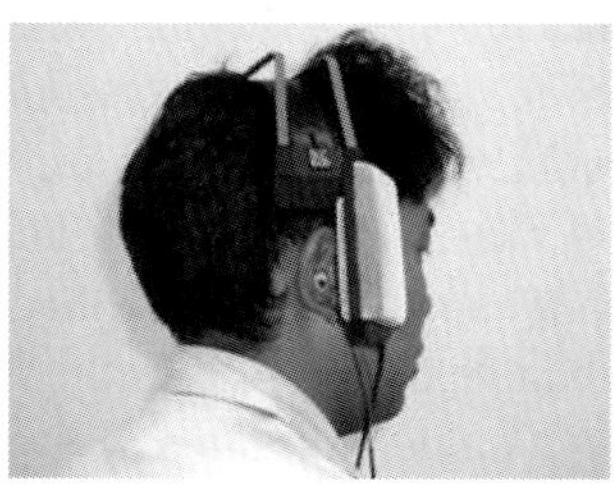

Fig. 5 A subject wearing ear microphones and ear speakers.

2.3. *Results of the tests*

Figure 6 shows the distributions of the responses of subject IT (a male of 30 years of age) for target elevation of 0,90, and 180°. The ordinate of each panel represents the perceived elevation, and the abscissa, the kind of stimulus. The 0°

is ahead of the listener, and the 180° is behind. Hereafter, the measured HRTF and parametric HRTF are expressed as the mHRTF and pHRTF, respectively. For the stimuli without an HRTF, the perceived elevation was not accurate, and the variance of responses was large. On the other hand, the subjects perceived the elevation of a sound source accurately at all the target elevations for the mHRTF. For the pHRTF(all), which is the parametric HRTF recomposed of all the spectral peaks and notches, the perceived elevation was as accurate as that for the mHRTF at all the target elevations. In other words, the elevation of a sound source can be perceived correctly when the amplitude spectrum of the HRTF is reproduced by the spectrum peaks and notches. For the pHRTF recomposed of only one spectral peak or notch, the variances of the responses were large at all the target elevations. One peak or notch did not provide sufficient information for localizing the elevation of a sound source. The accuracy of localization improved as the numbers of peaks and notches increased. Careful observation of the results indicates that the pHRTF recomposed of N1 and N2 provides almost the same accuracy of elevation perception as the mHRTF at most of the target elevations.

Fig. 7 shows the responses of subject IT to the mHRTF, pHRTF(all), and pHRTF(N1–N2) for seven target elevations. The ordinate of each panel represents the perceived elevation, and the abscissa, the target elevation. The diameter of each circle plotted is proportional to the number of responses within five degrees. For the pHRTF(all), the responses distribute along a diagonal line, and this distribution is practically the same as that for the mHRTF. For the pHRTF(N1–N2), the responses distribute along a diagonal line, in the case of subject IT.

Fig. 8 shows the responses of subject MK (a female of 22 years of age) to the mHRTF, pHRTF(all), pHRTF(N1–N2), and pHRTF(N1–N2–P1) for seven target elevations. For the pHRTF(all), the responses distribute along a diagonal line, and this distribution is practically the same as that for the mHRTF. For the pHRTF(N1–N2), the responses distribute along a diagonal line for the target elevations of 120, 150, and 180°, but the responses for the target elevations of 0, 30, 60, and 90° shift to the rear. For the pHRTF(N1–N2–P1), the responses generally distribute along a diagonal line, except for the target elevation of 90°.

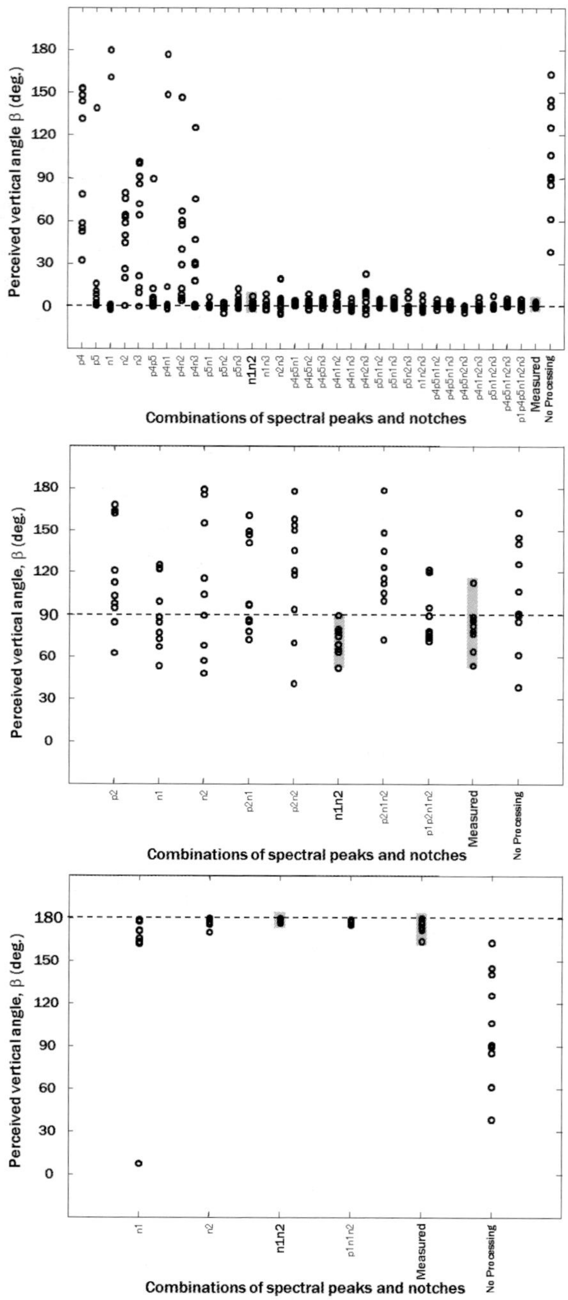

Fig. 6 Responses to stimuli of measured HRTFs and parametric HRTFs (0, 90, and 180deg.) (subject IT).

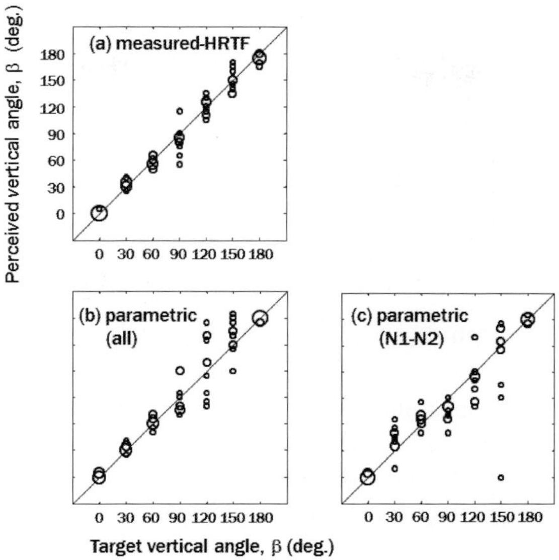

Fig. 7 Responses to stimuli of measured HRTFs and parametric HRTFs in the median plane (subject IT).

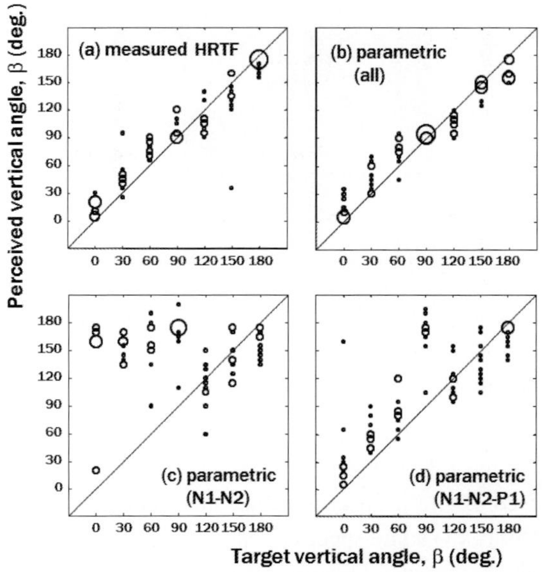

Fig. 8 Responses to stimuli of measured HRTFs and parametric HRTFs in the median plane (subject MK).

2.4. *Discussions*

The reason why some spectral peaks and notches markedly contribute to the perception of elevation is discussed. Fig. 9 shows the distribution of the spectral peaks and notches of the measured HRTFs of subject IT in the upper median plane. This figure shows that the frequencies of N1 and N2 change remarkably as the elevation of the sound source changes. Since these changes are non-monotonic, neither only N1 nor only N2 can identify the source elevation uniquely. It seems that the pair of N1 and N2 plays an important role as vertical localization cues.

The frequency of P1 does not depend on the source elevation. According to Shaw and Teranishi [7], the meatus-blocked response shows a broad primary resonance that contributes almost 10 dB of gain over the 4-6 kHz band, and the response in this region is controlled by a "depth" resonance of the concha. Consequently, the contribution of P1 to the perception of elevation cannot be explained in the same manner as the contributions of N1 and N2. It could be considered that the hearing system of a human being utilizes P1 as the reference information to analyze N1 and N2 in the ear-input signals.

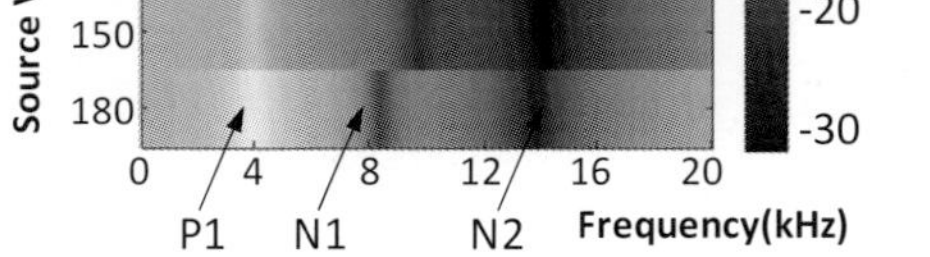

Fig. 9 Distribution of frequencies of N1, N2, and P1 in the upper median plane.

2.5. *Conclusions on the cues for vertical and front-back localization*

The authors carried out sound localization tests using a parametric HRTF model. The results show the following: (1) the perceived elevation for the parametric HRTF recomposed of all the spectral peaks and notches is as accurate as that for the measured HRTF; (2) some spectral peaks and notches play an important role in determining the perceived elevation, whereas some peaks and notches do not; (3) the parametric HRTF recomposed of the first and second notches (N1 and

N2) and the first peak (P1) provides almost the same accuracy of elevation perception as do the measured HRTFs.

Observations of the spectral peaks and notches of the HRTFs in the upper median plane indicate the following: (1) the frequencies of N1 and N2 change remarkably as the source elevation changes; (2) P1 does not depend on the source elevation.

From these results, some conclusions can be drawn: (1) N1 and N2 can be regarded as spectral cues; (2) the hearing system of a human being can utilize P1 as reference information to analyze N1 and N2 in ear-input signals.

3. What is an appropriate physical measure for individual differences of HRTFs?

Appropriate measures are necessary to extract the HRTFs that provide accurate localization to a listener from the HRTF database. *SD* (Spectral Distortion) defined by Eq. (4) has been used as a conventional measure to evaluate the individual differences of HRTFs. However, it is not proper to use *SD* because it calculates the differences in the entire range of spectral components.

$$SD = \sqrt{\frac{1}{N}\sum_{i=1}^{N}\left[20\log_{10}\frac{|HRTF_j(fi)|}{|HRTF_k(fi)|}\right]^2}\,. \tag{4}$$

Figure 10 shows four measured HRTFs of the same subject for the same direction (front) in the same anechoic chamber. Measurements were carried out in 1999, 2001, 2003, and 2005. These HRTFs show similar behavior, although they differ in detail. The calculated *SDs* for each HRTF pair range from 4.2 to 5.7 dB.

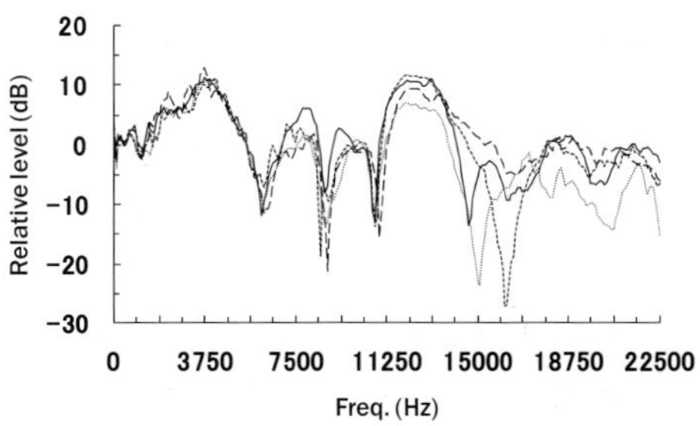

Fig. 10 Four measured HRTFs of the same subject for the same direction (front) in the same anechoic chamber.

Figure 11 shows two measured HRTFs of different subjects for the same direction (front) in the same anechoic chamber. These HRTFs differ in structure. The calculated *SD* is 7.2 dB. These examples show that it is not easy to distinguish the individual differences in *SD* from the measurement deviation.

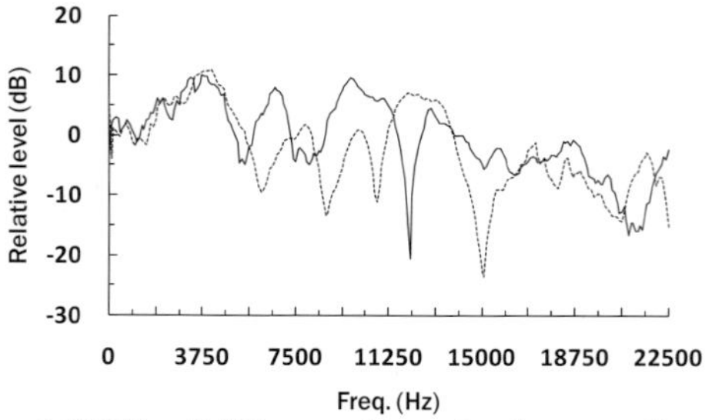

Fig. 11 Two measured HRTFs of different subjects for the same direction (front) in the identical anechoic chamber.

3.1. *Notch Frequency Distance (NFD)*

Individual differences in the N1 and N2 frequencies were observed since N1 and N2 can be regarded as spectral cues, as mentioned above. Fig. 12 shows the N1 and N2 frequencies of 50 subjects (100 ears) for the front direction. This figure indicates that the individual differences in N1 and N2 are very large. The N1 frequency ranges from 5.5 kHz to 10 kHz, and the N2 frequency ranges from 7 kHz to 12.5 kHz.

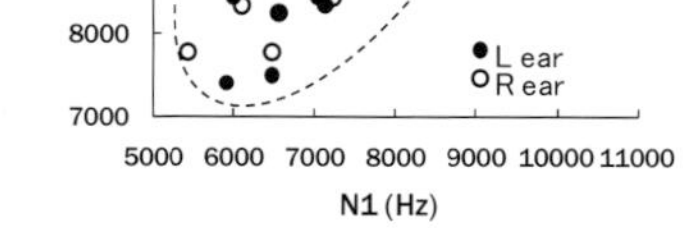

Fig. 12 N1 and N2 frequencies of 50 subjects (100 ears) for the front direction.

Thus, the authors propose Notch Frequency Distance (*NFD*) as a measure for the individual differences in HRTFs. *NFD* shows the distance between $HRTF_j$ and $HRTF_k$ in the octave scale, as follows:

$$NFD_1 = \log_2\left\{f_{N1}(HRTF_j)\big/ f_{N1}(HRTF_k)\right\} \text{ [oct.]}, \quad (5)$$

$$NFD_2 = \log_2\left\{f_{N2}(HRTF_j)\Big/ f_{N2}(HRTF_k)\right\} \text{ [oct.]}, \qquad (6)$$

$$NFD = \left|NFD_1\right| + \left|NFD_2\right| \text{[oct.]}, \qquad (7)$$

where f_{N1}, f_{N2} denote the frequencies of N1, N2, respectively.

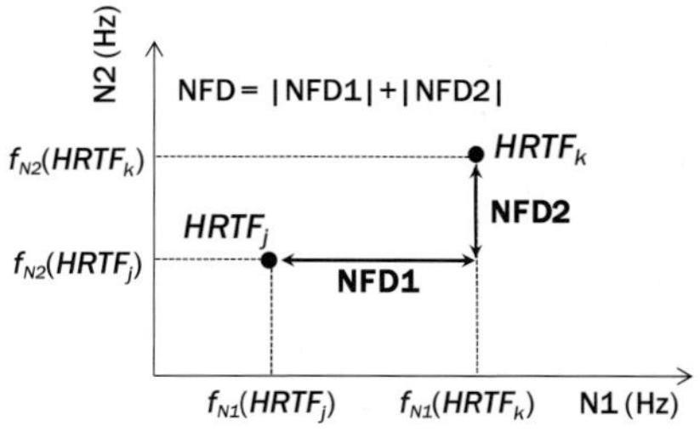

Fig. 13 Schematic explanation of *NFD*.

While calculated *NFDs* for each pair of four measured HRTFs of the same subject ranges from 0.00 to 0.05 octave, that of two measured HRTFs of different subjects is 0.28 octave. Therefore, it is considered that *NFD* can distinguish the individual differences of HRTFs from the measurement deviation.

3.2. *Acceptable range of NFD for accurate localization*

Localization tests were carried out to clarify the acceptance range of *NFD* for front localization. The HRTFs used were the subjects' own parametric HRTFs of the front direction recomposed of N1, N2, and P1, and the parametric HRTFs recomposed of N1, N2, and P1, in which the N1 and N2 frequencies were shifted by $\pm$0.1, $\pm$0.2, $\pm$0.4, and +0.6 octave from the subjects' own notch frequencies. In total, 26 parametric HRTFs, shown in Fig. 14, were prepared. The source signal was broadband (200 Hz-20 kHz) white noise.

The subjects sat at the center of the anechoic chamber. The ear microphones were put into the ear canals of each subject. Then, each subject wore the open-air headphones (SONY MDR-F1), and the stretched-pulse signals were emitted through them. The signals were received by the ear microphones, and the transfer functions between the open-air headphones and the ear microphones were obtained. Next, the ear microphones were removed, and stimuli were delivered through the open-air headphones. Stimuli $P_{l,r}(\omega)$ were created by Eq. (3). The duration of the stimulus was 1.2 s, including the rise and fall times of 0.1 s each. The subject could hear each stimulus over and over again. However, after the subject plotted the perceived elevation and moved on to the next stimulus, he could not return to the previous stimulus. The order of presentation of stimuli was randomized. The subjects responded ten times for each stimulus. The subjects were two males (ISY and IST). The task of the subjects was to mark down the perceived azimuth and elevation on the response sheet, on which a circle and an arrow indicating the median and horizontal planes were drawn.

Figure 15 shows the percent correct localization as the function of NFD for each subject. Percent correct localization is defined as the percentage of the responses that localized within 15° of the target direction (in front). The figures show that 80% correct localization is obtained when NFD is less than 0.2 and 0.1 octave for subjects ISY and IST, respectively. Therefore, the acceptable range of NFD for front localization is regarded to be 0.1 octave in this study.

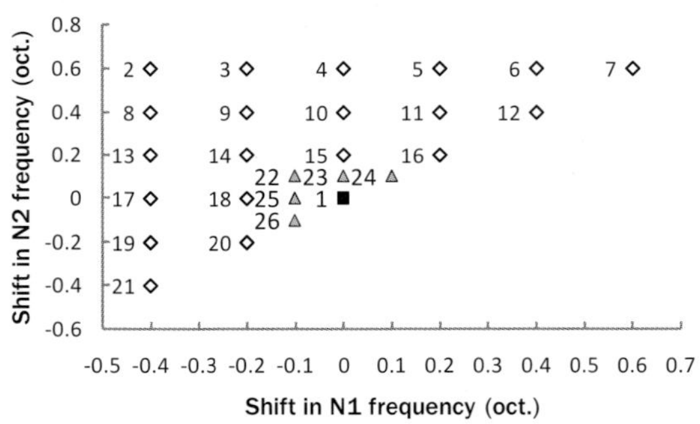

Fig. 14 Twenty-six parametric HRTFs with shifted N1 and N2 frequencies.

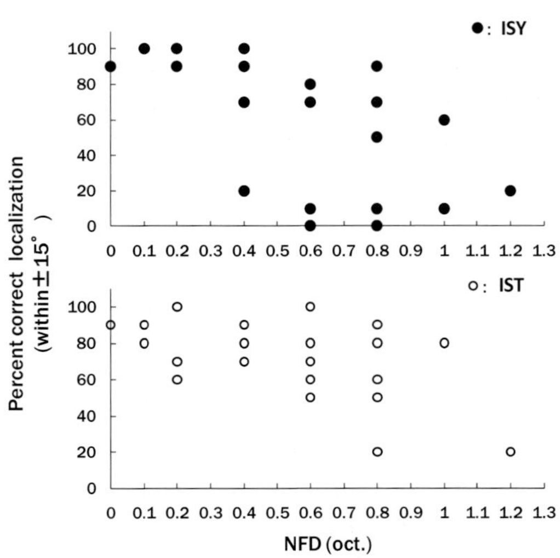

Fig. 15 Percent correct localization as the function of *NFD*.

4. A method to provide individualized HRTFs utilizing physical measures

Based on the abovementioned findings, the authors propose a method to provide individualized HRTFs for each listener. This method consists of the following three steps.

Step 1: Create the database of a minimal number of HRTFs for the front direction based on *NFD.*

Step 2: Select the appropriate parametric HRTFs for the front direction for each listener from the minimal database.

Step 3: Generate the individualized parametric HRTFs for various directions.

4.1. *Minimal parametric HRTF database for the front direction*

The HRTF database should be small in order to reduce the search time and burden of the listeners. The HRTF database, which is composed of the required minimum number of parametric HRTFs, is created by the following procedure.

First, the range of individual differences of N1 and N2 frequencies was obtained for many listeners for the front direction, at which the front-back localization error occurs frequently due to the individual differences. The distribution range of the N1 and N2 frequencies of 50 subjects (100 ears) for the front direction is shown in Fig. 12. One-hundred ears are considered a sufficient number of samples as a subgroup of the population.

Second, the distribution range was divided by the acceptable range of *NFD* for frontal localization (0.1 octave), as shown in Fig. 16.

Finally, pairs of N1 and N2 frequencies on grid points were extracted. Fig. 16 shows 38 extracted pairs of N1 and N2 frequencies. In this way, the minimum database consists of 38 parametric HRTFs composed of N1, N2 and P1.

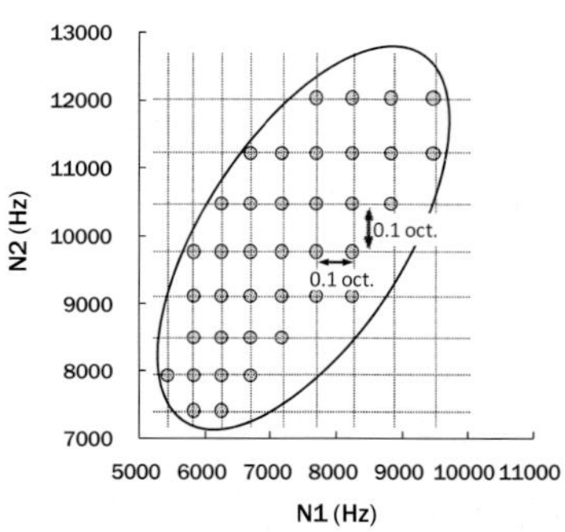

Fig.16 Extracted pairs of N1and N2; the distribution range is divided by the acceptance range of *NFD*.

4.2. *Selection of the appropriate parametric HRTFs for the front direction*

Thirty-eight parametric HRTFs were presented to a listener through open-air headphones. Before the listening tests, the transfer function of the apparatus was compensated, as mentioned in Section 3.2. The listener was asked to respond when he/she localized a sound image in front.

The parametric HRTF, by which a listener localized at the front, was selected as the appropriate one for the front direction. If the listener localized plural parametric HRTFs at the front, the parametric HRTF for which the listener perceived the most far and narrow sound images was selected as the appropriate one. The selection takes only 2 or 3 minutes.

4.3. *Generation of individual parametric HRTFs for various directions*

The individualized parametric HRTFs for directions in the horizontal and the median plane were generated from the selected parametric HRTF for the front direction.

4.3.1. *Generation of the individualized parametric HRTFs in the horizontal plane*

As shown in Fig. 17, the behavior of N1 and N2 frequencies as a function of the azimuth seems to be common among listeners, even though the frequencies of N1 and N2 of the front direction highly depend on the listener. Therefore, individualized N1 and N2 frequencies were obtained by regression equations, Eqs. (8) and (9), and by using the constant term given by the selected parametric HRTF in step 2.

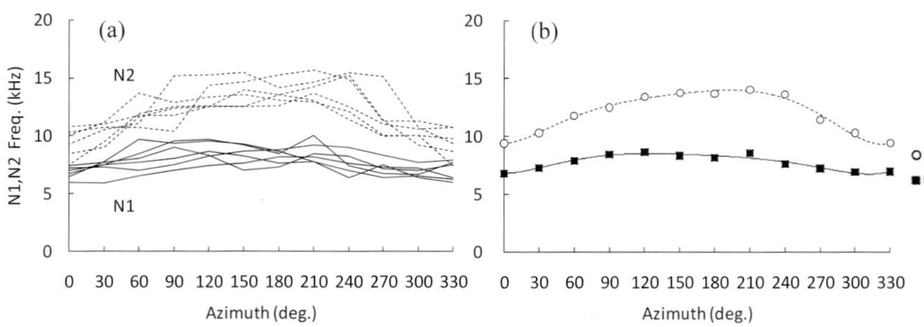

Fig. 17 N1 and N2 frequencies as a function of azimuth. (a): measured N1 and N2 frequencies of 6 subjects; (b): regression curves obtained from the mean values of the 6 subjects.

$$f_{N1}(azm) = 1.628\times10^{-10}\times azm^{6}-1.577\times10^{-7}\times azm^{5}+5.834\times10^{-5}\times azm^{4}$$
$$-1.011\times10^{-2}\times azm^{3}+7.152\times10^{-1}\times azm^{2}+1.839\times azm+6.809\times10^{3}\ \text{[Hz]}, \quad (8)$$

$$f_{N2}(azm) = 3.547\times10^{-10}\times azm^{6}-3.255\times10^{-7}\times azm^{5}+1.116\times10^{-4}\times azm^{4}$$
$$-1.796\times10^{-2}\times azm^{3}+1.236\times azm^{2}+8.557\times azm+9.380\times10^{3}\ \text{[Hz]}, \quad (9)$$

For Interaural Time Difference (ITD) and Interaural Level Difference (ILD), Eqs. (10) and (11) were used, respectively.

$$ITD(azm) = 1.0\times\sin(azm)\ \text{[ms]}, \quad (10)$$
$$ILD(azm) = 15.0\times\sin(azm)\ \text{[dB]}, \quad (11)$$

where *azm* denotes the azimuth in degrees.

4.3.2. *Generation of the individualized parametric HRTFs in the median plane*

As in the horizontal plane, the behavior of the N1 and N2 frequencies as a function of elevation seems to be common among listeners, even though the frequencies of N1 and N2 of the front direction highly depend on the listener (Fig. 18). The individualized N1 and N2 frequencies are obtained by the regression equations, Eqs. (12) and (13), and by using the constant term given by the selected parametric HRTF in step 2.

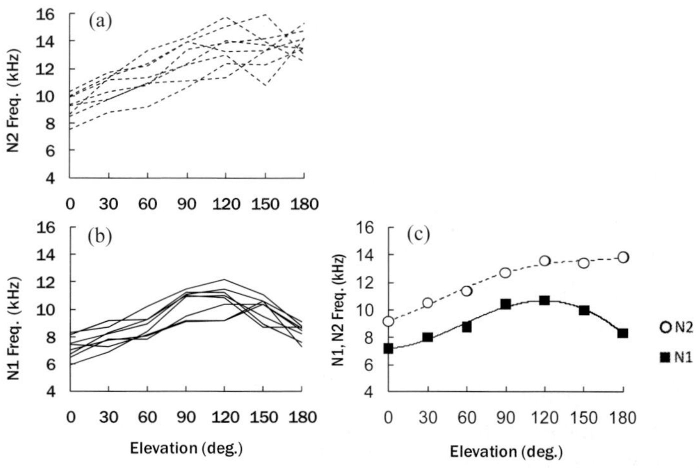

Fig. 18 N1 and N2 frequencies as a function of elevation. (a) measured N1 frequencies of 6 subjects; (b) measured N2 frequencies of 6 subjects; (c) regression curves obtained from the mean values of 6 subjects.

$$f_{N1}(elv)=1.001\times10^{-5}\times elv^{4}-6.431\times10^{-3}\times elv^{3}+8.686\times10^{-1}\times elv^{2}$$
$$-3.265\times10^{-1}\times elv+7.245\times10^{3}\ \text{[Hz]},\quad (12)$$

$$f_{N2}(elv)=1.310\times10^{-5}\times elv^{4}-5.154\times10^{-3}\times elv^{3}+5.020\times10^{-1}\times elv^{2}$$
$$+2.563\times10\times elv+9.244\times10^{3}\ \text{[Hz]},\quad (13)$$

where *elv* denotes the elevation in degree.

4.4. *Validity of the proposed individualization method*

To confirm the validity of the proposed individualization method, localization tests for steps 1 and 2 were carried out. The results show that the parametric HRTFs, which the subject localized at the front direction, have N1 and N2 near those of his/her own HRTF. Therefore, steps 1 and 2 could be considered to be valid. Examination of the total performance through steps 1 to 3 remains to be confirmed.

It is also important to utilize the existing measured HRTF database in the world. A method to find the appropriate measured HRTFs that correspond to the individualized parametric HRTFs will be a future subject of study.

5. Conclusions

To provide the appropriate HRTFs to a listener from the HRTF database quickly and easily, the authors proposed to utilize the differences in spectral cues to describe the individual differences of HRTFs and to create the database of a minimal number of HRTFs. In this study, the following three issues were discussed:
(1) the essence of spectral cues for vertical and front-back localization
(2) an appropriate physical measure for individual differences in HRTFs
(3) a method to provide individualized HRTFs by utilizing the physical measurements and the minimal HRTF database.

Systematic localization tests and observation of the measured HRTFs revealed the following: (1) the lowest first and second spectral notches (N1 and N2) above 4 kHz can be regarded as spectral cues; (2) the Notch Frequency Distance (NFD), which denotes the difference in the frequencies of N1 and N2, is a proper physical measure for individual differences of HRTFs; (3) the acceptance range of NFD for front localization is 0.1-0.2 octave; (4) a minimal HRTF database consisting of 38 parametric HRTFs for the front direction was obtained by dividing the distribution range of N1 and N2 frequencies by the acceptance range of NFD; (5) the appropriate parametric HRTFs for the front direction for each listener could be selected from the database of a minimal number of HRTFs by a brief localization test; (6) the individualized parametric HRTFs for various directions could be generated by the regression equation on the N1 and N2 frequencies.

Acknowledgments

The authors wish to thank Professor Masayuki Morimoto for meaningful discussion. The authors also thank Dr. Motokuni Itoh, Ms. Atsue Itagaki, Mr. Naokazu Gamoh for their cooperation in the localization tests.

References

1. M. Morimoto and Y. Ando, "On the simulation of sound localization," *J. Acoust. Soc. Jpn.(E)*. **1**, 167-174 (1980).
2. J. C. Middrebrooks, "Individual differences in external-ear transfer functions reduced by scaling in frequency", *J. Acoust. Soc. Am.* **106**, 1480-1492 (1999).
3. J. C. Middrebrooks, "Virtual localization improvrd by scaling nonindividualized external-ear transfer functions in frequency", *J. Acoust. Soc. Am.* **106**, 1493-1510 (1999).
4. J. C. Middrebrooks, E. A. Macpherson, and Z. A. Onsan, "Psychological customization of directional transfer functions for virtual sound localization", *J. Acoust. Soc. Am.* **108**, 3088-3091 (2000).
5. Y. Iwaya, "Individualization of head-related transfer functions with tournament-style listening test: Listening with other's ears," *Acoust. Sci. & Tech.* **27**, 340-343 (2006).
6. K. Roffler, A. Butler, "Factors that influence the localization of sound in the vertical plane", *J. Acoust. Soc. Am.* **43**, 1255-1259 (1968).
7. E. A. G. Shaw, R. Teranishi, "Sound pressure generated in an external-ear replica and real human ears by a nearby point source", *J. Acoust. Soc. Am.* **44**, 240-249 (1968).
8. J. Blauert, "Sound localization in the median plane", *Acustica* **22**, 205-213 (1969/70).
9. B. Gardner, S. Gardner, "Problem of localization in the median plane: effect of pinna cavity occlusion", *J. Acoust. Soc. Am.* **53**, 400-408 (1973).
10. J. Hebrank, D. Wright, "Spectral cues used in the localization of sound sources on the median plane", *J. Acoust. Soc. Am.* **56**, 1829-1834 (1974).

11. A. Butler, K. Belendiuk, "Spectral cues utilizes in the localization of sound in the median sagittal plane", *J. Acoust. Soc. Am.* **61**, 1264-1269 (1977).
12. S. Mehrgardt, V. Mellert, "Transformation character-istics of the external human ear", *J. Acoust. Soc. Am.* **61**, 1567-1576 (1977).
13. A. J. Watkins, "Psychoacoustic aspects of synthesized vertical locale cues", *J. Acoust. Soc. Am.* **63**, 1152–1165 (1978).
14. M. Morimoto, H. Aokata, "Localization cues of sound sources in the upper hemisphere", *J. Acoust. Soc. Jpn* (*E*). **5**, 165-173 (1984).
15. J. C. Middlebrooks, "Narrow-band sound localization related to external ear acoustics", *J. Acoust. Soc. Am.* **92**, 2607-2624 (1992).
16. K. Iida, M. Yairi, and M. Morimoto, "Role of pinna cavities in median plane localization", *Proc. 16th Int'l Cong. on Acoust.* 845-846 (1998).
17. B. C. J. Moore, R. Oldfield, G. J. Dooley, "Detection and discrimination of peaks and notches at 1 and 8 kHz", *J. Acoust. Soc. Am.* **85**, 820-836 (1989).
18. V. C. Raykar, R. Duraiswami, B. Yegnanarayana, "Extracting the frequencies of the pinna spectral notches in measured head related impulse responses", *J. Acoust. Soc. Am.* **118**, 364-374 (2005).
19. K. Iida, M. Itoh, A. Itagaki, M. Morimoto, "Median plane localization using a parametric model of the head-related transfer function based on spectral cues", *Applied Acoustics*, **68**, 835-850 (2007).
20. D. Hammershøi, H. Møller, "Sound transmission to and within the human ear canal", *J. Acoust. Soc. Am.* **100**, 408-427 (1996).

PRESSURE DISTRIBUTION PATTERNS ON THE PINNA AT SPECTRAL PEAK AND NOTCH FREQUENCIES OF HEAD-RELATED TRANSFER FUNCTIONS IN THE MEDIAN PLANE

H. TAKEMOTO, P. MOKHTARI, H. KATO and R. NISHIMURA
National Institute of Information and Communications Technology (NICT), 2-2-2, Hikaridai, Seika-cho, Soraku-gun, Kyoto, 619-0288 Japan

K. IIDA
Chiba Institute of Technology, 2-17-1, Tsudanuma, Narashino, Chiba, 275-0016, Japan

To reveal the mechanism generating spectral peaks and notches of the head-related transfer functions (HRTFs) in the median plane, head shapes were measured using magnetic resonance imaging. Then HRTFs were calculated from the shapes using the finite-difference time-domain method. Results showed that the pinna shape was the dominant factor for the basic peak–notch pattern. The simulation of pressure distribution patterns at a given frequency and elevation angle toward the sound source revealed that resonances in the pinna cavities, i.e., concha, cymba, and triangular and scaphoid fossae, contributed to generation of both peaks and the first notch. At the concha, a pressure anti-node developed at peaks, whereas a node developed at the notch. There were three peaks less than 10 kHz; the first, second, and third peaks had one, two, and three anti-nodes in the pinna cavities, respectively. Each peak frequency was stable across elevation angles, although the amplitude changed. The first notch frequency and the number of the anti-nodes changed with the elevation angle. Those anti-nodes appeared in pinna cavities other than the concha. Then they canceled the incoming wave at the concha because of the reverse phase.

1. Introduction

Various reports have described that spectral notches of head-related transfer functions (HRTFs) caused by the pinna act as important cues for localizing the elevation angle of sound sources in the median plane [1–8]. Iida *et al.* [9] revealed that the first (lowest) spectral peak (P1) as well as the first and second notches (N1 and N2) act as cues. They measured HRTFs of several subjects in the upper median plane. The P1 frequency was almost constant with the source elevation angle, although N1 and N2 frequencies changed systematically. The N1 and N2 frequencies increased gradually with the elevation angle, reached maxima at elevation angles of 120–150 deg; then decreased. They speculated that the human hearing system uses P1 as a fixed reference for changes in N1 and N2. Figure 1 presents the P1, N1, and N2 trajectories schematically.

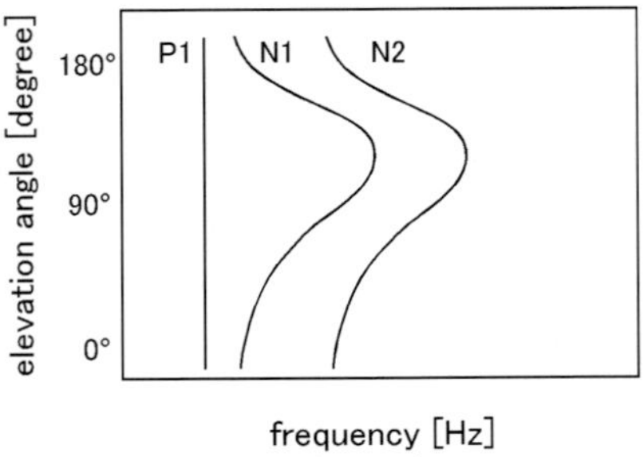

Figure 1. Schematic representation of P1, N1, and N2 trajectory patterns.

Some previous works have described the mechanism generating the spectral peaks and notches of HRTFs. According to Shaw and Teranishi [10], P1 resulted from the primary resonance of the concha. Hebrank and Wright [4] speculated that N1 resulted from the cancelation of the direct incoming wave by the reflected wave from the posterior wall of the concha. Lopez-Poveda and Meddis [11] and Raykar *et al.* [12] supported this speculation. Furthermore, Raykar *et al.* proposed a simple reflection model that can estimate notch frequencies based on the pinna shapes [12]. The model, however, cannot treat elevations behind the ear because the mechanism of the spectral notches is not clear at those elevation angles [12]. In short, the mechanism by which spectral peaks and notches are generated remains unclear.

We have developed an acoustic simulator based on the finite-difference time-domain (FDTD) method, which is a major method used for three-dimensional (3D) acoustic field analysis. This method can calculate acoustic characteristics much faster than other methods such as the finite element method (FEM) and the boundary element method (BEM), although this method requires higher spatial resolution for the analysis field. The most distinctive feature of the FDTD method is time domain analysis. This method is useful to visualize the propagation, reflection, and diffraction of sound waves around the analysis object. In addition, because this method uses a simple orthogonal grid system, it can load voxel data as an analysis model. Consequently, the simulator which we developed can directly analyze voxel-based data such as 3D magnetic resonance imaging (MRI) data of the human body [13]. Using the simulator, HRTFs are calculable from subjects' head MRI data. A comparison between measured and calculated HRTFs revealed that the calculation had adequate accuracy for examining the basic peak–notch pattern of HRTFs [14]. Using the simulator in the present study, we will examine how peaks and notches of HRTFs in the median plane are generated.

1.1. *Nomenclatures*

There are four main cavities in the pinna: the cavity of concha (concha), cymba conchae (cymba), triangular fossa, and scaphoid fossa. For concise descriptions hereafter, we will call all the cavities "the pinna cavities" and all the cavities other than the concha "the upper cavities" (Fig. 2).

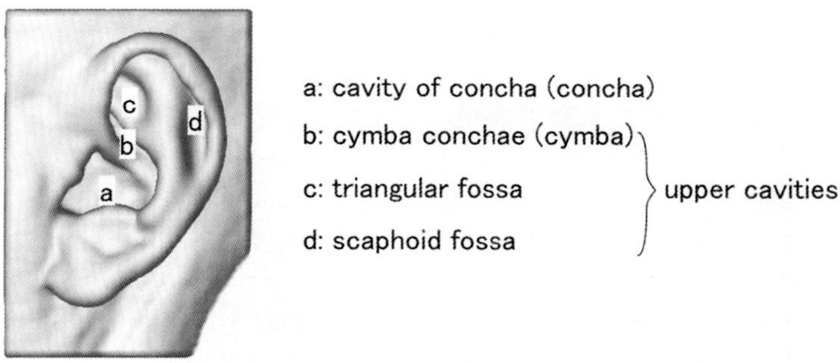

Figure 2. Anatomical names of pinna cavities.

2. Materials and Methods

2.1. *MRI data*

The head shapes of two males (M1 and M2) and two females (F1 and F2) were measured using MRI (Magnex Eclipse 1.5 T Power Drive 250; Shimadzu–Marconi) installed at the Brain Activity Imaging Center in ATR-Promotions Inc. The spatial resolution of MRI data for M1 and M2 was 1.2 mm; that for F1 was 1.1 mm, and that for F2 was 1.0 mm. The resolution depended on the head size. The echo time was 4.47 ms, and the repetition time was 12 ms. From scanned images, a 3D head shape for each subject was reconstructed.

The surface of the skin was segmented from the air by 3D image processing techniques such as binarization and region growing after the bilateral ear canals were occluded. Figure 3 depicts the whole head models for the four subjects on the same scale. From the whole head models, the left pinnae were extracted (Fig. 4) to evaluate head diffraction effects on HRTFs. Hereinafter, we simply refer to the whole head model as the "head model," and to the left pinna model as the "pinna model".

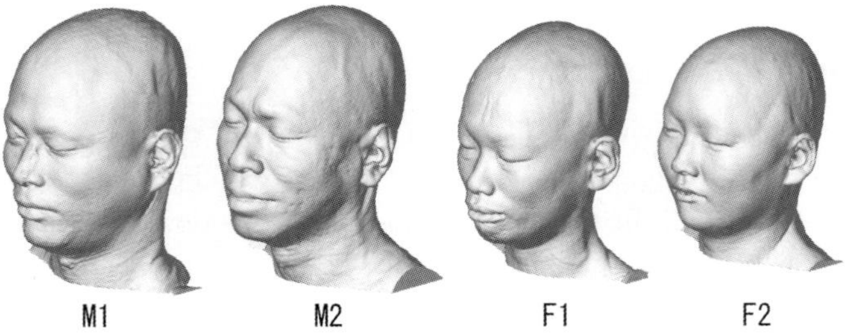

Figure 3. Four subjects' whole-head models on the same scale.

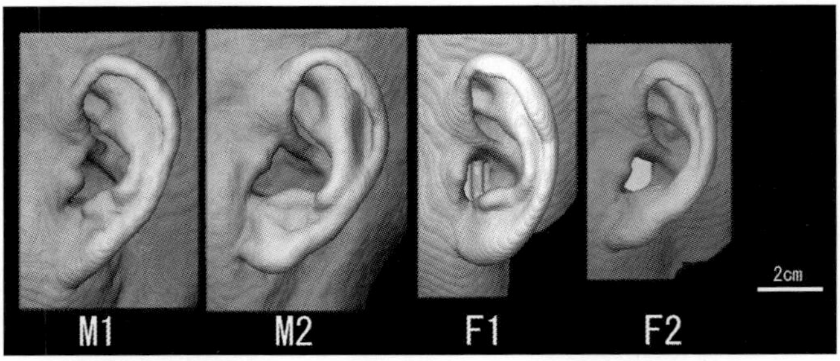

Figure 4. Four subjects' left pinna models on the same scale.

2.2. *FDTD method*

In the FDTD method, the particle velocity and pressure at each point within the analysis field were calculated using numerical solutions of the following two acoustic differential equations:

$$-\kappa\frac{\partial p}{\partial t}-\alpha p=\nabla u \text{ and} \tag{1}$$

$$-\rho\frac{\partial u}{\partial t}-\alpha^{*}u=\nabla p, \tag{2}$$

where p stands for the pressure, u signifies the particle velocity, α is the attenuation coefficient associated with compressibility of the medium, and ρ is the density of the medium. Also, κ, which is defined as $\kappa=1/\rho c$, denotes the compressibility of the medium, where c represents the sound velocity in the medium. In addition, α^{*} is the attenuation coefficient associated with density of the medium. It is generally zero in the analysis region, although the value is $\alpha\rho/\kappa$ in the absorbing boundary surrounding the analysis region, known as the Perfectly Matched Layer (PML) [15].

A simple surface impedance method that Yokota *et al.* [16] proposed was introduced to simulate the sound-wave reflection on the wall. This method cannot treat frequency-dependent losses. Consequently, the impedance uniformly affected the amplitudes of peaks and notches of HRTFs.

2.3. *HRTF calculation*

To examine the possibility for reducing the size of the analysis field, four subjects' HRTFs in the median plane were calculated using the FDTD method under the following three conditions (Fig. 5).

The first case was the "1 m head model." In this condition, using the head model, left-ear HRTFs were calculated at a distance 1 m from the center of the head. According to the reciprocity theorem, the sound source was placed at the entrance of the closed ear canal and observation points were placed on the circumference of a circle with a radius of 1.0 m at 10 degree intervals. The center of the circle was equal to that of the head.

The second condition was the "1 m pinna model." In this condition, the pinna model was used instead of the head model in the first condition. The center of the circle was the entrance of the closed ear canal. This condition was designed for evaluating diffraction effects of the head.

The third condition was the "0.1 m pinna model." In this condition, the radius of the circle was 0.1 m and the center of the circle was the same as that in the second condition. This condition was designed for evaluating the difference of the distance from the source point to the observation point.

Calculated HRTFs were mutually compared and the mean spectral distance was calculated for quantitative evaluation. The mean spectral distance (SD) was calculated as

$$SD = \frac{1}{N}\sum_{n=1}^{N}\left|H_n^1 - H_n^2\right|, \tag{3}$$

where N is the total number of frequency steps on a linear scale and H represents the transfer function (in decibels).

A Gaussian pulse was fed to the source point as a volume velocity and the pressure change was calculated during 8 ms at each observation point. The HRTFs were calculated up to 24 kHz. The elevation angle for the frontal direction was set at 0 deg and that for the overhead direction at 90 deg.

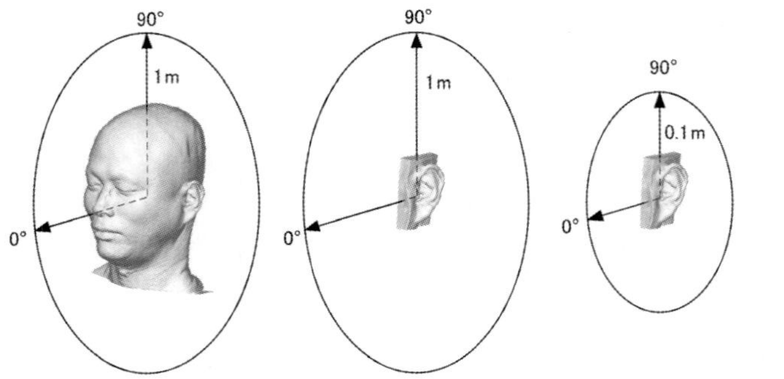

(a) 1 m head model (b) 1 m pinna model (c) 0.1 m pinna model

Figure 5. Three conditions for calculating HRTFs in the median plane.

2.4. *Simulation of pressure distribution pattern on the pinna*

To examine acoustic phenomena at frequencies of spectral peaks and notches, the time course of the instantaneous pressure distribution pattern on the pinna models was calculated. For a given elevation angle, the analysis field was excited by a sinusoidal source placed 0.1 m away from the entrance of the ear canal at a frequency equal to the HRTF peak or notch. After reaching a steady state, the pressure distribution pattern within the whole analysis field was recorded at a 200 kHz sampling rate during 0.5 ms. The pressure distribution pattern at each sampling step was visualized using a volume-rendering technique.

From the series of pressure distribution patterns, it was also possible to calculate the mean absolute pressure distribution pattern. This pattern could indicate more precisely which part of the pinna cavities resonated. In this study, however, instantaneous pressure distribution patterns were used for analysis because the mean absolute pressure distribution patterns lacked phase information.

3. Results and Discussion

3.1. *Comparison of HRTFs between head and pinna models*

Figures 6–9 present HRTFs calculated under the three different conditions for the four subjects and Table 1 presents mean spectral distances of HRTFs. The spectral distance between the 1 m head model and the 1 m pinna model was less than 1.91 dB. This spectral distance reflected diffraction effects of the head, which caused a fine structure resembling concentric rings, appearing mainly in the low-frequency region less than 5 kHz for all subjects. The spectral distance between the 1 m head model and the 0.1 m pinna model was less than 2.87 dB. These spectral differences resulted from distance effects observed mainly in the amplitudes of peaks and notches, in addition to head diffraction effects. The peak–notch pattern, however, was common across the three conditions for the four subjects. These facts suggest that the pinna shape is the primary factor for determining the peak–notch pattern of HRTFs in the median plane. Consequently, it is safe to say that the pinna model is adequate for examining how spectral peaks and notches in the median plane were generated.

Figures 6–9 also show that the peak–notch pattern is fundamentally the same among the subjects. Three spectral peaks, P1, P2, and P3, are visible at less than 10 kHz, although P3 is ambiguous for M1. Although the presence and strength of the peaks varied depending on the elevation angle, their frequencies were quite stable. On the other hand, the N1 frequency was changing with elevation angle to form a typical trajectory pattern presented in Fig. 1. Because the elevation angle increased, the frequency increased gradually, reached a maximum at approximately 120 deg, then decreased. No subject examined in this study had a clear trajectory pattern for N2.

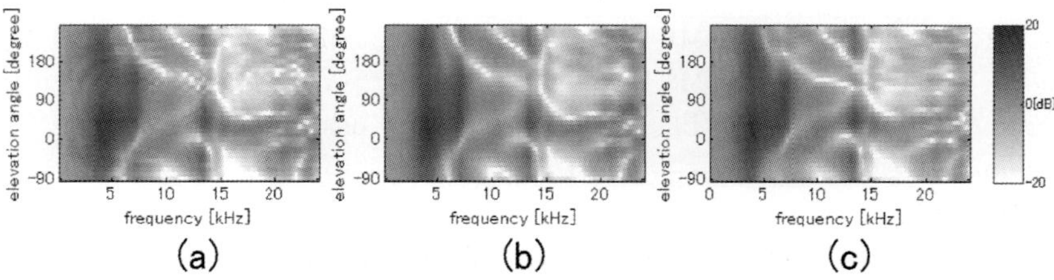

Figure 6. HRTFs for M1 in the median plane calculated under the three conditions: (a) 1 m head model, (b) 1 m pinna model, and (c) 0.1 m pinna model.

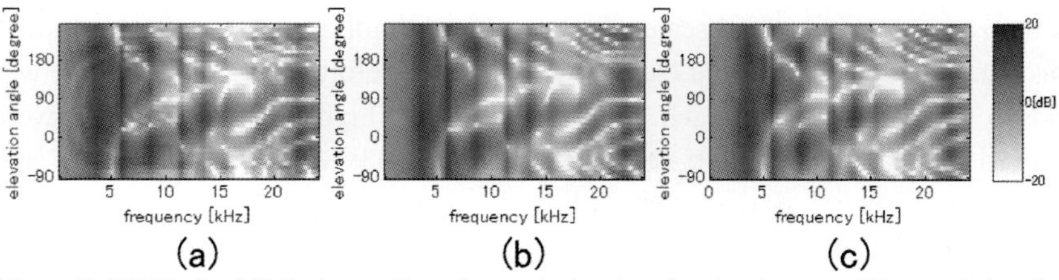

Figure 7. HRTFs for M2 in the median plane calculated under the three conditions: (a) 1 m head model, (b) 1 m pinna model, and (c) 0.1 m pinna model.

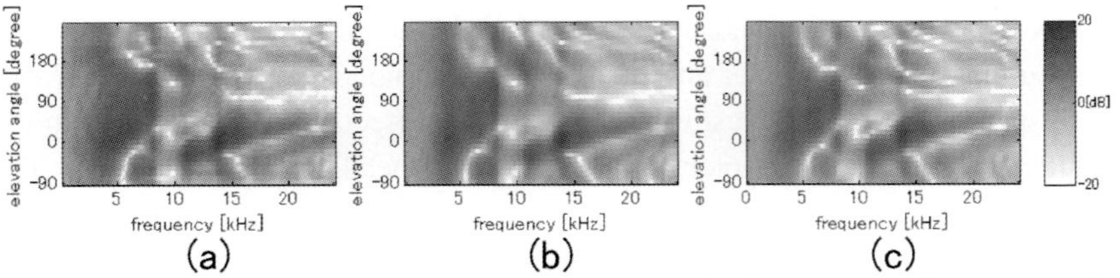

Figure 8. HRTFs for F1 in the median plane calculated under the three conditions: (a) 1 m head model, (b) 1 m pinna model, and (c) 0.1 m pinna model.

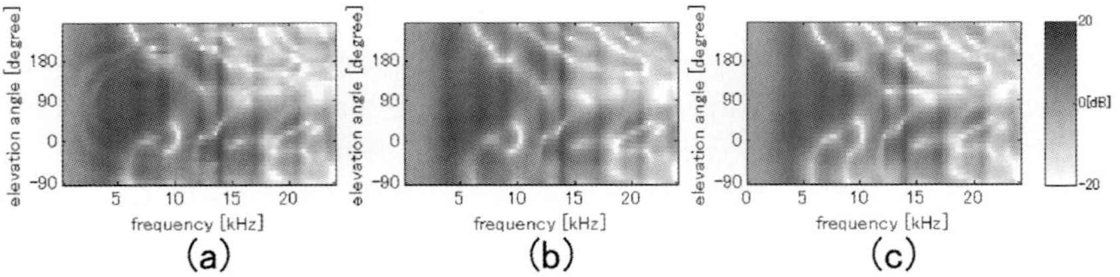

Figure 9. HRTFs for F2 in the median plane calculated under the three conditions: (a) 1 m head model, (b) 1 m pinna model, and (c) 0.1 m pinna model.

Table 1. Mean spectral distance of 1 m pinna model and 0.1 m pinna model relative to the 1 m head model [dB]

	1 m pinna model	0.1 m pinna model
1 m head model of M1	1.63	2.38
1 m head model of M2	1.91	2.87
1 m head model of F1	1.62	2.49
1 m head model of F2	1.78	2.76

3.2. *Pressure distribution patterns on the pinna at peak frequencies*

Figures 10–12 respectively represent instantaneous pressure distribution patterns on the pinna at P1, P2, and P3 frequencies. In these figures, the pinna models are visualized on a normalized scale. The pressure distribution pattern for each peak was common across different elevation angles. Therefore, an elevation angle was chosen for each peak of each subject. Voxels of high positive pressure were colored white, those of high negative pressure black. Consequently, white and black parts correspond to pressure anti-nodes with opposite phase.

Figure 10 represents pressure distribution patterns on the pinna at P1 frequency. A single pressure anti-node developed across the pinna cavities, which indicates that the pinna cavities resonated entirely in the same phase, i.e. the primary resonance mode occurred, whereas Shaw and Teranishi [10] reported that P1 was the primary resonance of the concha. Although the shape of the pinna differed substantially among the subjects, the P1 frequency was almost common among subjects.

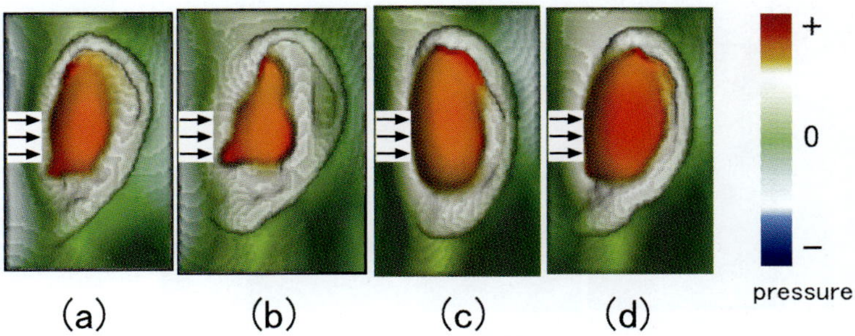

Figure 10. Instantaneous pressure distribution patterns on the pinna at P1 frequency. Arrows indicate the traveling direction of the incoming wave. The subject, elevation angle and excitation frequency of the source are as follows: (a) M1, 0°, 4 kHz; (b) M2, 0°, 3.5 kHz; (c) F1, 0°, 4 kHz; (d) F2, 0°, 4 kHz.

Figure 11 represents pressure distribution patterns on the pinna at P2 frequency. An anti-node developed at the concha (colored white), and another anti-node with reverse phase developed across the upper cavities (colored black). This pattern indicates that the second resonance mode occurred in the pinna cavities at P2 frequency.

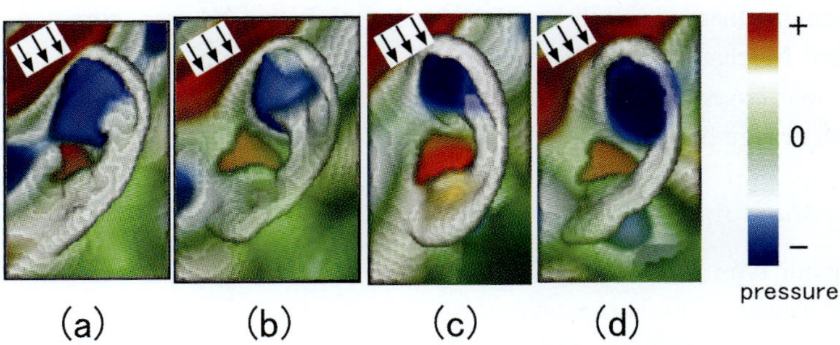

Figure 11. Instantaneous pressure distribution patterns on the pinna at P2 frequency. Arrows indicate the traveling direction of the incoming wave. The subject, elevation angle and excitation frequency of the source are as follows: (a) M1, 60°, 7 kHz; (b) M2, 60°, 6 kHz; (c) F1, 60°, 6 kHz; (d) F2, 60°, 6.75 kHz.

Figure 12 presents pressure distribution patterns on the pinna at P3 frequency. Because P3 of M1 was ambiguous, the pressure distribution pattern for its P3 is not shown. Three pressure anti-nodes were observed in the pinna cavities. Anterior parts of the concha and the triangular fossa resonated in phase (colored white), although the region from the cymba to the posterior part of the concha resonated with reverse phase (colored black). This pattern represents the third resonance mode of the pinna cavities.

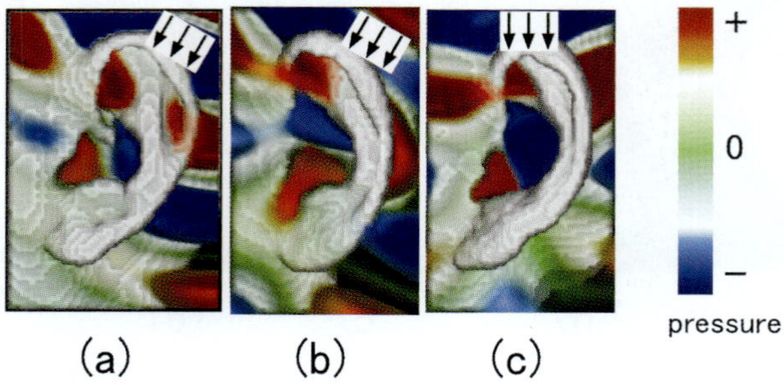

Figure 12. Instantaneous pressure distribution patterns on the pinna at P3 frequency. Arrows indicate the traveling direction of the incoming wave. The subject, elevation angle and excitation frequency of the source are as follows: (a) M2, 120°, 8 kHz; (b) F1, 120°, 8 kHz; (c) F2, 90°, 8.75 kHz.

3.3. *Pressure distribution patterns on the pinna at the first notch frequencies*

The pressure distribution pattern for N1 changed with source elevation angle, but the changing manner and the mechanism for generating N1 were common among subjects. Mechanisms of two types were observed: major and minor ones. The major one covered wider elevation angles, although the minor one covered a narrower range of angles.

3.3.1. *Major type: "counter" canceling*

The major type was "counter" canceling. The incoming wave was canceled by the resonance in the upper cavities because the phases were mutually reversed. Consequently, a pressure node developed just at the concha. When the sound source was located below the horizontal plane, i.e., at elevation angles from -90 to 0 deg or from 180 to 270 deg, the upper cavities resonated in the same phase. In other words, the first mode of resonance occurred in the upper cavities. Therefore, the primary resonance frequency of the upper cavities would be the lower limit of N1. Figure 13 presents typical patterns of this cancelation. The incoming wave (colored white) was canceled by the resonance in the upper cavities (colored black).

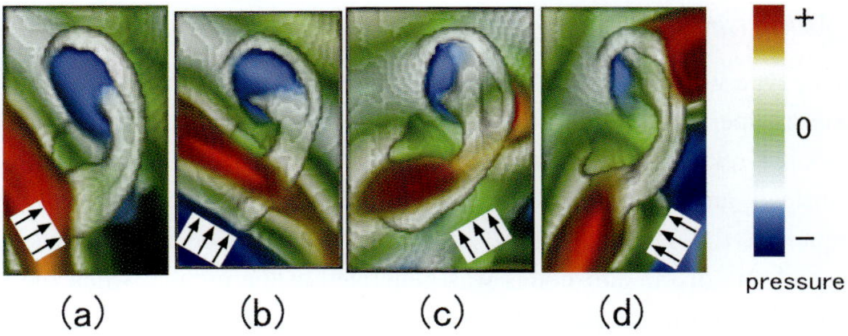

Figure 13. Instantaneous pressure distribution patterns on the pinna at N1 frequency. Arrows indicate the traveling direction of the incoming wave. The subject, elevation angle and excitation frequency of the source are as follows: (a) F1, -30°, 5.75 kHz; (b) M1, -60°, 6.75 kHz; (c) M2, 240°, 6 kHz, (d) F2, 210°, 7.25 kHz.

An extended version of this canceling was observed when the sound source was placed in the antero-superior direction. Figure 14 shows typical patterns of this canceling. In the extended version, in addition to the upper cavities, the posterior part of the concha connecting to the cymba was involved in the resonance which canceled the incoming wave. This fact supported the speculation that the incoming wave was canceled by the reflected wave from the posterior wall of the concha [4] because the reflection was a causal factor of resonance. At high elevation angles, 60 deg and 90 deg, two pressure anti-nodes were observed in the upper cavities and the posterior part of the concha, which indicates that the second resonance mode occurred in that region at high elevation angles; the anti-node at the posterior region of the concha dominated the cancellation of the incoming wave at the entrance of the ear canal (Figs. 14(c) and 14(d)).

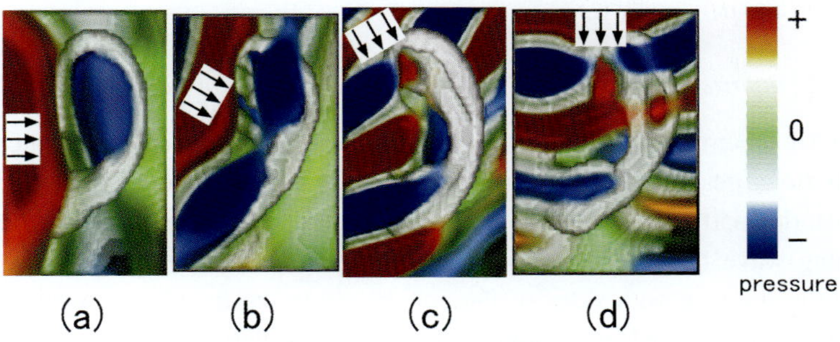

Figure 14. Instantaneous pressure distribution patterns on the pinna at N1 frequency. Arrows indicate the traveling direction of the incoming wave. The subject, elevation angle and excitation frequency of the source are as follows: (a) F2, 0°, 7 kHz; (b) M1, 30°, 10.25 kHz; (c) F1, 60°, 11.5 kHz, (d) M2, 90°, 9.5 kHz.

3.3.2. *Minor type: "intercept" canceling*

The minor type was "intercept" canceling. When the sound source was placed in the postero-superior direction, the upper cavities were on the same side as the sound source relative to the concha. Therefore, "counter" canceling could not occur. Instead, an anti-node at the cymba intercepted the incoming wave. At the N1 frequency, the second or third mode of resonance occurred in the upper cavities and one of the anti-nodes was generated at the cymba, when the sound source was located in the postero-superior direction. The phase of the anti-node at the cymba was opposite that of the incoming wave. Consequently, the incoming wave was canceled as it passed above the cymba. Figure 15 portrays the "intercept" canceling; Fig. 15(c) depicts the most typical pattern. The wavefront of the incoming wave (colored white) was divided by the anti-node at the cymba (colored black). Therefore, the pressure change was minimized at the concha.

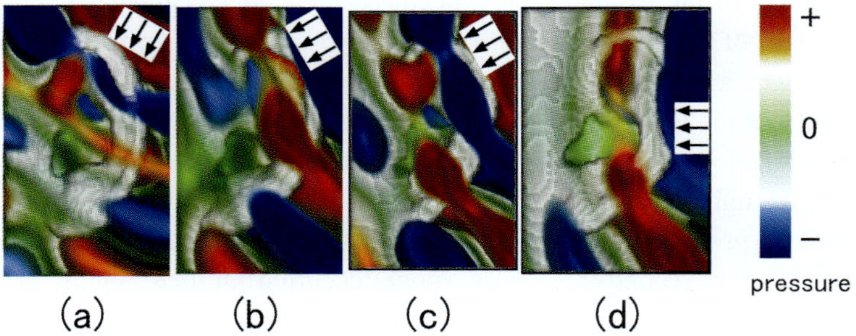

Figure 15. Instantaneous pressure distribution patterns on the pinna at N1 frequency. Arrows indicate the traveling direction of the incoming wave. The subject, elevation angle and excitation frequency of the source are as follows: (a) F1, 120°, 10.25 kHz; (b) F2, 150°, 10 kHz; (c) M1, 150°, 9.5 kHz, (d) M2, 180°, 6.75 kHz.

3.3.3. *Comparison of "counter" canceling and "intercept" canceling*

Figure 16 schematically presents mechanisms of the "counter" and "intercept" cancellations described above. In "counter" canceling, the upper cavities (and the posterior part of the concha in the extended version) resonated to cancel the incoming wave because of the reverse phase. In this case, two pressure anti-nodes appeared on both sides of the concha along the traveling sound wave; a pressure node emerged in the middle, i.e. at the concha. Therefore, the pressure change was minimized at the concha and the spectral notch was generated.

On the other hand, in "intercept" canceling, the typical pressure node placed between two anti-nodes was not generated at the concha. Instead, a part of the wavefront which led to the concha was canceled by a resonance with reverse phase at the cymba. Therefore, the pressure change of the incoming wave was reduced considerably after passing above the cymba. In other words, the incoming wave did not reach the concha to any great degree because the incoming wave was blocked by the resonance of the cymba.

At an elevation angle of around 90 deg, "counter" canceling changed to "intercept" canceling. The critical angle, however, was ambiguous because N1 was not clear around the angle. For that reason, no cancellation pattern was observed clearly in the pressure distribution pattern. At an elevation angle of around 180 deg, the critical elevation angle at which the "intercept" canceling changed to "counter" canceling was also ambiguous, although the N1 trajectory was more distinctive. This might be true because the cymba opened in the postero-inferior direction. The resonance of the cymba might affect the incoming wave over widely various elevation angles; therefore, "intercept" canceling would change continuously to "counter" canceling.

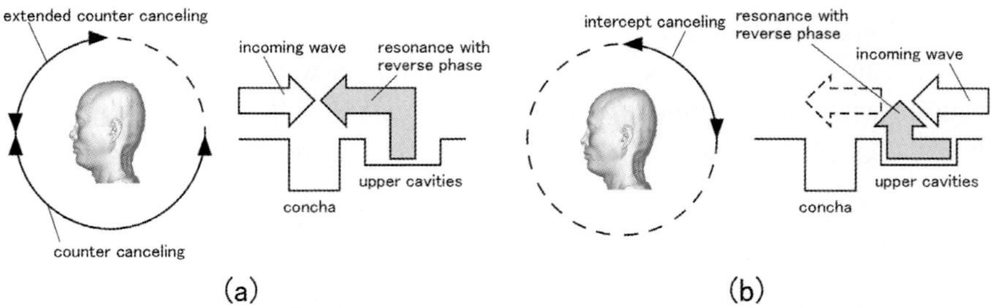

Figure 16. Mechanisms of generating N1. (a) "counter" canceling (b) "intercept" canceling.

4. Conclusion

The HRTFs in the median plane were calculated from head and pinna models for two males and two females under three conditions by an FDTD acoustic simulator. Comparisons of those HRTFs indicated that the basic pattern of peaks and notches could be attributed to the pinna shape, whereas diffraction effects of the head on HRTFs appeared in the low-frequency region less than 5 kHz. The trajectory of N2 was not distinct among the four subjects in this study.

Analyses of pressure distribution patterns at P1, P2, P3, and N1 were performed using pinna models. Various patterns of resonance occurred in the pinna cavities. At peaks, all the pinna cavities were involved in the resonance, although at N1, the concha was excluded. Because the number of anti-nodes determined the order of peaks, P1, P2, and P3 had one, two, and three anti-nodes, respectively. Those peak frequencies were distinct and stable across elevation angles. On the other hand, the N1 frequency and its associated number of anti-nodes changed with elevation angle, which indicates that N1 consists of different resonance modes of the upper cavities, varying from first order to third order. The N1 frequency and number of anti-nodes reached maxima at 120 deg elevation.

There were major and minor types of mechanism for generating N1. The major type, "counter" canceling, was observed when the sound source was placed in all the directions other than the postero-superior direction. In this type, a pressure anti-node, which was generated by resonances of the upper cavities and the posterior part of the concha, countered the incoming wave across the concha, and canceled the incoming wave at the concha. The minor type, "intercept" canceling, was observed when the sound source was placed in the postero-superior direction. In this type, the second or third resonance mode occurred in the upper cavities, and an anti-node at the cymba intercepted the incoming wave, passing above it to minimize pressure changes at the concha.

Although N1 was generated by the first, second, and third resonance modes of the upper cavities, N1 frequency changed with elevation angle gradually, rather than discretely as expected. In this study, we were unable to reveal the causal factor for continuous change. A possible factor is the ambiguous boundaries of the pinna cavities. The resonance of the upper cavities mainly occurred along a curved line from the triangular fossa to the posterior part of the concha via the cymba. The pressure node and anti-node were observed to shift gradually along this line; the resonance frequency changed with even small changes in elevation angle. The ambiguous boundaries of the pinna cavities would enable such a shift, although further analysis must be done to clarify this effect.

As described above, the resonance of the pinna cavities generated the peak–notch pattern of HRTFs. The resonance pattern is expected to be determined by the cavity shape and arrangement. Therefore, it is theoretically possible to estimate frequencies of peaks and notches from the morphological features of the pinna cavities. However, because the pinna shape is fairly complicated and varies greatly among individuals, it would be difficult to define the boundary of each cavity to measure its shape. Prior to such a study, it would be better to develop a simple model equivalent to the pinna cavities and examine the acoustic characteristics.

This study examined HRTFs only in the median plane and discussed the mechanism for generating peaks and notches. Results show that diffraction effects of the head on HRTFs were small and that the resonance of pinna cavities was the dominant factor for generating peaks and notches at frequencies higher than about 5 kHz. Although head diffraction effects are probably large outside the median plane, the basic mechanism for generating peaks and notches is expected to be general.

References

1. B. C. J. Moore, S. R. Oldfield and G. Dooley, "Detection and discrimination of spectral peaks and notches at 1 and 8 kHz," *J. Acoust. Soc. Am.* **85**, 820-836 (1989).
2. D. Wright, J. H. Hebrank and B. Wilson, "Pinna reflections as cues for localization," *J. Acoust. Soc. Am.* **56**, 957-962 (1974).
3. M. B. Gardner and R. S. Gardner, "Problem of localization in the median plane: Effect of pinna cavity occlusion," *J. Acoust. Soc. Am.* **53**, 400-408 (1974).
4. J. Hebrank and D. Wright, "Spectral cues used in the localization of sound sources on the median plane," *J. Acoust. Soc. Am.* **56**, 1829-1834 (1974).
5. P. Hofman, J. Van Riswick and A. Van Opstal, "Relearning sound localization with new ears," *Nat. Neurosci.* **1**, 417-421 (1998).
6. P. Poon and J. F. Brugge, "Sensitivity of auditory nerve fibers to spectral notches," *J. Neurophysiol.* **70**, 655-666 (1993).
7. P. Poon and J. F. Brugge, "Virtual-space receptive fields of single auditory nerve fibers," *J. Neurophysiol.* **70**, 666-676 (1993).
8. D. J. Tollin and T. C. T. Yin, "Spectral cues explain illusory elevation effects with stereo sounds in cats," *J. Neurophysiol.* **90**, 525-530 (2003).
9. K. Iida, M. Itho, A. Itagaki, and M. Morimoto, "Median plane localization using a parametric model of the head-related transfer function based on spectral cues," *Appl. Acoust.* **68**, 835-850 (2007).
10. E. A. G. Shaw and R. Teranishi, "Sound pressure generated in an external-ear replica and real human ears by a nearby point source," *J. Acoust. Soc. Am.* **44**, 240-249 (1968).
11. E. A. Lopez-Poveda and R. Meddis, "A physical model of sound diffraction and reflections in the human concha," *J. Acoust. Soc. Am.* **100**, 3248-3259 (1996).
12. V. C. Raykar, R. Duraiswami and B. Yegnanarayana, "Extracting the frequencies of the pinna spectral notches in measured head related impulse responses," *J. Acoust. Soc. Am.* **118**, 364-374 (2005).
13. H. Takemoto, P. Mokhtari and T. Kitamura, "Acoustic analysis of the vocal tract during vowel production by finite-difference time-domain method," *J. Acoust. Soc. Am.* **123**, 3233 (2008).
14. P. Mokhtari, H. Takemoto, R. Nishimura, and H. Kato, "Computer simulation of KEMAR's head-related transfer functions: verification with measurements and acoustic effects of modifying head shape and pinna concavity" *in the same volume* (2010).

15. J. P. Berenger, "A perfectly matched layer for the absorption of electromagnetic waves," *J. Comput. Phys.* **114**, 185-200 (1994).
16. T. Yokota, S. Sakamoto and H. Tachibana, "Visualization of sound propagation and scattering in rooms," *Acoust. Sci. & Tech.* **23**, 40-46 (2002).

SPATIAL DISTRIBUTION OF THE LOW-FREQUENCY HEAD-RELATED TRANSFER FUNCTION SPECTRAL NOTCH AND ITS EFFECT ON SOUND LOCALIZATION

M. OTANI*

Faculty of Engineering, Shinshu University,
4-17-1 Wakasato, Nagano, 380-8553, Japan
**E-mail: otani@cs.shinshu-u.ac.jp*

Y. IWAYA, T. MAGARIYACHI and Y. SUZUKI

Research Institute of Electrical Communication, Tohoku University,
2-1-1 Katahira, Aoba-ku, Sendai, 980-8577, Japan

Our previous work showed that the presence of pinnae produces a spectral notch in HRTFs for sound sources behind a listener (labeled N0). This spectral notch appears at frequencies lower than N1 and N2, as reported in previous works, as necessary cues for sound image localization. In this study, we examined the spatial distribution of N0 and subsequently conducted psychoacoustical experiments to clarify how N0 affects sound localization. Experimental results showed that N0 has little effect on front-back judgment, but that it has significant effects on sound localization in elevation.

Keywords: Head-related transfer functions; Spectral notch; Sound localization; ARMA modeling

1. Introduction

Human beings can perceive a sound position and thereby localize a sound source using head-related transfer functions (HRTFs).[1] The HRTFs represent acoustical transfer functions between a sound source and a listener's ears and include cues for sound localization. However, HRTFs have strong individuality because of individual variation in head and pinna shapes. Many researchers have been studying on the relationship between HRTFs and anthropometric measures such as head size, pinna shapes, and other physical characteristics of the human body.[2,3] Both interaural time difference (ITD) and interaural level difference (ILD) are well known as dominant localization cues in the horizontal plane. For elevation localization, however, it is difficult to explain sound localization mechanisms using ITD and ILD alone, especially in the median plane, on which interaural differences are extremely small. On the median plane, a spectral cue is a promising cue.[4] Therefore, researchers have investigated the spectral features of HRTFs. Previous works have suggested that the lowest spectral notch provides a localization cue.[5,6] Iida *et al.* reported that two distinct spectral notches called N1 and N2, which vary their frequencies with elevation angle of source position, provides necessary localization cues for the elevation angle.[7] For most source positions, N1 corresponds to the first notch, and N2 corresponds to the second lowest notch. However, we had found using numerical computer simulation of HRTFs that a pinna produces another spectral notch at frequencies lower than N1 for rear sources.[8] That notch is labeled as N0 for this discussion. For some cases, N0 can be the first notch. Figure 1 portrays HRTF spectra contours on the horizontal plane at a -30 degrees elevation angle. The figure shows that frequencies where N0 appears overlap those of N1, which indicates that N0 can provide a cue for localizing sound sources behind a listener.

Therefore, to examine the possibility that N0 can be a localization cue, we investigate the spatial distribution of N0 and effects of N0 on sound localization using a listening test in this study.

2. Distribution of N0 in measured HRTFs

2.1. *HRTF measurement*

Five subjects' HRTFs were measured. The measurement procedure is the following. The optimized Aoshima's time stretched pulse (OATSP) method[9] was used to measure impulse responses from sound sources to both ears. The TSP responses were convolved with an inverse TSP signal to obtain impulse responses. Similarly, impulse responses were measured at a head center without a head. HRTFs were calculated as IFFT of head-related impulse responses (HRIRs) by normalizing the impulse responses at both ears with head-center impulse responses. A loudspeaker array consisting of 18 loudspeakers located between -80 deg and 90 deg elevation angle on a hemicircle, rotates at a 5 deg interval (Fig. 2). The TSP length was 8,192 points. Measurements were performed using a 48-kHz sampling rate and 16-bit quantization. The TSP signals were DA-converted (MD-8D72-133; Pavec) and were amplified (MCA8050; Biamp Corp.). Amplified TSP signals were emitted using a loudspeaker (8 cm full range type, FE83E; Fostex Co.). The distance from each loudspeaker to center of the head was

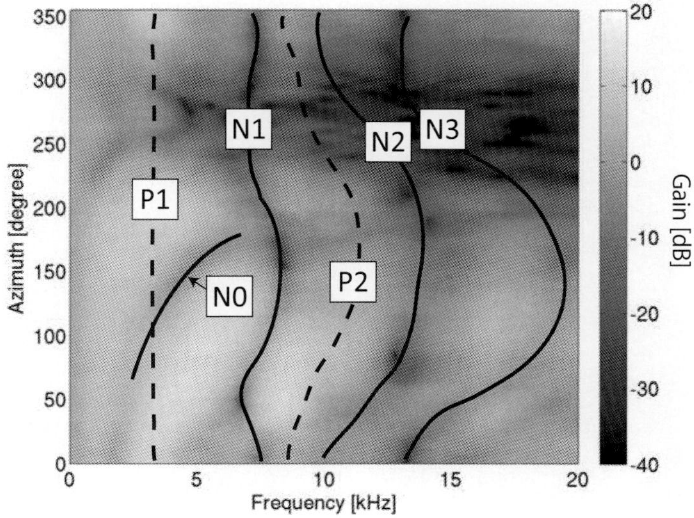

Fig. 1. HRTFs at -30 deg elevation angle. Colors represent gains in decibels. Lines show the spectral peaks and notches.

1.5 m. The TSP signals were detected using earplug type microphones, made with electret condenser microphones (FG3329; Knowles) and silicone impression material placed in both ear canals. Acquired TSP signals that were sent to microphone preamplifiers (MA-2016C; Thinknet) were low-pass filtered (FV-665; NF Corp.). They were AD-converted (MD-8D72-133; Pavec). For each source position, four TSP responses were measured and averaged to obtain an impulse response.

2.2. *Detection of spectral notches*

To examine the spatial distribution of N0, spectral notches were detected by application of auto regressive moving average (ARMA) modeling and dynamic programming (DP).[10] Using the ARMA model, transfer functions can be represented as an infinite impulse response filter, written as

$$H(z) = \frac{\prod_{i=1}^{Q}(1 - q_i z^{-1})}{\prod_{i=1}^{P}(1 - p_i z^{-1})},$$

where P and Q respectively denote the orders of poles and zeros. In this study, P and Q were set as 30 because a correlation coefficient reportedly existing between measured HRTF and ARMA-modeled HRTF saturates for

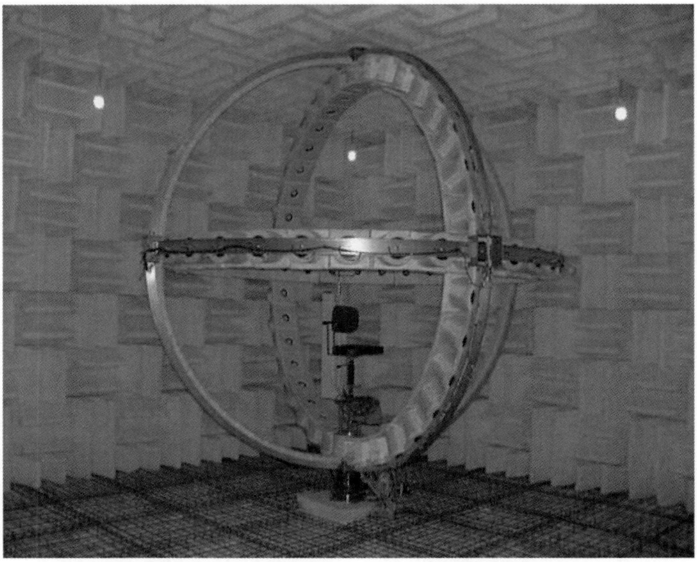

Fig. 2. Spherical loudspeaker array for HRTF measurements in an anechoic chamber at Research Institute of Electrical Communication, Tohoku University

orders larger than 30.[11] Next, DP was applied to find neighboring zeros, which produce N0.

2.3. *Spatial distribution of N0*

Figure 3 demonstrates a spatial distribution of frequency of N0 appearance for both ears of the five subjects (10 ears in total). Black color shows that N0 appears more frequently, and the maximum value is 10. A perspective is located behind the head. The left hemisphere corresponds to the ipsilateral source positions. The results for the right ears are symmetrically mirrored with respect to the median sagittal plane so that results for both the ears can be demonstrated in a single sphere. The figure shows that N0 appears mainly at 130–180 deg of azimuth and from -60 to 10 degrees of elevation.

Figure 4 depicts the HRTF spectrum variation of one subject for elevation angle of -20 deg as an example of measurement. The abscissa and ordinate respectively represent the frequency in Hertz and horizontal angle in degrees. The angle of 0 deg corresponds to a frontal source position and positive values represent counter-clockwise rotation. The color bar shows the gain in decibels. Circles in the figures represent zeros, which correspond to the detected N0. Other elevation angles were also analyzed. The results

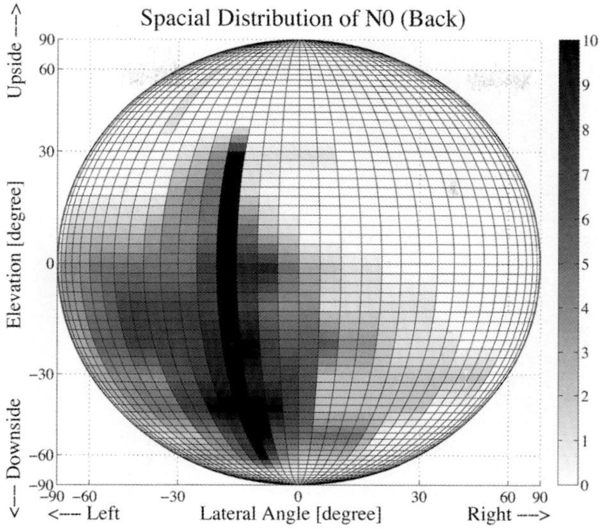

Fig. 3. Spatial distribution of appearing instances of N0 for both the ears of five subjects (10 ears in total). Black color represents that N0 appears more frequently. The left hemisphere shows the ipsilateral source positions. The figure shows that N0 appears more frequently for lower rear sound sources on the ipsilateral side: behind the ears.

show that N0 clearly appears at 0 deg elevation and at lower elevation. In particular, at elevation angles between 0 deg and -40 deg, the frequencies of N0 (from 5 kHz to 7 kHz) overlap those of N1 at some horizontal directions. Therefore, at an elevation where N0 appears clearly, N1 is not always the first spectral notch in some cases, especially for the lower rear position.

3. Effects of N0 in sound localization

3.1. *Methods*

Because of the existence of N0, HRTF spectral features for lower rear source positions are distinctively different from those for other source positions. In other words, N0 and N1 are, respectively, the first and second notches in such source positions. For controlling virtual sound image direction with high accuracy, it can be inferred that the effect of N0 should be taken into account. Therefore, the effects of existence of N0 were examined with a sound localization test.

The test was performed in a soundproof room. Four listeners with normal hearing ability participated in the test. Each set of the listener's own

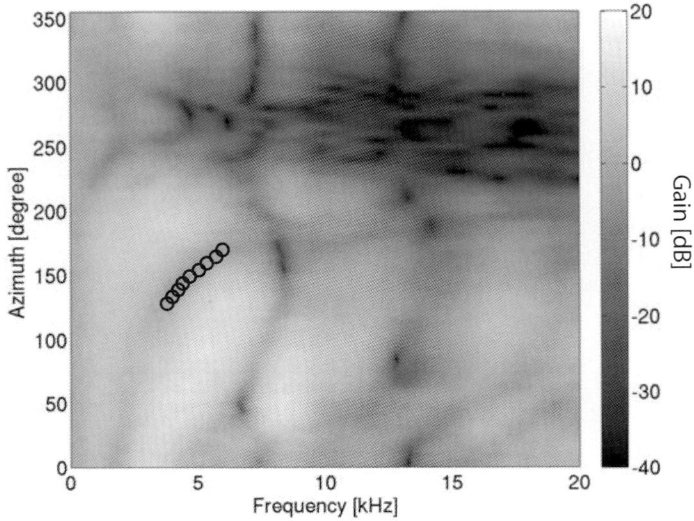

Fig. 4. Detected spectral notches N0 on the horizontal plane for -20-deg elevation angle. Circles in the figure represent N0.

HRTFs was measured using the loudspeaker array described above. A virtual sound image was synthesized by convolving each listener's HRTFs with the sound source signal. The position of a virtual sound image was selected randomly from 10 positions, which included the combinations among two elevation (-10 deg and -30 deg) and five rear horizontal angles (-120, -150, -180, -210, and -240 deg) (Fig. 5). The sound source signal was pink noise with 48-kHz sampling frequency and 2.7-s duration. In fact, HRTFs of three types were used in the experiment: the original set of listener's own HRTFs (*Original* condition), a set in which N0 was eliminated from frequency spectra (*noN0* condition), and a set in which N0 had been deepened by 10 dB from the original level (*Deep* condition). The stimuli were presented as binaural signals via a set of headphones. We conducted another listening test for the horizontal perceptual angle in the same manner in advance and confirmed that no statistically significant differences existed among the three HRTF types. In this experiment, listeners were asked to report the perceived elevation with precision of 10 deg.

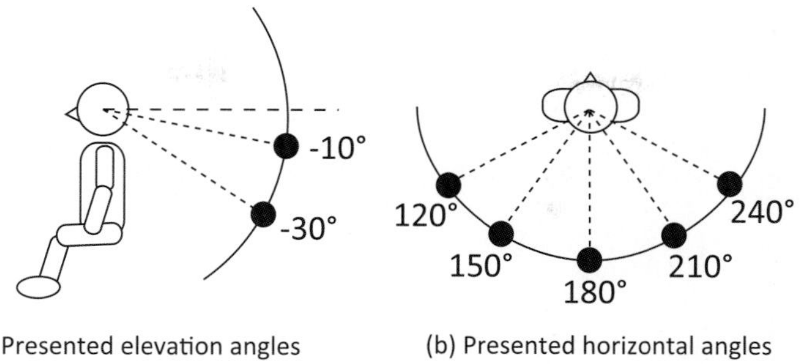

Fig. 5. Positions of virtual sound sources presented in localization test: (a) presented elevation angles and (b) presented horizontal angles. 10 positions in all (combination among (a) and (b) were used in the experiment.)

3.2. *Results and discussion*

Figures 6–9 present results of perceived elevation for each type of HRTF for four listeners. Sound images were presented at -10 deg elevation. For 3 out of 4 listeners, the perceived elevation in *noN0* condition was significantly higher than other conditions. On the other hand, the perceived elevation was lowest in the *Deep* condition. Irrespective of presented horizontal angles and listeners, these tendencies could be confirmed in other listeners' results. This indicates that N0 is an important spectral cues to perceive the elevation of lower rear sound sources. Blauert identified frequency band that influcences the direction of auditory images, known as directional bands.[12] In our study, N0 appears at lower rear position of the listener's head, and the frequency position of N0 is distributed from 5 kHz to 7 kHz. This frequency distribution of N0 includes the directional band, which yields elevated perception. If we can consider that a spectral notch (N0) weakens the directional band, listeners might perceive lower elevation because of N0 depth.

4. Conclusion

As described in this paper, we investigated spectral features of HRTFs for the lower rear source position, as characterized with the distinctive spectral notch N0. The results of the analysis and the listening test confirmed that N0 is an important cue for localizing the elevation of lower rear sound sources.

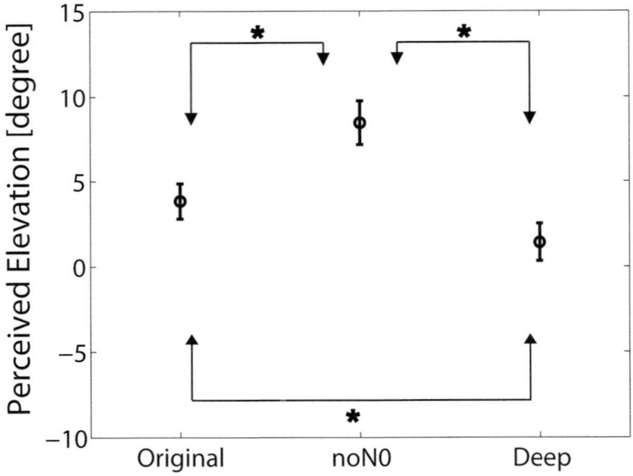

Fig. 6. Perceived elevation angles for each condition of listener JM. The presented elevation angle was -10 deg; results were averaged among all horizontal angles. Asterisks (*) denote that significant differences exist between the corresponding pairs of condition($p < 0.05$).

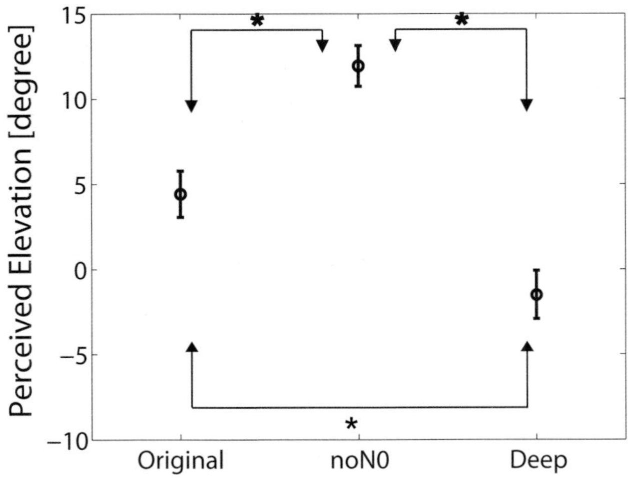

Fig. 7. Perceived elevation angles for each condition of listener SF. The presented elevation angle was -10 deg; and results were averaged among all horizontal angles. Asterisks (*) denote a significant differences existing between the corresponding pairs of condition($p < 0.05$).

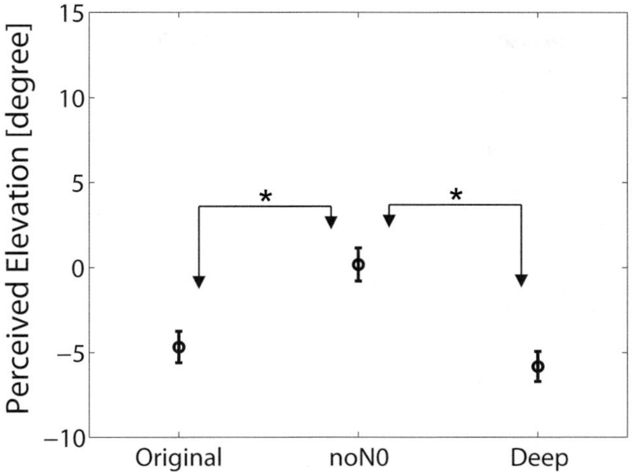

Fig. 8. Perceived elevation angles for each condition of listener YM. The oresented elevation angle was -10 deg; results were averaged among all horizontal angles. Asterisks (*) denote that a significant differences existing between the corresponding pairs of condition($p < 0.05$).

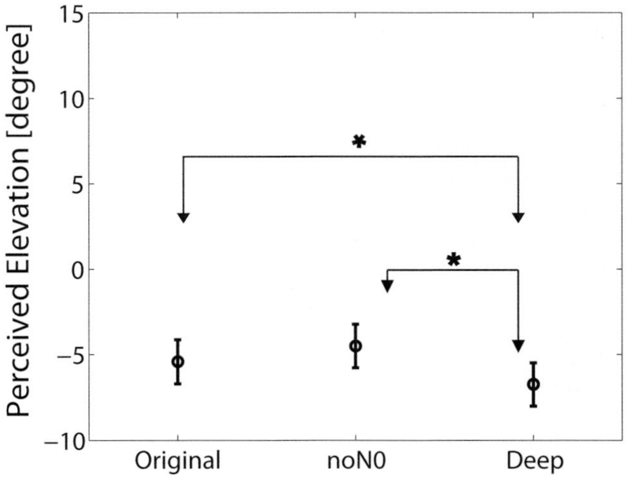

Fig. 9. Perceived elevation angles for each condition of listener TM. The presented elevation angle was -10 deg; results were averaged among all horizontal angles. Asterisks (*) denote a significant differences existing between the corresponding pairs of condition($p < 0.05$).

5. Acknowledgment

This work was supported by a Grant-in-Aid for Scientific Research (C) (Grant No. 20500110) of JSPS, Japan.

References

1. J. Blauert, *Spatial Hearing.* (The MIT Press, Cambridge, 1997)
2. Y. Iwaya and Y. Suzuki, Numerical analysis of the effects of pinna shape and position on the characteristics of head-related transfer functions, *J. Acoust. Soc. Am.* Vol.123, 3279, presented in Acoustics '08 Paris (2008)
3. K. Watanabe, K. Ozawa, Y. Iwaya, Y. Suzuki, and K. Aso, Estimation of interaural level difference based on anthropometry, *J. Acoust. Soc. Am.* Vol.122, 2832–2841 (2007)
4. F. Asano, Y. Suzuki, and T. Sone, Role of spectral cues in median plane localization, *J. Acoust. Soc. Am.* Vol.88, 159–168 (1990)
5. A. Huan and B. May, Sound orientation behavior in cats: II mid-frequency spectral cues for sound localization, *J. Acoust. Soc. Am.* Vol.100, No.2, 1070–1080 (1996)
6. J. Middlebrooks, Spectral shape cues for sound localization, in *Binaural and spatial hearing in real and virtual environments*, eds. T.R. Anderson and R.H. Gilkey (Lawrence Erlbaum Associates, Mahwah, 1997), ch. 4
7. K. Iida, M. Itoh, A. Itagaki, and M. Morimoto, Median plane localization using a parametric model of the head-related transfer function based on spectral cues, *Applied Acoustics* Vol.68, 835–850 (2007)
8. S. Sekimoto, R. Ogasawara, Y. Iwaya, Y. Suzuki, and S. Takane, Numerical investigation of effects of head sizes and ear positions on head-related transfer functions, in *Proc. Japan–China Joint Conf. on Acoust. 2007*, P-1-12, Sendai (2007)
9. Y. Suzuki, F. Asano, H.Y. Kim, and T. Sone, An optimum computer-generated pulse signal suitable for the measurement of very long impulse responses, *J. Acoust. Soc. Am.* Vol.97, 1119–1123 (1995)
10. L. Rabiner and B.H. Juang, *Fundamentals of Speech Recognition* (Prentice Hall, New Jersey, 1993)
11. S. Sekimoto, Y. Iwaya, and Y. Suzuki, Directional and individual variation of distinctive notches in HRTFs, in *Proc. AES Japan Conf.* P15, Fukuoka (2006) in Japanese
12. J. Blauert, Sound localization in the median plane, *Acustica* Vol.22, 205–213 (1969/70)

COMPUTER SIMULATION OF KEMAR'S HEAD-RELATED TRANSFER FUNCTIONS: VERIFICATION WITH MEASUREMENTS AND ACOUSTIC EFFECTS OF MODIFYING HEAD SHAPE AND PINNA CONCAVITY

P.MOKHTARI[†], H.TAKEMOTO, R.NISHIMURA and H.KATO

National Institute of Information and Communications Technology (NICT),
2-2-2 Hikaridai, Seikacho, Kyoto 619-0288 Japan
[†]E-mail: parham@nict.go.jp

The Finite-Difference Time-Domain (FDTD) method was used to simulate Head-Related Transfer Functions (HRTFs) of KEMAR (Knowles Electronics Manikin for Acoustic Research). Compared with KEMAR's measured HRTFs available in the CIPIC database, the mean spectral mismatch on a linear frequency scale up to 14 kHz was 2.3 dB; this was better than the 3.1 dB mismatch between KEMAR's left- and right-ear measured HRTFs. FDTD simulations were then run to clarify acoustic consequences of smoothing away facial features, morphing the head shape towards a sphere, and either exaggerating or reducing the degree of concavity/convexity in the folds and cavities of the pinna.

1. Introduction

Head-Related Transfer Functions (HRTFs) have long been central to the principles and applications of spatial hearing. It is therefore important to develop and evaluate methods for accurate numerical simulation of HRTFs, both as a means of bypassing the difficulties of acoustic measurements and to gain deeper insights into the physical correlates of HRTF features.

HRTF simulations have been reported with the Boundary Element Method (BEM) [1]-[6] and the Finite-Difference Time-Domain (FDTD) method [7]-[10]. However, few studies compared simulation results extensively and quantitatively with acoustic measurements: Katz [1] made comparisons up to 6 kHz and at six spatial locations; Mokhtari *et al.* [9] did so up to 20 kHz in the front hemisphere with measurements on human subjects; Iwaya *et al.* [4] did so qualitatively in the horizontal plane; and Takemoto *et al.* [10] did so up to 20 kHz for one, frontal location. More extensive results were reported in two recent studies using accelerated BEM simulations: Kreuzer *et al.* [5] compared HRTFs in horizontal and median planes with measurements on a human subject up to 16 kHz, but showed reasonable congruence only up to about 7 kHz; and Gumerov *et al.* [6] made side-by-side qualitative comparisons of manikins' HRTFs in a selection of 2D planes, concluding that reasonably accurate simulations can be achieved up to 20 kHz given a suitably accurate 3D mesh of the manikin's head and pinnae.

Compared with BEM, advantages of the FDTD method include direct use of volumetric data without having to generate a surface mesh, and simulation across a wide frequency range in a single run. Here, with a spatial resolution of 2 mm and a minimum of 12 voxels per wavelength to ensure accuracy and stability, we evaluate FDTD simulation of KEMAR's [11] HRTFs against measurements in the CIPIC database [12], up to 14 kHz and across a wide array of locations.

2. Head Geometry Data and Acoustic Simulation Methods

To avoid the possibility of errors caused by involuntary head movements when measuring HRTFs of human subjects, we used the KEMAR manikin [11]. Right-half head shape data were kindly provided by Dr. Yuvi Kahana who had measured (with laser scanners) 3D coordinates of points on the surface of KEMAR's head and neck, and at much finer resolution KEMAR's "DB60" right pinna [3]. We used a semi-automatic procedure to optimally align the pinna on the right-half head, and the combined data were reflected about the median plane to obtain a left-right symmetric whole head with pinnae. The surface data were then volumetrized on a 3D grid of uniform resolution 2 mm, by setting the physical properties at each voxel (i.e., sound speed and material density) to either *air* (outside the manikin) or *water* (inside the manikin). The ear canals were blocked by replacing air voxels therein with water.

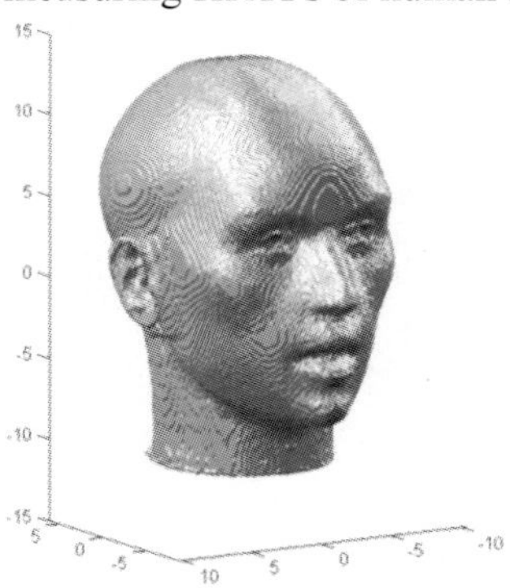

Figure 1. A view of KEMAR with DB60 pinnae. Coordinates are in cm; origin at head center.

FDTD simulations on the volumetric data were run with the computation domain bounded by an optimum Perfectly Matched Layer (oPML) [13] and with the bottom of KEMAR's neck inserted partially into the lower oPML [14]. Computation and memory load were reduced by placing a point source at the (right) pinna according to the acoustic reciprocity principle; and by including the Kirchhoff-Helmholtz integral equation within the FDTD algorithm [15] to obtain pressure signals (of duration 5 ms) at 1250 spatial locations as specified in the CIPIC data [12], a distance 1 m from the head center. FDTD temporal and spatial derivatives were accurate to 2nd and 4th order, respectively.

3. Verification of Simulated HRTFs with Measurements

For evaluation of the simulation results, reference HRTFs were calculated from the right-ear impulse responses (of duration 4.5 ms) of subject 165 ("KEMAR with small pinnae") in the CIPIC data [12]. Differences between corresponding pairs of measured and simulated HRTFs were then quantified by an objective Spectral Distance (SD), defined as the mean absolute difference (in dB) between a pair of HRTF log-magnitude spectra on a linear frequency scale from 500 Hz to 14 kHz. As simulated HRTFs depend somewhat on the precise position of the reciprocal source at the pinna [9][10][5], the mean SD over all spatial locations was used as a criterion to find the best source position (i.e., that which yielded the lowest mean SD) close to the blocked entrance of the ear canal. To match the CIPIC data, spatial locations around the manikin's head were specified in terms of inter-aural polar coordinates, where *azimuth* is the lateral angle away from the median plane with -90° at left and +90° at right, and *elevation* is the angle along each cone-of-confusion with 0° at the front, 90° above, and 180° behind.

Figures 2(a) and (b) show all 1250 simulated right-ear HRTFs superimposed on the right-ear CIPIC measurements. Qualitatively, a good spectral match was obtained over most spatial locations, not only in regard to the frequencies and relative amplitudes of major peaks and notches, but also the gradual rise in overall amplitude towards the ipsilateral (right) side. These observations were confirmed quantitatively by an overall mean SD of only 2.3 dB between simulated and measured HRTFs on a linear frequency scale (an even lower mean SD of 1.8 dB was obtained on a log-frequency scale, owing to a general rise in mismatch with increasing frequency). This best match was achieved with the source placed within the cavum concha and close to the tragus, 7 mm from the closed entrance of the ear canal and 1 mm from the anterior wall of the concha.

Considering the general characteristics of the HRTFs, Takemoto *et al.* [16] recently showed that the first three peaks are resonances of the entire set of pinna cavities including the concha, cymba, triangular and scaphoid fossae. The first main notch at around 7-8 kHz and rising in frequency with elevation angle in the median plane, was attributed to various types of acoustic interference causing cancellation of the incoming sound wave at the concha [16]. These and other spectral features were here simulated fairly accurately, although there appear a number of mismatches in the depth (or bandwidth) of some notches.

The spatial distribution of SDs shown in Fig. 3, indicates that the smallest spectral mismatches occurred across a broad range of locations above KEMAR's head. Figure 2 confirms that the relatively broad peaks and shallow notches of these overhead HRTFs were indeed obtained accurately with simulation.

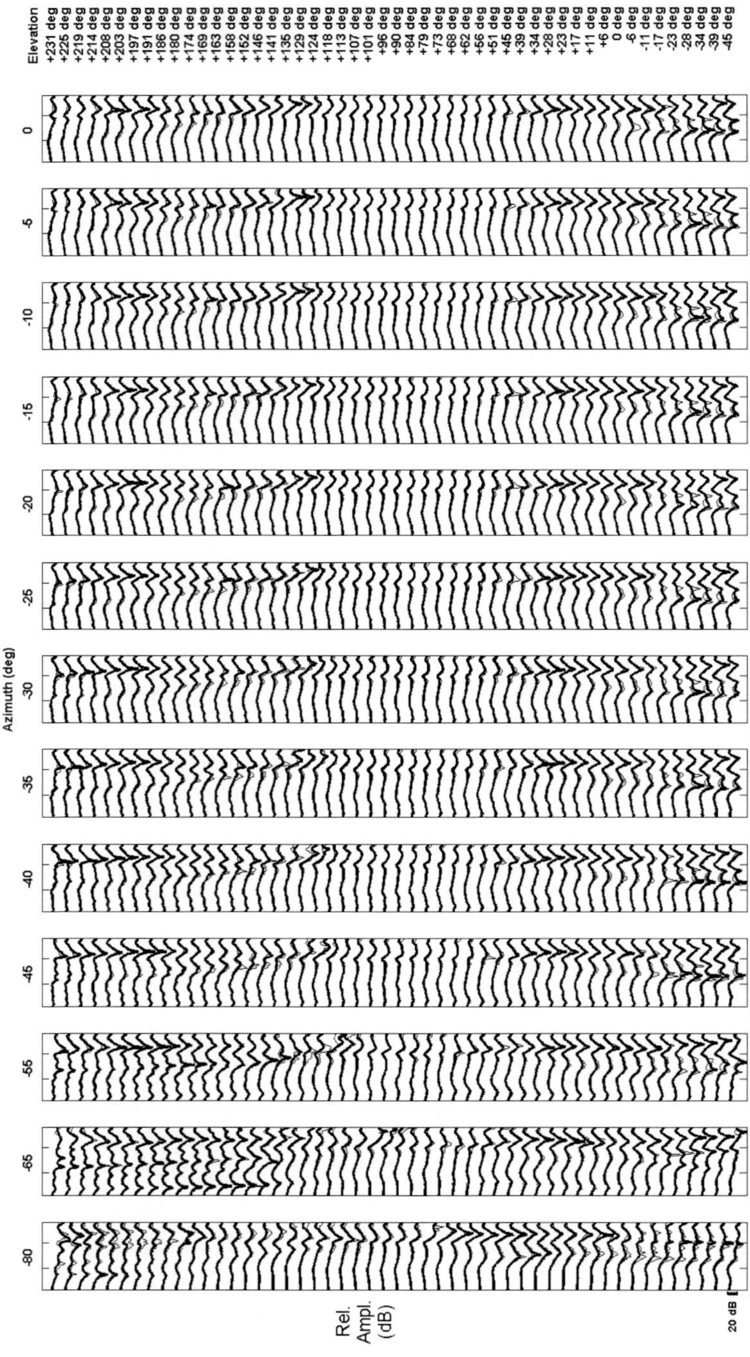

Figure 2(a). KEMAR's right-ear contralateral and median-plane HRTFs. <u>Bold lines</u>: simulation. <u>Thin lines</u>: CIPIC measurements.

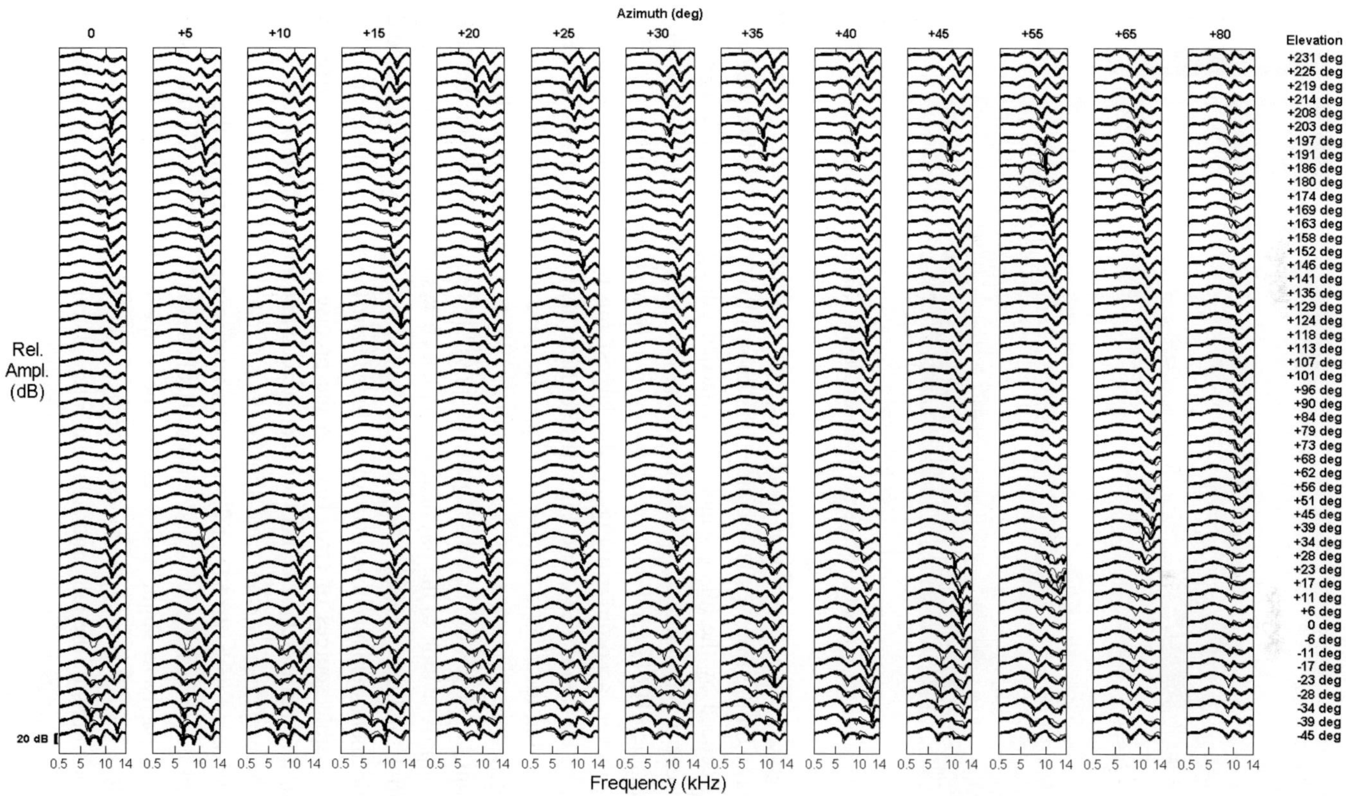

Figure 2(b). KEMAR's right-ear ipsilateral and median-plane HRTFs. Bold lines: simulation. Thin lines: CIPIC measurements.

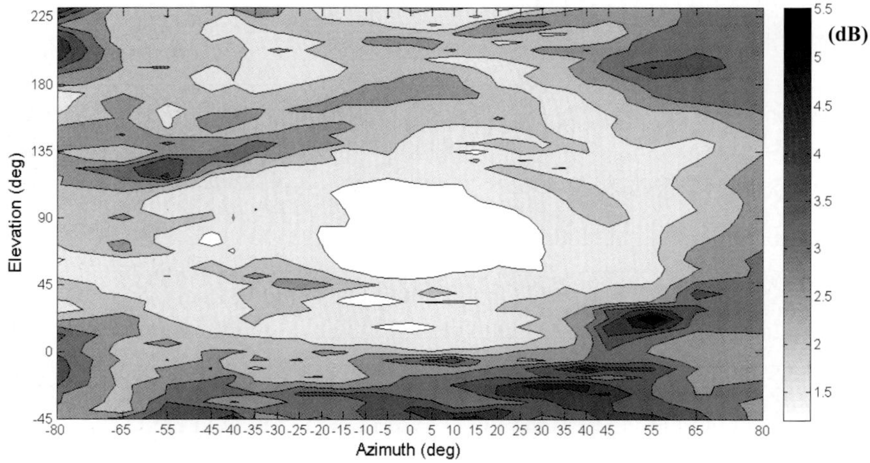

Figure 3. Spatial distribution of spectral distances between simulated and measured HRTFs.

Figure 3 also shows that most differences occurred at lower elevations, especially on the front ipsilateral side. It is known that HRTFs at low elevations are influenced by the torso, which was attached during the measurements [17] but absent in our simulations. The ripples at low frequencies visible in some of the measured HRTFs in Fig. 2 are also known to be torso effects [18].

However, according to Fig. 3, the largest mismatch occurred at azimuth +55° and elevation +23°; and Fig. 2(b) confirms that for only a few HRTFs at and near that location, a number of higher-frequency peaks and notches appear in the measurements but not in simulation. While it is presently difficult to ascertain the origin of such discrepancies, possible causes include the relatively rough approximation to the physically complex curves of the pinna even at 2 mm resolution, and the presence of environmental factors that may have rendered the acoustic measurements less than ideal. Indeed, in the face of these and many other possible sources of error (e.g., in the precise alignment of the head and pinnae, and the exact placement of the sources and microphones), the overall match between measurement and simulation is remarkably good.

Compared with our SD of 2.3 dB, it is interesting to note that the mean SD between the measured right- and left-ear HRTFs was 3.1 dB. While this can be attributed partly to slight differences in the shape of the left and right pinnae fitted to KEMAR during the measurements [12][17] and perhaps to differences in the positions of the left- and right-ear microphones, it is encouraging that our simulation results were objectively closer to the intended right-ear HRTFs, than the two sets of measurements were to each other.

4. Head and Pinna Shape Modifications

Having validated the simulation results against acoustic measurements, five additional simulations were run after selectively modifying the shape of KEMAR's head and pinnae. As in the study by Iwaya *et al.* [4], these preliminary experiments were carried out towards a better understanding of the causal links between head and pinna geometry and acoustic HRTF features.

First, the right-ear HRTFs were obtained by simulation after occluding all the main cavities of the left pinna. Compared with the original simulation results, the HRTFs were found to be unaffected (mean SD of 0.0 dB). This indicates that the contralateral pinna cavities are acoustically inconsequential.

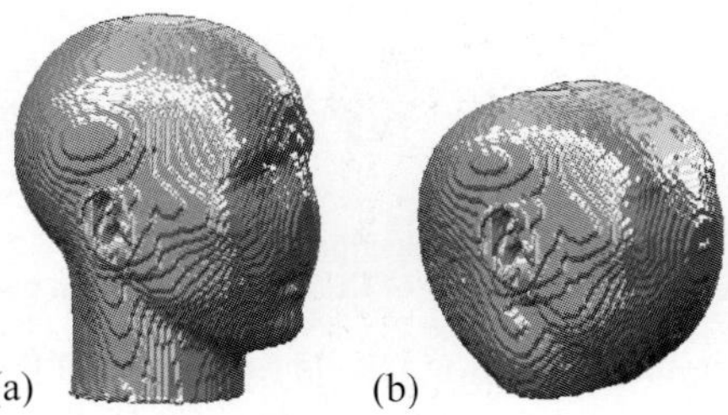
(a) (b)

Figure 4. KEMAR with modified head shapes: (a) smoothed face, (b) quasi-spherical head.

Second, as shown in Fig. 4(a) KEMAR's facial features, including the eyes, nose and mouth, were smoothed away by morphing the face towards a sphere of radius 10.3 cm, with the degree of morphing gradually fading to merge with the original head shape near the forehead, cheekbones, and the bottom of the chin. The resulting HRTFs (dotted lines in Fig. 5) were only slightly different from the

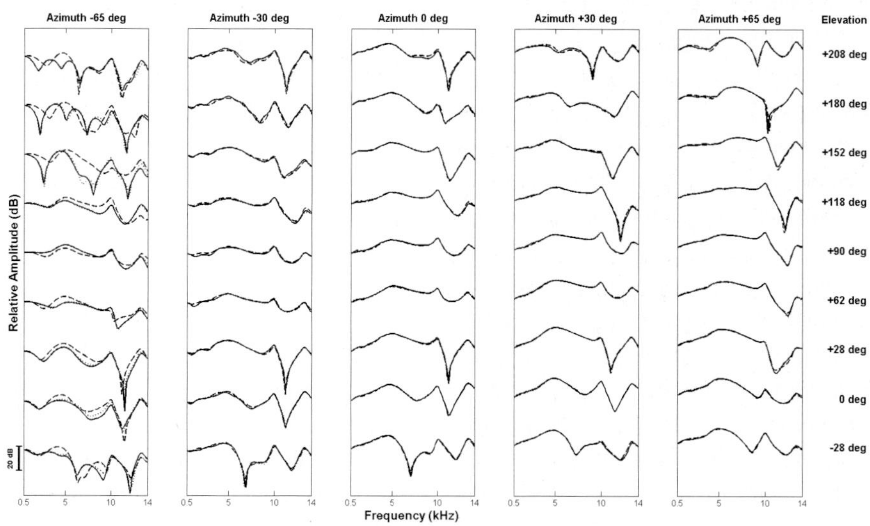

Figure 5. Simulated HRTFs showing the effects of modifying only the head shape. Solid lines: unmodified KEMAR. Dotted lines: smoothed face (as in Fig. 4(a)). Dashed lines: quasi-spherical head (as in Fig. 4(b)).

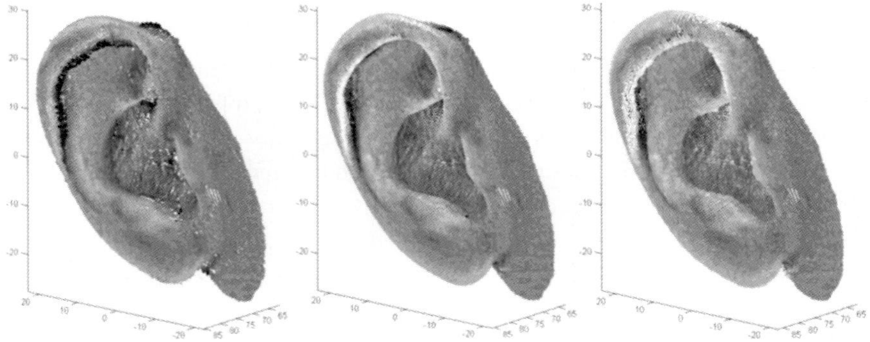

Figure 6. Original (center panel) and modified versions (left panel: reduced features, right panel: exaggerated features) of KEMAR's DB60 right pinna. Coordinates in mm; origin at head center.

original simulation results (mean SD of 0.4 dB), with small differences in certain mid-frequency features in front and behind, but only on the far contralateral side.

Third, as shown in Fig. 4(b) KEMAR's head was morphed towards a sphere of radius 9 cm, with a perfectly circular cross-section in the median plane and a gradual reduction in the degree of morphing towards the sides of the head approaching the pinnae, and with the original pinnae retained. HRTFs simulated with this quasi-spherical head (dashed lines in Fig. 5) differed from the original simulation (mean SD of 1.2 dB) almost entirely on the far contralateral side, where indeed the modified head shadow would be expected to have the largest influence.

For the fourth and fifth simulations, KEMAR's original head shape was restored and only the pinnae were modified by either exaggerating or reducing the depth of cavities and the thickness of folds. For exaggeration, each point on the pinna surface was shifted inward/outward along its normal vector by a small distance (maximally 1 mm) proportional to a measure of the local concavity/ convexity at that point (see right panel of Fig. 6). Conversely, shifting all surface points in the opposite direction (maximally 2 mm) rendered a pinna with slightly occluded cavities and thinner folds (see left panel of Fig. 6). Compared with the original simulation results, the pinna-related peaks and notches of the resulting HRTFs were found to be shifted down in frequency as expected in the case of exaggerated pinna features with deeper resonant cavities (dotted lines in Fig. 7, mean SD of 3.1 dB); and shifted up in frequency as expected for the opposite case of reducing the pinna's local curvature and depths of cavities (dashed lines in Fig. 7, mean SD of 2.0 dB). In contrast, the upper-left of Fig. 7 shows that the low-frequency notches on the contralateral side that were caused mainly by the head shape (indeed, the HRTF features that were affected the most in Fig. 5 by morphing to a quasi-spherical head) were unaffected by these pinna-only

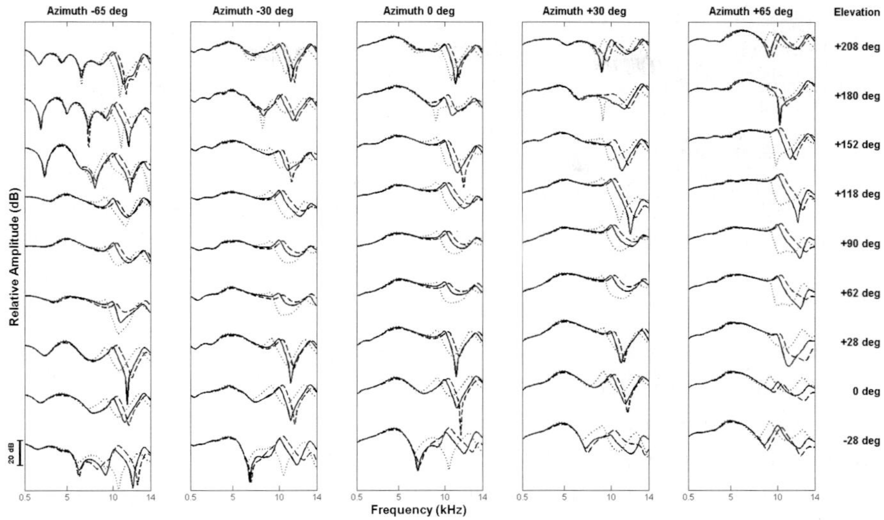

Figure 7. Simulated HRTFs showing the effects of modifying only the pinna features. Solid lines: unmodified KEMAR. Dashed lines: reduced features (cf. left panel of Fig. 6). Dotted lines: exaggerated features (cf. right panel of Fig. 6).

modifications. These results suggest that acoustic effects of head and pinna shape are, to a certain extent, separable in the spatio-frequency domain. The results in Fig. 7 also confirm [16] that the pinna cavities' resonances generate not only peaks but also notches of HRTFs.

5. Conclusions

Complementing the recent BEM studies of Kreuzer *et al.* [5] and Gumerov *et al.* [6], our FDTD results provide a promising validation of HRTFs obtained by numerical simulation. The close match between simulated and measured HRTFs was made possible thanks to the high fidelity of both the 3D head geometry data and our FDTD simulation methods. In turn, the validation results enabled a preliminary investigation into the relations between the head and pinna geometry and the acoustic features of HRTFs.

Clearly, this is one small step towards clarifying the physical factors underlying HRTF features such as the first few peaks and notches that are known to be critical spectral cues in human spatial hearing [19]. Accurate acoustic simulation can now be used as a powerful tool to investigate the mechanisms for generating HRTF features [16], and to identify contributions of the detailed pinna surface geometry to the overall HRTF pattern [20]. Furthermore, simulation can be applied to head and pinna geometries of a number of individuals in order to learn how and why HRTFs vary from person to person. Such knowledge will lead to more effective HRTF personalization needed for truly realistic, virtual 3D audio.

Acknowledgment

We are grateful to Dr. Yuvi Kahana for kindly providing his measurements of the KEMAR head and DB60 pinna surfaces.

References

1. B. F. G. Katz, "Boundary element method calculation of individual head-related transfer function. II. Impedance effects and comparisons to real measurements," *J. Acoust. Soc. Am.* **110**(5), 2449-2455 (2001).
2. M. Otani and S. Ise, "Fast calculation system specialized for head-related transfer function based on boundary element method," *J. Acoust. Soc. Am.* **119**(5), 2589-2598 (2006).
3. Y. Kahana and P. A. Nelson, "Boundary element simulations of the transfer function of human heads and baffled pinnae using accurate geometric models," *J. Sound and Vibration* **300**(3-5), 552-579 (2007).
4. Y. Iwaya and Y. Suzuki, "Numerical analysis of effects of pinna's shape/ position on characteristics of head-related transfer functions," *J. Acoust. Soc. Am.* **123**(5, Pt.2), 3297 (2008).
5. W. Kreuzer, P. Majdak and Z. Chen, "Fast multipole boundary element method to calculate head-related transfer functions for a wide frequency range," *J. Acoust. Soc. Am.* **126**(3), 1280-1290 (2009).
6. N. A. Gumerov, A. E. O'Donovan, R. Duraiswami and D. N. Zotkin, "Computation of the head-related transfer function via the fast multipole accelerated boundary element method and its spherical harmonic representation," *J. Acoust. Soc. Am.* **127**(1), 370-386 (2010).
7. T. Xiao and Q. H. Liu, "Finite difference computation of head-related transfer function for human hearing," *J. Acoust. Soc. Am.* **113**(5), 2434-2441 (2003).
8. M. Nakazawa and A. Nishikata, "Development of sound localization system with tube earphone using human head model with ear canal," *IEICE Trans. Fundamentals* **E88-A**(12), 3584-3592 (2005).
9. P. Mokhtari, H. Takemoto, R. Nishimura and H. Kato, "Comparison of simulated and measured HRTFs: FDTD simulation using MRI head data," *Audio Engineering Society 123rd Convention*, New York, USA, Paper 7240, 12 pp. (2007).

10. H. Takemoto, R. Nishimura, P. Mokhtari and H. Kato, "Comparison of head related transfer functions obtained by numerical simulation and measurement of physical head model," *Autumn Meet. Acoust. Soc. Japan*, Paper 1-8-11, 607-610 (in Japanese) (2008).
11. M. D. Burkhard and R. M. Sachs, "Anthropometric manikin for acoustic research," *J. Acoust. Soc. Am.* **58**(1), 214-222 (1975).
12. V. R. Algazi, R. O. Duda, D. M. Thompson and C. Avendano, "The CIPIC HRTF database," *IEEE Workshop on Applications of Signal Processing to Audio and Acoustics*, 99-102 (2001).
13. P. Mokhtari, H. Takemoto, R. Nishimura and H. Kato, "Optimum loss factor for a perfectly matched layer in finite-difference time-domain acoustic simulation," *IEEE Trans. Audio, Speech & Language Processing* **18**(5), 1068-1071 (2010).
14. P. Mokhtari, H. Takemoto, R. Nishimura and H. Kato, "Computer simulation of HRTFs for personalization of 3D audio," *IEEE 2nd International Symp. on Universal Communication*, Osaka, 435-440 (2008).
15. P. Mokhtari, H. Takemoto, R. Nishimura and H. Kato, "Efficient computation of HRTFs at any distance by FDTD simulation with near to far field transformation," *Autumn Meet. Acoust. Soc. Japan*, Paper 1-8-12, 611-614 (2008).
16. H. Takemoto, P. Mokhtari, H. Kato, R. Nishimura and K. Iida, "Pressure distribution patterns on the pinna at spectral peak and notch frequencies of head-related transfer functions in the median plane," *in the same volume*, (2010).
17. R. O. Duda and V. R. Algazi, *personal communication* (2009).
18. V. R. Algazi, R. O. Duda, R. Duraiswami, N. A. Gumerov and Z. Tang, "Approximating the head-related transfer function using simple geometric models of the head and torso," *J. Acoust. Soc. Am.* **112**(5), 2053-2064 (2002).
19. K. Iida, M. Itoh, A. Itagaki and M. Morimoto, "Median plane localization using a parametric model of the head-related transfer function based on spectral cues," *Applied Acoustics* **68**, 835-850 (2007).
20. P. Mokhtari, H. Takemoto, R. Nishimura and H. Kato, "Acoustic sensitivity to micro-perturbations of KEMAR's pinna surface geometry," *International Congress on Acoustics*, Sydney, Australia, 8 pp. (2010).

ESTIMATION OF WHOLE WAVEFORM OF HEAD-RELATED IMPULSE RESPONSES BASED ON AUTO REGRESSIVE MODEL FOR THEIR ACQUISITION WITHOUT ANECHOIC ENVIRONMENT

S. TAKANE*

Department of Systems Science and Technology, Akita Prefectural University,
84-4 Ebinokuchi, Tsuchiya, Yurihonjo, Akita 015-0055, Japan
**E-mail: takane@akita-pu.ac.jp*

The method of estimating Head-Related Impulse Responses (HRIRs) from impulse responses involving the reflection and/or noise was investigated. This method is based on the Auto Regressive (AR) model of HRIR. The AR coefficients are estimated using the conventional Linear Prediction (LP) algorithm. The data used for estimation are the part of the response which is regarded as the direct component of the impulse response from the sound source to the ear, *i.e.*, the part of the HRIR. A computer simulation in which the method is applied to the estimation of the HRIRs of the Head-and-Torso Simulator (HATS) showed the following: (1) SDR was improved in some directions examined when the order of AR coefficients is half of the cutout points. Otherwise the improvement of SDR was not observed. Although the reason for such occasional improvement remains unclear, the improvement demonstrates that the proposed method can better estimate the part of the HRIRs lost by reflection and/or noise; (2) The number of samples used for the computation of AR coefficients greatly affects the estimation accuracy. Ideally, the whole waveform of the HRIR is useful for estimation of the AR coefficients, which shows that the proposed method brings about accurate estimation in an ideal situation.

Keywords: HRIR; Auto Regressive model; Linear prediction; Reflection; Noise

1. Introduction

Individual Head-Related Transfer Functions (HRTFs) or their inverse Fourier transforms, designated as Head-Related Impulse Responses (HRIRs) in whole directions are ideally required to synthesize 3D sound imagery accurately. A free field such as an anechoic chamber must be prepared for acquisition of the HRTFs based purely on their measurement. Such a field is difficult to prepare because of the various costs that are necessary for its construction. If the acoustic engineers and/or researchers intend to measure the HRTFs of a certain subject, then they must take that person to the free field constructed somewhere.

The problems described above often complicate the measurement of impulse responses from the sound source to the subject's ears conducted in an ordinary sound field with reflections and/or noise. Acquisition of the HRIRs is possible with this measurement if the direct component can be extracted from the measured response. However, the reflection components frequently overlap the direct component. In this condition, the entire waveform of the HRIR cannot be obtained using a simple time window. Takane *et al.* discussed this problem based on the characteristics of poles and zeros of the measured transfer functions; the objective accuracy of the extracted HRIRs was not sufficient.[1] A method based on Linear Prediction (LP) from a part of the impulse response, measured using its reflection component cut out, was also proposed by Takane *et al.*[2] The estimated HRIRs were evaluated via hearing experiments.[3] However, the performance of the estimation done using proposed method is not confirmed because there exist some parameters affecting the performance. As described in this paper, two such parameters —the order of the coefficients and number of samples for estimation of the coefficients— are subjects of the performance evaluation. The effectiveness and the limitations of the proposed method are investigated.

2. Some theoretical discussion of our proposed method

2.1. *Outline of the proposed method*

In this section, an outline of the proposed method for the estimation of HRIR from a part of the impulse response measured in an ordinary sound field is described.[2]

The procedures of the proposed method can be listed as follows:

(1) The initial delay of the obtained impulse response is extracted.[4,5] The remaining part consists of the direct and the reflection components.
(2) After the point at which the reflection component seems to appear, the response is cut out. This point is defined as a "cutout point," denoted as N_C hereinafter. The value N_C corresponds to the maximum number of sample points available for the estimation of the AR coefficients to be stated next. The processed impulse response is denoted as $h_C(n)$, of which sample length is N_C.
(3) The following Auto Regressive (AR) model is assumed as

$$h_C(n) = -\sum_{j=1}^{N_{AR}} a_j^{(M)} h_C(n-j) \quad (0 \le n < N_C), \tag{1}$$

where $a_j^{(M)} (j = 1, \cdots, N_{AR})$ is a set of AR coefficients, and (M) denotes how many samples of $h_C(n)$ are used for computation of the AR coefficients.
(4) The cutout part of the impulse response is predicted using the coefficients

calculated in (3) and the following equation:

$$\hat{h}(n) = -\sum_{j=1}^{N_{\mathrm{AR}}} a_j^{(M)} \tilde{h}(n-j) \quad (n \geq N_{\mathrm{C}}), \tag{2}$$

where $\hat{h}(n)$ is the estimated part of the impulse response, and $\tilde{h}(n)$ is defined as follows:

$$\tilde{h}(n) = \begin{cases} h_{\mathrm{C}}(n), & 0 \leq n < N_{\mathrm{C}}, \\ \hat{h}(n), & n \geq N_{\mathrm{C}}. \end{cases} \tag{3}$$

The sequence $\tilde{h}(n)$ is the estimated result, except for restoring the initial delay extracted in (1).

Equation (1) resembles the Linear Prediction (LP), when one regards the sequence $h_{\mathrm{C}}(n)$ as a signal. Here it is a part of the impulse response, thereby this model is simply called the "AR model" hereinafter.

The above-described set of procedures simplify the following discussion because the reflection component is not used, meaning that the performance of the procedures is unaffected by the property of the reflection components, except for its delay relative to the direct one. Although the reflection component must be cut out to execute this method effectively, the appearance point of that component might not necessarily be known accurately. The only limitation is that some samples corresponding to the direct component are available.

2.2. *Theoretical discussion on validity of the proposed method*

In many studies, the HRTF or HRIR is represented by the ARMA model (see[12] for example). The general input–output relation of the ARMA model is represented as

$$y(n) = -\sum_{k=1}^{N_{\mathrm{AR}}} a_k y(n-k) + \sum_{k=0}^{N_{\mathrm{MA}}} b_k x(n-k), \tag{4}$$

where $x(n)$ is the input signal, $y(n)$ is the output signal, $a_k(\mathrm{k}=1,\cdots,N_{\mathrm{AR}})$ and $b_k(\mathrm{k}=0,\cdots,N_{\mathrm{MA}})$ respectively denote the AR and the MA coefficients. The HRIR $h(n)$ is obtained as the output signal with a unit impulse as the input. This means the following relation:

$$h(n) = -\sum_{k=1}^{N_{\mathrm{AR}}} a_k h(n-k) + \sum_{k=0}^{N_{\mathrm{MA}}} b_k \delta(n-k), \tag{5}$$

where $\delta(n)$ indicates the unit impulse. The following relation is valid if the relation $n > N_{\mathrm{MA}}$ stands:

$$h(n) = -\sum_{k=1}^{N_{\mathrm{AR}}} a_k h(n-k), \quad n > N_{\mathrm{MA}}. \tag{6}$$

This equation shows that the AR model is valid after the N_{MA}-th samples. The ARMA model of the HRIR has larger order of MA coefficients than that of AR coefficients in some studies (ex.[5]), and the proposed AR model is not valid for the estimation of HRIRs if N_{MA} is larger than the cutout point, N_{C}. To make use of this relation, it is necessary that the HRIR can be modeled under the condition that the order of MA coefficients, N_{MA}, is small.

As an example, the left-ear HRIR and HRTF (source direction: front) of a HATS (Samrai; Koken Co. Ltd.) are compared between the original and the ARMA-modeled. Fig. 1 shows the results. In this figure, the order of AR and MA coefficients is: $N_{\mathrm{AR}} = 40$ and $N_{\mathrm{MA}} = 10$. There is apparently little difference in the HRIRs (Fig. 1(a)), although there is slight difference between the original and the modeled HRTFs (Fig. 1(b)) less than 1 kHz. The Signal-to-Distortion Ratio (SDR), which is defined as follows, is calculated as the simple measure of the similarity of two signals in time domain:

$$\mathrm{SDR}(h,\hat{h}) = 10\log_{10}\left[\frac{\sum_{n=0}^{N-1} h(n)^2}{\sum_{n=0}^{N-1}\left\{h(n)-\hat{h}(n)\right\}^2}\right] \quad [\mathrm{dB}], \tag{7}$$

where h and $\hat{h}$ denote the original and the modeled response. For responses in Fig. 1, the SDR is about 25.8 dB. The results show that the HRIR can be modeled with acceptable accuracy when N_{MA} is small.

However, this result does not mean that the AR-modeling and the estimation of the HRIR is well performed because the AR coefficients in Eq. (6) must be estimated from a part of the impulse response in the case dealt with in this paper. This result is only the least precondition under which the proposed method works.

3. Effects of parameters on the performance of the proposed method

To perform the proposed method in the best way, the following two main parameters seem to be important:

- **Order of AR coefficients (N_{AR}):** This clearly determines how many past samples are related to the current samples. Considering that the HRIR has

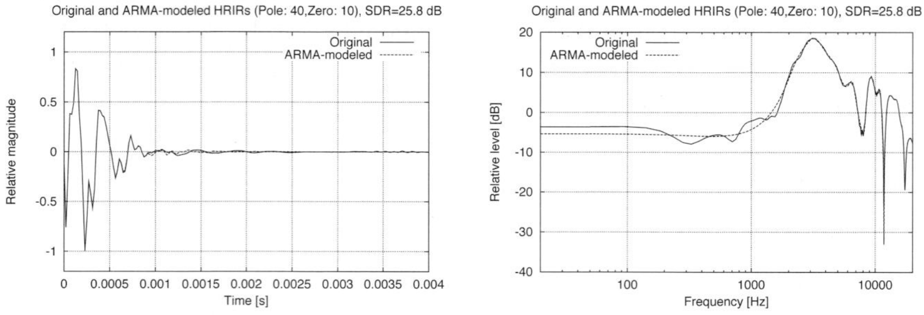

(a) HRIR (b) HRTF

Fig. 1. Example comparison of HRIR and HRTF between original and ARMA-modeled with their orders, N_{AR} and N_{MA}, set to 40 and 10, respectively (Initial delay extracted, source direction: front, left ear, HATS: Samrai, Koken Co. Ltd.).

duration of several milliseconds (except the initial delay), N_{AR} greater than the number of samples corresponding to this duration is sufficient and capable of modeling any type of response.

- **Samples used for estimation of AR coefficients (M):** It is likely that this parameter strongly affects the performance of the estimated AR coefficients. In this study, this must be small compared to the restoration of the audio signals.[6]

The proposed method is examined with the set of HRIRs of a HATS, changing the parameters listed above. In this article, the AR coefficients were computed using the Burg method[7–9] implemented in GNU Octave.[10,11]

3.1. *Used set of HRIRs*

A set of HRIRs of a HATS (Samrai, Koken Co. Ltd.) measured in the anechoic room of the author's laboratory was used. According to the proposed method, the part of the impulse response including the component other than the direct one is cut out, and the remaining part is used with the AR model. Therefore, a set of the HRIRs measured in free field is useful for the examination of the effectiveness of the proposed method.

The right-ear HRIR of the sound source on the horizontal plane was measured in the anechoic room of the authors' laboratory with the sampling frequency of 48 kHz. Distance from the sound source to the HATS was fixed at 1.5 m, and the azimuth was changed from 0 to 357.5 deg in 2.5 deg interval, where 0 deg corresponds to the front of the subject, 90 deg corresponds to the right (ipsilateral side), and 270 deg corresponds to the left (contralateral side). An HRTF is usually defined as the ratio of sound pressure at the listener's ear position to that at the point corresponding to the center of the listener's head (with the absence of the listener); HRIR is defined as its inverse Fourier transform.[12] However, the impulse response from the sound source to the listener's ear is used as the HRIR in this paper for convenience' sake.

3.2. *Results and discussion*

3.2.1. *Effect of N_{AR}*

Figures 2(a)–2(b) show the azimuthal change in SDR with the order of AR coefficients varied in various cutout points (N_C). In each panel, "Zero-padded" shows the SDR between the original response and the one with zero samples padded after the cutout points. Therefore, no estimation is conducted. It is also readily apparent that the higher SDR in the "zero-padded" case is achieved with the cutout point N_C increased because this means that the remaining number of samples is increased. Seeing Fig. 2, the SDR is almost the same or slightly improved or degraded, compared with that of the zero-padded response at most of the examined order of AR coefficients.

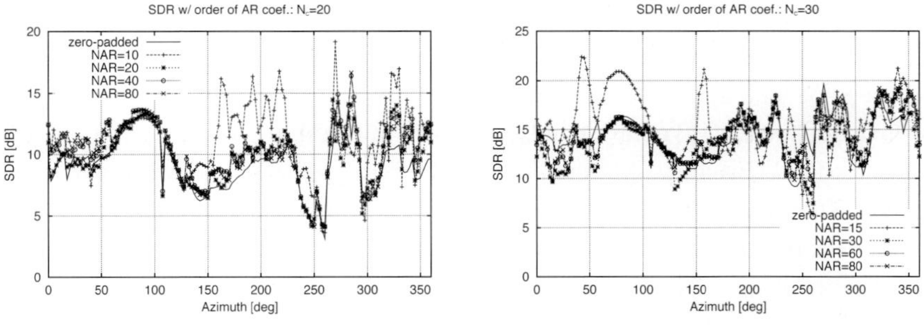

(a) $N_C = 20$ (b) $N_C = 30$

Fig. 2. Change of SDR in azimuth ($N_{AR} = 20, 30$).

At some directions, however, the improvements in SDR are visible when the order of the AR coefficient is half of the cutout points N_C. As an example of this case, the HRIR and the HRTF with source direction equal to 40 deg, $N_C = 30$, $N_{AR} = 15, M = 30$ are shown in Fig. 3. Improvement of SDR is about 7 dB in this case. To discuss the reason for this improvement, Spectral Distortion (SD) was calculated; it is defined as shown below.

$$\mathrm{SD}(H, \hat{H}, N_l, N_h) = \sqrt{\frac{1}{N_h - N_l + 1} \sum_{k=N_l}^{N_h} \left(20 \log_{10} \left| \frac{\hat{H}(k)}{H(k)} \right| \right)^2} \quad [\mathrm{dB}], \qquad (8)$$

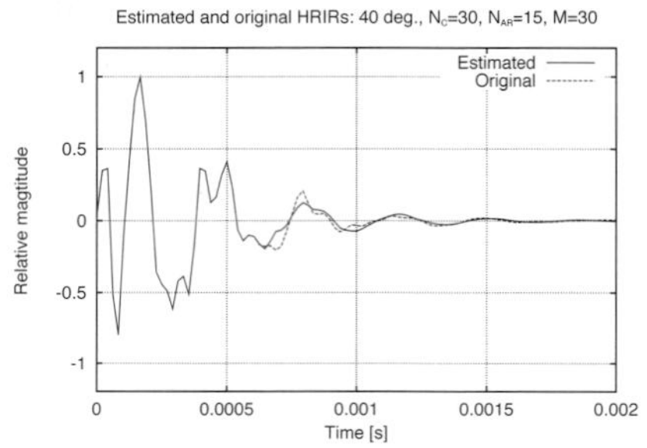

(a) HRIR

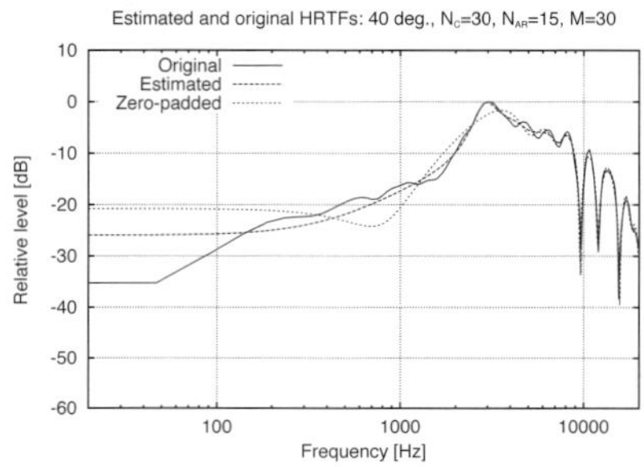

(b) HRTF

Fig. 3. Comparison between original and estimated characteristics (HRIR and HRTF, azimuth: 40 deg, $N_{\mathrm{C}} = 30$, $N_{\mathrm{AR}} = 15$, $M = 30$).

where H and $\hat{H}$ respectively denote the original and the estimated HRTFs, and the N_l and N_h respectively denote the sample number corresponding to the lower and the higher frequency. The SD is the quantity computed in frequency domain; its value close to 0 dB means that the corresponding frequency components of the estimated HRTF are close to those of the original HRTF. This calculation enables discussion of the estimation accuracy in the frequency domain. The SD of the zero-padded and the estimated HRTFs with the original one as the reference were calculated in 1/3 octave bands, as shown in Fig. 4. Figure 4(a) shows that the estimated HRTF achieve SD smaller than the zero-padded one in whole frequency bands. This means that the estimation method brings about higher accuracy not only in the time domain but also in frequency domain. The reason for this improvement when $N_{\mathrm{AR}} = N_{\mathrm{C}}/2$ occurs is currently unclear. Furthermore, improvement cannot be observed at some azimuthal angles. Figure 4(b) presents an example of the estimation method not properly working. The SD of the estimated HRTF has its larger value in a low-frequency region less than 1 kHz, although the low SD is achieved in the high-frequency region. Almost all responses with and without the estimation commonly have the same tendency of the larger SD in a low-frequency region and smaller in the high-frequency region. That seems to imply two points:

- the cutting procedure produces the error mainly in a low-frequency region,
- it is generally inadequate for the simple AR model to restore the cutout part accurately, at least in the simulating conditions used for the study explained here.

Frequency characteristics of HRTFs have some peaks, but they are not so sharp generally. The AR model properly works when the plant generates the periodical response,[9] but the results in this article shows that the HRTFs are not completely fittable to the AR model, although this model works well in some azimuths.

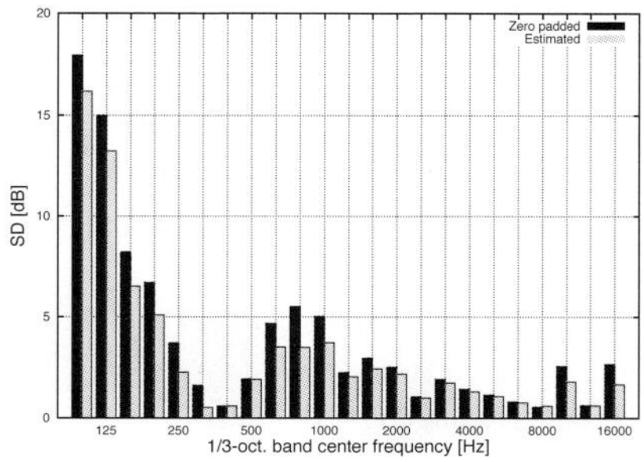

(a) Azimuth: 40 deg

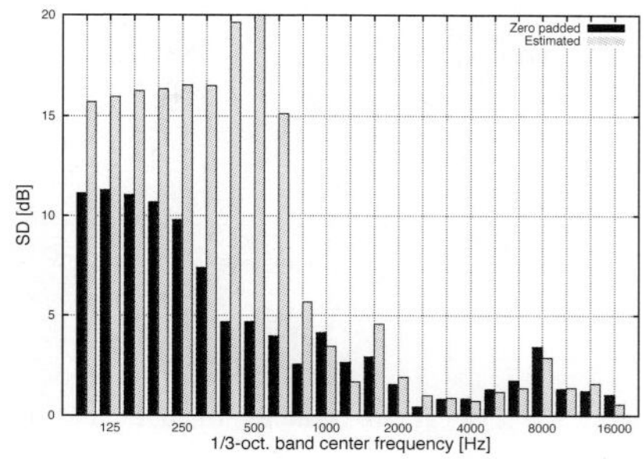

(b) Azimuth: 0 deg

Fig. 4. SD of estimated and original HRTFs calculated in 1/3 octave band ($N_C = 30$, $N_{AR} = 15$, $M = 30$).

3.2.2. *Effect of M*

Here the effect of samples used for the estimation of AR coefficients, M is discussed. Fig. 5 describes the change of SDR in azimuth with various values of M in the case of $N_C = 20$ and $N_{AR} = 10, 80$. Fig. 5(a) is the case of $N_{AR} = 10$. This panel shows that the tendency does not change greatly with the increase of M. However, in the case of $N_{AR} = 80$, the SDR is greatly improved as M becomes large, as shown in Fig. 5(b). This result reflects that the good estimates of AR coefficients are obtainable when the whole of the response is available. In this sense, our previous results[2,3] were almost the best ones obtainable using the proposed method. However, this seldom has availability in practical use because it requires no estimation when the whole of the response is obtained. Practical effectiveness of the proposed method should be evaluated using the results presented in Fig. 2, in which the improvement in SDR seems slight.

3.2.3. *Some comments on subjective evaluation*

Nishino *et al.* discussed the relation of the interpolation accuracy of the HRTFs to this influence in the subjective evaluation with SD (in overall frequency range).[4] Then they stated that the SD of 2 dB was sufficient. According to this criterion, the prediction accuracy with the proposed method seems insufficient. On the other hand, Hanazawa *et al.* discussed the relation of the subjective evaluation to the

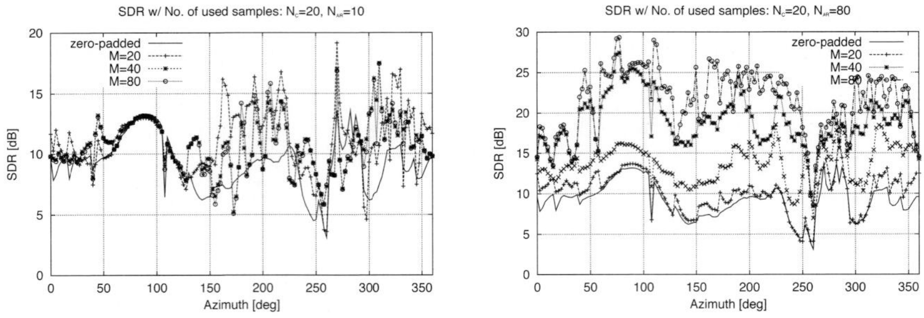

(a) $N_{\mathrm{AR}} = 10$ (b) $N_{\mathrm{AR}} = 80$

Fig. 5. Change of SDR in azimuth with various values of M ($N_{\mathrm{C}} = 20$).

interpolation accuracy with SDR.[13] They concluded that the SDR of 7 dB was the threshold to perceive the difference between the original and the interpolated HRIRs. According to this criterion, the HRIRs in some azimuth satisfies it, as shown in Fig. 2. The relation between the accuracy of the prediction results, such as SD and SDR, and its influence in subjective evaluation is not clear from these two studies. It must also be noted that those studies were aimed at the evaluation of their interpolation methods, whereas the target for evaluation in this article is the estimation method outlined in **2**. The previously described feature that the estimated results involves the estimation error in low-frequency region might also greatly affect the subjective evaluation considering the property of the auditory system,[12] but it might be possible that the estimation improves the subjective features to some degree. Consequently, the subjective evaluation of the HRIRs is regarded as necessary to clarify the effectiveness and limitations of the method introduced here.

4. Concluding remarks

As described in this paper, the estimation method of HRIRs from the impulse responses measured in ordinary sound field, as proposed in our previous study based on the AR model of the HRIRs, was investigated in two parameters affecting the performance. The results obtained using the set of HRIRs of a Head-and-Torso Simulator (HATS) showed that: (1) SDR was improved in some directions when the order of AR coefficients is a half of the cutout points; otherwise, the improvement of SDR was not observed. Although the reason for such occasional improvement is not clear, it demonstrates that the proposed method can better estimate the part of the HRIRs lost by reflection and/or noise; (2) The number of samples used for the computation of AR coefficients greatly affects the estimation accuracy. Ideally, the whole waveform of the HRIR can be used for the estimation, which indicates that the proposed method brings about the accurate estimation in ideal situation. Analysis of the estimated results in greater detail and the subjective evaluation of the estimated HRIRs are the main subjects of future works.

References

1. S. Takane and T. Sone, "A fundamental study on the extraction of Head-Related Transfer Functions from binaural room transfer functions," Proc. RADS04 (Room Acoustics: Design and Science 2004), No. 20, 1-4(2004).
2. S. Takane, K. Abe and S. Sato, "Acquisition of individual HRTFs from measured transfer functions in ordinary sound field," Proc. Japan–China Joint Conference of Acoustics, 1-6(2007).
3. S. Takane, M. Nabatame, K. Abe, K. Watanabe and S. Sato, "Subjective evaluation of HRIRs linearly predicted from impulse responses measured in ordinary sound field," Proc. Audio Eng. Soc. Japan Conference (Poster No. 18), 1-8(2008).
4. T. Nishino, S. Kajita, K. Takeda and F. Itakura, "Interpolating head related transfer functions in median plane," Proc. 1999 IEEE Workshop on Applications of Signal Processing to Audio and Acoustics, New Paltz, NY, 17-20(1999).
5. K. Watanabe, S. Takane and Y. Suzuki, "A novel interpolation method of HRTFs based on the Common-Acoustical-Pole and Zero model," Acta Acustica united with Acustica, **91**, 958-966(2005).
6. S. J. Godsil and P. J. W. Rayner, *Digital audio restoration* (Springer Verlag, London, 1998).
7. D. S. G. Pollock, *A handbook of time-series analysis, signal processing and dynamics* (Academic Press, London, 1999).
8. J. .P. Prokias and D. G. Manolakis, *Digital signal processing – principles, algorithms, and applications,* The fourth edition (Prentice Hall, New Jersey, 2007).
9. P. P. Vaidyanathan, *The theory of linear prediction* (Morgan & Claypool Publishers, California, 2008).
10. GNU Octave, `http://www.octave.org/`
11. Octave-Forge, `http://octave.sourceforge.net/`
12. J. Blauert, *Spatial Hearing*, Revised edition (MIT Press, Cambridge, 1999).
13. K. Hanazawa, H. Yanagawa and M. Matsumoto, "Subjective evaluations of interpolated binaural impulse responses and their interpolation accuracies," Proc. Spring Research Meeting of the Acoust. Soc. Jpn., 3-Q-25, 677-678(2006.3), in Japanese.

ANALYSIS OF MEASURED HEAD-RELATED TRANSFER FUNCTIONS BASED ON SPATIO-TEMPORAL FREQUENCY CHARACTERISTICS

Y. MORIMOTO

Graduate School of Information Science, Nagoya University,
Nagoya, Aichi, Japan
E-mail: y-morimoto@sp.m.is.nagoya-u.ac.jp

T. NISHINO

EcoTopia Science Institute, Nagoya University,
Nagoya, Aichi, Japan
E-mail: nishino@nagoya-u.jp

K. TAKEDA

Graduate School of Information Science, Nagoya University,
Nagoya, Aichi, Japan
E-mail: kazuya.takeda@nagoya-u.jp

A head-related transfer function (HRTF) is an acoustic transfer function between a sound source and the entrance of the ear canal. Since an HRTF is defined as an acoustic function of time and the sound source's location, the spatio-temporal frequency characteristics of HRTFs can be visualized and analyzed by multi-dimensional Fourier transform in time and space. In our experiments, we investigate the basic property of spatio-temporal frequency characteristics based on the coordinate system of measuring HRTFs on the horizontal plane and analyze the measured HRTFs. Moreover, the reverberation and pinnae effects of the spatio-temporal frequency characteristics are examined. As a result, the spatio-temporal spectrum components of HRTFs were mostly concentrated in specific frequency bands, and the spectrum components of other factors appeared clearly.

Keywords: HRTF; Spatio-temporal frequency; Fourier transform; Visualization

1. Introduction

We can perform sound localization using such acoustic features as interaural level differences, interaural time differences, and frequency characteristics.[1] These acoustic features between a sound source and the entrance of the ear canal are represented by an acoustic transfer function called the head-related transfer function (HRTF). HRTF controls a sound image by convolving with the source signal. There have been many HRTF studies, and many application areas also exist, such as for 3D sound reproduction systems.

HRTFs are usually obtained by measurement with a head-and-torso simulator or a human, and measured HRTFs are generally visualized by a figure whose axes correspond to the angle of the sound source and temporal frequency. Most previous works employed frequency analysis in the time domain to emphasize the time variation, and these conventional figures illustrate the differences in HRTFs among the sound source directions. However, the acoustic transfer function defined in the wave equation is a function of time and space. Based on this definition, not only the time variation but also the space variation should be analyzed. Space variation is caused in HRTFs when the sound source or the listener moves. By analyzing HRTFs over both variations, the relation between HRTF and space can be clarified to simplify the extraction of factors attributed to space.

In this study, we analyze measured HRTFs with spatio-temporal frequency analysis, which is frequency analysis in time and space.[2] Even though such analysis is predominantly used in visual neurology fields, some studies have applied it to acoustic waves.[3] For example, a broadband beamforming approach with a filter in the spatio-temporal frequency domain has been described by Nishikawa *et al.*[4] and an efficient method for coding the sound field using spatio-temporal frequency analysis has been proposed by Pinto and Vetterli.[5] Ajdler *et al.*[6] calculated the specific frequency band concentrated by HRTF spectrum components in the spatio-temporal frequency domain for interpolating HRTFs. To apply the same analysis as the previous work to HRTFs measured under various conditions, we first describe the spatio-temporal frequency characteristics for the sampling data on the coordinate system of ordinary HRTF measurements by using a spherical-head model. Then, we analyze the measured HRTFs and investigate the similarities and differences with their spatio-temporal frequency characteristics for the numerical results. Such factors as reverberation and pinnae that might influence the spatio-temporal frequency characteristics are also examined.

2. Spatio-temporal frequency analysis

A head-related impulse response (HRIR), which is an impulse response between a sound source and an ear and the time domain representation of an HRTF, can be defined by the sound source direction and the time. HRIRs on the horizontal plane are used in this study, and their spatio-temporal frequency analysis is calculated by a two-dimensional Fourier transform on the time and the source direction:

$$\underline{H}(\phi_l, f_k) = \sum_{m=0}^{M-1} \sum_{n=0}^{N-1} h(\theta_m, t_n) e^{-j2\pi\left(\frac{lm}{M} + \frac{kn}{N}\right)}, \quad (1)$$

$$\underline{H}(\phi_{M-l}, f_{N-k}) = \underline{H}^*(\phi_l, f_k) \quad (-M/2 \leq l < M/2,\ 0 < k \leq N/2). \quad (2)$$

$\underline{H}(\phi_l, f_k)$ is the spatio-temporal frequency spectrum. f_k is the k-th temporal frequency, and we call ϕ_l the l-th azimuthal frequency. ϕ_l denotes the periodicity of HRIRs when the sound source direction was changed and its unit is [rad^{-1}], based on the definition of the Fourier transform. M is the number of angular sampling points, and it means dividing the circumference into M points. N is the length of the HRIRs. A complex conjugate is denoted by $*$ in Eq. (2), which indicates that the spatio-temporal frequency characteristics have origin symmetry.

2.1. *Analysis of impulse responses with spherical-head model*

Impulse responses obtained by spatial sampling on the coordinate system of HRTF measurements are analyzed numerically. A sound source is usually arranged on the circumference of a circle whose center is the middle point between both tragions. Although HRTFs are usually measured as a transfer function from the sound source to the entrance of the ear canal, a sound source and a received point can be replaced mutually by the reciprocity theorem. Therefore, HRTF can be regarded as a transfer function between a sound source located at the entrance of the ear canal and a received point on the circumference, as illustrated in Fig. 1, which shows the set-up of the spatial sampling of the sound field with respect to the right ear's HRTFs.

Figure 2 shows the spatio-temporal frequency spectrum of the impulse responses that were calculated with the spherical-head model. The horizontal and vertical axes represent azimuthal frequency ϕ [rad^{-1}] and temporal frequency f [kHz]. The gray-scale indicates magnitude spectrum $20\log_{10}|\underline{H}(\phi, f)|$ [dB]. In this calculation, a sound source and received points were on the horizontal plane, and distance r between them was 1.0 [m], and distance s between a sound source and the center of the sphere was 0.09 [m]. The sampling frequency was 48 [kHz], the sound source interval was 5 [°], and the impulse response length was 512 points (about 0.011 [s]). In Fig.

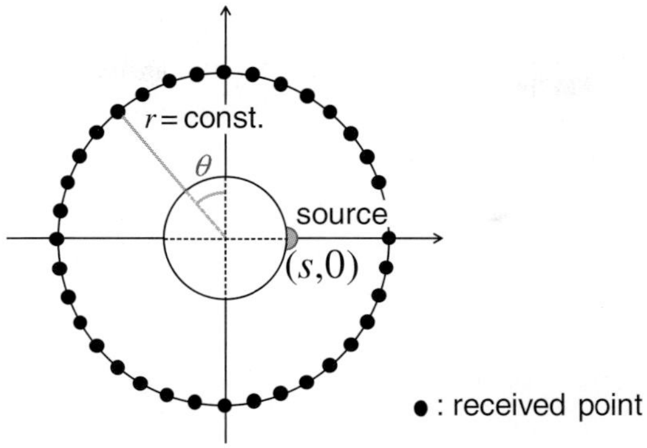

Fig. 1. Set up of spatial sampling for HRIRs.

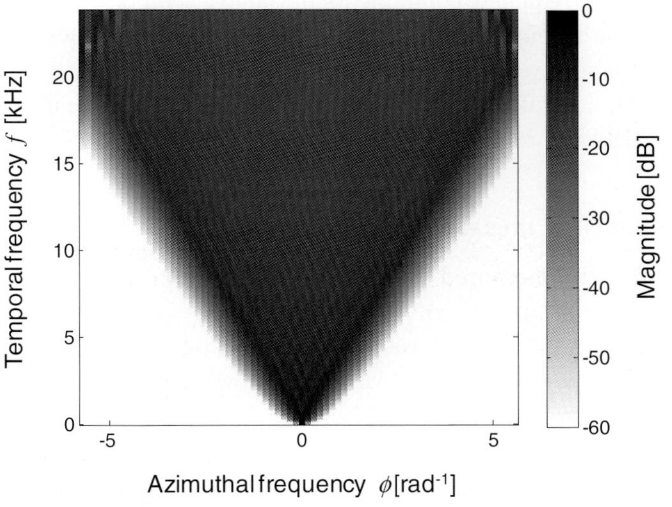

Fig. 2. Spatio-temporal frequency spectrum of impulse responses using spherical-head model.

2, the spectrum has origin symmetry, and most of the components are concentrated in the inverse triangle because of the spatial coordinate system. In Fig. 1, in an HRIR measurement coordinate system, the received points are equally spaced concerning angle θ to the origin, but they are unequally spaced concerning angle θ' to the source. This causes nonlinear variations of the observed signals along with the space, i.e., the source direction. Con-

Table 1. Measurement conditions of HRTF databases.

	Database A	Database B
Head-and-torso simulator	KEMAR	B&K 4128
Distance of sound source [m]	1.2	1.2
Sampling frequency [kHz]	48	44.1

sidering the added element of diffraction by the sphere, the signals are also combined by various space-variation components. Consequently, the azimuthal frequency of certain temporal frequency components is not unique and is spread over a wide frequency band. The azimuthal frequency bandwidth depends on the temporal frequency because of the relation between the sound wavelength and the temporal frequency. The sound wavelength of the low temporal frequency is long, and the variability of the signals is small when the sound source direction is changed. Therefore, there are fewer high-azimuthal frequency components in the low temporal frequency. However, since the sound wavelength is short in the high temporal frequency, azimuthal frequency components exist in the wide bandwidth from low- to high-azimuthal frequency. The spatio-temporal frequency characteristics of measured HRTFs are expected to resemble Fig. 2.

2.2. *Analysis of measured HRTFs*

We analyzed the measured HRTF databases in the spatio-temporal frequency domain. Database A was measured with a head-and-torso simulator (KEMAR) in a soundproof chamber whose reverberation time was 150 [ms]. Database B[7,8] was also measured with a head-and-torso simulator (B&K 4128) in a soundproof chamber whose size was smaller than that of database A. The measurement conditions are shown in Table 1. Differences between databases A and B were the soundproof chamber, the head-and-torso simulator and the sampling frequency. The sound source interval and the impulse response length were the same as in the sphere model.

Figures 3 and 4 show the spatio-temporal frequency spectra obtained from the measured HRTFs. Even though they have the expected triangular shape, the spectrum components are not exactly distributed as in Fig. 2 because of the measurement environment and the shape of the head and the pinnae. Figure 4 shows many components in the whole region, not only in the triangle area. This was probably caused by the reverberation and background noise. HRTFs should essentially be measured in the free sound

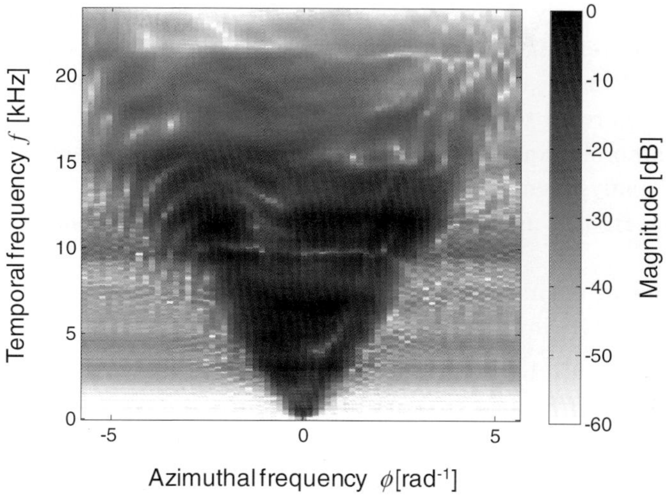

Fig. 3. Spatio-temporal frequency spectrum of measured HRTFs: Database A.

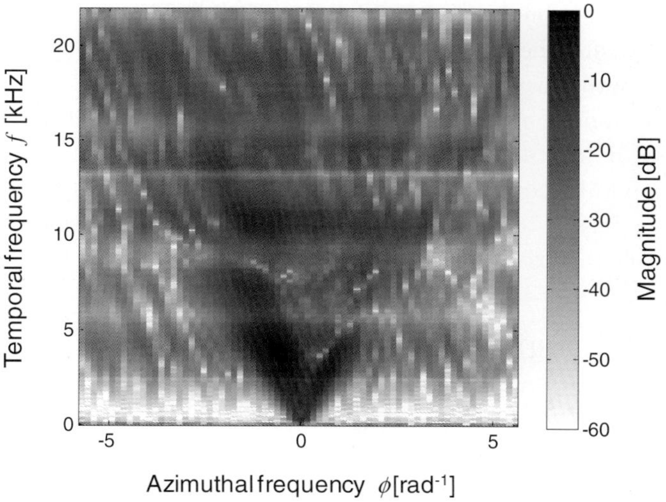

Fig. 4. Spatio-temporal frequency spectrum of measured HRTFs: Database B.

field and should only consist of direct sound waves and diffraction waves caused by the body. However, achieving such required conditions is difficult because reflection waves occur from the equipment located near the direct sound wave path. Noise and such variations in the measurement environment as room temperature also exert an influence. We consider that spatio-temporal frequency analysis can display such extra factors that conventional analysis cannot, and this method is also useful for comparing and confirming HRTFs under different measurement environments.

3. Factors affecting the spatio-temporal frequency characteristics

The effects of reverberation and pinnae on the spatio-temporal frequency characteristics of HRTFs are investigated.

3.1. *Reverberation*

To examine the influence of reverberation on spatio-temporal frequency characteristics, two binaural room impulse responses that were measured under different reverberation conditions in a variable reverberation room were compared. The distance from the sound source (BOSE Acoustimass) to a head-and-torso simulator (B&K 4128) was 1 [m]. The sampling frequency was 48 [kHz], and the measurement direction interval was 5 [°]. The impulse response length was 65,536 points (about 1.365 [s]). $h_0(\theta, t)$ and $h_1(\theta, t)$ denote the impulse responses measured when the reverberation time was set up as 151 [ms] and 459 [ms], respectively. $\underline{H}_0(\phi, f)$ and $\underline{H}_1(\phi, f)$ are spatio-temporal frequency spectra obtained by applying spatio-temporal frequency analysis to both conditions. To compare both spatio-temporal frequency characteristics, the spectral difference calculated by Eq. (3) was used:

$$\underline{\mathrm{SD}}(\phi, f) = 20\log_{10}\frac{|\ \underline{H_1}(\phi, f)\ |}{|\ \underline{H_0}(\phi, f)\ |} \quad [\mathrm{dB}]. \tag{3}$$

The resultant SD is shown in Fig. 5. Black represents the large spectral difference. Since the black portion is concentrated outside the triangle area, as shown in the numerical result (Fig. 2), the reverberations have different spatio-temporal frequency characteristics from direct sound waves.

3.2. *Pinnae*

HRTFs have individuality because their pinnae shape is different. Therefore, analyzing the pinnae influence is important. We analyzed two HRIRs

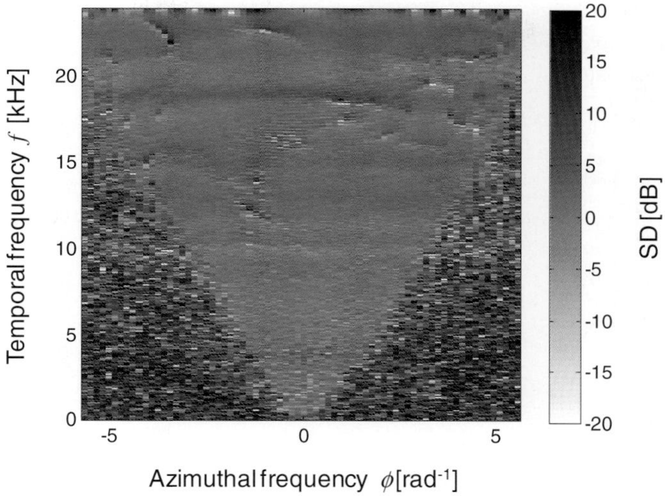

Fig. 5. Spectral difference between $\underline{H}_0(\phi, f)$ and $\underline{H}_1(\phi, f)$ whose reverberation times differ.

measured with and without pinnae. They were measured in the soundproof chamber when the distance between a sound source and a head-and-torso simulator (KEMAR) was 1 [m]. The sampling frequency was 48 [kHz], the sound source interval was 5 [°], and the impulse response length was 512 points (about 0.011 [s]).

The spatio-temporal frequency spectra of HRTFs with and without pinnae are shown in Figs. 6 and 7, respectively. The distributions in the triangle area are different. In Fig. 6, the components are especially concentrated in the triangle area of the absolute temporal frequency lower than about 8 [kHz]. In contrast, in Fig. 7, the same feature is not shown, and the spectrum is spread out and wavy regardless of the temporal frequency. This phenomenon is caused by the sound collection effect of pinnae.[9]

4. Conclusions

In this study, we described spatio-temporal frequency analysis for sampling data on a coordinate system of HRTF measurements and analyzed the measured HRTFs in the spatio-temporal frequency domain. Because of the coordinate system of spatial sampling and the wave property, all of the spatio-temporal frequency spectrum components of the transfer functions

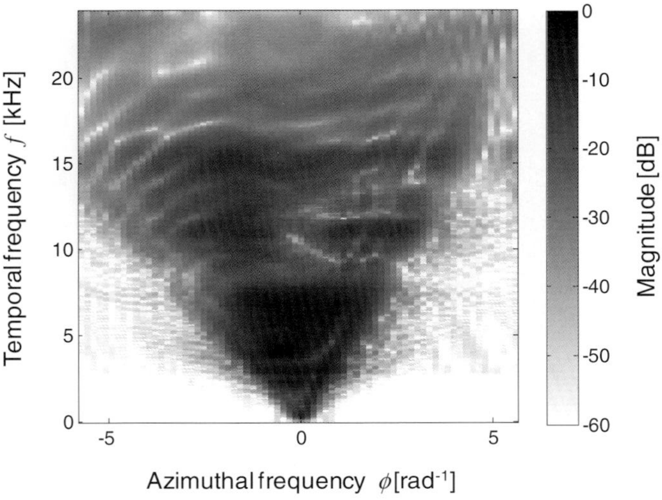

Fig. 6. Spatio-temporal frequency spectrum of HRTFs with pinnae.

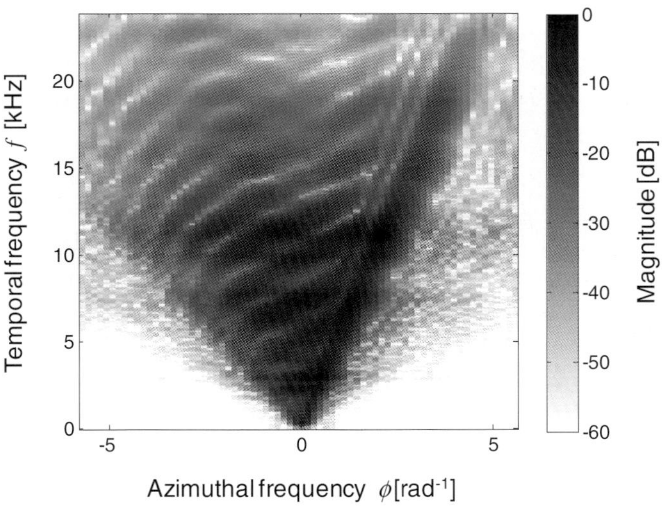

Fig. 7. Spatio-temporal frequency spectrum of HRTFs without pinnae.

were concentrated in a triangle area in the spatio-temporal frequency domain. On the spatio-temporal frequency spectrum of the measured HRTFs, the influence of the actual measurement, such as reverberations and background noise, also appeared. We also examined the reverberation in the measurement environment and the pinnae influences. The reverberation components were distributed on different areas from the numerical results, and the sound collection effect of the pinnae was observed in the triangle of the temporal frequency lower than 8 [kHz]. Future works include investigations of other factors of HRTF measurement and applications using such spatio-temporal frequency characteristics for dereverberation.

References

1. J. Blauert, *Spatial Hearing (revised ed.)* (MIT Press, 1996).
2. Y. Morimoto, T. Nishino and K. Takeda, Analysis of head related transfer functions based on the spatio-temporal frequency characteristic, in *Proc. Audio Eng. Soc. 14th Reg. Conv.*, (Tokyo, Japan, 2009).
3. D. H. Johnson and D. E. Dudgeon, *Array signal processing: concepts and techniques* (Simon & Schuster, 1992).
4. K. Nishikawa, T. Yamamoto, K. Oto and T.Kanamori, Wideband beamforming using fan filter, in *Proc. IEEE ISCAS'92*, (San Diego, California, U.S.A., 1992).
5. F. Pinto and M. Vetterli, Wave field coding in the spacetime frequency domain, in *IEEE International Conference on Acoustics, Speech, and Signal Processing (ICASSP 2008)*, (Las Vegas, Nevada, U.S.A., 2008).
6. T. Ajdler, C. Faller, L. Sbaiz and M. Vetterli, Sound field analysis along a circle and its applications to HRTF interpolation, *J. Audio Eng. Soc.* **56**, 156 (2008).
7. T. Nishino, HRTF database `http://www.sp.m.is.nagoya-u.ac.jp/HRTF/`.
8. T. Nishino, M. Ikeda, K. Takeda and F. Itakura, Interpolating head related transfer functions, in *Proc. Western Pacific Regional Acoustics Conference (WESTPRAC VII)*, (Kumamoto, Japan, 2000).
9. K. Sugiyama, Sound collection effect of a pinna of an artificial head, *Acoust. Sci. & Tec.* **24**, 311 (2003).

INFLUENCE ON LOCALIZATION OF SIMPLIFYING THE SPECTRAL FORM OF HEAD-RELATED TRANSFER FUNCTIONS ON THE CONTRALATERAL SIDE

K. WATANABE*, R. KODAMA, S. SATO, S. TAKANE, and K. ABE

Faculty of Systems Science and Technology, Akita Prefectural University
84-4 Ebinokuchi, Tsuchiya, Yuri-Honjo, Akita 015-0851, Japan
**E-mail: kwatanabe@akita-pu.ac.jp*
www.akita-pu.ac.jp

We investigated the influence on localization of simplifying HRTFs on the contralateral side. For the study described in this report, the spectral form of HRTFs are flattened in a region higher than a certain frequency. The results of a localization test suggest that the contralateral-side HRTFs at frequencies of 4 kHz or lower influenced front–back perception.

Keywords: head-related transfer function, HRTF, simplifying, contralateral side, localization

1. Introduction

Sound localization is known to be controllable by synthesizing HRTFs.[1,2] Simplification of head-related transfer functions (HRTFs) is important for effective implementation of their synthesis from a computational perspective. The frequency resolution of the auditory system shows that the detailed spectral form of the HRTFs is not evaluated in the high-frequency region, which might enable simplification of the HRTFs to some degree. In particular, the HRTFs on the contralateral side to a sound source have a complicated spectral form with peaks and dips, and lower energy than those on the ipsilateral side. Associating the simplification with these points can support an investigation of how much simplification of the contralateral-side HRTFs is possible. It remains unclear, however, whether the peaks and dips involved in the contralateral-side HRTFs are necessary as localization cues. Because binaural cues play dominant role in auditory localization, simplification of the HRTFs on the contralateral side might be possible if the HRTFs on the ipsilateral side and the binaural cues are accurately reproduced.[3]

2. Simplification of Contralateral-side HRTFs

2.1. *HRTF measurement conditions*

As described in this paper, a set of HRTFs for each subject was measured using a time stretched pulse (TSP)[4] with a sampling frequency of 48 kHz. Figure 1 shows the measurement system. Each subject was seated in an anechoic chamber. A loudspeaker (Q1; KEF) was mounted at a position 1.5 m from the center of the subject's interaural axis. The subject's ear canals were blocked,[5] and miniature microphones (KE4-211-2; Sennheiser GmbH and Co.) were used to pick up the TSP signals. A headrest was mounted on the chair to constrain the subject's head. The chair was rotated to measure HRTFs for any source azimuth.

2.2. *Method of simplifying HRTFs*

A set of HRTFs is measured in the horizontal plane for each subject. For descriptions in this paper, the source of 0° is defined as front; the others are defined in a clockwise direction. Consequently, the contralateral side for right ear corresponds to the source directions which are greater than 180°, whereas that for the left ear corresponds to those less than 180°.

The HRTFs on the contralateral side are simplified by flattening the magnitude characteristics in a frequency region higher than a certain frequency. This method is similar to one that Zhang and Hartmann used.[6] This frequency is designated as the *boundary frequency* in this paper. The level of the flattened region is set to retain the interaural level difference (ILD). The interaural time difference (ITD) is also kept by adjusting each time delay included in the simplified head-related impulse response (HRIR) to that in the original one. Consequently, the simplification method only deteriorates the spectral cues in the contralateral-side HRTFs. An example of the method is portrayed in Fig. 2. As shown in Fig. 2(b), the spectral form in the frequency region higher than the boundary frequency, 8 kHz, is flattened. The relative level in that region is −2.8 dB. It is determined to be equal to the average level of the original HRTF in the frequency region to which the flattening is applied. Consequently, the ILD is retained in all directions.

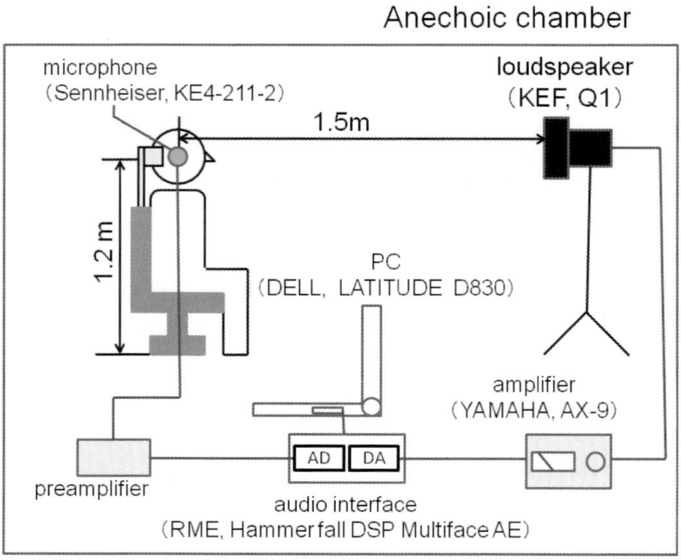

Fig. 1. HRTF measurement system.

3. Evaluation Experiment

3.1. *Outline of the experiment*

A localization test was conducted to evaluate the influence of the simplified HRTFs. Subjects were seated in an anechoic chamber and asked to listen to a stimulus presented with headphones (HD-650; Sennheiser GmbH and Co.). Each stimulus was a sound signal that was convolved with the original HRTFs or the HRTFs with that on the contralateral side simplified. Subjects were asked to report the perceived direction in the horizontal plane.

3.2. *Conditions*

Four male subjects with normal hearing acuity participated in the experiment. A set of HRTFs was measured for each subject. Source directions, including 0°, were spaced at intervals of 30° in the horizontal plane; the number of source directions was 12. Each source signal was pink noise with a length of 1 s convolved with an HRTF or a simplified HRTF. The boundary frequencies used to simplify the HRTFs were 20 (original), 8, 4, 2, 1, 0.5, and 0 (flattened for overall frequency range) kHz. Figures 3(a) and 3(b) show characteristics of stimuli convolved with the original HRTF and

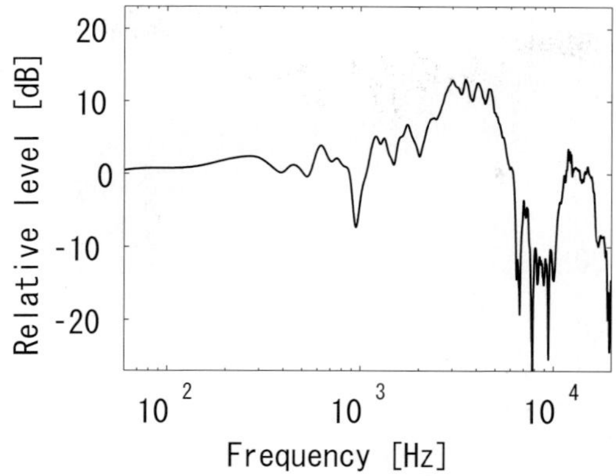

(a) original HRTF

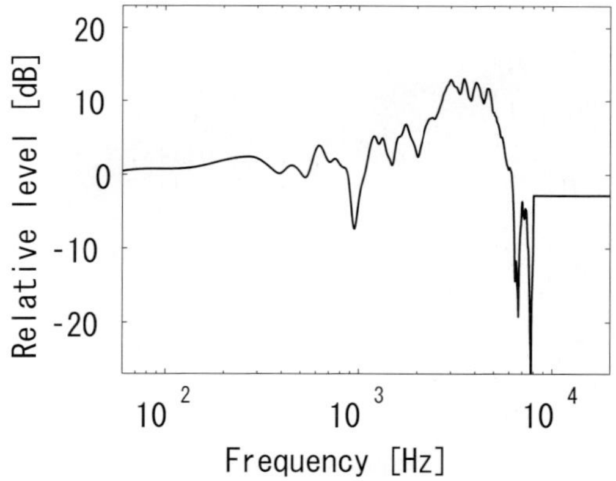

(b) simplified HRTF (boundary frequency: 8 kHz)

Fig. 2. Example of simplification of the HRTFs on the contralateral side (30°, left ear). The level in the flattened region is −2.8 dB, which is the same value as the average level of the original HRTF in the frequency region corresponding to the flattened one.

the simplified HRTF, respectively. In Fig. 3(b), the spectral form is the same as that shown in Fig. 3(a) for frequencies lower than the boundary frequency, 8 kHz, whereas the pink noise form remains in the frequency region higher than that. Simplification was not applied to the HRTFs for the two source directions of 0° and 180° because the contralateral side could not be defined. Each stimulus was evaluated five times in random order.

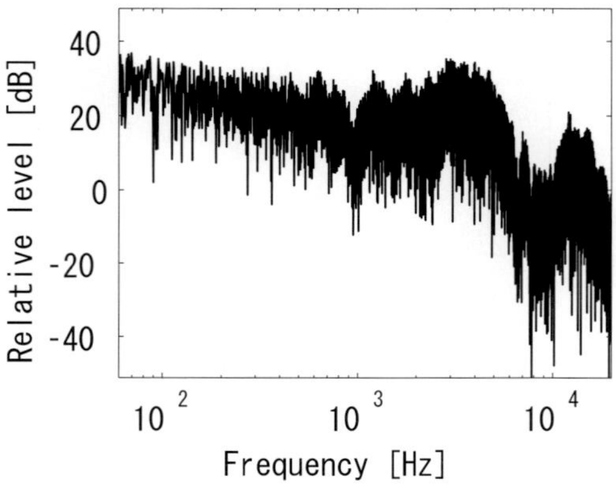

(a) pink noise convolved with the original HRTF

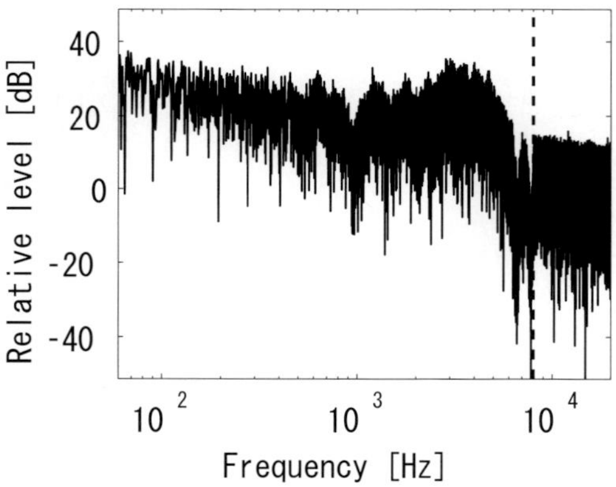

(b) pink noise convolved with the simplified HRTF (boundary frequency: 8 kHz)

Fig. 3. Examples of frequency characteristics of stimuli (source direction is 30°).

4. Results

4.1. *Results of localization tests*

Results for all subjects in each condition are presented in Figs. 4(a)– 4(d). The abscissa shows the simulated direction; the ordinate shows the subject's perceived direction. The boundary frequencies are indicated at the top of each panel.

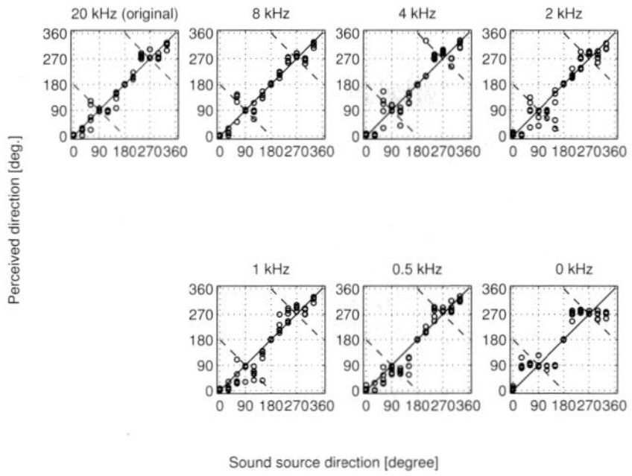

(a) Subject 1

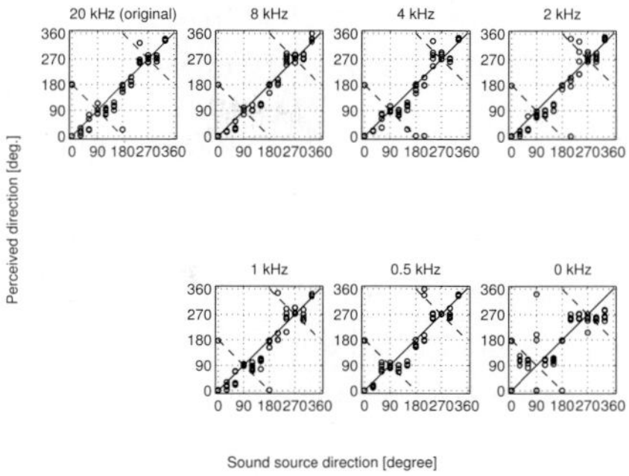

(b) Subject 2

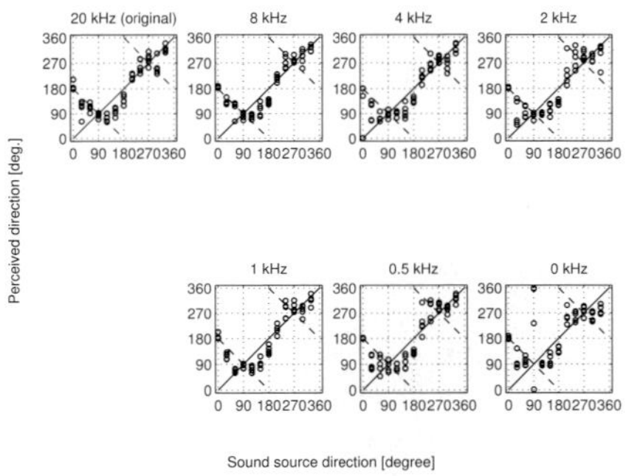

(c) Subject 3

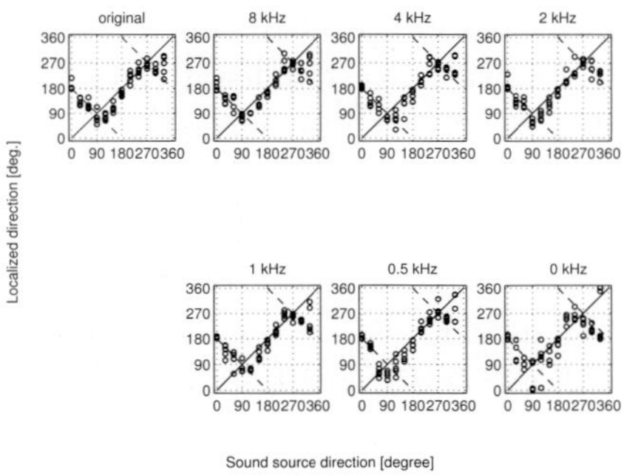

(d) Subject 4

Fig. 4. Scatter plots of simulated directions versus a subject's perceived directions.

In the "20 kHz" condition, the responses of subjects 1 and 2 fall near the diagonal line, whereas those of subjects 3 and 4 show that they tended to perceive a sound image presented in the frontal direction as being in the rear. Therefore, the tendency of localization differs among subjects. However, a similar tendency for the simplification of HRTFs is apparent across subjects. The results showed that the lower the boundary frequency is, the more variable the perceived directions are. In the "0 kHz" condition, the perceived directions are almost directly to the side (90° or 270°) for all subjects. To evaluate the results of localization for the simplified HRTFs, localization errors and front–back confusion rates are analyzed.

4.2. *Analysis of variance for localization error and front–back confusion*

As described in this paper, localization error is defined as the absolute value of the difference between a source direction and a perceived direction from which any front–back confusion is extracted in advance. The errors are calculated for all source directions. The front–back confusion rates are calculated for all source directions except 90° and 270°.

Figure 5 shows localization errors for each subject averaged over the source directions. The abscissa is the simplified HRTF condition; the ordinate is the average localization error. Effects of the contralateral-side HRTF simplification, source direction, and subject were analyzed using analysis of variance (ANOVA). The main effect of the HRTF simplification was statistically significant ($F(6, 207) = 6.08, p < 0.01$). The main effect of source direction was also significant ($F(9, 207) = 6.08, p < 0.01$). However, the main effect of subject was insignificant. Because the main effect of the simplification was significant, the least significant difference (LSD) test was applied for multiple comparisons. The results showed that the difference between the "20 kHz" and "0 kHz" conditions was significant (MSe = 1994.16, level of significance 5 %).

Figure 6 shows front–back confusion rates for each subject for all source directions. The abscissa is the simplified HRTF condition; the ordinate is the front–back confusion rate. No trend is apparent for the variation of confusion rates with the simplification condition because great differences exist among individuals. The effects of the contralateral-side HRTF simplification and source direction were analyzed using ANOVA. The main effect of the HRTF simplification was not significant. Regarding results for subjects 3 and 4 shown in Figs. 4(c) and 4(d), front–back confusion is apparent even for the stimulus convolved with the original HRTFs. That is considered to contribute to the insignificance of the effect of the HRTF simplification. However, the main effect of source direction was significant ($F(7, 168) = 5.73, p < 0.01$). The results shown in Fig. 4 reveal that the confusion of back to front directions is small for all subjects in the "20 kHz" condition. Consequently, the analysis will be conducted only for back source directions.

4.3. *Analysis of variance for front–back confusion*

Figure 7 shows front–back confusion rates calculated for each simplified HRTF condition and each subject when the sound sources are back direc-

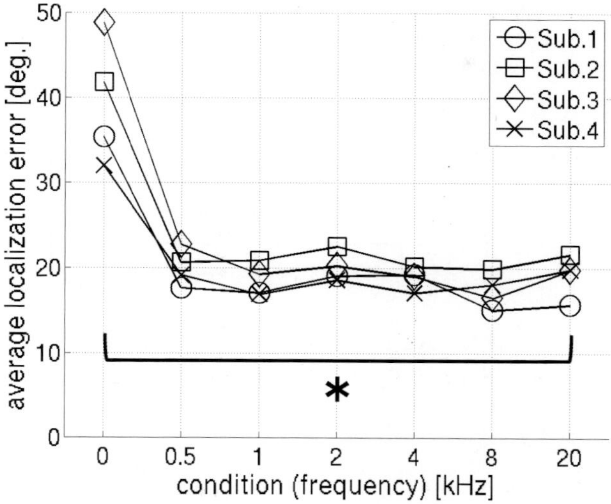

Fig. 5. Average localization error. The asterisk denotes the pair for which results showed a significant difference for the effect of the contralateral-side HRTF simplification in the LSD test.

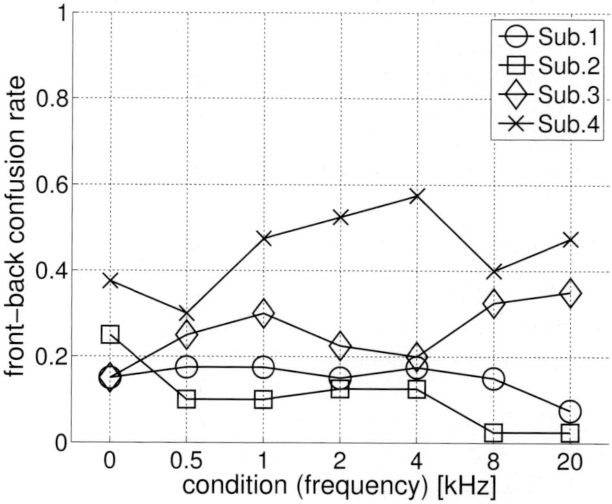

Fig. 6. Front–back confusion rates for all source directions.

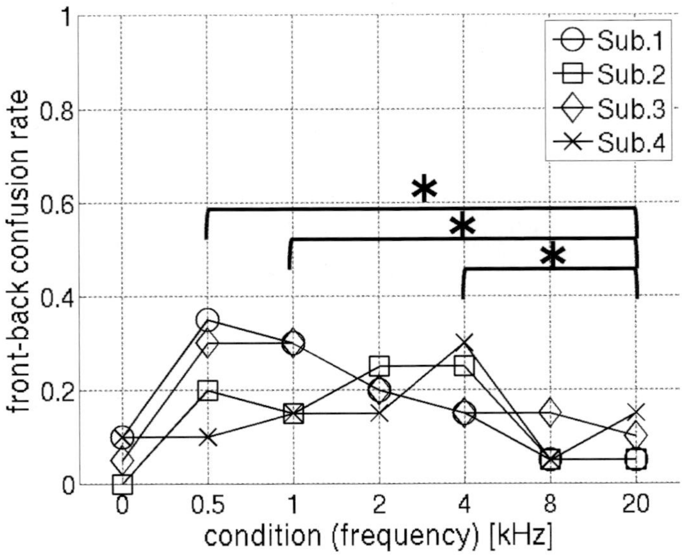

Fig. 7. Front–back confusion rates for back to front directions. The asterisk denotes the pair for which results showed that a significant difference was found for the effect of the contralateral-side HRTF simplification in the LSD test.

tions (120°, 150°, 210°, and 240°). In this figure, the front–back confusion rates are small when the boundary frequencies are 20, 8, and 0 kHz for all subjects. To evaluate the effect of the simplification statistically, ANOVA was applied to those data. The main effect of the simplification was significant ($F(6, 84) = 3.62, p < 0.01$). The main effect of source direction was also significant. Results of the LSD test demonstrated that the front–back confusion rates for the boundary frequency of 20 kHz are significantly different from those for 4, 1, and 0.5 kHz (MSe = 0.03, level of significance 5 %). Those combinations are denoted by asterisks in Fig. 7.

5. Discussion

The results portrayed in Fig. 5 show that the spectral forms of the HRTFs on the contralateral side in the frequency region higher than 0.5 kHz do not influence horizontal localization, ignoring front–back confusion. Because the ILDs and ITDs were retained in the simplified HRTFs, subjects were able to use them as cues for the source azimuth.[7] Localization was greatly degraded in the "0 kHz" condition, in which the simplification of the HRTFs on the contralateral side was applied to the overall frequency range. Although it remains unclear what caused that degradation from this experiment, using only the ipsilateral-side HRTFs as spectral cues might cause lateralization. Then, based on the results shown in Fig. 7, front–back confusion rates significantly increase when the boundary frequency of the simplification is 4 kHz or lower, which suggests that the spectral forms of the HRTFs on the contralateral side in the frequency region lower than 4 kHz are important for front–back discrimination. The front–back confusion rates are small when the boundary frequency is 0 kHz. It is considered that subjects' localizations in that condition were lateral directions for which the rates were not calculated.

6. Conclusion

In this study, the HRTFs were simplified only on the contralateral side to the sound source direction. The localization test results showed that the spectral form of the HRTFs on the contralateral side at frequencies of 0.5 kHz or higher does not influence horizontal localization, ignoring the effect of front–back confusion. Considering front–back confusion, results suggest that the frequency range of 4 kHz or lower is important.

7. Acknowledgement

This work was partially supported by Grant-in-Aid for Young Scientists (B) (21700140).

References

1. J. Blauert, *Spatial Hearing* (The MIT Press, Cambridge, MA, 1983).
2. D. R. Begault, *3-D sound for virtual reality and multimedia* (AP Professional, Cambridge, 1994).
3. E. A. Macpherson, and A. T. Sabin, Binaural weighting of monaural spectral cues for sound localization, *J. Acoust. Soc. Am.* **121**(6), 3677–3688 (2007).
4. Y. Suzuki, F. Asano, H. Y. Kim, and T. Sone, An optimum computer-generated pulse signal suitable for the measurement of very long impulse responses, *J. Acoust. Soc. Am.* **97**(2), 1119–1123 (1995).
5. H. Møller, Fundamentals of binaural technology, *Applied Acoustics* **36**, 172–218 (1992).
6. P. X. Zhang and W. M. Hartmann, On the ability of human listeners to distinguish between front and back, *Hearing Research* **260**, 30–46 (2010).
7. Lord Rayleigh, On our perception of sound direction, *Philosophical Magazine* **13**(74), 214–232 (1907).

3D SOUND TECHNOLOGY: HEAD-RELATED TRANSFER FUNCTION MODELING AND CUSTOMIZATION, AND SOUND SOURCE LOCALIZATION FOR HUMAN–ROBOT INTERACTION

Y. PARK

Center for Noise and Vibration Control, Department of Mechanical engineering, KAIST, Korea

S. HWANG

Vibration and Noise Research Part, Marine Research Institute, Samsung Heavy Industries, Korea

B. KWON

Center for Noise and Vibration Control, Department of Mechanical Engineering, KAIST, Korea

Humans depend mainly on visual and auditory cues to capture spatial information about their environment. Frequently, although auditory cues are given minimal attention in designing virtual environments, they can play an important role in the production of virtual environments because spatial information provided by vision is limited to the viewing direction. The ability to identify the location of a sound source is known as auditory localization. Technologies that replicate the sounds from a real environment through an artificially created environment that exploits human auditory localization are known as three-dimensional (3D) sound technology. This chapter presents a description of the modeling of head-related transfer functions (HRTFs)—the core technology for synthesizing 3D sound—based on principal components analysis (PCA). An HRTF-customization method for moving sound sources based on subjective tuning of a few parameters is also introduced. Applications of 3D sound technology to robots are introduced. Robot auditory systems including speech and speaker recognition techniques and robot artificial ears, have been implemented in an actual robot platform, and the performance is verified with experiments in a household environment.

1. Virtual Auditory Display Based on HRTFs

In a headphone-based simulation, if sounds are filtered with head-related transfer functions (HRTFs) and delivered to a listener via headphones, then a virtual acoustic environment can be produced: the listener feels the spatialized sounds appear to originate from the desired directions in the surrounding three-dimensional (3D) space. Systems or techniques generating spatialized sounds and conveying them to a listener are designated as virtual auditory displays (VADs). Because virtual sound sources can be generated by real-time convolution of the audio source with the HRTFs corresponding to the desired source positions, the HRTFs or head-related impulse responses (HRIRs), which are the time domain counterparts of the HRTFs, play a key role in rendering for high-fidelity VADs.

1.1.1. *Modeling and customization of HRTFs*

A typical VAD requires a large library containing the HRTFs corresponding to source positions that are densely distributed in 3D space. In other words, many HRTFs must be measured empirically and stored to generate well-spatialized sounds by VAD. For that reason, large amounts of memory size are necessary for rendering of high-fidelity VAD. To save memory, it is necessary to model the HRTFs using only a few parameters that preserve the perceptually relevant features of the HRTFs [1–3]. Typical VAD systems mainly depend on the non-individualized HRTFs measured from a dummy head microphone system. However, many earlier reports have described that non-individual HRTFs might cause high error rates in sound localization [4, 5]. Individualized HRTFs can enhance the localization performance of virtual sources, but it is impractical to measure the individual HRTFs of every listener because of the requirements of heavy and expensive equipment, as well as the long measurement time. Therefore, it is important to develop an HRTF customization method that provides a listener with proper sound cues without measurement of individual HRTFs [6–8].

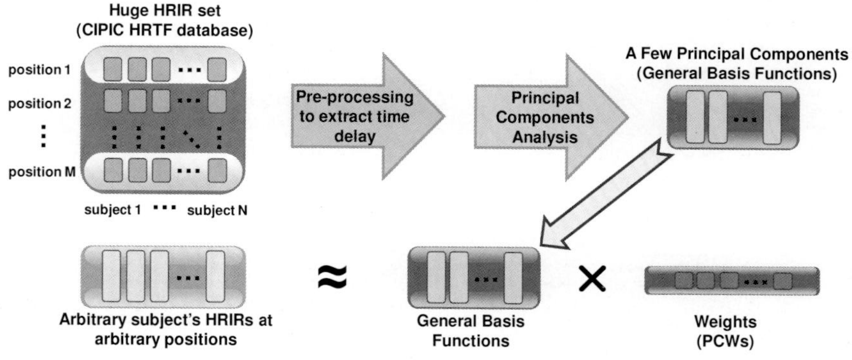

Figure 1. Diagram showing HRIR modeling based on general basis functions.

To achieve efficient modeling and customization of HRTFs, we introduced general basis functions in the time domain [9]. Principal component analysis (PCA) was performed using the median-plane HRIRs in the CIPIC HRTF database [10]; 12 principal components (PCs) were extracted. These PCs are general basis functions that can represent not only inter-subject variation but also inter-position variation in HRIRs [9, 11]. An HRIR can be reproduced as a weighted linear combination of the general basis function as

$$HRIR(position, subject) \cong \sum_{k=1}^{12} w_k \mathbf{v}_k + \mathbf{u}\,, \qquad (1)$$

where w stands for the respective weight of each PC (PCW), $\mathbf{v}$ signifies the PC and $\mathbf{u}$ represents the empirical mean. A diagram of HRIR modeling based on general basis functions is presented in Fig. 1. The HRIR is a function of both the source position and subject, and $\mathbf{v}_k$ (k=1,2,...12) are the general basis functions. Consequently, only w depends on both the source position and subject. In other words, modeling of an arbitrary subject's HRIRs is done by determining w to minimize the modeling error in the least-squares sense. Results show that an arbitrary subject's HRIRs, measured using somewhat different measurement conditions, techniques, and source positions from the CIPIC HRTF database are reproducible using the 12 general basis functions obtained from PCA of the CIPIC HRTF database with 7.7% modeling error in the least-squares sense [9]. Figure 2 presents the distributions of the responses of a representative subject with the measured and modeled individual HRIRs from 12 PCs, 8 PCs, and 4 PCs. Each column corresponds to the responses with the measured or modeled individual HRIRs as denoted at the top of each column. In each panel, the horizontal and vertical axes respectively show the target and perceived elevation. The circle radius is directly proportional to the response frequency within 5°.

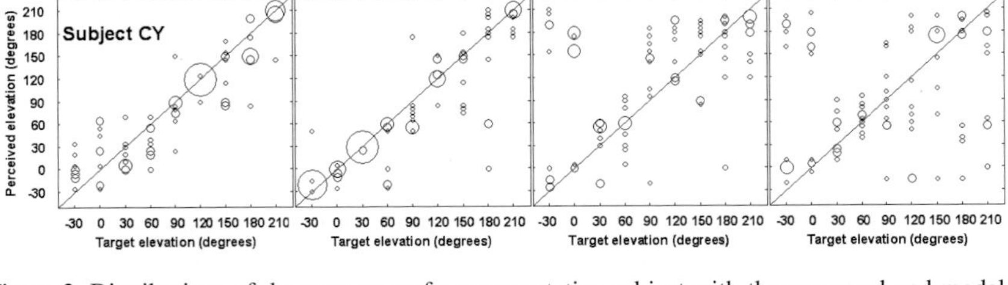

Figure 2. Distributions of the responses of a representative subject with the measured and modeled HRIRs. Each column corresponds to the responses of subject CY with the measured or modeled HRIRs, as denoted at the top of each panel.

The subject perceived the elevation of a sound source accurately with the measured HRIRs. Front–back confusion did not occur. In the second column, the subject also perceived the elevation accurately with the modeled HRIRs from 12 PCs. In the third column, the responses with the modeled HRIRs from eight PCs are shown as more scattered than those in the first and second columns. The subject reported that front–back confusion occurred frequently at low frontal elevations. The last column shows that when the modeled HRIRs from four PCs were presented, all subjects reported inaccurate responses. Student's *t*-tests were performed to infer whether a difference in localization error between the measured HRIR and the modeled HRIR is significant or not. The subjective listening tests showed no significant difference in localization error between the measured HRIR and the modeled HRIRs from the 12 PCs [9]. The memory requirement can be dramatically reduced—up to five-fold—by modeling the HRIR based on the general basis functions, compared to the memory capacity necessary to store the HRIRs as FIR filters [12].

An arbitrary subject's HRIRs can be modeled using the general basis functions.

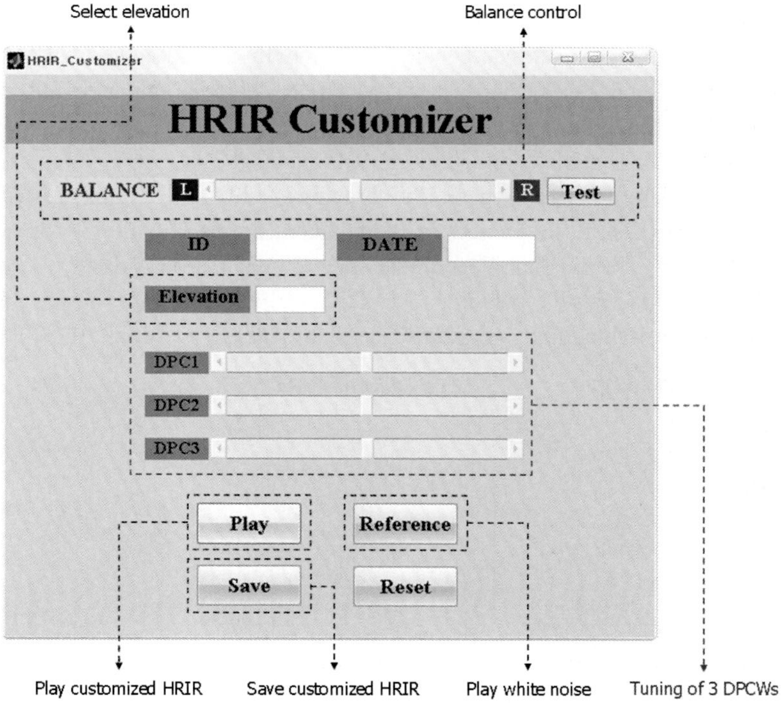

Figure 3. MATLAB^{TM} GUI for HRIR customization.

Therefore, it is intuitively reasonable to infer that one can make customized HRIRs if the 12 PCWs are selected properly to provide sound cues for the vertical perception and front–back discrimination. Customization is conducted by letting a subject tune the weight of each basis function while listening to a broadband stimulus filtered with the resulting customized HRIRs [9]. A MATLAB™ graphical user interface (GUI), as depicted in Fig. 3, was used for customization. Tuning of PCWs can continue until satisfactory vertical perception and front–back discrimination are achieved. Subjective localization listening tests were performed by three subjects to evaluate the localization performance. Figure 4 shows a response distribution of a representative subject. The subject exhibited accurate localization performance for stationary sounds synthesized by measured (individual) or customized HRIRs, whereas their localization performance was inaccurate with Kemar HRIRs. All subjects reported dramatically improved performance for the vertical perception and the front–back discrimination with the customized HRIRs compared to those with the Kemar HRIRs [9]. No statistically significant difference was found in localization error between the measured HRIR and the customized HRIR for two subjects [9].

In many practical applications, generation of moving sound sources in VAD is important. In such cases, the customized HRIRs corresponding to all source positions on the trajectory of interest must be obtained. In other words, customization of spatially contiguous HRIRs along the trajectory of interest is necessary. Therefore, we expand the previous study to customize spatially contiguous HRIRs in the trajectory simultaneously by the tuning of a few parameters [13]. The median plane in the upper hemisphere was divided into two sectors by considering the perceptual sensitivity to HRIRs [14]. The weights on the specific basis functions were tuned at each of three positions (0°, 70°, and

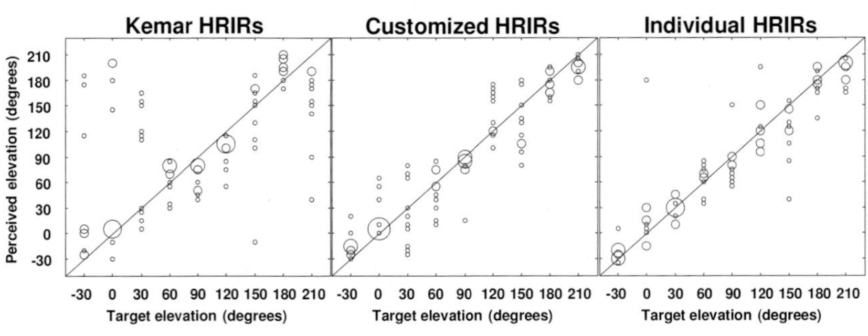

Figure 4. Subjective listening test results of a representative subject using stationary sounds synthesized by the individual, customized, and Kemar HRIRs.

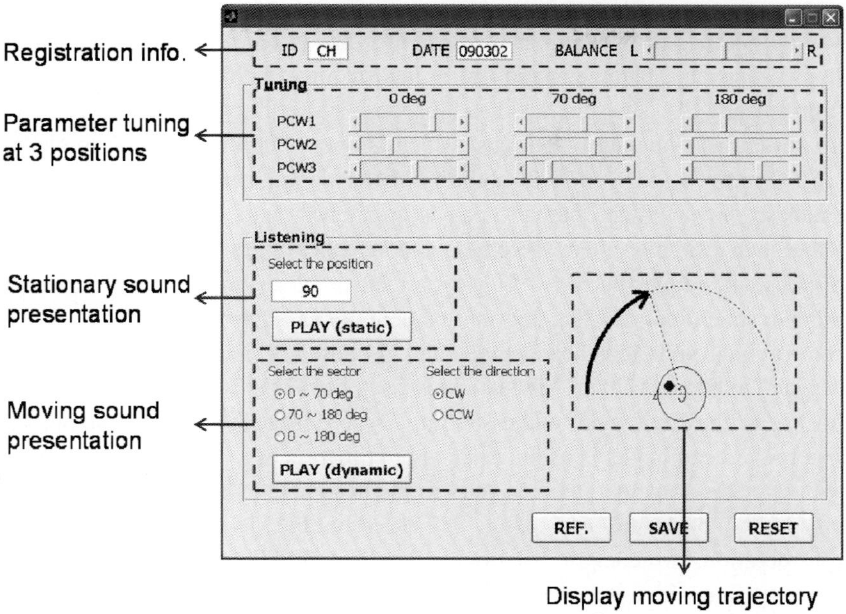

Figure 5. MATLAB™ GUI for customization of spatially continuous HRIRs.

180°), which are the endpoints of sectors. The customization process is based on the MATLAB™ GUI, as depicted in Fig. 5. Subjects tune the parameters using the slider bars at three static source positions: 0°, 70°, and 180° of elevation. Subjective listening tests were performed to evaluate the performance for synthesizing moving sounds on three sets of HRIRs including the individual, customized, and non-individualized (Kemar) HRIRs. When the moving sounds were presented to four subjects, all subjects more accurately perceived the center position of the target trajectory with the customized HRIRs than with the Kemar HRIRs. The error for the center position with the customized HRIRs was comparable to that with the individual HRIRs. No statistically significant difference was found in the errors for the center position between the individual and customized HRIRs for all subjects, although a statistically significant difference between the individual and Kemar HRIRs was found for three subjects.

1.1.2. *HRTF modeling based on a spheroidal head model*

Mainly because of its simplicity, researchers have used a spherical head with a single adjustable parameter (sphere radius) to model diffraction and reflection from a head [15]. The spherical head response is useful to construct a head block for structural modeling of HRTFs, which is one mode of HRTF customization [16]. Simply by extraction of anthropometric parameters of the CIPIC HRTF database, however, the human head is apparently better fitted as an ellipsoid rather than a simple sphere. To obtain a more general head response rather than a response of simple sphere, we suggest a mathematical form to calculate analytic prolate spheroidal HRTFs for an incident point source. Schematics for the conventional spherical head model and proposed spheroidal head model are depicted in Fig. 6. Matlab^{TM} pseudo-code for spheroidal HRTF can be downloaded freely at the following address: (http://sdac.kaist.ac.kr/data/spheroid.php). A spheroidal head model is useful as an approximation of a human's head response with five adjustable parameters: head width, head height, head tilt, downward ear offset, and backward ear offset. To ascertain the necessity of head height consideration, two analytic solutions—spherical HRTF and spheroidal HRTF—are compared with a measured HRTF from the CIPIC HRTF database.

By varying head dimensions and downward ear offset, interaural time difference (ITD) patterns and notch patterns are optimally matched to measured patterns. From the perspective of ITD, however, two analytic HRTFs show almost identical ITD-matching performance to the measured ITD information [17]. The reason is that the ITD information is closely dependent on azimuthal source changes rather than the elevational source changes. Therefore, consideration of

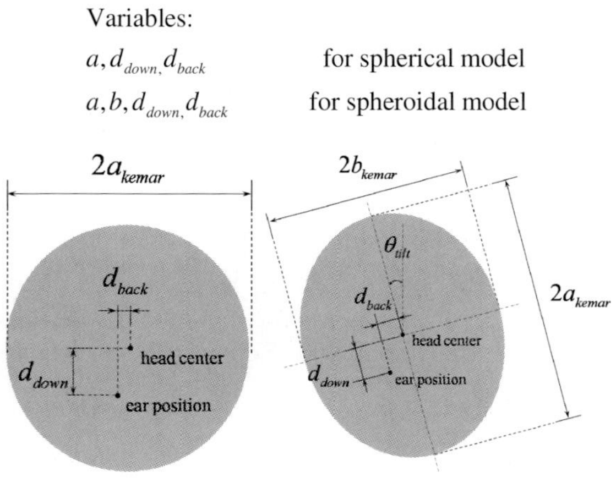

Figure 6. Head parameters of two head models: spherical head model (left), spheroidal head model (right).

the head height dimension is apparently unnecessary in ITD matching. Prolate spherical models might be used to match the notches and peaks in the frequency domain more accurately to the individual HRTFs [18].

2. 3D Sound Localization for Human–Robot Interaction

If the objective of a VAD system is to generate spatialized sound for humans, then the purpose of robot auditory systems is to provide hearing ability to a robot. It is important for a robot to acquire spatial information such as where a sound source is and what a sound means, for human–robot interaction (HRI). The techniques to perceive such information are named "sound-source localization" and "speech recognition". Sound source localization is concerned with estimating the location of sources using the measurement of acoustic signals by microphone arrays. Speech recognition is the technology that makes it possible for a machine to identify human speech components. This section specifically addresses sound source localization. Speech recognition is used to enhance the performance of sound source localization.

2.1.1. *Robot artificial ear*

In the intelligent robot industry, the ultimate goal of creating a humanoid robot is to make robots mimic humans in all respects, including appearance. For that reason, the robot industry is demanding a robot auditory system shaped with a human-like external ear, particularly pinnae, to achieve more natural HRI. Therefore, we developed a robot artificial ear as a first prototype using two

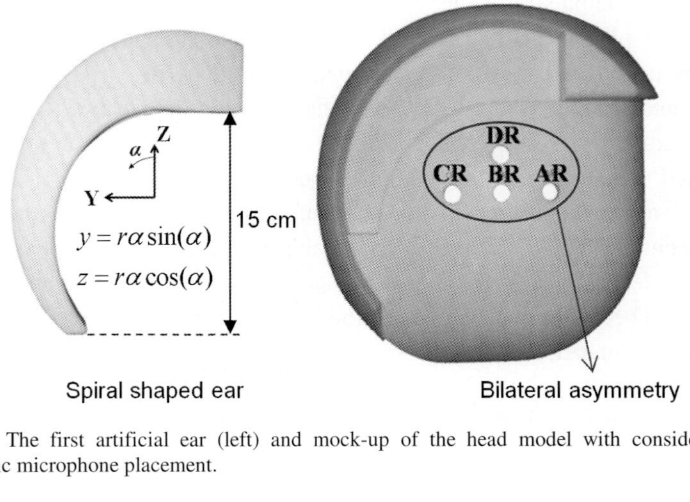

Figure 7. The first artificial ear (left) and mock-up of the head model with consideration of asymmetric microphone placement.

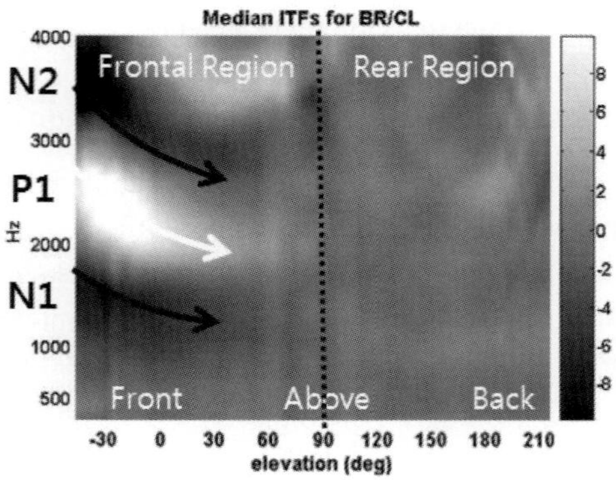

Figure 8. Magnitude response of median ITFs using asymmetric placement of microphones at BR and CL.

microphones as portrayed in Fig. 7 [19]. Our design of the first artificial ear is based on the assumption, proved experimentally and analytically by prior research [20–22], that spectral notches are obtained by cancellation involving a direct wave from the sound source and reflected wave from the interior surface of concha.

The first artificial ear was designed to create direction-dependent spectral features in the voice frequency range. When two microphones are placed symmetrically in the head like a human's left and right ear positions, no interaural differences for the sources in median plane and spectral modifications of two ear outputs are the same. Therefore, we considered asymmetric microphone placement on both sides. For consideration of eight microphone positions in left and right sides, we made four holes in each side. Based on experimental results, BR and CL positions were chosen because it shows more useful features for sound-source localization [19, 23]. When microphones are positioned at BR and CL, the interaural transfer function (ITF) [24] is depicted in Fig. 8 for sound sources in the median plane. However, both the memory requirement of HRTF database for the localization algorithm and the rather large pinnae made the proposed robot artificial ear not readily applicable to robot platforms. Therefore, we re-designed the artificial ear, as the second prototype, which has a smaller size resembling that of human pinnae but which uses two additional microphones [25, 26]. The two designed artificial ears and head models are presented in Fig. 9. The second artificial ear is smaller than that of the first prototype. To apply the second artificial ear to differently sized and shaped robot platforms, we used a cleansing method for elimination of reflected

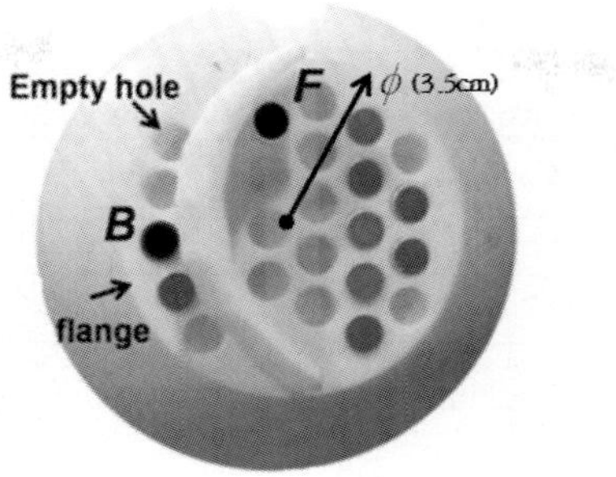

Figure 9. This second artificial ear prototype was attached to a spherical head model.

waves caused by the robot platform, i.e. a robot shoulder. The feasibility of sound source localization using the second artificial ear for multi-platform application was examined using three robot platforms of different heights [26].

2.1.2. *Sound source localization using spatially mapped generalized cross correlation (GCC) function*

A novel sound source localization method based on time delay of arrival (TDOA) is proposed for application to robot auditory systems using microphone arrays [27]. The main concept of the proposed method is to transform the cross-correlation function in the time domain into the spatially mapped GCC function

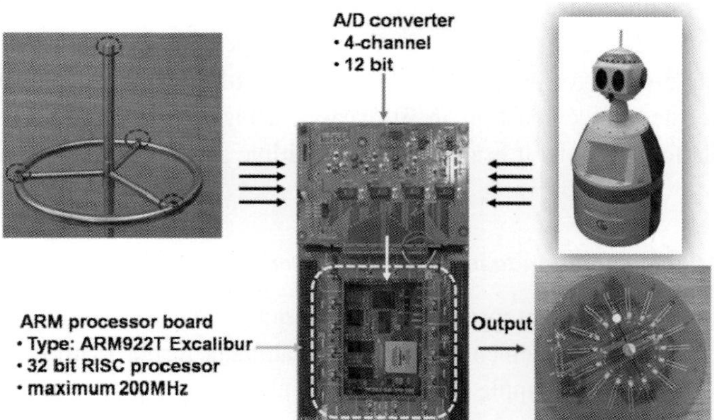

Figure 10. Implementation system consists of microphone arrays on the robot platform (Infotainment Robot Platform Ver.1), A/D converter (manufactured in person), ARM processor in SoC and output device.

Table 1 Experimental results

(% of CE: ±10° error bound)

SNR(dB)	MEE (°) (Conventional)	MEE (°) (Proposed)	% of CE (Conventional)	% of CE (Proposed)
-4	85	58	14	45
2	31	10	49	91
7	20	5	73	99
12	15	4	74	99

in spatial domain using an appropriate mapping function. In the general TDOA approaches using multiple microphones, various error criteria [28] employing all TDOAs must be used to estimate the true source location. However, because the spatially mapped GCC functions of each microphone pair represent the likelihood of the source direction in 3D space, those of all microphone pairs can be summed or multiplied easily in the same spatial domain. Then the true source location is estimated by finding the maximum peak location of a summed or multiplied single function. Because of the light computational load of this procedure, it is possible for the proposed method to be embedded on an ARM processor for a system-on-a-chip (SoC) implementation, as presented in Fig. 10. Actual environmental experiments to estimate the azimuth angle of a source with three microphones show better localization performance than that of the conventional TDOA method under various SNR conditions. Experimental results are presented in Table 1. The mean value of Estimation Error (MEE) is the average of differences between the actual and estimated source locations; the percentage of Correct Estimation (CE) is the percentage of estimation results under the given error bound to all estimation results. Moreover, this method can estimate the source location in 3D spaces using only three microphones and robot platform effects; it is applicable to multiple source localization as well [29, 30].

2.1.3. ***Implementation to an actual platform***

Figure 11 (a) shows that a robot auditory system combining the abovementioned two techniques is implemented on the Infotainment Robot Platform Ver.1 [31]. For HRI, a light and simple algorithm for real time, isolated word recognition is applied additionally to the robot platform. This can enable estimation of a specified persons' location, isolating the voice characteristics of the person. By discriminating human speech and non-speech sound, it is possible to reduce the unnecessary computing time and power. Microphone arrays and artificial ears for azimuth and elevation angle estimation of the sound source are installed on the robot platform. Real-time implementation of the unified system is realized using MATLAB™ software and an NI-DAQ board. The overall procedure of the robot auditory system is presented in Fig. 12. First, signals over the specific power are detected and are then examined to test whether it is speech sound or not. If they are speech sounds, then this system recognizes the prespecified words and estimates the speech sound's direction simultaneously. The unified auditory system can perfectly estimate the azimuth angle of source within ±10° error bound and the elevation angle within a ±20° error bound.

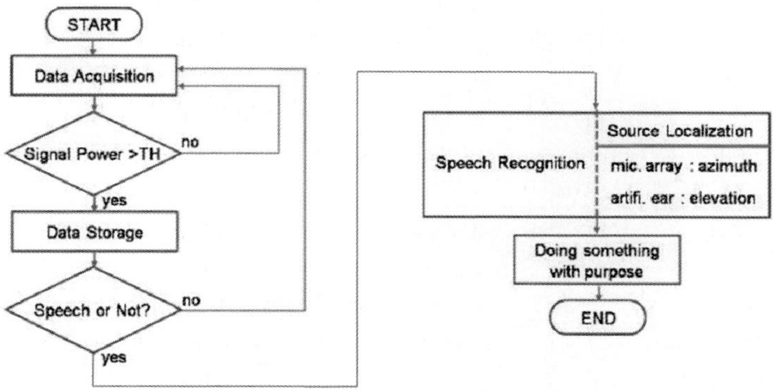

Figure 12. Flow chart of the unified robot auditory system.

Acknowledgments

This work was supported by the BK21 program and the Korea Science and Engineering Foundation through the National Research Laboratory Program (R0A-2005-000-10112-0) funded by the Ministry of Education, Science and Technology.

References

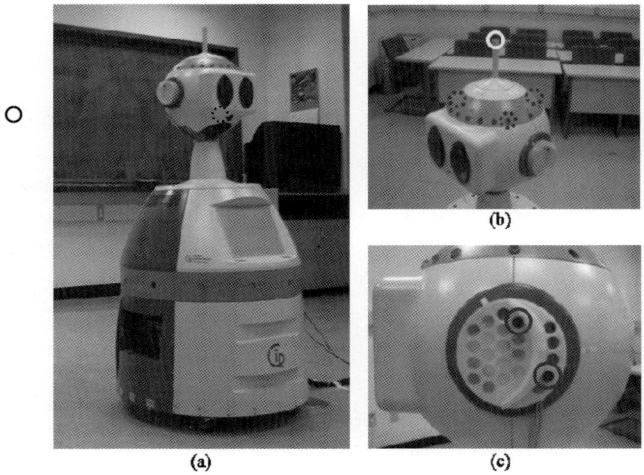

Figure 11. (a) Robot platform installed auditory system, (b) microphone positions for speech recognition,   and for localization,   (c) robot artificial ear.

[1] D. A. Durant and G. H. Wakefield, *IEEE Trans. on Speech and Audio Processing*, **10**, 18 (2002)
[2] D. W. Grantham, J. A. Willhite and K. D. Frampton, *J. Acoust. Soc. Am.*, **117**, 3116 (2005)
[3] K. Iida, M. Itoh, A. Itagaki and M. Morimoto, *Applied Acoustics*, **68,** 835 (2007)
[4] E. Wenzel, M. Arruda, D. Kistler and F. Wightman, *J. Acoust. Soc. Am.*, **94**, 111 (1993)
[5] H. Møller, M. F. Sørensen, D. Hammershøi and C. B. Jensen, *J. Audio Eng. Soc.*, **43**, 300 (1995)
[6] K. Shin and Y. Park, *IEICE Trans. on Fundamentals*, **E91-A**, 345 (2008)
[7] J. C. Middlebrooks, *J. Acoust. Soc. Am.*, **106**, 1480 (1999)
[8] D. N. Zotkin, R. Duraiswami and L. S. Davis, *IEEE Trans. on Multimedia,* **6,** 553 (2004)
[9] S. Hwang, Y. Park and Y. Park, *Acta Acustica united with Acustica*, **94**, 965

(2008)
[10] CIPIC HRTF database files, *CIPIC Interface Laboratory, U.C. Davis.* (2001) (http://interface.cipic.ucdavis.edu/)
[11] S. Hwang and Y. Park, *J. Acoust. Soc. Am.*, **123**, EL65 (2008)
[12] S. Hwang, Y. Park and Y. Park, *J. KSNVE*, **18**, 448 (2008)
[13] S. Hwang, *Ph.D. dissertation, Dep. Mech. Eng., KAIST* (2009)
[14] S. Hwang, Y. Park and Y. Park, *J. Mech. Sci. and Tech.* (2009)
[15] R. Duda and W. Martens, *J. Acoust. Soc. Am.* **104**, 3048 (1998)
[16] C. Brown and R. Duda, *IEEE Trans. on Speech and Audio Processing*, **6**, 476 (1998)
[17] H. Jo, Y. Park and Y. Park, In *Proc. of KSNVE* 35-05 (2008)
[18] H. Jo, Y. Park and Y. Park, In *Proc. ICCAS*, 251 (2008)
[19] Y. Park and S. Hwang, In *Proc. of 16th IEEE ICRHIC*, 405 (2007)
[20] H. Nanakashima and T. Mukai, In *Proc. of ICSMC*, 3534 (2005)
[21] F. Keyrouz and A. A. Saleh, In *Proc. of ICCP,* 97 (2007)
[22] E. Lopez-Poveda and R. Meddis, *J. Acoust. Soc. Am.*, **100**, 3248 (1996)
[23] S. Hwang, Y. Park and Y. Park, In *Proc. of ICCAS*, 1906 (2007)
[24] J. Blauert, *Spatial hearing*, revised edition, MIT Press, 1997
[25] S. Lee, S. Hwang, Y. Park and Y. Park, In *Proc. ICCAS*, 246 (2008)

[26] S. Lee, Y. Park and Y. Park, In *Proc. KACC*, 358 (2009)
[27] B. Kwon, Y. Park and Y. Park, *J. KSNVE*, **19**, 355 (2009)
[28] M. S. Brandstein and H. F. Silverman, *Computer Speech and Language*, **11**, 91 (1997)
[29] B. Kwon, Y. Park and Y. Park, In *Proc. of ICCAS*, 241 (2008)
[30] B. Kwon, Y. Park and Y. Park, In *Proc. of ICCAS*, 1773 (2009)
[31] S. Hwang, *Master's thesis, Dept. Mech. Eng., KAIST* (2006)

Section 3

Capturing and Controlling the Spatial Sound Field

A STUDY ON 3D SOUND IMAGE CONTROL BY TWO LOUDSPEAKERS LOCATED IN THE TRANSVERSE PLANE*

K. IIDA, T. ISHII, AND Y. ISHII†

Faculty of Engineering, Chiba Institute of Technology,
Tsudanuma, Narashino, Chiba, Japan
†E-mail: kazuhiro.iida@it-chiba.ac.jp
http://www.iida-lab.it-chiba.ac.jp/

The ordinary trans-aural system involves two loudspeakers positioned in the frontal horizontal plane. However, listeners often perceive a sound to be located at the rear direction at the front. This front-back error appears to be caused by small difference in listening position, because the characteristics of the cross-talk cancel filter highly depend on the position to be controlled. In the present study, the analysis of the transfer function between the loudspeakers positioned in various directions in the transverse plane and the entrances of the ear canals of the subjects was conducted using the following two physical measures: (1) the flatness of the amplitude spectrum of the transfer functions between the loudspeakers and the ipsi-lateral ears (direct component), and (2) the level of the amplitude spectrum of the transfer functions between the loudspeakers and the contra-lateral ears (cross-talk components) relative to that of direct components. In addition, sound localization tests were carried out. The results of the analysis and the sound localization tests showed that accurate sound image control could be achieved by two loudspeakers positioned in the transverse plane at 100 - 110°.

1. Introduction

Accurate sound localization is accomplished when the listener's own Head-Related Transfer Functions (HRTFs) are reproduced at his/her eardrums. Some trans-aural systems, which are designed to reproduce 3D sound images using two loudspeakers positioned in the frontal horizontal plane, have been proposed [1,2]. However, listeners often perceive a sound to be located at the rear direction at the front. This front-back error seemed to be caused by a small difference in listening position, because the characteristics of the cross-talk cancel filter depend strongly on the position to be controlled.

Morimoto and Ando [3] demonstrated that accurate sound image control in the horizontal and median planes were achieved with the trans-aural system, in

* A part of this work is supported by the "Academic Frontier" Project for Private Universities: matching fund subsidy from MEXT (Ministry of Education, Culture, Sports, Science and Technology).

which two loudspeakers are positioned in the upper transverse plane (T30 in Fig. 1) of the head-fixed subjects. One of the reasons for their success appears to be that the transfer functions between the loudspeakers and the ipsi-lateral ears do not have remarkable spectral peaks or notches. The flatness of the amplitude spectrum of the transfer function between the loudspeakers and the ipsi-lateral ears leads to the robustness of the cross-talk cancel filter.

In the present study, an analysis in terms of two physical measures, namely, 1) the flatness of the amplitude spectrum of the transfer functions between the loudspeakers and the ipsi-lateral ears (direct component), and 2) the level of the amplitude spectrum of the transfer functions between the loudspeakers and the contra-lateral ears (cross-talk components) relative to that of direct components, and sound localization tests were carried out for various loudspeaker arrangements in the transverse plane.

2. Analysis of the transfer function between the loudspeakers and the entrances of the ear canals of the subjects

2.1. *Method*

The transfer functions between the loudspeakers and the entrances of the ear canals of the subjects were measured in an anechoic chamber. The loudspeakers were positioned in the horizontal plane and the transverse plane (Fig. 1). The subjects were three males (IST, ISY, and UEO) with normal hearing sensitivity. The subjects were asked to face forward without having their heads fixed.

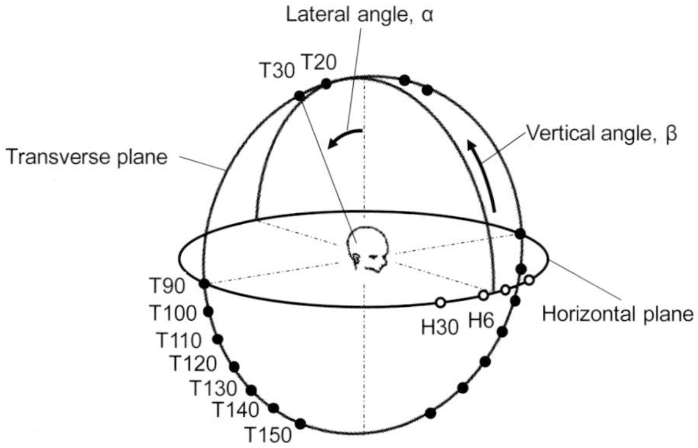

Fig.1 Loudspeaker arrangements in the horizontal plane (H30 and H6) and in the transverse plane (T20, T30, and T90 - T150).

2.2. *Results*

Figure 2 shows the amplitude spectrum of the direct components and cross-talk components of the measured transfer functions of subject IST. Figure 3 shows the standard deviation of the amplitude spectrum of the direct components (200-17,000 Hz) and the mean amplitude difference between the direct components and the cross-talk components (200-17,000 Hz) for subject IST. These figures indicate that the direct components are relatively flat when the loudspeakers are positioned near the zenith (T20 and T30), as compared with other directions, and that the levels of the cross-talk components are remarkably low when the loudspeakers are located at T110, T120, and T130. These results imply that two loudspeakers located in the transverse plane could provide more accurate 3D sound image control than those in the horizontal plane.

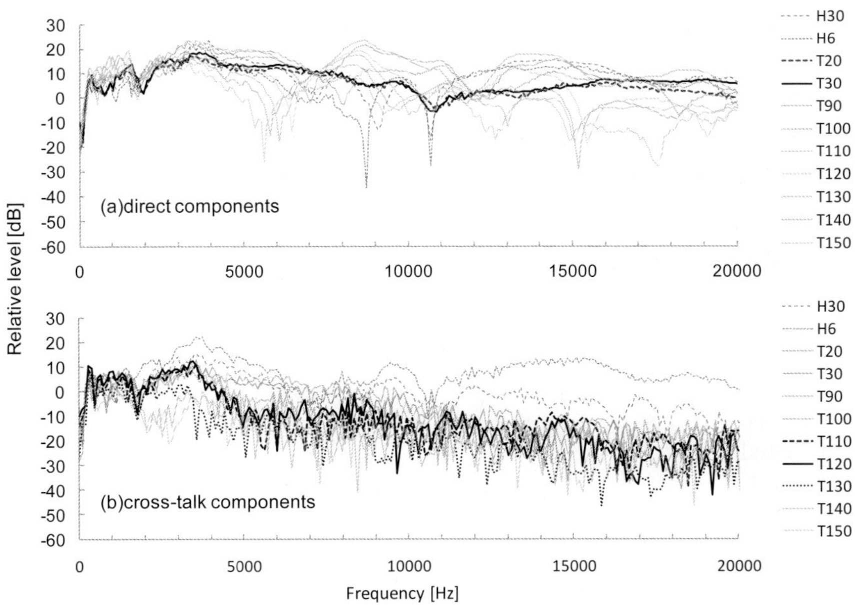

Fig.2 (a) Amplitude spectrum of the direct components, and (b) cross-talk components of the measured transfer functions of subject IST.

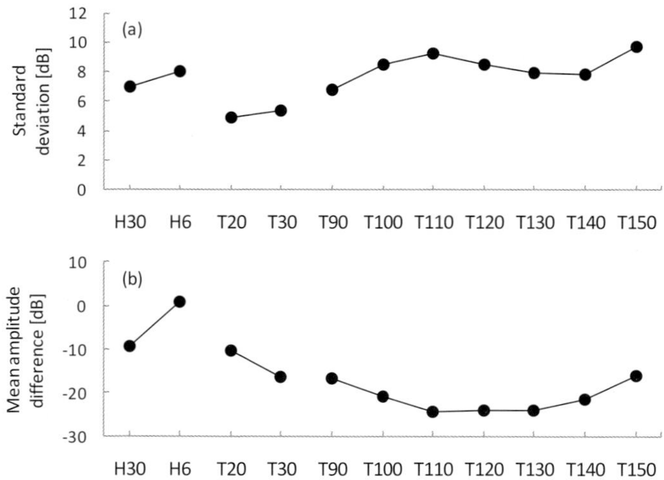

Fig.3 (a) Standard deviation of the amplitude spectrum of the direct components (200-17,000 Hz) and (b) mean amplitude difference between the direct components and the cross-talk components (200-17,000 Hz) of subject IST.

3. Localization tests

3.1. *Method*

Sound localization tests were performed using trans-aural systems with the HRTFs of each subject and the measured transfer functions between the loudspeaker and the entrances of the ear canals of the subject. The sound source was a wide-band white noise (200-17,000 Hz). The sampling frequency was 48,000 Hz. Eleven loudspeaker arrangements, as shown in Fig. 1, were considered.

The target directions were 12 directions in the horizontal plane (0 - 330°) and seven directions in the upper median plane (0 - 180°) in 30° steps. The experiment was carried out in a darkened anechoic room. The subjects were three males who participated as the subjects for the transfer function measurements. The subjects were asked to face forward without having their heads fixed. The subjects were asked to plot the perceived azimuth and elevation on the circle on a response sheet. At the beginning of the localization tests, the transfer functions between the loudspeakers and the entrances of the ear canals were measured again in order to confirm the reproducibility of the transfer functions by comparison with the transfer functions measured in Section 2.

3.2. *Results*

Figures 4 through 9 show the responses of each subject.

3.2.1. *Responses for the target directions in the horizontal plane*

For H30 and H6, front-back errors were observed for the responses of subjects IST and UEO. These subjects never perceived a sound image at the rear.

For the transverse plane arrangements, all of the subjects localized a sound image approximately in the target direction. In particular, the perceived lateral angles agreed well with the target lateral angles for T30 and T90-T130.

3.2.2. *Responses for the target directions in the median plane*

For H30 and H6, subject IST perceived all of the stimuli at the front, whereas the responses of subject ISY were relatively accurate.

For the transverse plane arrangements, subject IST localized a sound image in approximately the target directions, except for T140 and T150. The responses of subject ISY reveal that he localized sound images accurately for all of the transverse arrangements. The responses of subject UEO were not so accurate, except for T90 and T130. These results indicate that T90-T130 were relatively accurately localized by all of the subjects.

3.2.3. *Localization error*

Figure 10 shows the mean localization error, *e*, defined by the following equation:

$$e = \left|\overline{S - R}\right|, \qquad (1)$$

where *S* indicates the target direction, and *R* is the perceived direction.

The figure shows that the mean localization error for the target directions in the horizontal plane reaches a minimum at T100 (8.3°), and for the target directions in the median plane reaches a minimum at T110 (23.5°). Based on these results, T100 and T110 are considered to be the proper loudspeaker arrangements for the target direction for both the horizontal and median planes. These results agree with the results of the analysis of the measured transfer function between the loudspeakers and the entrances of the ear canals, as described in Section 2.

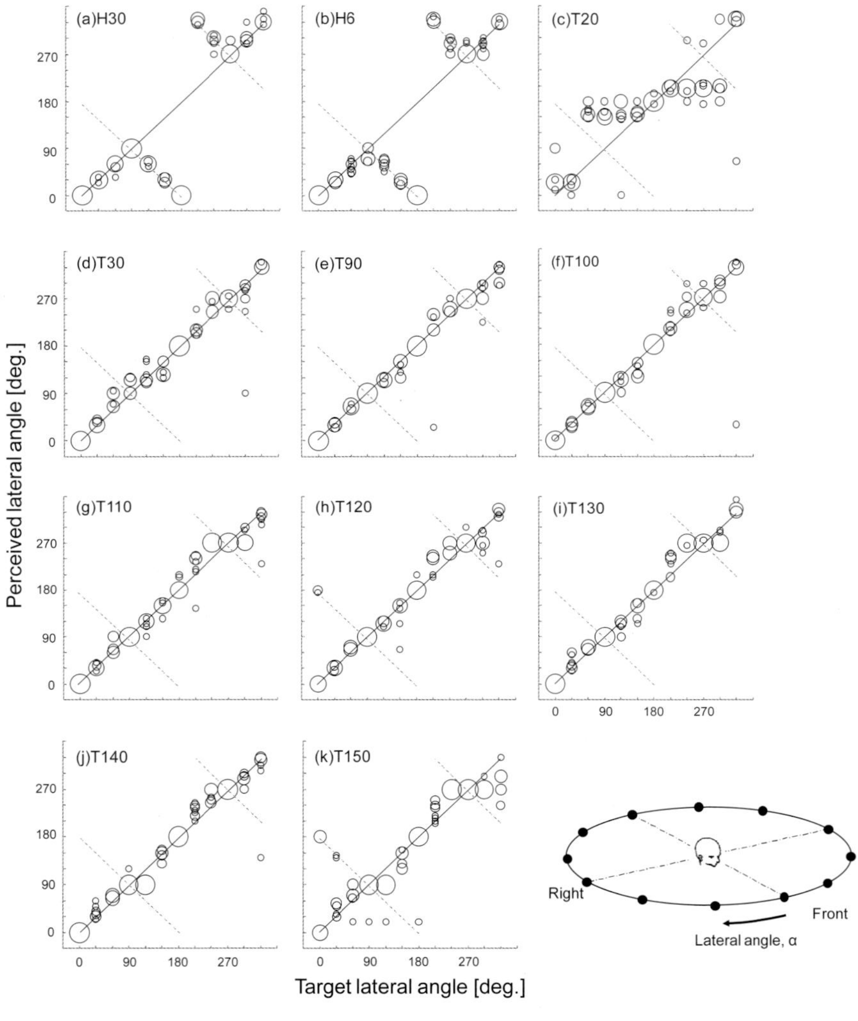

Fig.4 Localization responses for the stimuli in the horizontal plane of subject IST.

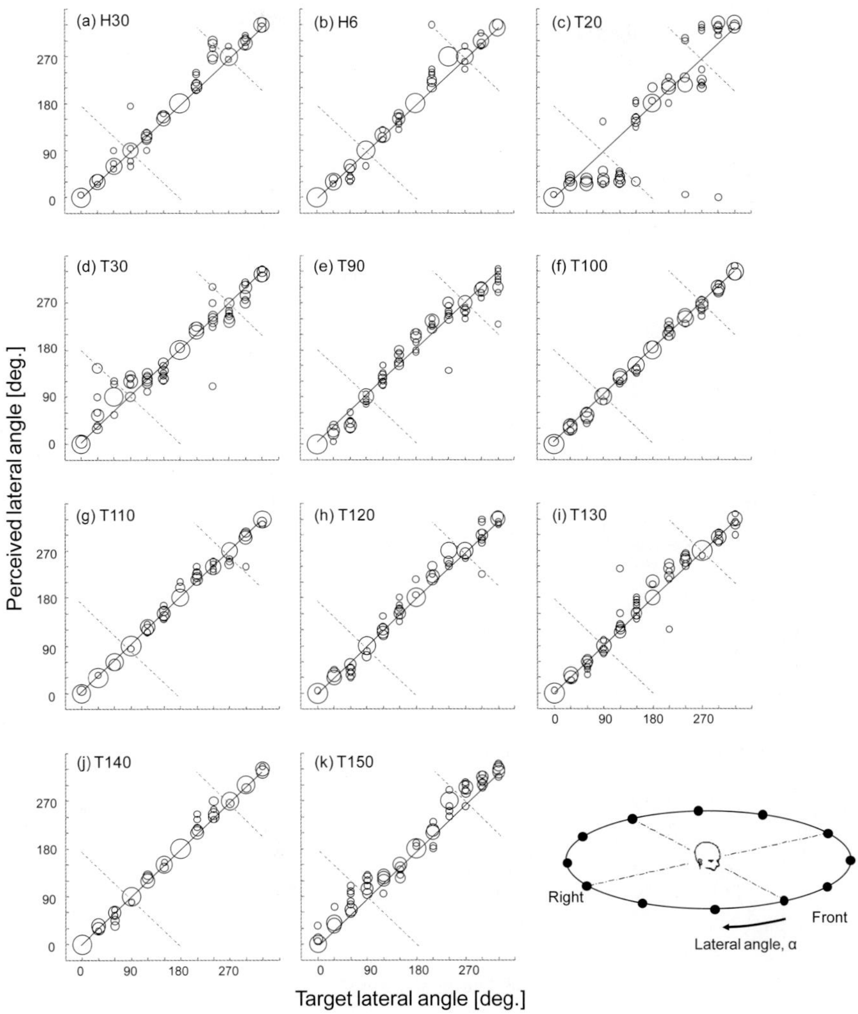

Fig.5 Localization responses for the stimuli in the horizontal plane of subject ISY.

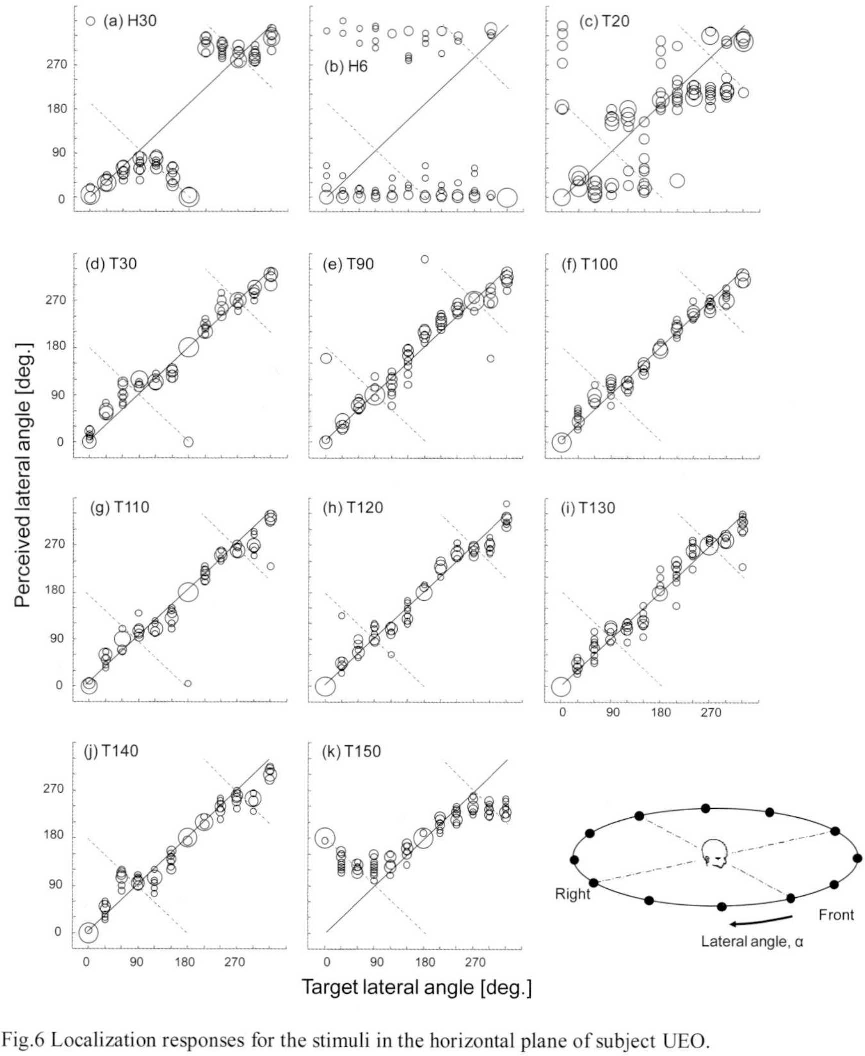

Fig.6 Localization responses for the stimuli in the horizontal plane of subject UEO.

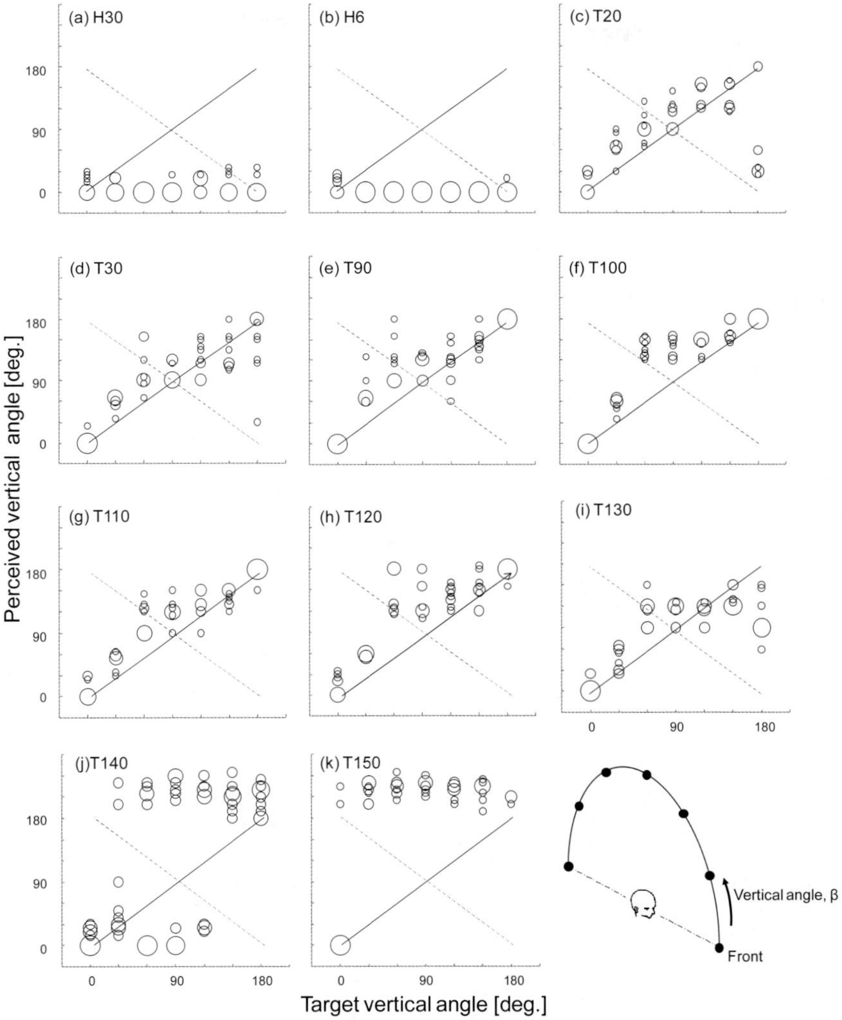

Fig.7 Localization responses for the stimuli in the median plane of subject IST.

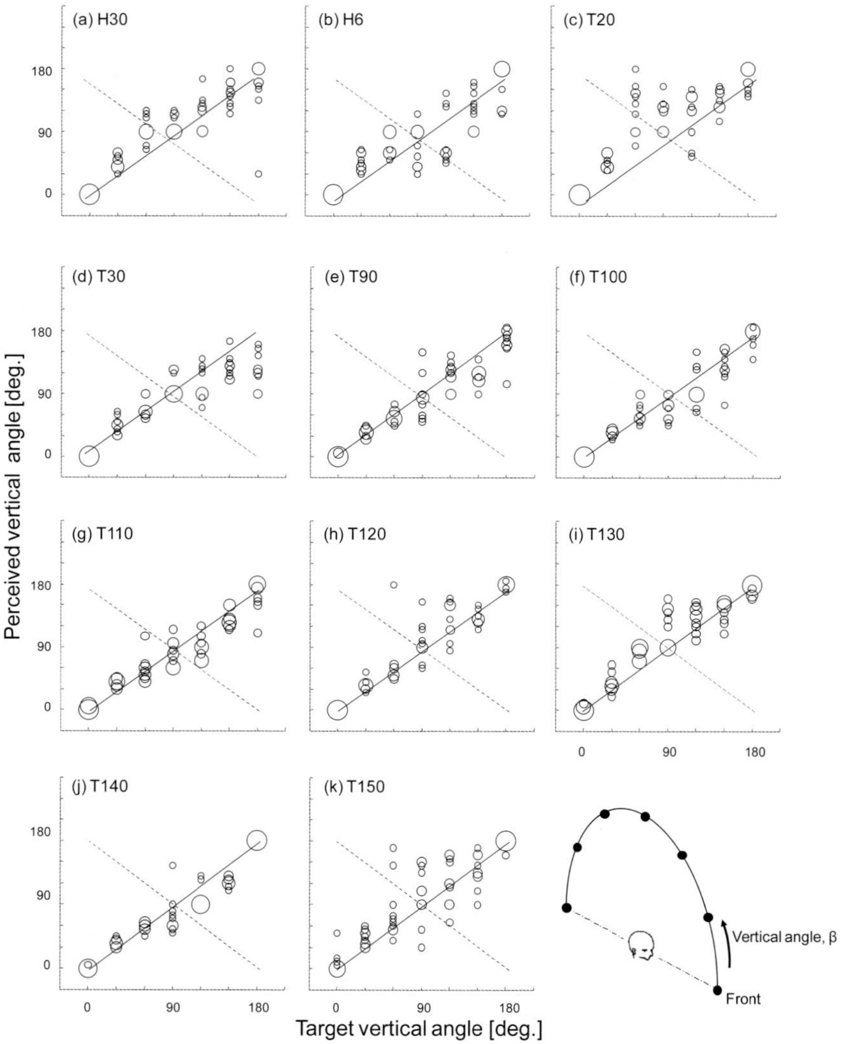

Fig.8 Localization responses for the stimuli in the median plane of subject ISY.

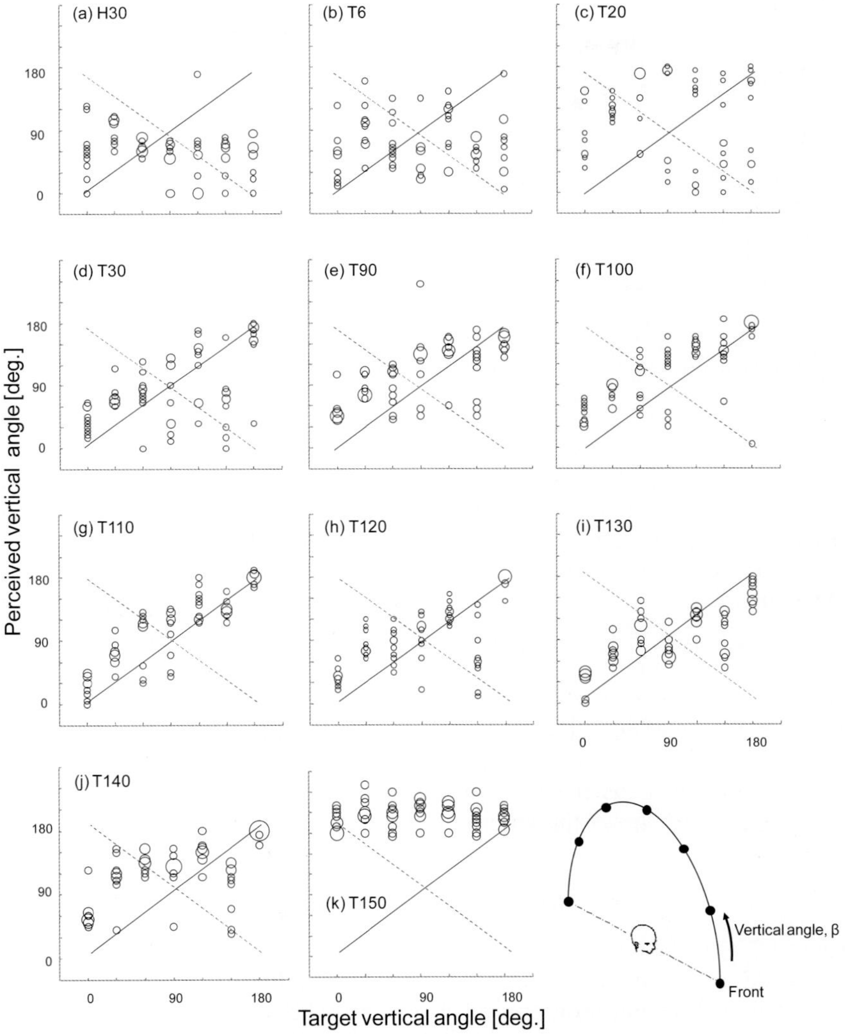

Fig.9 Localization responses for the stimuli in the median plane of subject UEO.

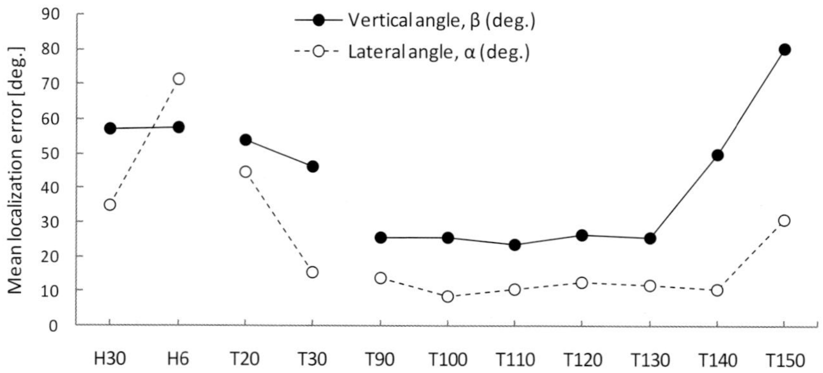

Fig.10 Mean localization error for the target directions in the horizontal and median planes.

4. Discussion

The reason why localization accuracy for the transverse loudspeaker arrangement is better than that for the horizontal loudspeaker arrangement is discussed in this section.

Figure 11(a) shows the spectrum of the signal obtained by the following equation:

$$C(\omega)\times C^{-1}(\omega)\times \mathrm{HRTF}(\omega), \tag{2}$$

where $C(\omega)$ is the transfer function between the loudspeakers and the entrances of the ear canals of the subject. The spectrum of the signal of both H6 and T100 is approximately same as that of the target HRTF, because this is the ideal condition, i.e., the transfer function is constant.

Figure 11(b) shows the spectrum of the signal obtained by the following equation:

$$C'(\omega)\times C^{-1}(\omega)\times \mathrm{HRTF}(\omega), \tag{3}$$

where $C'(\omega)$ is the transfer function between the loudspeakers and the entrances of the ear canals of the subject measured at the localization tests. A small difference in listening position caused a remarkable cancellation error for H6. For T100, significant spectral notches of the target HRTFs (N1 and N2 [4]) were reproduced, although a certain amount of cancellation error was observed. The transverse loudspeaker arrangement could be considered to be robust with respect to small differences in listening position.

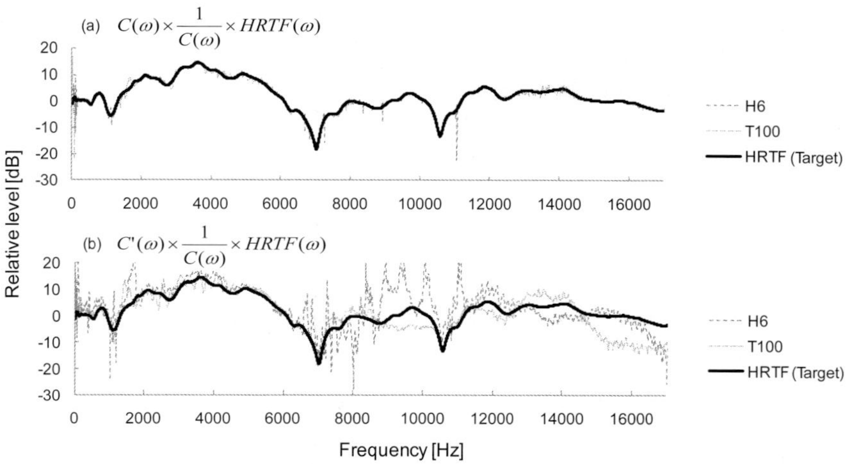

Fig. 11 Simulated HRTF by a trans-aural system for (a) the ideal condition (the transfer function is constant), and for (b) the case in which the transfer function is not constant (small difference in listening position).

5. Conclusion

We investigated the accuracy of 3D sound image control by trans-aural systems for various loudspeaker arrangements in the transverse plane, as compared with conventional arrangements. The results of the analysis of the transfer functions between the loudspeakers and the entrances of the ear canals of the subjects and sound localization tests revealed that accurate sound image control could be achieved by two loudspeakers positioned in the transverse plane at 100 - 110°.

Acknowledgments

The authors wish to thank Mr. T. Ikemi and Mr. Y. Yamamoto for their cooperation in the localization tests.

References

1. M. R. Schroeder and B. S. Atal,"Computer simulation of sound transmission in room", IEEE Intern. Conv. Rec. 11, pp.150-155 (1963).
2. O. Kirkeby, P. A. Nelson, and H. Hamada,"The stereo dipole: A virtual source imaging system using two closely spaced loudspeakers", J. Audio Eng. Soc., 45, pp.387-395 (1998).
3. M. Morimoto and Y. Ando,"On the simulation of sound localization," J. Acoust. Soc. Jpn.(E), 1(3): 167-174 (1980).

4. K. Iida, M. Itoh, A. Itagaki, and M. Morimoto,"Median plane localization using parametric model of the head-related transfer function based on spectral cues," Applied Acoustics, 68: 835-850 (2007).

SELECTIVE LISTENING POINT AUDIO BASED ON BLIND SIGNAL SEPARATION AND 3D AUDIO EFFECT

T. NISHINO

EcoTopia Science Institute, Nagoya University,
Nagoya, Aichi, Japan
E-mail: nishino@nagoya-u.jp

M. OGASAWARA

Graduate School of Information Science, Nagoya University,
Nagoya, Aichi, Japan
E-mail: ogasawara@sp.m.is.nagoya-u.ac.jp

K. NIWA

NTT Cyber Space Laboratories, NTT Corporation,
Musashino, Tokyo, Japan
E-mail: niwa.kenta@lab.ntt.co.jp

K. TAKEDA

Graduate School of Information Science, Nagoya University,
Nagoya, Aichi, Japan
E-mail: kazuya.takeda@nagoya-u.jp

We propose a novel sound field reproduction method called selective listening point audio. The proposed system uses blind source separation and stereophonic technology. In this system, multichannel acoustic signals captured at distant microphones are decomposed to virtual sound sources based on frequency-domain independent component analysis. The spatial sound is constructed at the selected listening point by convolving head-related transfer functions with a local signal mixture that is produced with virtual sound sources and classification information. In our system, an imperfect separation does not cause serious problems because the source signals are remixed in the target signal. We examined signal separation using not only linear and boundary microphone arrays but also a dodecahedral microphone array, and the signal separation performances were satisfactory. From the subjective results, the resultant spatial sound provided by our proposed method was as high as the original sounds.

Keywords: Selective listening point audio (SLP audio), Head-related transfer function (HRTF), Blind source separation, Frequency-domain independent component analysis (FD-ICA), Dodecahedral microphone array (DHMA)

1. Introduction

In most audio and visual contents, we can only view images and listen to sounds at fixed points. However, if the viewpoint and listening point could be freely selected, the user's degree of freedom would improve. Making images from an arbitrary viewpoint[1,2] has been attempted in broadcasting and movies, but not making the sounds from an arbitrary listening point. For sounds, a head-related transfer function (HRTF) can be used to control the sound images; however, the sound source signals and their locations in the environment are needed. When we use the sound signals in the actual sound environment, it is difficult to obtain sound signals and their locations because signals are mixed and reverberation and background noise exist. Therefore methods to observe or obtain the sound signals and their locations in the actual sound environment are needed.

We propose a novel sound field reproduction method called selective listening point audio (SLP audio).[3,4] SLP audio is a spatial sound reproduction system characterized by four requirements: 1) the microphones must be placed at distant locations from the sound sources, 2) the system must work on the condition that the number and the locations of the sound sources are unknown, 3) each sound source may move independently, and 4) the reproduced sound signals can be presented with ordinary equipment such as earphones, headphones, and a stereo loudspeaker system. Figure 1 shows a block diagram of the SLP audio system. Our proposed method can decompose multiple microphone signals into a set of virtual sound source information, i.e., the location and associated signals, which is the natural generalization of a typical 3D sound field representation. After decomposition, the local sound field at the selected listening point can be flexibly presented.

In this paper, we introduce SLP audio and its applications. Section 2 describes the selective listening point audio. Two kinds of microphone array systems and signal separation methods using these array systems are described in Section 3. Experiments and results are shown in Section 4. Required components and technologies for future SLP audio are discussed in Section 5. Section 6 introduces demonstration software, and Section 7 concludes this paper.

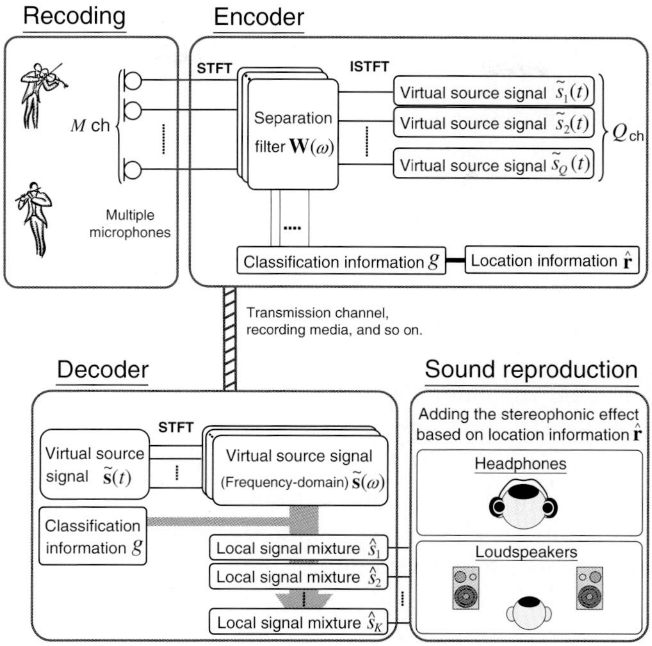

Fig. 1. Block diagram of selective listening point audio system[4] consisting of four parts: 1) recording with multiple distant microphones, 2) encoder based on blind source separation, 3) decoder based on spatial audio technique, and 4) sound reproduction with ordinary audio device.

2. Selective listening point audio

One of the simplest ways to define the 3D sound field is to specify the locations of the sound sources and the corresponding source signals:

$$\Omega = \{\mathbf{r}_n, s_n(t)\}, \qquad n = 1, \cdots, N, \tag{1}$$

where $\mathbf{r}_n$ and $s_n(t)$ denote the location and the signal of the n-th sound source. Given listening position $\mathbf{r}^{(R)}$, target sound $y(t)$ can be calculated by

$$y(t) = \sum_{n=1}^{N} h(\mathbf{r}_n, \mathbf{r}^{(R)}) * s_n(t). \tag{2}$$

Typically in the binaural audio case, column vector $\mathbf{h}(\mathbf{r}_\alpha, \mathbf{r}_\beta) = [h^{(\text{left})}(\mathbf{r}_\alpha, \mathbf{r}_\beta), h^{(\text{right})}(\mathbf{r}_\alpha, \mathbf{r}_\beta)]^T$ is used for the transfer function. In our study, since we used a binaural system based on an HRTF, the main problem of an SLP audio system is decomposing the multi-channel signals captured through M distant microphones into source information Ω.

Potentially, blind signal separation (BSS) can be used for part of the decomposing by finding estimated source information $\hat{\Omega}$. In particular, frequency-domain ICA[5] combined with advanced methods for solving permutation ambiguity[6] is powerful under realistic acoustic conditions. However, since the assumption on the number of sources is crucial in BSS, accurate estimation of the independent sources is difficult in such applications as SLP where the number of sound sources varies widely.

In a previous study, we evaluated the performance of SLP audio using BSS assuming prior knowledge of the number and locations of the sound sources. We found that imperfect separation does not cause serious problems in an SLP audio application because the source signals are remixed in the target signal. Therefore, to achieve an SLP audio system, we extended the BSS algorithm to operate it without any prior knowledge of the sound sources and built a decomposing algorithm that converts the multi-channel signals into estimated source information $\hat{\Omega}$.

3. Recoding system

The SLP audio system needs many microphones and microphone arrays to obtain sound signals and their locations. In our study, two different arrays were examined. One of them surrounds target sound sources, and the other is surrounded by target sources.

3.1. *Linear and boundary microphone arrays*

Figures 2 and 3 show a linear microphone array and a boundary microphone array, respectively. Multiple arrays were arranged to surround sound sources. Not only combinations of the linear and boundary microphone arrays but also one or the other array can be used.

We used a linear microphone array with four microphones installed at intervals of 2 cm and a boundary microphone array made from ABS resin with seven microphones at intervals of 1.4 cm. In section 4.2, we used seven linear microphone arrays and four boundary microphone arrays. A signal separation method using these arrays, proposed by Niwa *et al.*,[4] was applied to the recorded signals.

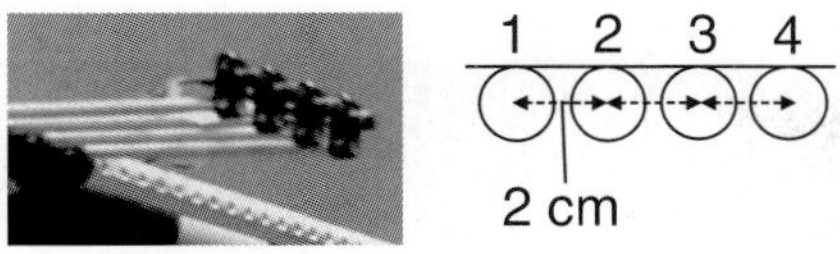

Fig. 2. Linear microphone array.

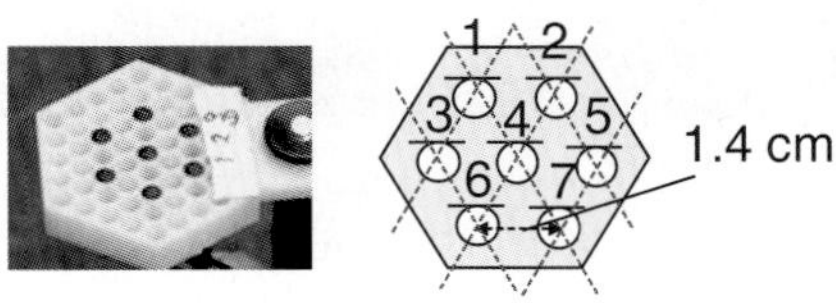

Fig. 3. Boundary microphone array.

3.2. *Dodecahedral microphone array*

Figure 4 shows our developed dodecahedral microphone array (DHMA) device.[7] Its diameter is 8 cm, and the interval between adjacent faces is 36°. Microphones can be installed on ten faces, except for the top and bottom faces, and 16 holes appear on each face. The distance is 7 mm between the centers of the holes on the same face. The top and bottom faces are used for installing the microphone stand. The observed signals at each face have different acoustic features such as sound pressure level, arrival time, and influence of diffraction waves. Our proposed method uses these features to group the frequency components of the separated signals obtained by FD-ICA.

3.2.1. *Solving permutation problem using dodecahedral microphone array*[7]

Figure 5 shows an outline of the entire separation process. Our proposed method uses amplitude and phase information observed by the developed array to solve the permutation problem by grouping the frequency features and relating them to the sound source location. This is equivalent to grouping transfer functions $\mathbf{w}^{+}(f)$ between the sound source and the microphone. This transfer function corresponds to each column vector of the pseudo-inverse matrix of separation filter $\mathbf{W}(f)$, which is obtained by FD-ICA. $\mathbf{w}^{+}(f)$ is grouped by the k-means algorithm to calculate the centroid, and then the order of the transfer function is determined in every frequency bin. k-means clustering needs a cost function. The cost function

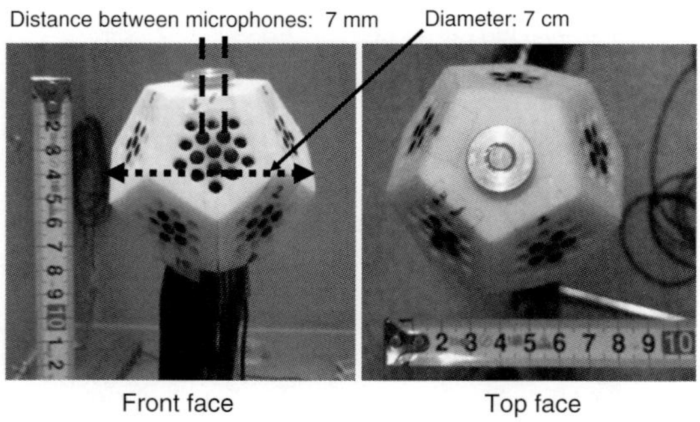

Fig. 4. Developed dodecahedral microphone array made from ABS resin. Ten faces, except the top and bottom, are available for installing microphones, and the maximum number of microphones is 160. Here, six microphones are installed around the center of each face.

of the conventional method evaluates the similarities of the amplitude and the phase at the same weight,[8] but we propose a new cost function based on the human sound localization ability. A human's sound localization cue was described by the duplex theory[9] in the early stages, and the ITD effect is large at frequencies below 1.6 kHz, where the wavelength corresponds to the order of the head size. For frequencies above 1.6 kHz, ILD is a primary factor.[10,11] Therefore, the proposed method emphasizes the importance of phase information for low frequencies and amplitude for high frequencies.

First, the k-means algorithm clusters transfer function $\mathbf{w}^+(f)$ for all frequencies. The similarity between q-th transfer function $\mathbf{w}_q^+(f)$ and k-th centroid $\mathbf{c}_k$ is evaluated by the following cost function:

$$\mathcal{J}\left(\mathbf{w}_q^+(f), \mathbf{c}_k\right) = a(f)\,\mathcal{D}_a + b(f)\,\mathcal{D}_p, \tag{3}$$

where $\mathcal{D}_a$ and $\mathcal{D}_p$ are the similarities of amplitude and phase, respectively.[7] In the case of the surrounding microphone array, the cost function used the similarity of phase $\mathcal{D}_p$ only. However, since DHMA can observes differences in the sound pressure level and phase, the cost function represented by Eq.(3) is able to solve the permutation problem. Here, $a(f)$ and $b(f)$ are weight functions defined by

$$a(f) = \left\{\frac{f}{F_s/2}\right\}^n, \quad b(f) = 1 - a(f), \tag{4}$$

where f is the frequency in Hz and F_s is the sampling frequency. Preliminary experiments determined the appropriate n for the weighting function. The permutation error rate was evaluated with the three sound source conditions while changing n. The preliminary experiment results are shown in

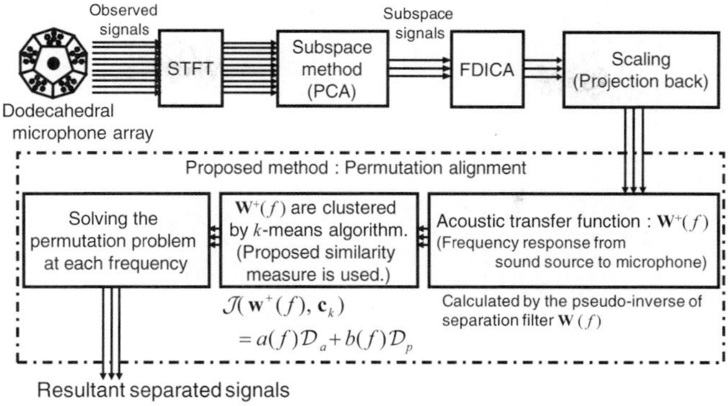

Fig. 5. Block diagram of separation procedure with dodecahedral microphone array. Our proposed part is shown in the center (permutation alignment scheme). Dimension of observed signals is reduced by a subspace method using principle component analysis (PCA).[12]

Fig. 6. $n = 1$ is an appropriate value by considering both the balance of the information on the amplitude and phase and the separation performance. Therefore, the weight function for $n = 1$ is used in the following experiment. The distances between the centroid and the transfer function that correspond to all sources are evaluated for each frequency. Finally, permutation matrix $\mathbf{\Pi}(f)$ is estimated:

$$\mathbf{\Pi}(f) = \underset{\mathbf{\Pi}}{\operatorname{argmax}} \sum_{k=1}^{N} \mathcal{J}\left(\mathbf{w}^{+}_{\mathbf{\Pi}_k}(f), \mathbf{c}_k\right), \tag{5}$$

where N is the number of sound sources.

4. Experiments

The performances of the proposed methods were evaluated through sound source separation experiments.

4.1. *Signal separation using dodecahedral microphone array*

4.1.1. *Experimental conditions*

In this experiment, from three to twelve sound sources were arranged. Speech and musical instrument signals were used. All sound sources and DHMA were located on the same horizontal plane, which was 130 cm from the floor. Figure 7 shows the arrangement of the 12-source case. All sound sources were positioned at equal intervals. Other experimental conditions

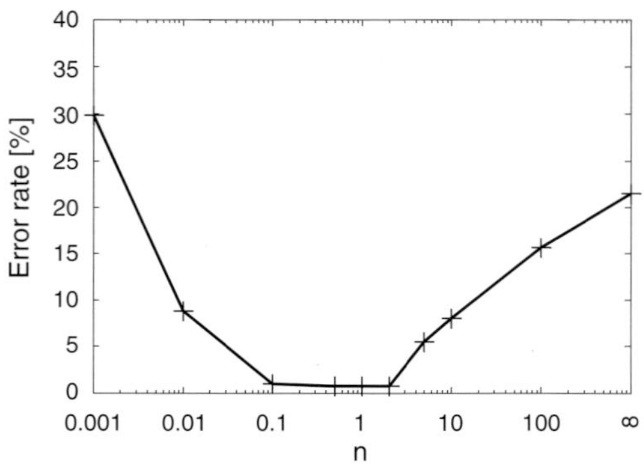

Fig. 6. Preliminary experimental results to determine appropriate n for weighting function (Eq. (4)). Permutation error rate was evaluated with three sound source conditions while changing n.

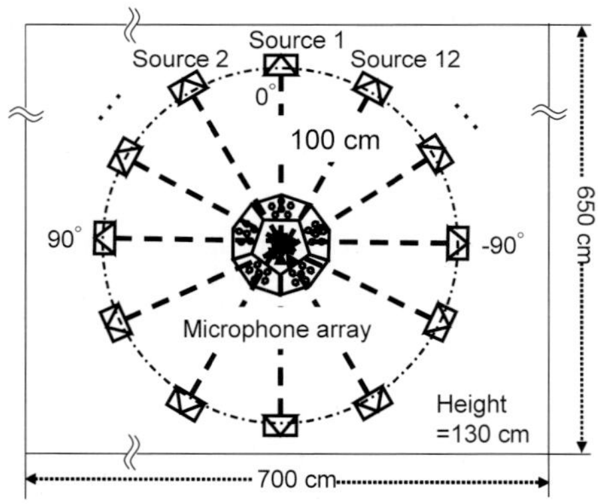

Fig. 7. Experimental arrangement for 12 sound sources under multiple-sound-source condition.

are shown in Table 1. The number of sound sources is given throughout the experiments. We assumed that the source directions were unknown. Test signals were generated by convolving the dry source and measuring the acoustic transfer functions. The experiments were performed in a sound-proof chamber with a reverberation time of 138 msec.

4.1.2. *Results*

The performances were compared with those of the ideal condition and the conventional method.[8] In the ideal condition, permutation was solved by taking the correlation with the original source signal for each frequency

Table 1. Experimental conditions for signal separation using dodecahedral microphone array.

Sampling frequency	40 kHz
Frame length	1024 points (64 ms)
Frame shift	256 points (16 ms)
Window function	Hanning
FFT point	1024
Background noise level	17.7 dB(A)
Sound pressure level (1 m)	75.4 dB(A)
Number of microphones	60
Number of sound signals	speech (3, 4, 5, 6, 8, 10, 12) musical instrument (4, 4, 5, 6)
Signal duration	5 s

bin. The conventional method uses time delays and amplitude differences in equal weights to evaluate the similarity between acoustic transfer functions. Separation performances were evaluated by the improvement scores of the signal-to-interference ratio (SIR).[7,8]

$$\text{SIR improvement}_n = \text{OutputSIR}_n - \text{InputSIR}_n \quad [\text{dB}], \tag{6}$$

$$\text{InputSIR}_n = 10\log_{10}\left[\frac{\sum_t x_{mn}(t)^2}{\sum_t\{\sum_{s\neq n} x_{ms}(t)\}^2}\right] \quad [\text{dB}], \tag{7}$$

$$\text{OutputSIR}_n = 10\log_{10}\left[\frac{\sum_t y_{nn}(t)^2}{\sum_t\{\sum_{s\neq n} y_{ns}(t)\}^2}\right] \quad [\text{dB}], \tag{8}$$

where x_{ms} is an input signal from source signal s observed by microphone m and y_{ns} is an output signal from source signal n processed by separation filter w_s.

Figures 8 and 9 show the average SIR improvement scores of the speech and musical instrument signals, respectively, as a function of the number of sound sources. The separation of the proposed method was better than the conventional method, and the separation performances of the speech signals were especially close to the ideal condition for up to six sound sources. The performances of the musical instrument signals were also better than the conventional method. These results indicate that the proposed method is superior to the conventional method and its separation is accurate. However, since the number of sound sources was given in the experiments, an improved method must include a method of estimating the number of sound sources.

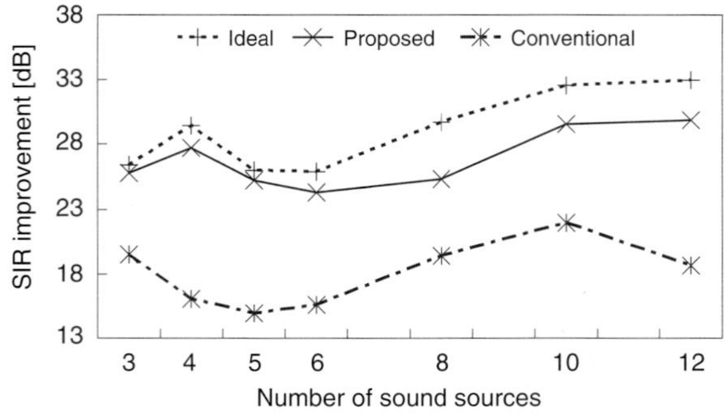

Fig. 8. Average SIR improvement scores of speech signals.

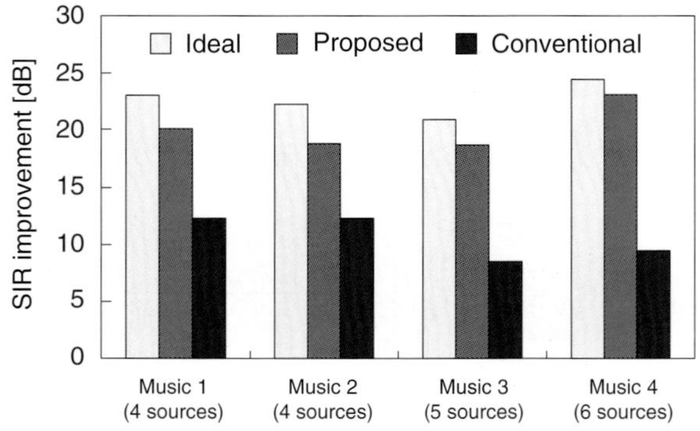

Fig. 9. Average SIR improvement scores of musical instrument signals.

4.2. *Combining sounds with images*

From the results of the previous study[4] and those of Section 4.1, signal separation performances were good. In this experiment, we investigate the effectiveness of the SLP audio by evaluating a system that combines sounds with images.

4.2.1. *Experimental conditions*

The proposed system was evaluated using a musical performance called "Variation on a theme by Joseph Haydn" by Brahms. A small orchestra with 10 players performed in a lecture room whose reverberation time (RT_{60}) was 423 ms. Figure 10 shows the recording conditions. Eleven microphone arrays consisting of non-directional microphones surrounded the players,

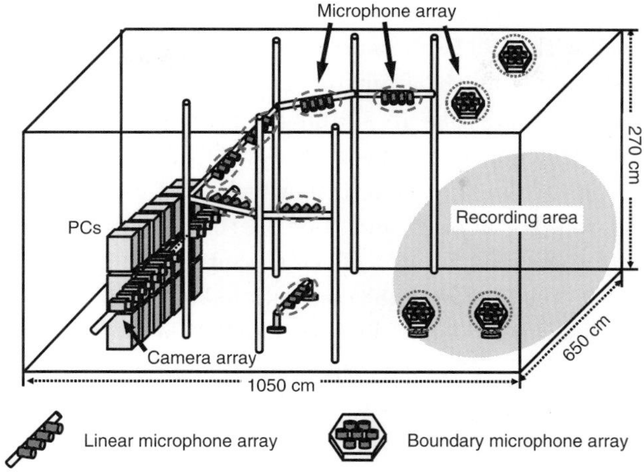

Fig. 10. Recording conditions: Seven linear microphone arrays and four boundary microphone arrays were used. Background noise level was 45.9 dB(A).

Table 2. Recording and analysis parameters.

Number of microphone arrays	11
Number of microphone	56
Sampling frequency	48 kHz
Number of virtual sources, Q	10
Number of clusters, K	24

and the recording system[13] simultaneously recorded 100-ch images and audio. The analysis parameters are shown in Table 2. The obtained estimated local signal mixtures were convolved with HRTFs in the appropriate direction. The measured HRTFs were used to add a spatial impression to the estimated local signal mixtures. The HRTFs in the non-measured directions were obtained by an interpolation method.[14] Spatial impressions such as sound source distance and direction were added by the HRTFs. However, only HRTFs on the horizontal plane were used in this experiment.

4.2.2. *Results*

A very simple subjective test was used as a preliminary evaluation of whether the reproduced signals were suitable for images. Three subjects evaluated the signals with a subjective score: good (3), poor (2), and bad (1). All subjects were males aged 22 – 24 years. The images at the three locations shown in Fig. 11 were evaluated. The reproduced sound signals were convolved with HRTFs measured with a dummy head (B&K, 4128).[15] The observed signals at the front of "Loc 2" were used for comparison. The duration of stimuli was 15 s, and the silence between stimuli was 3 s. Stimuli were transduced by headphones (SONY, MDR-7506). Figure 12 shows the results of the subjective test, which indicate that the SLP audio system works well. For "Loc 1" and "Loc 3," the proposed method provided suitable sound signals for images. Since the interaural differences were not coordinated with the images in the conventional method, the results of "Loc 1" and "Loc 3" seemed reasonable. For "Loc 2," the proposed method's results are inferior to those of the conventional method. The noises brought from the signal processing and the unsuitable HRTFs for the subjects are probably the main causes.

5. Discussion

Our previous studies and this paper's results suggest that an exciting sound reproduction or communication system can be developed by signal separation and spatial audio techniques. However, many other issues need further study. The optimal array arrangement and performance under more reverberant and/or noisy conditions were insufficiently investigated. The array arrangement and the number of microphones and arrays are especially important problems, although they may depend on the target contents. Dealing with a non-stationary sound field, e.g., moving sources, is also important.

On the other hand, better spatial sound reproduction needs suitable HRTFs and such sound effects as reverberation and depth perception. In our study, HRTFs were used for the 3D sound reproduction; however, we did not address such HRTF problems as individuality and sound localization. Obtaining suitable HRTF for users is not easy. Therefore, improving the measurement method,[16,17] developing the equipment,[18] and comprehending the characteristics of HRTFs are very important tasks. Establishing a subjective evaluation method is also necessary when the image and sound are simultaneously presented to the subject, as is developing a reproduction system that is connected with head motion.[19,20]

6. Demonstration software

We developed simple demonstration software to appreciate the musical performance at the selected location. Figure 13 shows the image window and

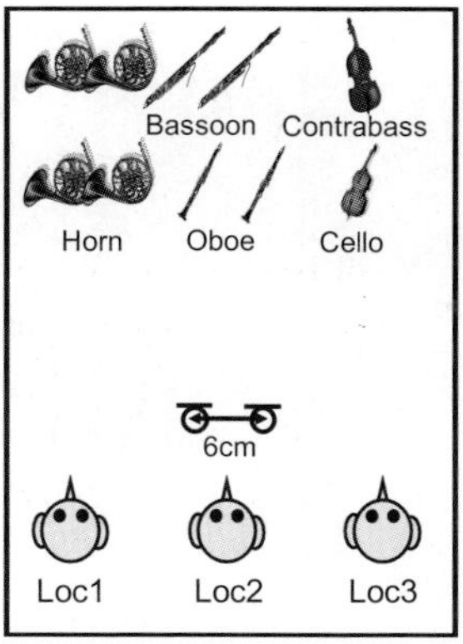

Fig. 11. Experimental conditions of subjective test. Two microphones located at the front of "Loc 2" recorded stereo signals for comparison.

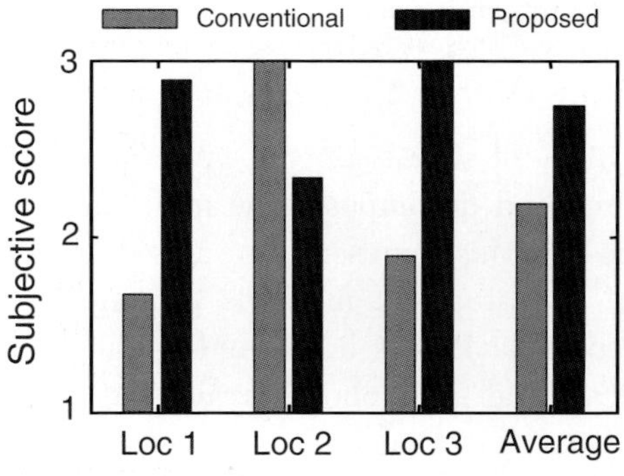

Fig. 12. Results of subjective test.

control panel. The square represents the user's location, which can be moved to an arbitrary point, and the circles are estimated source information $\hat{\Omega}$. There are many circles at the top of the control panel, and these are related to players. The two circles located at the bottom of the control panel consist of reflection waves. However, since no stereophonic effect was used in this software to increase execution performance, it must be improved.

7. Conclusions

In this paper, we introduced and evaluated a new spatial audio scheme: a selective listening point audio system. In this system, a 3D acoustic field

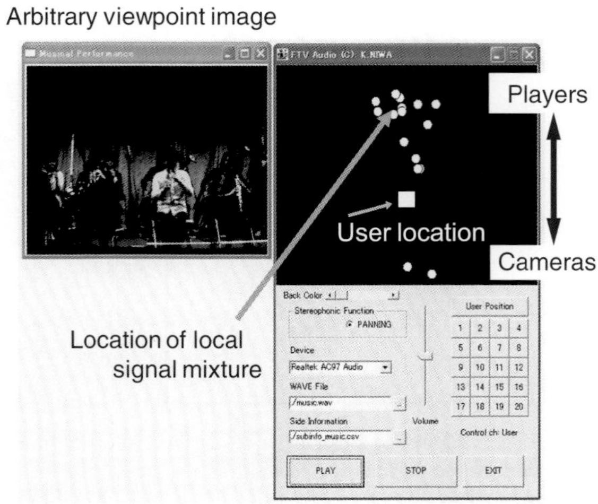

Fig. 13. SLP audio software. Left window displays an image at an arbitrary viewpoint. Right window has a control panel that can move the user's location. Square in right window is user's location and circles are local signal mixtures.

is represented by a set of signal sources with their locations and associated signals. Our method decomposes the multi-channel signals recorded at distant locations into this representation based on BSS technologies.

For evaluation, the proposed methods decomposed signals captured through not only combinations of linear and boundary microphone arrays but also the dodecahedral microphone array. The performances of signal separation were better than that of the conventional method. Subjective evaluation showed its effectiveness and revealed that the spatial impression of the resultant spatial sound was as high as the reference sounds. Future works include investigating the number of microphones and their arrangement, reducing computational complexity, and developing a real-time SLP audio system.

References

1. http://www.ri.cmu.edu/events/sb35/tksuperbowl.html.
2. T. Fujii and M. Tanimoto, Free-viewpoint TV system based on the ray-space representation, *SPIE ITCom* **4864-22**, 175 (2002).
3. K. Niwa, T. Nishino and K. Takeda, Encoding large array signals into 3D sound field representation for selective listening point audio based on blind source separation, in *IEEE International Conference on Acoustics, Speech, and Signal Processing (ICASSP 2008)*, (Las Vegas, Nevada, U.S.A., 2008).
4. K. Niwa, T. Nishino and K. Takeda, Selective listening point audio based on

blind signal separation and stereophonic technology, *IEICE Trans. on Inf. & Sys.* **E92-D**, 469 (2009).
5. P. Smaragdis, Blind separation of convolved mixtures in the frequency domain, *Neurocomputing* **22**, 21 (1998).
6. H. Saruwatari, S. Kurita, K. Takeda, F. Itakura, T. Nishikawa and K. Shikano, Blind source separation combining independent component analysis and beamforming, *EURASIP J. Applied Signal Processing* **2003**, 1135 (2003).
7. M. Ogasawara, T. Nishino and K. Takeda, Blind source separation based on acoustic pressure distribution and normalized relative phase using dodecahedral microphone array, in *The 17th European Signal Processing Conference (EUSIPCO2009)*, (Glasgow, Scotland, 2009).
8. H. Sawada, S. Araki, R. Mukai and S. Makino, Solving the permutation problem of frequency-domain bss when spatial aliasing occurs with wide sensor spacing, in *IEEE International Conference on Acoustics, Speech, and Signal Processing (ICASSP 2006)*, (Toulouse, France, 2006).
9. L. Rayleigh, On our perception of sound direction, *Philos. Mag.* **13**, 214 (1907).
10. G. G. Harris, Binaural interactions of impulsive stimuli and pure tone, *J. Acoust. Soc. Am.* **32**, 685 (1960).
11. J. Blauert, *Spatial hearing (revised edition)* (The MIT Press, 1996).
12. M. Wax and T. Kailath, Detection of signals by information theoretic criteria, *IEEE Trans. Acoustics, Speech, and Signal Processing* **33**, 387 (1985).
13. T. Fujii, K. Mori, K. Takeda, K. Mase, M. Tanimoto and Y. Suenaga, Multipoint measuring system for video and sound - 100-camera and microphone system, in *2006 IEEE International Conference on Multimedia and Expo*, (Toronto, Ontario, Canada, 2006).
14. T. Nishino, S. Mase, S. Kajita, K. Takeda and F. Itakura, Interpolating HRTF for auditory virtual reality, in *The Third Joint Meeting ASA and ASJ*, (Honolulu, Hawaii, U.S.A., 1996).
15. http://www.sp.m.is.nagoya-u.ac.jp/HRTF/.
16. D. N. Zotkin, R. Duraiswami, E. Grassi and N. A. Gumerov, Fast head-related transfer function measurement via reciprocity, *J. Acoust. Soc. Am.* **120**, 2202 (2006).
17. K. Fukudome, T. Suetsugu, T. Ueshin, R. Idegami and K. Takeya, The fast measurement of head related impulse responses for all azimuthal directions using the continuous measurement method with a servo-swiveled chair, *Applied Acoustics* **68**, 864 (2007).
18. S. Hosoe, T. Nishino, K. Itou and K. Takeda, Development of micro-dodecahedral loudspeaker for measuring head-related transfer functions in the proximal region, in *IEEE International Conference on Acoustics, Speech, and Signal Processing (ICASSP 2006)*, (Toulouse, France, 2006).
19. S. Yairi, Y. Iwaya and Y. Suzuki, Development of virtual auditory display software responsive to head movement, *Trans. VR Soc of Jpn.* **11**, 437 (2006).
20. M. Otani, T. Hirahara and S. Ise, Pan-spatial dynamic virtual auditory display, *Trans. VR Soc of Jpn.* **12**, 453 (2007).

SWEET SPOT SIZE IN VIRTUAL SOUND REPRODUCTION: A TEMPORAL ANALYSIS

Y. LACOUTURE PARODI* and P. RUBAK

Department of Electronic Systems, Section of Acoustics
Aalborg University Aalborg East, 9220, Denmark
**E-mail: ylp@es.aau.dk*
http://www.es.aau.dk/sections/acoustics/

The influence of head misalignments on the performance of binaural reproduction systems through loudspeakers is often evaluated based on the amplitude ratio between the crosstalk and the direct signals. The changes in magnitude give us an idea of how much of the crosstalk is leaked into the direct signal and therefore a sweet spot performance can be estimated. However, as we move our heads, the time information of the binaural signals is also affected. This can result in ambiguous cues that can destroy the virtual experience. In this paper, we present an analysis in the time domain of the influence of head misalignments. Using the interaural cross-correlation we estimated the interaural time delay and defined a sweet spot. The analysis is based on measurements carried out on 21 different loudspeaker configurations, including two- and four-channel arrangements. Results show that closely spaced loudspeakers are more robust to lateral displacements than wider span angles. Additionally, the sweet spot as a function of head rotations increases systematically when the loudspeakers are placed at elevated positions.

Keywords: Virtual acoustics; Crosstalk cancellation; Sweet spot; Interaural time delay; Stereo dipoles.

1. Introduction

The reproduction of an authentic auditory event is possible if the sound signals at the ears matches the sound pressures of the real environment. This is the central idea of binaural techniques and it is based on the assumption that the sound pressures at the ears control our perception of any auditory event. Virtual auditory events can be rendered through headphones or loudspeakers.

One of the biggest challenges of binaural reproduction through loudspeakers is to avoid that the signals that are to be hear in one ear are also hear in the other. This problem can be solved by introducing the appropriate filters into the reproduction chain. These filters are usually designed for a fixed position.

Head movements are known to add important dynamic cues to the localization of sound sources. However, in binaural reproduction systems through loudspeakers, when the listener moves the head, the transfer functions of the acoustical paths from the loudspeakers to the ears do not longer correspond to the transfer functions used to design the filters. This can result in leakages from the contralateral path into the ipsilateral path and thus, the virtual image can be destroyed.

The maximum amount of displacement allowed such that the errors introduced do not significantly affect the virtual reality effect had been the focus of different studies in the past.[1–4] In most instances, these analysis were conducted based on magnitude ratios between the crosstalk and the direct signals, given that with the spectral information we can easily observe how much of the signal from the contralateral path leaks into the ipsilateral path. However, spectral information might not be sufficient to assess the space region in which the errors introduced are negligible. As we move our heads, the time information of the signals changes accordingly, introducing delay errors into the desired binaural signal. This can also produce conflicting cues and therefore destroy the spatial perception.

The robustness of temporal cues were discussed by Takeuchi *et. al.* in Ref. 1. In their analysis, they employed a free field model and the head related transfer functions (HRTF) from a head and torso simulator. They only analyzed two loudspeaker configurations: two-channel configurations with 10° and 60° span angles placed on the horizontal plane. In this paper we present a temporal analysis of the sweet spot for 21 different loudspeaker configurations, including two- and four-channel arrangements placed at different elevations. The analysis is based on measurements carried out at the acoustical laboratories at Aalborg university and is intended as a complement of the channel separation analysis presented in Ref. 4.

2. Crosstalk Cancellation

The purpose of a crosstalk cancellation network is to cancel the signals that arrive from the contralateral path, so that the binaural signals are reproduced at the ears in the same way they would be reproduced through headphones. Figure 1 illustrates a simplified diagram of a binaural reproduction system through loudspeakers. The blocks C_{ji} represents a set of optimal filters and the functions h_{ji} describes the acoustical paths from the jth loudspeakers to the ith ear. The signal d_i contains the binaural signal that is to be reproduced at the ith ear and v_i is the signal that is actually reproduced at the ith ear.

Perfect crosstalk cancellation is achieved when $d_i = v_i$. In other words, we need a set of filters C_{ji} such that $\mathbf{H} \cdot \mathbf{C} = \mathbf{I}$. Here $\mathbf{C}$ is an $n \times 2$ matrix containing the crosstalk cancellation filters, where n is the number of loudspeakers. $\mathbf{H}$ is an $2 \times n$ matrix which contains the transfer functions describing the acoustical paths from the sources to the ears and which we will refer to as the plant matrix. $\mathbf{I}$ is the

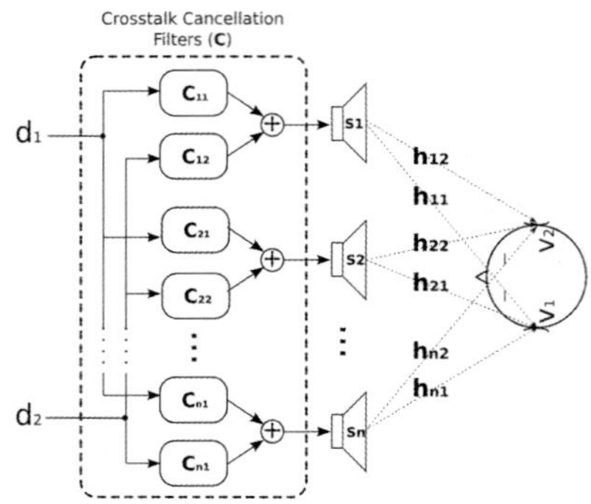

Fig. 1. Simplified diagram of a crosstalk cancellation system.

identity matrix. The problem is basically to find the inverse of **H**.

The plant matrix **H** is generally singular and therefore not invertible. Besides, when the reproduction system consists on more than two loudspeakers, the equation system becomes overdetermined and a direct inversion is not feasible. Thus, it is necessary to model the system such that we can obtain an approximation that is closed enough to the required solution.

There exists a number of methods to obtain the optimal inverse filters **C**. In this study, we implemented three different crosstalk cancellation techniques. The first one, which we will refer as the generic crosstalk canceler (GCC), applies the exact matrix inverse definition. It obtains the filters by inverting directly a matrix composed of the minimum-phase section of the plant transfer functions h_{ji}. It models the interaural transfer functions (ITF) as the ratio between the minimum phase component of the ipsilateral and contralateral transfer functions. The all-pass section of the transfer functions are approximated to a frequency independent delay. This is based on the assumption that the phase of the all-pass section is approximately linear at low frequencies. Given that it is based on a direct matrix inverse, this methods is only applicable for two-channel arrangements.

The other two methods are based on least square approximations. These methods do not try to invert the plant matrix directly but seek for the best approximation which result in minimum errors. One of the methods we implemented is the so called fast deconvolution method, which is based on the fast fourier transform. We will refer to it in the results as LS_f. The other method calculates the optimal filters in the time domain, using matrices that contain digital FIR filters. We will refer to this method as LS_t. Frequency dependent regularization was incorporated. A detailed description of the methods and the implementation can be found in Refs. 5 and 4.

3. Changes in the Interaural Time Delay

In Ref. 4 we presented an analysis of the sweet spot size as function of lateral displacements and head rotations. This analysis was done for 21 different loudspeaker configurations, including two- and four-channel arrangements. In that study, we made use of the absolute and relative sweet spot definitions described in Ref. 3. The first one is defined as as the maximum displacement from the nominal center position that results in an average channel separation index larger than -12dB. The second one is defined as the maximum displacement from the nominal center position that results in a channel separation index degradation of 12dB with respect to the nominal center position. This is,

$$\text{Absolute sweet spot} := max\left\{d|\ \overline{CHSP}_d \leq -12dB\right\} \quad (1)$$

$$\text{Relative sweet spot} := max\left\{d|\ \overline{CHSP}_d \leq \overline{CHSP}_o + 12dB\right\}, \quad (2)$$

where the variable d corresponds to either a lateral displacement or a head rotation from the nominal center position. $\overline{CHSP}_d$ is the channel separation index at the position d and $\overline{CHSP}_o$ is the channel separation index at the nominal center position. The channel separation index is defined as the magnitude ratio between the contralateral and ipsilateral signals.[4]

In general, we observed that a wider control area is obtained when the loudspeakers are closely spaced and at elevated positions. Additionally, results showed us that the two-channel configurations tend to be more robust and result in wider control area than when using the four-channel configurations. However, questions were still open on whether such movements influence the phase information in a similar manner and whether this phase changes depend only on the loudspeakers positions or also on the method used to calculate the filters.

In an ideal situation, if we send a pair of impulses at the same time through our binaural reproduction system illustrated in Fig. 1, the reproduced signals v_i will be the same impulses with the same delay. Thus, the interaural time delay (ITD) of the binaural reproduction system will equal $0\mu s$. When the listener moves the head this ITD changes accordingly to the movement. This time difference is introduced into the desired binaural signal, changing the original ITD and therefore generating ambiguous cues.

Several methods to determine the ITD can be found in the literature. In Ref. 6, Minnaar *et. al.* described and compared different methods to calculate the ITD. It is suggested that determining the ITD by calculating the group delay of the excess phase components evaluated at 0Hz is numerically more robust and consistent than other methods proposed in the literature. However, if the group delay at 0Hz is incorrect due to for example a high-pass filtering, results from this methods are not longer consistent. In Ref. 1, Takeuchi *et. al* propose to use the interaural cross-correlation (IACC) to determine the ITD and analyze the temporal changes. According to Minnaar's results,[6] the IACC consistently overestimated the ITD values. Nevertheless, the relative variation with respect to angle was similar to the variations observed with the other methods. One advantage of the IACC method is that it is less sensitive to noise as opposed to methods such as the leading-edge. Furthermore, there are indications that the nervous system calculates the ITD by means of a cross-correlation.[7] Thus, it can be considered a better approximation of the auditory system. Based on these arguments and given that the analysis presented here was done with high-pass filtered signals, we decided to use the IACC to estimate the ITD changes of our reproduction system as a function of head movements. The discrete time IACC is defined as:

$$\Psi(m) = \begin{cases} \sum_{n=0}^{N-m-1} p_1(n)p_2(n+m) & m \geq 0 \\ \Psi^*(-m) & m < 0 \end{cases} \tag{3}$$

where $p_i(t) = R_i^{contra}(t) + R_i^{ipsi}(t)$ is the linear combination of the ipsilateral and contralateral signals, when an impulse is sent to the ith ear and N is the length of the signals.[1] Here the ITD of the binaural reproduction system can be estimated as the delay corresponding to the maximum in the cross-correlation function.

Now, we need to define a threshold for the ITD discrimination in order to assess the sweet spot quantitatively. Experiments presented in Ref. 8 suggest that the audibility threshold for the ITD is $10\mu s$. In Ref. 9 larger values were obtained when evaluated with naïve listeners. However, these results showed a large variance between subjects. Thus, $10\mu s$ can be considered a strict but safe limit. Following this line, we defined the sweet spot as the maximum head misaligment allowed, such that the ITD difference between the nominal center position and the new position does not exceed $10\mu s$. In other words,

$$\text{ITD sweet spot} := max\{d|\ |ITD_o - ITD_d| \leq 10\mu s\}, \tag{4}$$

where ITD_o and ITD_d are the ITDs at the nominal center position and at a laterally displaced or rotated position respectively. In order to distinguish from the other sweet spot definitions mentioned above, we will refer to this definition as the ITD sweet spot. Since we are not assuming symmetry, the total sweet spot is calculated by adding the maximum leftward misalignment and the maximum rightward misalignment.

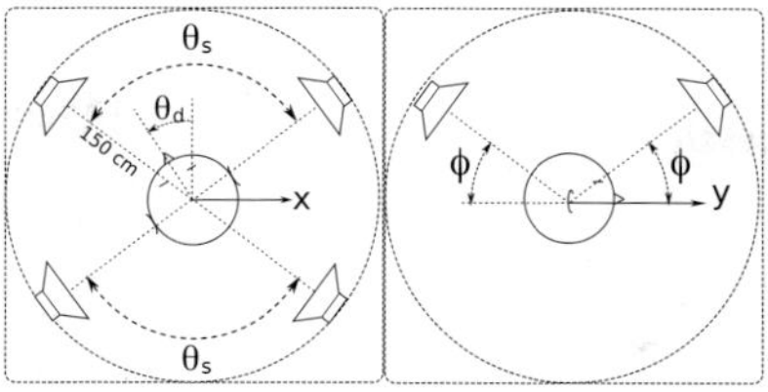

Fig. 2. General diagram of the measured loudspeaker configurations. θ_s corresponds to the span angle and ϕ corresponds to the elevation angle. Only the loudspeakers in front of the listener were used to evaluate the two-channels configuration.

4. Measurements

We carried out the measurements in an anechoic chamber at the acoustics laboratories at Aalborg University. We measured 21 different loudspeaker configurations, including two- and four-channel arrangements. The loudspeakers were placed at three different span angles: $\theta_s = \{12°, 28°, 60°\}$. Each of these configurations were measured at four different elevations: $\phi = \{0°, 30°, 60°, 90°\}$. Only two loudspeakers were measured at an elevation angle of 90°. Note that the mentioned loudspeaker positions are not expressed in spherical coordinates, strictly speaking. The angle θ_s refers to the span between loudspeakers in the front and in the back. ϕ is the angle between the horizontal plane and the plane formed by each pair of loudspeakers and the center of the manikin's head. Figure 2 illustrates the general diagram of the measured configurations.

To design the filters, we used the transfer functions of each of the aforementioned loudspeaker configurations measured with the artificial head Valdemar designed at Aalborg university.[10] For this purpose, the manikin was placed in the geometrical center of the arcs formed by the frontal and rear loudspeakers, facing towards the middle point between the two frontal loudspeakers. We refer to this position as the nominal center position. The filters were calculated using the three different crosstalk cancellation techniques mentioned before. Details of the measurement setup can be found in Ref. 4.

To evaluate the effect of head misalignments, we measured the channel separation when the manikin was placed at positions corresponding to lateral displacements, frontal displacements and head rotations. The lateral displacements x and frontal displacements y were measured from -20 cm to 20 cm with a resolution of 2 cm, where $x = 0$ and $y = 0$ correspond to the nominal center position. The head rotations θ_d were measured between -30° and 30° with a resolution of 2°, where $\theta_d = 0°$ corresponds to the nominal center position and the negative angles to clockwise rotations.

4.1. *Results*

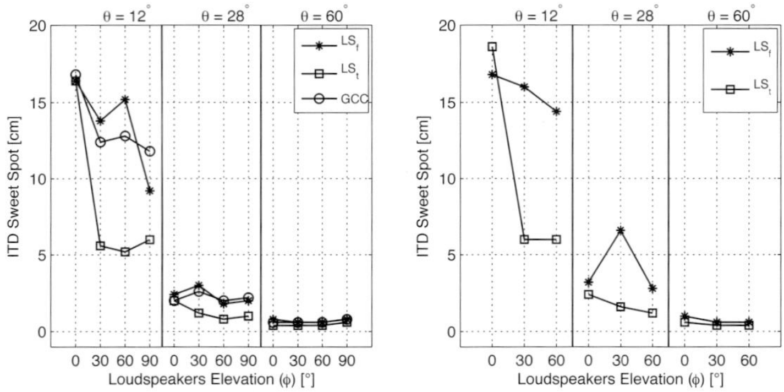

(a) Two-channel configurations (b) Four-channel configuration

Fig. 3. ITD sweet spot size for lateral displacements as function of loudspeaker configuration. Each column corresponds to each measured span angle θ_s and the x–axis to each measured elevation ϕ. Only the two-channel configuration was measured at 90° elevation.

The ITD does not change significantly with frontal displacements, hence we only present the results obtained with lateral displacements and head rotations. Figure 3 shows the ITD sweet spot size for the two- and four-channel configurations. The plots are split into three sections, each one corresponding to one of the different span angles. The x-axis corresponds to the different elevation angles ϕ.

We can notice in Fig. 3(a) that the loudspeakers set with 28° and 60° span angles in two-channel configurations are significantly less robust to lateral displacements than the 12° span angle configuration. In our previous study, we found that the absolute sweet spot decreases gradually with wider span angles (see App. A Fig. 1(a)). However, the ITD sweet spot suggests us that only the 12° span angle is robust to lateral displacements, especially when it is placed on the horizontal plane.

We can also observe that the three different methods yield different results. In general, the LS_t results in narrower controlled areas than the LS_f and the GCC approaches. Even though there is no redundancy when inverting a two-channel system, the different numerical approximations used yield different filter coefficients. Thus, different phase errors are introduced by each method in the signals that reach the ears.

Figure 3(b) shows the ITD sweet spot size for the four-channel configurations as a function of lateral displacements. The results follow a trend similar to the two-channel case. The LS_f shows to be more robust to lateral displacements than the LS_t, especially at 30° elevation. We can also observe a slight improvement in controlled area when compared with the two-channel configuration.

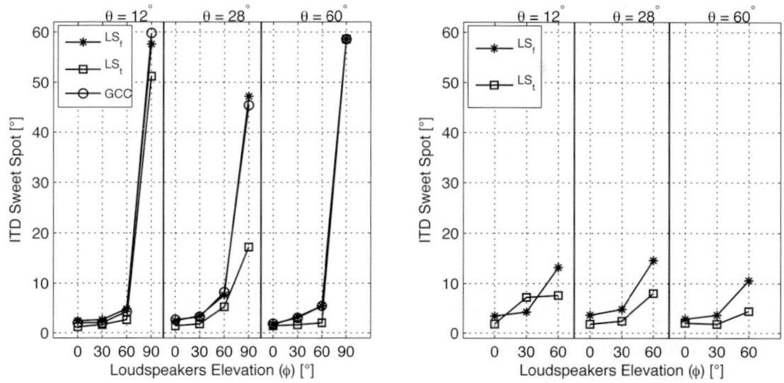

(a) Two-channel configurations

(b) Four-channel configurations

Fig. 4. ITD sweet spot size for head rotations as function of loudspeaker configuration. Each column corresponds to each measured span angle θ_s and the x–axis to each measured elevation ϕ. Only the two-channel configuration was measured at 90° elevation.

Regarding head rotations, we can see in Fig. 4(a) that there is a dramatic improvement in sweet spot when the loudspeakers are placed at 90° elevation for the two-channel case. In general, the ITD sweet spot increases with elevation. We obtained a similar trend with the absolute sweet spot (see App. A Fig. 1(c)). Nevertheless, the ITD sweet spot shows to be narrower for elevations angles below 60°.

Looking at the results obtained with the four-channel configurations we can see that it follows the same tendency as the two-channel configurations (see Fig. 4(b)). Then again, there is an improvement in sweet spot size if compared with the two-channel configurations, especially at 60° elevation.

5. Conclusions

Different studies have evaluated the effect of head misalignments in binaural reproduction systems. Yet, these evaluations are often based on magnitude ratios between the contralateral and ipsilateral signals. In this paper, we intended to extend the results described in a previous analysis presented in Ref. 4 in which the absolute and relative sweet spot as a function of lateral displacements, frontal displacements and head rotations was evaluated for 21 different loudspeaker configurations. Here, we presented an evaluation of the robustness to movements from the time domain perspective. We defined the ITD sweet spot as the maximum movement such that the ITD difference between the nominal center position and the new position does not exceed $10\mu s$. This is a rather strict limit, but based on some studies of the minimum audible ITD, we consider it a safe criteria.[8,9]

We observed that when evaluating the sweet spot in the time domain, a narrower control area is usually obtained in comparison to the results obtained in the sweet spot definitions based on magnitude ratios (see App. A Fig. A1). Only the 12° span angle showed to be sufficiently robust with respect to lateral displacements and actually resulted in larger values than those observed with the absolute sweet spot.

The controlled area with respect to head rotations of the two-channel configurations increases with elevation. Especially at 90° elevation, the ITD does not vary significantly with large rotations. In contrast, when we place the loudspeakers on the horizontal plane, small head rotations result in ITD changes larger than $10\mu s$. This is expected, since the slope of the ITD as a function of head rotations decreases with elevation.

The four-channel configurations showed to be more robust to head rotations than the two-channel case. This is surprising, since when analyzing the absolute sweet spot, the performance of the four-channel configurations showed to be poorer than the two-channel setups (Fig. A1). Another aspect we noted in the results obtained with the two- and four-channel cases, is that the ITD sweet spot does not vary significantly with the different span angles. Only elevation shows an improvement in the controlled area.

In this paper we also evaluated three different crosstalk cancellation techniques. We expected that the variations of the ITD depend only on the placement of the loudspeakers. Yet, the methods not always yielded the same results. In general, the LS_t resulted in narrower controlled area than the LS_f and the GCC. That is especially noticeable with the 12° span angle configuration and lateral displacements. Even though the LS_t is known to make an efficient use of the available coefficients,[11] these results suggest that it is less robust to phase errors than the LS_f. The results presented here show that not only the loudspeaker placement influences the phase variations of the binaural system, but also the numerical errors introduced by the different approximations used by each method. This should be taken into consideration when designing optimal crosstalk cancellation filters.

It is well known that the human auditory system employs two mechanism to discriminate the location of a sound source. The first one extracts the interaural time differences and it is known to work up to 1.6kHz. The second one uses the interaural sound pressure level differences and it is known to be dominant for signals with frequencies above 1.6kHz. Even though they function independently up to a certain extend, there is evidence that they interact with each other. For example, trading experiments had shown that up to a certain extend an auditory event can be displaced by either a time or a level difference.[7] So far, we have analyzed the influence of head misalignment in the frequency and time domain separately and that give us a pretty good idea of how the performance of the different configurations change when varying specific parameters. However, it should be possible to find a model that combines both results and predicts the controlled area more accurately. In this way, the results can be better understood from the human localization point of view. This question is the topic of future research.

Appendix A. Absolute Sweet Spot

In order to give the reader a better understanding of the results described in this paper, we include here the results of the absolute sweet spot presented in Ref. 4.

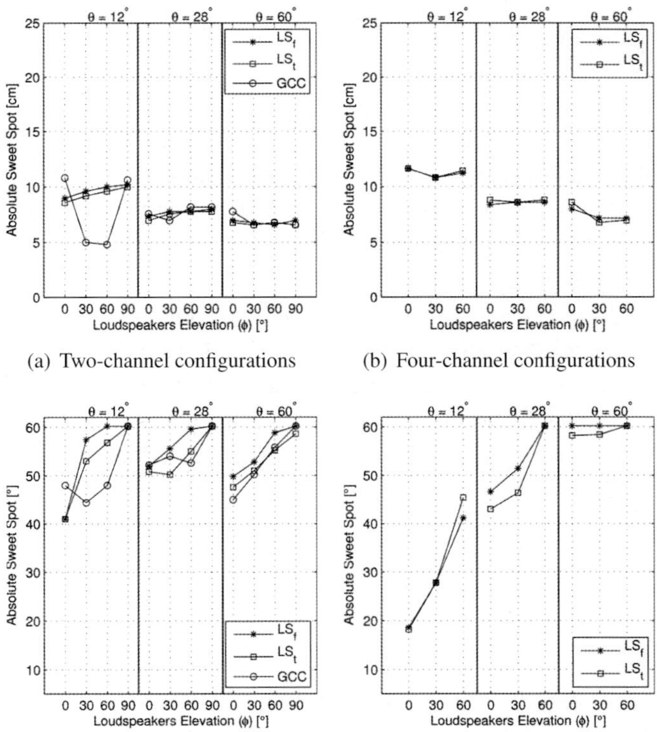

(a) Two-channel configurations (b) Four-channel configurations

(c) Two-channel configurations (d) Four-channel configurations

Fig. A1. Absolute sweet spot at the left ear for lateral displacements [(a) and (b)] and head rotations [(c) and (d)] as a function of loudspeaker configuration. Each column corresponds to each measured span angle θ_s and the x–axis to each measured elevation ϕ. Only the two-channel configuration was measured at 90° elevation.

References

1. T. Takeuchi and P. A. Nelson, Robustness to Head Misalignment of Virtual Sound Imaging Systems, *Journal of the Acoustic Society of America* **109**, 958(March 2001).
2. J. Rose, P. Nelson, B. Rafaely and T. Takeuchi, Sweet Spot Size of Virtual Acoustic Imaging Systems at Asymmetric Listener Locations, *Journal of the Acoustic Society of America* **112**, 1992(November 2002).
3. M. R. Bai and C. Lee, Objective and Subjective Analysis of Effects of Listening angle on Crosstalk Cancellation in Spatial Sound Reproduction, *Journal of the Acoustic Society of America* **120**, 1976(October 2006).
4. Y. Lacouture Parodi and P. Rubak, Preliminary Evaluation of Sweet Spot Size in Virtual Sound Reproduction Using Dipoles, in *126th Convention of the Audio Engineering Society*, (Munich, Germany, 2009).
5. Y. Lacouture Parodi, Analysis of Design Parameters for Crosstalk Cancellation Filter Applied to Different Loudspeaker Configurations, in *125th Convention of The Audio Engineering Society*, (San Francisco, C.A, 2008).
6. P. Minaar, J. Plogsties, S. Krarup, F. Christensen and H. Møller, The Interaural Time Difference in Binaural Synthesis, in *108th Convention of The Audio Engineering Society*, (Paris, France, 2000).
7. J. Blauert, *Spatial Hearing*, 3rd edn. (Hirzel-Verlag, 2001).
8. R. G. Klump and H. R. Eady, Some Measurement of Interaural Time Difference Thresholds, *The Journal Acoust. Soc. Am.* **28**, 859 (September 1956).
9. P. F. Hoffmann and H. Møller, Audibility of Differences in Adjacent Head-Related Transfer Functions, *Acta Acustica united with Acustica* **94**, 945 (2008).
10. F. Christensen, C. B. Jensen and H. Møller, The Design of VALDEMAR - An Artificial Head for Binaural Recordings Purposes, in *109th Convention of The Audio Engineering Society*, (Los Angeles, CA, 2000).
11. O. Kirkeby and P. A. Nelson, Digital Filter Design for Inversion Problems in Sound Reproduction, *Audio Engineering Society* **47**, 583(July/August 1999).

PSYCHOACOUSTIC EVALUATION OF DIFFERENT METHODS FOR CREATING INDIVIDUALIZED, HEADPHONE-PRESENTED VIRTUAL AUDITORY SPACE FROM B-FORMAT ROOM IMPULSE RESPONSES

A. KAN, C. T. JIN* and A. VAN SCHAIK

Computing and Audio Research Laboratory,
School of Electrical and Information Engineering,
University of Sydney, Australia, 2006
**E-mail: craig.jin@sydney.edu.au*

We evaluate a new technique for synthesizing individualized binaural room impulse responses for headphone-rendered virtual auditory space (VAS) from B-format room impulse responses (RIRs) recorded with a Soundfield microphone, and a listener's anechoic head-related impulse responses (HRIRs). Traditionally, B-format RIRs are decoded for loudspeaker playback using either Ambisonics or Spatial Impulse Response Rendering. For headphone playback, virtual loudspeakers are commonly simulated using HRIRs. However, the number and position of loudspeakers should not really be a factor in headphone playback. Hence, we present a new technique for headphone-rendered VAS which is not limited by the number and position of loudspeakers and compare its performance with traditional methods via a psychoacoustic experiment.

Keywords: Virtual auditory space; Binaural room impulse response; Soundfield microphone; Room impulse response

1. Introduction

A virtual auditory space (VAS) is an auditory display that conveys three-dimensional acoustic information to a listener such that a virtual sound source in the VAS will be perceived to be the same as that of a naturally-occurring sound source in an equivalent real-world space. A VAS can be presented to a listener using loudspeakers or headphones. For headphone-presented VAS, a binaural room impulse response (BRIR) is typically recorded at the ears of a listener for every sound source position of interest in the room or listening space. The BRIR completely characterizes the acoustical transformation of the sound signal from its source position to the listener's ears. This transformation arises from reflections and scattering due to the room and the listener's ears, head and physique, and provides acoustic information to the listener about the source location and also the room's physical characteristics. Recording BRIRs may not always be easily achieved or even possible because it clearly requires that each listener must travel to the acoustic space of interest to have the measurements taken. A possibly more flexible method would be to record the components of the acoustical transformation that arise from the room and the listener separately and then to recombine these components together again to synthesize the BRIR. This paper examines various techniques to achieve this separation and recombination of acoustic information for the synthesis of individualized VAS.

Consider now the two separate components of a BRIR. First, a head-related impulse response (HRIR), or in the frequency domain a head-related transfer function (HRTF), characterizes the directionally-dependent acoustical transformation of a sound signal from a location in the free-field to the listener's ears. These are typically recorded for a listener[1] in an anechoic room, i.e. a room without reflections, and therefore characterize the acoustic properties of a listener's ear. Secondly, the acoustical transformation of a sound signal from its source location in a room to a listening position is characterized by a room impulse response (RIR) and can be recorded using a Soundfield microphone.[2] The advantage of using a Soundfield microphone is that the directional characteristics of the RIR are encoded within its B-format signals, which consist of an omni-directional pressure signal, $W(t)$, and three orthogonal figure-of-eight, pressure-gradient signals, $X(t)$, $Y(t)$ and $Z(t)$, oriented in the directions of the Cartesian axes.

Because the methods for decoding B-format signals have traditionally been designed for loudspeaker playback, we will first review the application of B-format RIRs for loudspeaker playback and then consider common adaptations of this technique for headphone presentation which will ultimately use a listener's recorded HRIRs. There are two primary methods for loudspeaker playback of B-format signals: Ambisonic decoding and Spatial Impulse Response Rendering (SIRR). With the Ambisonic technique, a monaural sound source signal is first filtered with the B-format RIRs to produce a vector of B-format signals, b. Ambisonic decoding then solves a least-mean square optimization problem[3,4] based on the location of the loudspeakers to obtain a decoding matrix M_d. Given the decoding matrix, the vector of loudspeaker feeds, l, are obtained using $l = M_d b$. It should be noted that with a limited number of loudspeakers, the size of the listening area and frequencies at which the sound field can be accurately reconstructed is limited due to spatial aliasing. Above the spatial-aliasing frequency (typically around 400 Hz), the loudspeaker gains can be modified in order to maximize the high frequency energy coming from the direction of a sound source; we will refer to this as "Ambisonic - $\max_{\rm re}$". To improve the robustness of the sound field across a larger listening area, an additional decoding correction can be added[5] such that the loudspeakers are played "in-phase", that is, the decoding prevents loudspeakers from playing signals out of phase, particularly from those loudspeakers that are diametrically opposite to the sound source location. We will refer to this method of Ambisonic decoding as "Ambisonic - in-phase".

An alternative method for loudspeaker playback using B-format RIRs is SIRR.[6] SIRR assumes that perfect reconstruction of the original sound field is not necessary to reproduce the spatial impression of a room, but rather the same spatial impression can be generated by recreating the time-frequency features of a sound field. To achieve this, SIRR applies an energy analysis to the B-format RIRs in the time-frequency domain in order to determine the direction of arrival and the diffuseness of the energy at each time-frequency tile. The time-frequency analysis is usually performed using a short-time Fourier transform (STFT). The information derived from the energy analysis is then used to create a set of decoding filters for a loudspeaker array. A monaural source signal is then filtered with the decoding filters to generate loudspeaker signals that preserve the direction of arrival, diffuseness and spectrum of the sound field when played back over the array of loudspeakers.

In our view, the primary drawback with SIRR is that the diffuse sound field is rendered somewhat arbitrarily. One of the main contributions of this work is that we have developed a technique along the lines of SIRR that better preserves the diffuse sound field when rendered via headphones. Before describing our new method, we first review SIRR in some detail.

The SIRR energy analysis is based on the concept of sound intensity which describes the transfer of energy in a sound field. For a given time-frequency tile, the active intensity, $\mathbf{I}_a(k,\omega)$, and diffuseness, $\psi(k,\omega)$, of the B-format RIR are given by:

$$\mathbf{I}_a(k,\omega) = \frac{\sqrt{2}}{Z_0} Re\left\{W^*(k,\omega)\mathbf{V}(k,\omega)\right\} \tag{1}$$

and

$$\psi(k,\omega) = 1 - \frac{\sqrt{2}\left\|Re\left\{W^*(k,\omega)\mathbf{V}(k,\omega)\right\}\right\|}{\left|W(k,\omega)\right|^2 + \left|\mathbf{V}(k,\omega)\right|^2/2} \tag{2}$$

where $W(k,\omega)$ and $\mathbf{V}(k,\omega)$ are the STFT (k is the time-frame index and ω is the frequency variable) of $W(t)$ and $\mathbf{V}(t) = X(t)\mathbf{e}_x + Y(t)\mathbf{e}_y + Z(t)\mathbf{e}_z$, respectively, where $\mathbf{e}_x$, $\mathbf{e}_y$ and $\mathbf{e}_z$ are the unit vectors in the directions of the Cartesian co-ordinate axes; * denotes the complex conjugation, $|.|$ denotes the absolute value of a complex number, $\|.\|$ denotes the norm of a vector, and Z_0 is the acoustic impedance of air (typically 413.2 Nsm^{-3} at 20°C). The quantity ψ takes a value between 0 and 1. A value of $\psi = 1$ indicates an ideal diffuse sound field (no net transport of energy), and a value of $\psi = 0$ signifies the sound field consists only of a directional component. From the intensity vector, the direction of arrival of the net flow of energy, i.e. the azimuth, $\theta(k,\omega)$, and elevation, $\phi(k,\omega)$, can be calculated as:

$$\theta(k,\omega) = \tan^{-1}\left[\frac{-I_y(k,\omega)}{-I_x(k,\omega)}\right], \phi(k,\omega) = \tan^{-1}\left[\frac{-I_z(k,\omega)}{\sqrt{I_x^2(k,\omega) + I_y^2(k,\omega)}}\right] \tag{3}$$

where $I_x(k,\omega)$, $I_y(k,\omega)$, $I_z(k,\omega)$ are the components of the active intensity in the directions corresponding to the Cartesian co-ordinate axes.

After performing the energy analysis of the B-format RIRs as described above, an STFT representation of the decoding filters for the loudspeaker array is determined as follows. It should be noted that for each time window, zero-padding is used prior to the Fourier transform to prevent time-domain aliasing. For each time-frequency tile, the omni-directional signal, $W(k,\omega)$, is split into directional and diffuse components according to the diffuseness estimate $\psi(k,\omega)$. The directional component is given by: $\sqrt{1-\psi(k,\omega)}W(k,\omega)$ and the diffuse component by: $\sqrt{\psi(k,\omega)W^2(k,\omega)}$. At each time-frequency tile, the directional component is distributed among the decoding filters using a vector-based amplitude panning (VBAP) technique,[7] while the diffuse component is added to all of the decoding filters using a technique that distributes the total diffuse energy in a decorrelated manner among all of the loudspeakers. Pulkki et al.[8] suggests a decorrelation method for SIRR whereby random panning of the diffuse energy from different loudspeakers is used at the low frequencies (< 800 Hz), with a smooth transition (800 - 1200 Hz) into a phase randomization method at the high frequencies. Time-domain decoding filters for the loudspeakers are then obtained by applying an inverse STFT to the STFT representation of the decoding filters with appropriate overlap-and-add processing.

To render the signals from Ambisonics or SIRR over headphones as VAS, it is common to use a virtual loudspeaker technique in which the loudspeaker signals are filtered with the HRIRs corresponding to the di-

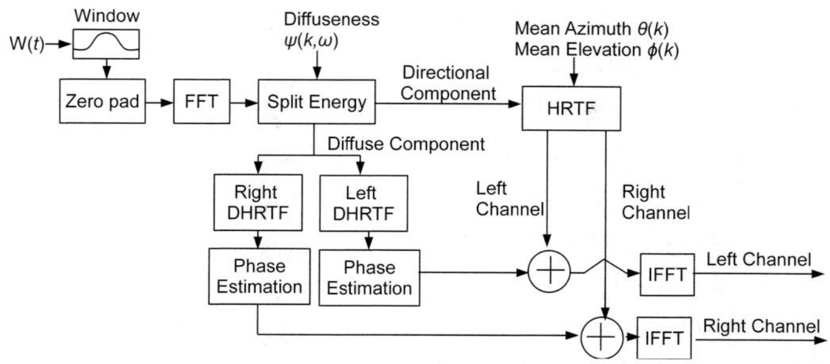

Fig. 1. Synthesis of a BRIR using BSFR.

rection of the loudspeaker relative to the listener and summed together to create left and right headphone signals.[9] In reality, however, the limitations on the number and position of the loudspeakers should not be a factor when reproducing the sound field over headphones. For example, the quality of an Ambisonic decoding varies with the order of the decoding and also with the number of loudspeakers. With too many loudspeakers, Ambisonics solves an under-determined system of equations to determine the decoding matrix and the quality of the reproduction suffers. On the other hand, with too few loudspeakers the directional resolution of sound sources will suffer. SIRR partly overcomes the problems associated with using a large number of loudspeakers in an Ambisonic decoding. It achieves this by using VBAP, but the diffuse or ambient sound can be incorrectly reproduced. In the following, we propose a new method called binaural sound field rendering (BSFR) for using B-format RIRs to generate an individualized VAS for headphone playback which is not limited by the number and position of the loudspeakers.

2. Binaural sound field rendering

BSFR is a method for synthesizing individualized BRIRs from B-format RIRs and a set of anechoic HRIRs. Fig. 1 shows the steps for BSFR. BSFR begins with exactly the same steps as SIRR and applies an energy analysis on the B-format RIRs in the STFT domain to determine the directional and diffuse components of the omni-directional signal, $W(k,\omega)$. The STFT of the desired BRIR is then determined as follows. At each time window, $W(k,\omega)$, is split into directional and diffuse components according to the diffuseness estimate $\psi(k,\omega)$. The directional component of the BRIR is then obtained by: $\sqrt{1-\psi(k,\omega)}W(k,\omega)HRTF_{lr}(k,\omega,\theta,\phi)$ where $\psi(k,\omega)$ is the estimated diffuseness, $W(k,\omega)$ is the omni-directional channel of the Soundfield RIR, $HRTF_{lr}(k, \omega, \theta, \phi)$ is the complex-valued HRTF corresponding to the direction of the active intensity vector at a particular frequency bin and the subscript 'lr' denotes the left or right ear. The real-valued magnitude spectra of the diffuse component of the BRIR is obtained by: $\sqrt{\psi(k, \omega)W^2(k, \omega)}DHRTF_{lr}$ where $DHRTF_{lr}$ is the real-valued magnitude of the directionally-averaged or diffuse-field HRTF for the left or right ear. It is calculated separately for the left and right ears from HRTFs recorded for an evenly distributed set of sound source directions around the listener using:

$$DHRTF_{lr} = 10^{\left(\frac{1}{N}\sum_{i=1}^{N} 20\log_{10}(|HRTF_{lr}(\theta_i, \phi_i)|)\right)/20} \tag{4}$$

where N is the number of HRTFs and θ_i and ϕ_i are the azimuth and elevation co-ordinates, respectively, corresponding to the direction of the HRTF. In order to estimate the phase of the diffuse component of the BRIR, a spectrogram inversion method[10] was used. This method iteratively estimates the phase at a particular time window while minimizing the difference in the magnitude response between the magnitude-only diffuse field BRIR and the estimated complex-valued diffuse field BRIR. Additionally, phase continuity between time windows is maintained by taking into account the magnitude spectra from past, present and future time windows during the phase estimation process. The use of the spectrogram inversion method for synthesizing the diffuse field BRIR gives a natural sounding reproduction of the diffuse sound field without the need for decorrelation methods. The diffuse field BRIR estimated by our method is naturally decorrelated at the two ears since the diffuse-field HRTF for the left and right ears are different and hence lead to different phase estimates for the final left and right ear signals. Finally, the directional and diffuse-field parts of the BRIR are added together and the time-domain BRIR is obtained by applying an inverse STFT with appropriate overlap-and-add processing.

3. Listening Test

A listening test was conducted to evaluate the different methods mentioned above for generating headphone-rendered VAS. Subjects rated the VASs generated by these methods against a reference VAS generated from their own BRIRs. In the following, we first describe the methods employed in recording the B-format RIRs used in this listening test, and the BRIR and HRIRs of each subject. Details on how the different methods are applied to these recordings to generate test stimuli are then given. Finally, a description of the listening test will be presented.

Subject BRIRs and a B-format RIR were recorded in a room 7.52 x 12.14 x 2.72 m^3 in size. A Tannoy V6 loudspeaker, driven by an Ashley 4400 power amplifier, was used to provide the stimulus. The loudspeaker was located 2.7 m away from the recording position at a height of 1.5 m. A silent computer equipped with an RME Multiface sound card was used to play and record the audio signals at a 48 kHz sampling rate. Since the output transfer function of the loudspeaker did not have constant gain across frequency, a compensation filter was used so that the output transfer function of the loudspeaker was flat within 3 dB between 300 Hz and 20 kHz. A 6 s long logarithmic sine sweep from 10 Hz to 20 kHz, filtered with the compensation filter, was used as stimulus for the recordings and the impulse responses recovered from the recorded sweep via deconvolution.[11] A Soundfield microphone was used for recording the B-format RIR. Subject BRIRs were recorded using a "blocked ear canal" method.[1] The subjects faced the loudspeaker for the BRIR recordings.

HRIRs were also recorded for each of the subjects in an anechoic chamber using the "blocked ear canal" method. HRIRs were recorded for 393 different sound source directions, around the subject's head. HRIRs for any sound source direction were then obtained by interpolation of the 393 HRIR recordings using a spherical thin-plate spline interpolation method.[12] 14 subjects participated in the listening test. Of the 14 subjects, 7 subjects had extensive experience, 5 subjects had some previous experience and 2 subjects had no previous experience in listening tests.

Test stimuli were generated using the four different methods described above. For Ambisonic - $\max_{re}$, Ambisonic - in-phase and SIRR decoding, a cubic configuration of virtual loudspeakers was used where the loudspeakers were placed at the corners of the cube. For SIRR and BSFR, 3 ms sine-squared windows with 50% overlap were used for the energy analysis. The same windows were used for the synthesis with 1.5 ms zero-padding before and after each window. The diffuse-field HRTF for BSFR was calculated by averaging the 393 recorded HRTFs for each subject separately. Additionally, a reference sound was created by filtering anechoic sound stimuli with the measured BRIRs and a low quality anchor stimuli was created by filtering the anechoic sounds with the anechoic HRIR of the subject for a sound source in front of the listener and low-passed filtered at 3.5 kHz. A total of 8 anechoic sounds were chosen for the listening test (see Table 1). In order to achieve a consistent perceived loudness across the test stimuli generated by

Table 1. The different sound excerpts are shown along with the name (key) with which the sound will be identified.

Music for Archimedes			**Denon Professional Test CD**		
No.	Description	Key	No.	Description	Key
4	Female Speech - English	voice	23	Symphony No. 4 in E-flat (Bruckner)	orch
12	Guitar Capriccio Arabe	guitar	25	The Marriage of Figaro (Mozart)	figaro
27	Xylophone Sabre Dance	xylo	27	Pizzicato Polka (Strauss)	strings
37	Bb Trumpet Over the Rainbow	trumpet	30	Violin solo	violin

the different methods, a loudness model[13] was used to calculate a single gain adjustment factor for each of the test stimuli separately.[14] The calculated gain adjustment factor was then applied to the corresponding left and right ear sound signals of the test stimuli.

The listening test was conducted in a sound-attenuating booth to reduce external sound interference. Sound stimuli were presented using Etymotic ER-1 headphones from an RME Multiface soundcard attached to a computer located outside the booth. An adapted version of the multi-stimulus test with hidden reference and anchor (MUSHRA) paradigm[15] was used. In the standard MUSHRA paradigm, a subject is asked to rate how close each test stimuli, generated by the different methods, is to a reference sound using a scale from 0 to 100. The scale is divided into 5 equal intervals, where [0-19] = bad, [20-39] = poor, [40-59] = fair, [60-79] = good, and [80-100] = excellent. However, during preliminary listening tests, it was determined that making one rating for each test stimuli was too difficult since the stimuli generated by the different VAS methods differed from the reference in more than one perceptual aspect. Hence, subjects were instructed to first rate the test stimuli on three perceptual attributes separately, prior to making an overall rating of the test stimuli. The three perceptual attributes were: (1) the quality of the reverberation in the sound, that is, whether the test sound sounded like it was in the same room as the reference; (2) the quality of the sound source, that is, how similar the sound source was to the reference and whether there were noticeable timbral difference or changes in the sound source width; and (3) the position of the sound source, that is, how close the sound source was in position compared to the reference sound. Sliders were provided on a graphical user interface for the subject to make the ratings for each trial. After rating each test stimuli on the perceptual attributes, the subject was then asked to make

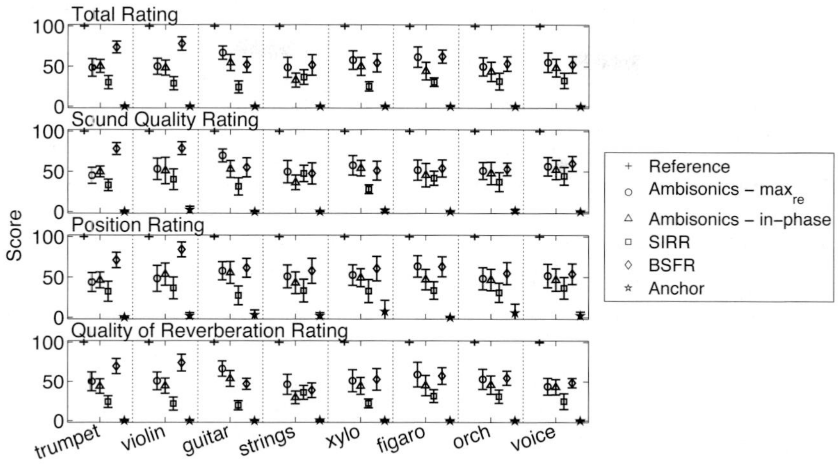

Fig. 2. Mean ratings with the 95% confidence interval of the mean are shown for each test sound separately.

an overall rating of the test stimuli. For the overall ratings, subjects were required to rate one of the sounds in each trial at a score of 100 and one at a score of 0, while for the perceptual attributes, subjects were not required to rate any of the stimuli at a particular score. A comment box was also provided for subjects to leave comments about the sound stimuli.

4. Results

The mean overall rating given by subjects for the different VAS generation methods are shown in Fig. 2. A number of observations can be made from the overall ratings: (1) For most of the sounds, subjects gave similar scores to Ambisonic - $\max_{re}$ and Ambisonic - in-phase decoding methods. This is to be expected since the subject always remains in the "sweet spot" when the Ambisonic reconstruction of the sound field is presented over headphones. (2) For two of the sounds, trumpet and violin, BSFR was on average rated significantly higher than the other methods; and (3) the scores for SIRR were lower compared to the other methods for most sounds.

To test the significance of the above observations, a Kruskal-Wallis non-parametric ANOVA was conducted on the mean overall ratings to test the hypothesis that there are no statistically significant differences in the ratings for the four different methods. The analysis was done for each of the test stimuli separately and the results are shown in Table 2. The analysis revealed significant differences in the ratings for all test sounds except for

Table 2. The χ^2 and p-value results of a Kruskal-Wallis non-parametric ANOVA conducted on the overall ratings for each of the test sounds. The number of degrees of freedom for all tests is 3.

	trumpet	violin	guitar	strings	xylo	figaro	orch	voice
χ^2	26.20	26.21	26.49	6.64	26.26	20.77	11.34	9.40
p	< 0.005	< 0.005	< 0.005	0.08	< 0.005	< 0.005	0.01	0.02

the strings. A post-hoc analysis (Tukey HSD) was conducted to investigate these differences and the analysis revealed the higher ratings for BSFR for the trumpet and violin sounds and the low ratings for SIRR for most test stimuli to be statistically significant. The ratings for the two Ambisonic methods showed no statistically significant differences.

Some understanding of how subjects may have arrived at their overall ratings can be obtained by studying the ratings for the sound quality and position of the sound source, and the quality of the reproduced reverberation. The mean ratings for each perceptual parameter is shown in Fig. 2. It can be observed that for the trumpet and violin sounds, BSFR was, on average, rated as being better at reproducing the sound quality and position of the sound source. Also, it can be observed that SIRR was rated significantly lower for most stimuli when judged for its ability to reproduce the reverberant qualities of the sound field.

5. Discussion and Conclusions

A listening test was conducted to evaluate a number of different methods for generating an individualized, headphone-rendered VAS from B-format RIRs. The results show that there is a noticeable difference in the VAS generated from the different methods compared to the VAS generated using the subjects' measured BRIRs. Anecdotally, subjects commented that the VAS generated by the different methods were acceptable, even reasonable, except for SIRR where most subjects commented that the reproduced sound field was too reverberant. This is due to the fact that there is no control over the amount of decorrelation applied in the decorrelation method used in SIRR. In the case of the trumpet and violin sounds, there is an improvement in the generated VAS when using BSFR. Furthermore, the BSFR method was anecdotally reported to provide a better frontal image. Subject ratings in these sounds for the three perceptual attributes indicate that there was improved position localization and timbral qualities of the sound sources when using BSFR. In summary, while the B-format RIRs do not provide complete information to synthesize perceptually-accurate BRIRs, the BSFR method provides a technique that is not limited by the position or number of loudspeakers and seems to recreate the characteristics of the sound field reasonably well.

References

1. H. Møller, Fundamentals of binaural technology, *Applied Acoustics* **36**, 171 (1992).
2. A. Farina and R. Ayalon, Recording concert hall acoustics for posterity, in *AES 24th International Conference on Multichannel Audio*, (Banff, Alberta, Canada, 2003).
3. J. Daniel, J.-B. Rault and J.-D. Polack, Ambisonics encoding of other audio formats for multiple listening conditions, in *105th Audio Engineering Society Convention*, September 1998.
4. M. Gerzon, Practical periphony: The reproduction of full-sphere sound, in *AES Preprint 1571*, (65th Convention of the Audio Engineering Society, London, February 25-28 1980).
5. D. G. Malham, Experience with large area 3-D ambisonic sound systems, in *Proceedings of the Institute of Acoustics*, (5)1992.
6. J. Merimaa and V. Pulkki, Spatial Impulse Response Rendering I: Analysis and Synthesis, *Journal of the Audio Engineering Society* **53**, 1115 (December 2005).
7. V. Pulkki, Virtual sound source positioning using vector based amplitude panning, *Journal of the Audio Engineering Society* **45**, 456 (1997).
8. V. Pulkki and J. Merimaa, Spatial Impulse Response Rendering II: Reproduction of Diffuse Sound and Listening Tests, *Journal of the Audio Engineering Society* **54**, 3 (February 2006).
9. D. McGrath and A. Reilly, Creation, manipulation and playback of soundfields with the huron digital audio convolution workstation, in *Signal Processing and Its Applications, 1996. ISSPA 96., Fourth International Symposium on*, Aug 1996.
10. X. Zhu, G. Beauregard and L. Wyse, Real-time signal estimation from modified short-time fourier transform magnitude spectra, *Audio, Speech, and Language Processing, IEEE Transactions on* **15**, 1645 (July 2007).
11. A. Farina, Simultaneous measurement of impulse response and distortion with a swept-sine technique, in *Proceedings of the 108th AES Convention*, 2000.
12. C. Jin, Spectral analysis and resolving spatial ambiguities in human sound localization, PhD thesis (2001).
13. D. Robinson, Replay gain - a proposed standard `http://replaygain.hydrogenaudio.org/` (July, 2001).
14. A. Q. Li, Spatial hearing through different ears: A psychoacoustic investigation, Masters thesis (2007).
15. ITU-R BS.1534-1:2003, *Method for the subjective assessment of intermediate quality level of coding systems.*

Effects of microphone arrangement on the accuracy of a spherical microphone array (SENZI) in acquiring high-definition 3D sound space information

Shuichi Sakamoto[1], Jun'ichi Kodama[1], Satoshi Hongo[2],
Takuma Okamoto[3], Yukio Iwaya[1] and Yôiti Suzuki[1]

[1]*Research Institute of Electrical Communication and Graduate School of Information Sciences, Tohoku University,*
2-1-1, Katahira, Aoba-ku, Sendai, 980-8577, Japan
E-mail: {saka, kodama, iwaya, yoh}@ais.riec.tohoku.ac.jp
www.ais.riec.tohoku.ac.jp/index.html

[2]*Dept. of Design and Computer Applications, Sendai National College of Technology,*
48 Nodayama, Natori, 981-1239, Japan
E-mail: hongo@sendai-nct.ac.jp

[3]*Research Institute of Electrical Communication and Graduate School of Engineering, Tohoku University,*
2-1-1, Katahira, Aoba-ku, Sendai, 980-8577, Japan
E-mail: okamoto@ais.riec.tohoku.ac.jp

We propose a three-dimensional sound space sensing system using a microphone array on a solid, human-head-sized sphere with numerous microphones, which is called SENZI (Symmetrical object with ENchased ZIllion microphones). It can acquire 3D sound space information accurately for recording and/or transmission to a distant place. Moreover, once recorded, the accurate information might be reproduced accurately for any listener at any time. This study investigated the effects of microphone arrangement and the number of controlled directions on the accuracy of the sound space information acquired by SENZI. Results of a computer simulation indicated that the microphones should be arranged at an interval that is equal to or narrower than 5.7° to avoid the effect of spatial aliasing and that the number of controlled directions should be set densely at intervals of less than 5° when the microphone array radius is 85 mm.

Keywords: Microphone array, Dummy head recording, Head-related transfer function (HRTF), Tele-existence

1. Introduction

Sensing technologies of three-dimensional (3D) sound space information are indispensable partners of 3D sound reproduction technologies. Comprehensive and accurate sensing of 3D spatial audio information is therefore a key to realization of high-definition 3D spatial audio systems. Although several research efforts have addressed sensing topics, few technologies reflect listeners' head movements in the sensing. Many authors [1–3] have described that listeners' head movements are effective to enhance the localization accuracy as well as the perceived realism in human spatial hearing perception.

In this context, a few methods have been proposed to realize sensing of three-dimensional sound space information considering the listener's movement [4–6]. All of these methods apply special objects to sense sound space information. These objects are set at the recording place. Then recorded information is transmitted to a distant place. However, these methods are insufficient to sense accurate 3D audio space information and to provide appropriate sound information to plural listeners individually and simultaneously.

As another approach to sense and/or to reproduce accurate sound information, ambisonics, especially higher-order ambisonics, have been specifically examined [7,8]. In this technology, 3D sound space information is encoded and decoded on several components with specific directivities based on spherical harmonic decomposition. However, even with higher-order ambisonics of the highest order available such as five, the directional resolution might be insufficient compared with human resolution of spatial hearing. A sensing system matching human performances is highly desired but it remains unclear how many orders are necessary to yield directional resolution that is sufficient to satisfy perceptual resolution.

Consequently, we have proposed a system that can sense accurate 3D sound space information and/or transmit it to a distant place using a microphone array on a human-head-sized solid sphere with numerous microphones on its surface. We designate this spherical microphone array as SENZI (Symmetrical object with ENchased ZIllion microphones) [9]. The system can sense 3D sound space information comprehensively: information from all directions is available, over locations and over time if once recorded, for any listener orientation and head/ear shape with correct binaural cues. However, the accuracy of the acquired 3D sound space information necessarily depends on the arrangement of microphones that are set on the solid sphere [10].

In this study, we investigated the effect of the microphone arrangement on the accuracy of the acquired sound space information of SENZI.

2. System Outline[9]

2.1. *System Concept*

Figure 1 presents a scheme of the proposed system. The microphone array named SENZI is made from an acoustically hard sphere with plenty of microphones. The SENZI is set at the recording place and sound signals inputted to all microphones on SENZI are used for synthesizing a listener's head related transfer function (HRTF). Calculated signals are typically presented to a listener binaurally, for example, via headphones. It is noteworthy, however, that SENZI can output suitable signals to any spatial sound reproduction system. Individual HRTFs are synthesized using digital signal processing. Moreover, listener's head movement can be reflected to the output signal processing for any listener with various head and ear shape and for any time for any place if the input signals are once recorded. Therefore, it is even possible to present individualized 3D sound space information to many listeners simultaneously using this system.

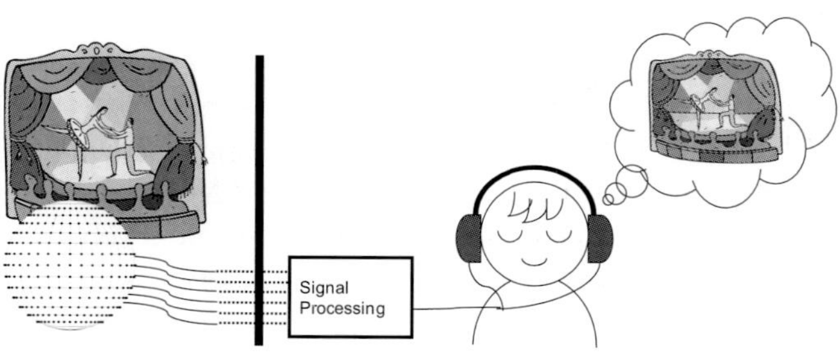

Fig. 1. Concept of the proposed system.

2.2. *Calculation method of HRTFs for individual listeners*

In the proposed system, to calculate and synthesize a listener's HRTFs using input signals from spatially distributed multiple microphones, each input signal from each microphone is simply weighted and summed to synthesize a listener's HRTF. Moreover, the weight is changed according to the 3D head movement of a human who is in a different place. Therefore, 3D sound space information is acquired accurately corresponding to any head movement. Moreover, it should be noted that this is possible for any listener for any time if input signals are once recorded.

As the simplest set of circumstances, we first assume the case in which sounds come only from the horizontal plane. Let $\boldsymbol{H}_{\text{listener}}$ signify the listener's HRTF for one ear. For a certain frequency f, $\boldsymbol{H}_{\text{listener},f}(\theta)$ is expressed according to the following equation:

$$\boldsymbol{H}_{\text{listener},f}(\theta) = \sum_{i=1}^{n} z_{i,f} \cdot \boldsymbol{H}_{i,f}(\theta) + \varepsilon. \tag{1}$$

In that equation, $\boldsymbol{H}_{i,f}(\theta)$ is the transfer function of the i-th microphone from a sound source as a function of the direction of the sound source (θ). Actually, $\boldsymbol{H}_{i,f}(\theta)$ and $z_{i,f}$ are all complex. Equation 1 reflects that the listener's HRTF is calculable from these transfer functions. Equation 1 cannot be solved; a residual ϵ remains if the number of sound source directions differs from the number of microphones n. In fact, ϵ varies according to the weighting coefficient $z_{i,f}$. Therefore, a set of optimum $z_{i,f}$ is calculated using the pseudo-inverse matrix. The coefficients $z_{i,f}$ are calculated for each microphone at each frequency in this method. Calculated $z_{i,f}$ is constant irrespective of the direction of a sound source. This feature of our method is extremely important because one important advantage of the system is that the sound source position need not be considered when sound-space information is acquired.

When $z_{i,f}$ is calculated, we must select directions (θ), that are incorporated into the calculation. The selected directions are designated as "*controlled directions*" hereinafter. However, in a real environment, sound waves come from all directions, including directions that are not incorporated into calculations. These directions are called "*uncontrolled directions*." To synthesize accurate sound information for all directions including "*uncontrolled directions*," the number of microphones, the arrangement of the microphones on the object, and the shape of the object should be optimized.

3. Accuracy of acquired sound space information for uncontrolled directions

3.1. *Experimental method*

We analyzed the accuracy of synthesized HRTFs of all the directions including at *uncontrolled directions*, when the transfer functions of the SENZI were adjusted at *controlled directions*.

For this study, the HRTFs of a dummy head (SAMRAI; Koken Co. Ltd.) were used as the target characteristics to be realized using this system. They were calculated using the boundary element method (BEM). Calculated

HRTFs are depicted in Fig. 2. The frequency range for synthesis was set as 0–20 kHz in 93.75 Hz steps.

Table 1 presents verified conditions that were considered. In SENZI, microphones were set at steps of 20° (conditions **a**, **b** and **c**), 10° (conditions **d**, **e** and **f**), or 5° (condition **g**) from 0° (in front of the listener) to 359° in the horizontal plane and at steps of 20° from −60° to 60° in the vertical plane. The *controlled directions* were set at steps of 20° (condition **a**), 10° (conditions **b**, **d**), 5° (conditions **c**, **e**, **g**), or 2° (condition **f**) from 0° (in front of the listener) to 359° in the horizontal plane and at steps of 20° from −40° to 80° in the vertical plane. After calculating the weighting coefficient $z_{i,f}$ in each condition, the transfer functions of 2,520 (360×7) directions (including both *controlled* and *uncontrolled directions* were synthesized using $z_{i,f}$ in all conditions.

The error of the synthesized HRTFs in terms of the spectral distortion (SD) was calculated as follows [9]:

$$\varepsilon_f(\theta) = \left| 20\log_{10} \left| \frac{\boldsymbol{H}_{\text{listener},f}(\theta)}{\boldsymbol{H}_{\text{synthesized},f}(\theta)} \right| \right| \quad [\text{dB}]. \tag{2}$$

3.2. *Results and discussion*

Figures 3 to 6 show examples of the SD between the HRTFs of the dummy-head and the synthesized HRTFs. From these figures, a certain boundary

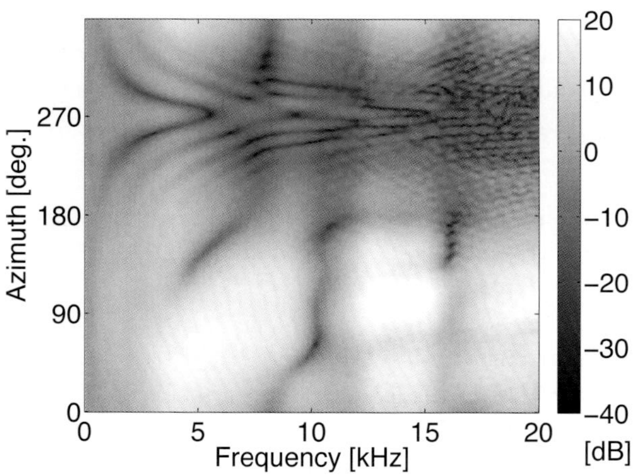

Fig. 2. HRTF of SAMRAI (0 deg elev. angle).

Table 1. Conditions considered

Condition	Number of microphones	Number of controlled directions
a	126 (18×7)	126 (18×7)
b	126 (18×7)	252 (36×7)
c	126 (18×7)	504 (72×7)
d	252 (36×7)	252 (36×7)
e	252 (36×7)	504 (72×7)
f	252 (36×7)	1260 (180×7)
g	504 (72×7)	504 (72×7)

frequency is visible between the frequency range where sound space information is synthesized accurately and that where a large synthesis error is observed. That is expected to be attributable to the effect of spatial aliasing from the intervals between microphones [11]. When the microphones are set at $\theta[rad]$ steps on the SENZI, the interval between each microphone d is calculated using the following equation:

$$d = r\theta, \tag{3}$$

where r is the radius of the SENZI.

Because the radius of the SENZI in this study was 85 mm, d was 29.5 mm in conditions **a**, **b** and **c**, and 14.8 mm in conditions **d**, **e** and **f**. Therefore, the frequency at which the half-wavelength is equal to d is 5.8 kHz in conditions **a**, **b** and **c**, and 11.6 kHz in conditions **d**, **e** and **f**. These values correspond to the boundary frequency of the error in Figs. from 3 to 6. Therefore, microphones should be set at intervals of 5.7° to avoid the effect of spatial aliasing when the radius of the SENZI is 85 mm.

Figure 7 portrays the average SD for "*controlled directions*" and "*controlled directions* and *uncontrolled directions*." In this figure, the accuracy of synthesized HRTFs calculated with *controlled* and *uncontrolled directions* is much lower than that calculated only with *controlled directions* in conditions **a**, **d**, and **g**. In these conditions, the number of microphones was equal to the number of *controlled directions*. Therefore, the calculated coefficients $z_{i,f}$ were fitted strictly at the *controlled directions*, but they were not fitted at *uncontrolled directions*.

For quantitative consideration, we simplified the situation to one in which microphones and *controlled directions* were set to 0° elevation angle, i.e. the horizontal plane. In this situation, we investigated the relation between the intervals of *controlled directions* and the accuracy of synthesized

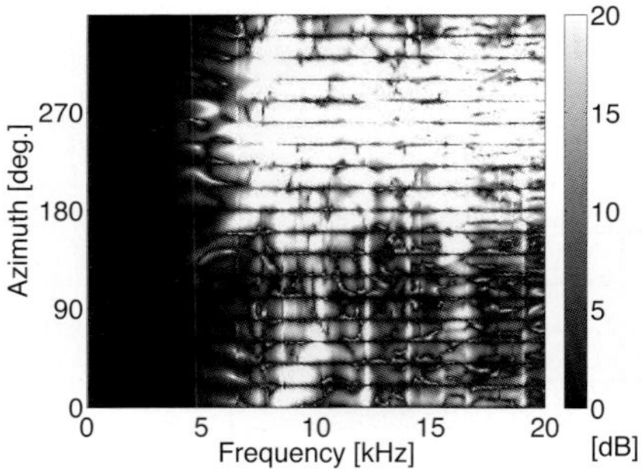

Fig. 3. Spectral distortion of condition **a** (0 deg elev. angle).

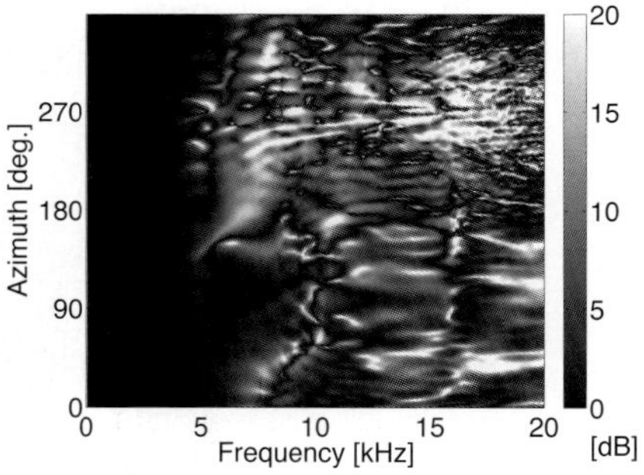

Fig. 4. Spectral distortion of condition **c** (0 deg elev. angle).

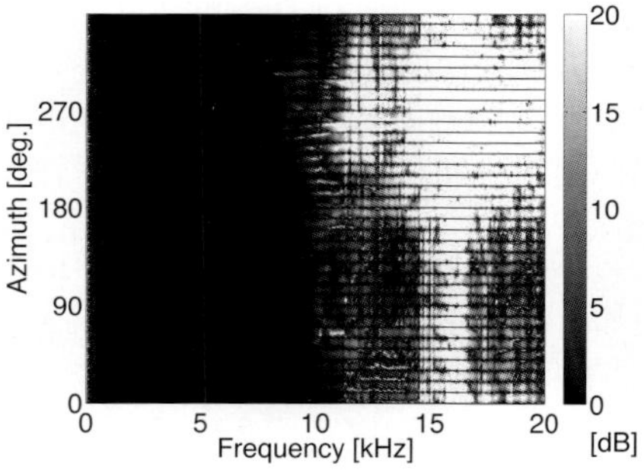

Fig. 5. Spectral distortion of condition **d** (0 deg elev. angle).

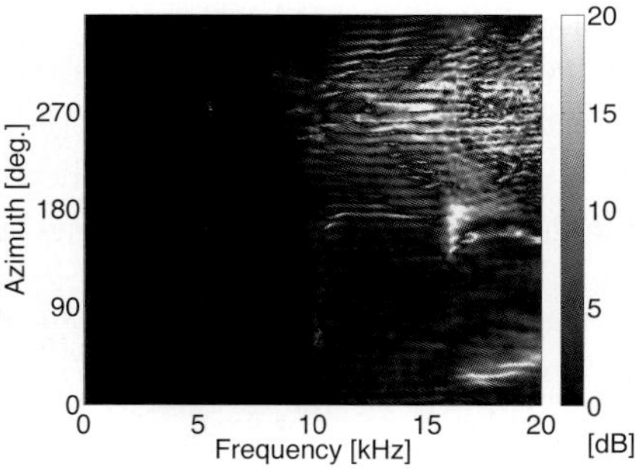

Fig. 6. Spectral distortion of condition **e** (0 deg elev. angle).

HRTFs. The microphones were set at steps of 90°, 45°, 30°, 15°, or 10° from 0° (in front of the listener) to 359° in the horizontal plane of the solid sphere. The intervals of the *controlled directions* were changed from 1° to 90°, with the intervals of *controlled directions* always narrower than that of microphones. The sampling frequency was 48 kHz. The range for synthesis was set as 0–20 kHz in steps of 93.75 Hz. Then, SD was calculated for all directions including *controlled* and *uncontrolled directins*.

Figure 8 portrays the relation between the average SD and the intervals of *controlled directions* for each microphone arrangement. Results show that the average SD decreases when the intervals of *controlled directions* becomes narrow. Moreover, the average of SD is almost constant when the *controlled directions* are set densely at intervals of less than 5°. Results show that the *controlled directions* should be set at 5° intervals.

Figures 9 to 12 show the SD of each condition with 36 microphones.

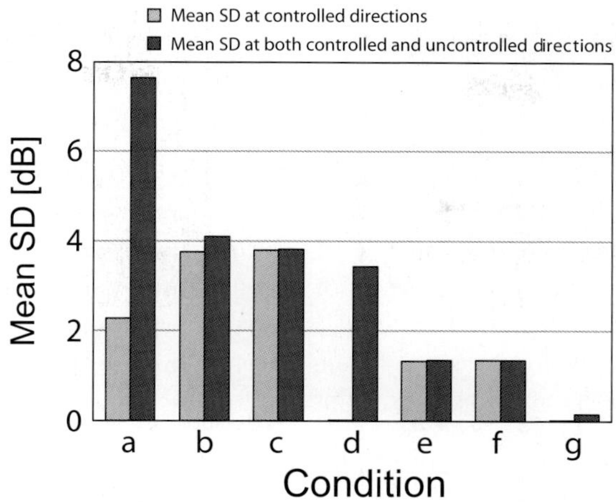

Fig. 7. Average of spectral distortion in all conditions.

In this case, the intervals of microphones are 10° (14.8 mm); thus spatial aliasing frequency is around 11.5 kHz. These figures show that the error in the head shadow region around 270° decreases as the intervals of *controlled directions* become narrow and that the error becomes constant when the *controlled directions* are set densely at intervals of less than 5°.

In summary, if the microphones and the "controlled directions" are arranged appropriately, then SENZI can accurately synthesize any HRTF up to around the spatial aliasing frequency for the shadow region and up to even several kilohertz higher frequency for the sunny-side region.

4. Summary

In this study, we investigated the effects of microphone arrangement and the number of *controlled directions* on the accuracy of the comprehensive 3D sound space information acquisition system called SENZI. The simulation results indicate that the microphones should be arranged at intervals of 5.7° or narrower to avoid the effect of spatial aliasing. Furthermore, the number of *controlled directions* should be set densely at intervals of less than 5° when the radius of the microphone array is 85 mm.

Acknowledgements

This work was supported by Strategic Information and Communications R&D Promotion Programme (SCOPE) No. 082102005 from the Ministry of

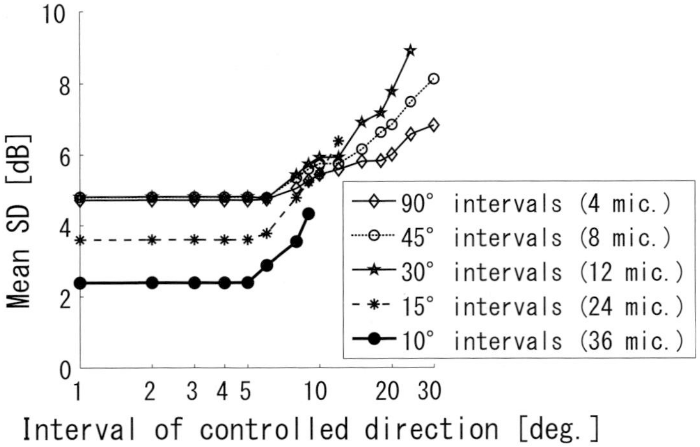

Fig. 8. Relation between the average of spectral distortion and the number of *controlled directions* for each microphone arrangement.

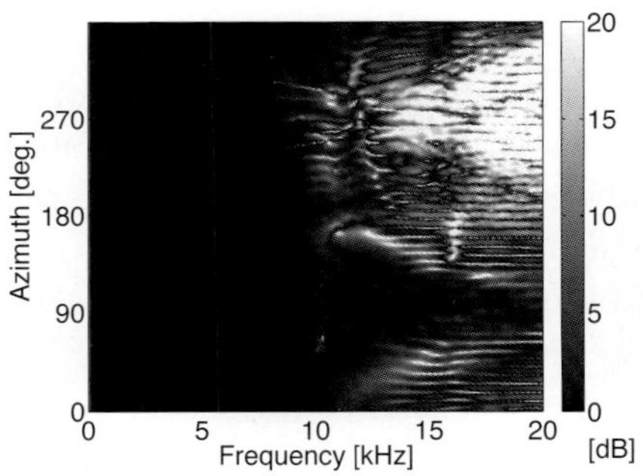

Fig. 9. Spectral distortion of 36 microphones and 8 deg intervals of *controlled directions* (45 directions).

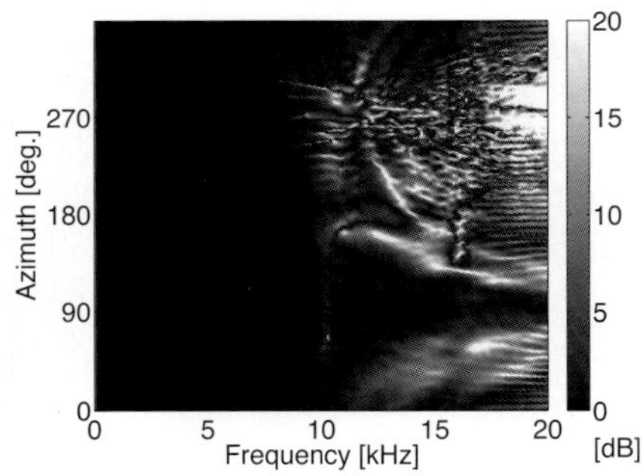

Fig. 10. Spectral distortion of 36 microphones and 6 deg intervals of *controlled directions* (60 directions).

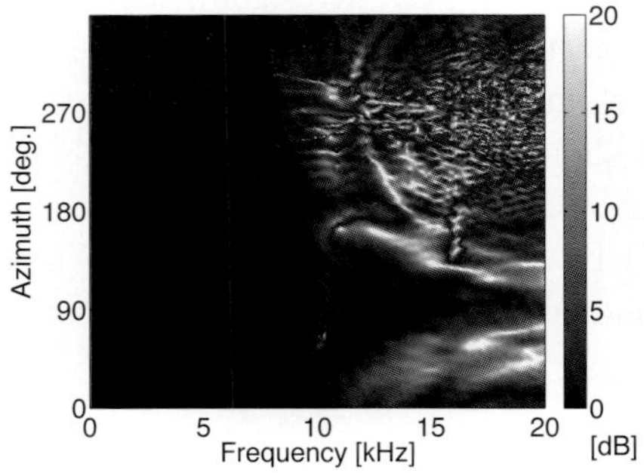

Fig. 11. Spectral distortion of 36 microphones and 5 deg intervals of *controlled directions* (72 directions).

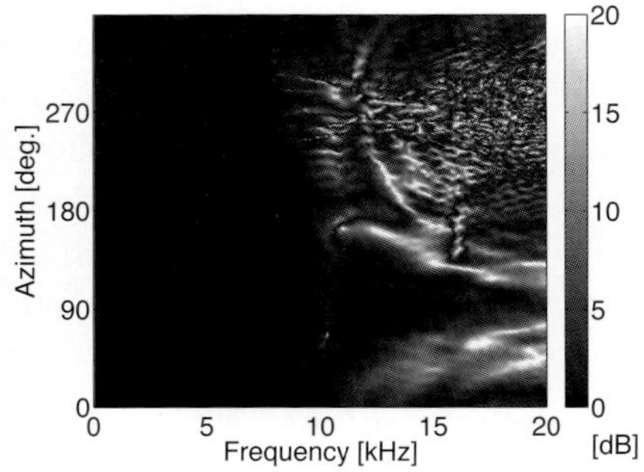

Fig. 12. Spectral distortion of 36 microphones and 4 deg intervals of *controlled directions* (90 directions).

Internal Affairs and Communications (MIC) Japan and a grant for Tohoku University Global COE Program CERIES from MEXT Japan. We thank Dr. Makoto Otani (Shinshu University) for assistance in calculating the HRTFs.

References

1. H. Wallach, On sound localization, *J. Acoust. Soc. Am.* **10**, 270 (1939)
2. W. R. Thurlow and P. S. Runge, Effect of induced head movement in localization of direction of sound, *J. Acoust. Soc. Am.* **42**, 480 (1967)
3. Y. Iwaya, Y. Suzuki and Kimura D, Effects of head movement on front-back error in sound localization, *Acoust. Sci. & Tech.* **24**, 322 (2003)
4. I. Toshima, H. Uematsu and T. Hirahara, A steerable dummy head that tracks three-dimensional head movement, *Acoust. Sci. & Tech.* **24**, 327 (2003)
5. V. R. Algazi, R. O. Duda and D. M. Thompson, Motion-Tracked Binaural Sound, *J. Audio Eng. Soc.* **52**, 1142 (2004)
6. J. B. Melick, V. R. Algazi, R. O. Duda and D. M. Thompson, Customization for personalized rendering of motion-tracked binaural sound, *Proc. 117th AES Convention* **6225**, 1 (2004)
7. D.H. Cooper and T. Shige, "Discrete-Matrix Multichannel Stereo," *Journal of Audio Engineering Society*, 20(5), pp. 346–360, 1972
8. R. Nicol and M. Emerit, "3D-sound reproduction over an extensive listening area: A hybrid method derived from holophony and ambisonic," Proc. AES 16th International Conference, 16(39), pp. 436–453, 1999
9. S. Sakamoto, S. Hongo, R. Kadoi and Y. Suzuki, SENZI and ASURA: New high-precision sound-space sensing systems based on symmetrically arranged numerous microphones, *Proc. Second International Symposium on Universal Communication (ISUC2008)*, 429 (2008)
10. J. Kodama, S. Sakamoto, M. Otani, S. Hongo, Y. Iwaya and Y. Suzuki, Numerical investigation of reproduction accuracy for ASURA (A Symmetrical and Universal Recording Array): Geometry and microphone position effects, *Proc. ASJ Spring meeting* (in Japanese), 1-9-6, 1457-1456 (2008)
11. S. U. Pillai, *Array signal processing* (Springer-Verlag, New York, 1989)

PERCEPTION-BASED REPRODUCTION OF SPATIAL SOUND WITH DIRECTIONAL AUDIO CODING

V. PULKKI[1], M-V. LAITINEN[1], J. VILKAMO[1], J. AHONEN[1], T. LOKKI[2] and T. PIHLAJAMÄKI[1] *

[1] *Dept Signal Processing and Acoustics,*
[2] *Dept of Media Technology*
Aalto University, Finland
**E-mail: ville.pulkki@tkk.fi*
www.acoustics.hut.fi

This article presents a review of Directional audio coding (DirAC), which is a perceptually motivated technique for spatial audio processing. DirAC analyzes, in short time windows, the sound spectrum together with direction and diffuseness in frequency bands of human hearing. It then uses this information in synthesis. The applications of DirAC are also discussed, which include capturing, coding, and resynthesis of spatial sound, teleconferencing, directional filtering, and virtual auditory environments. The subjective evaluation results presented elsewhere are also summarized.

Keywords: Spatial audio, sound reproduction, perceptual signal processing

1. Introduction

The spatial properties of sound perceivable by humans are the directions and distances of sound sources in three dimensions, and the effect of the room on sound. In addition, the spatial arrangement of sound sources affects the timbre, which corresponds to perceived sound color. The directional resolution of spatial hearing is limited within auditory frequency bands.[1] In principle, all sound within one critical band can be perceived only as a single source with broader or narrower extent. In some special cases, binaural narrow-band sound stimulus can be perceived as two distinct auditory objects, but the perception of three or more concurrent sources is generally not possible, which differs from visual perception, where already one eye can detect the directions of numerous visual objects sharing the same color.

The limitations of spatial auditory perception imply that such spatial realism that is needed in visual reproduction is not needed in audio. In other words, the spatial accuracy in reproduction of acoustical wave field can be compromised without decreasing perceptual quality. A recent technology for spatial audio, Directional audio coding (DirAC),[2] explores the possibilities of exploiting the frequency-band resolution of human sound perception in audio. In this paper, the basic DirAC processing principles are overviewed. Then different applications are presented.

2. Directional Audio Coding

Directional audio coding (DirAC)[2] has been proposed recently as a signal processing method for spatial sound. It is applicable for spatial sound reproduction for any multi-channel loudspeaker layout, or for headphones. Other applications for it have been suggested: teleconferencing and perceptual audio coding.

In DirAC, it is assumed that at one instant and at one critical band, the spatial resolution of auditory system is limited to decoding one cue for direction and another for inter-aural coherence. It is further assumed that if the direction and diffuseness of sound field is measured and reproduced correctly, a human listener will perceive the directional and coherence cues correctly.

The concept of DirAC is presented in Fig. 1. In the analysis phase, the direction and diffuseness of the sound field are estimated in auditory frequency bands depending on time, forming metadata transmitted together with a few audio channels. In the "low-bitrate" approach shown in the figure, only one channel of audio is transmitted. The audio channel might also be compressed further to obtain a lower transmission data rate. The version with more channels is shown as the "high-quality version", for which the number of transmitted channels is three for horizontal reproduction, and four for 3D reproduction. In the high-quality version, the analysis might be conducted in the receiving end.

2.1. *Division into frequency bands*

In DirAC, both analysis and synthesis are performed in the frequency domain. Several methods exist for dividing the sound into frequency bands, each with distinct properties. The most commonly used frequency transforms include short time Fourier transform (STFT), and quadrature mirror filterbank (QMF). In addition to these, there is full liberty to design a filterbank with arbitrary filters that are optimized to any specific purpose.

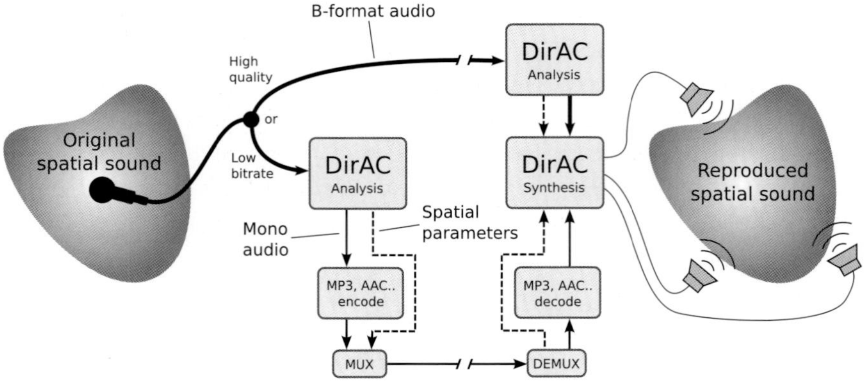

Fig. 1. Two typical approaches of DirAC are high quality (top) and low bitrate (bottom).

Irrespective of the selected time-frequency transform, the design goal is to mimic the resolution of human spatial hearing.

In the first implementations of DirAC, a filterbank with arbitrary subband filters alternatively with STFT with 20 ms time windows was used.[2] The even time resolution at all frequencies is a disadvantage of STFT implementation, which might produce some artifacts at high frequencies with some critical signals because of overly long temporal windows. The filterbank implementation solves this problem, but it might present constraints in computational complexity. In[3] a linear-phase filterbank was used; in[4] a multi-resolution version of STFT was used, where the input sound was divided into a few frequency channels and processed with different STFTs having window lengths suited to each frequency band.

However, based on informal listening during the development of the technique, it is quite clear that the choice of time-frequency transfer is not a critical issue for audio quality in DirAC reproduction. The differences are typically audible only with input material having about 100–1000 Hz modulations in signal envelopes at high frequencies, an example of such sound is the snare drum sound.

2.2. *Directional analysis*

The target of directional analysis, which is shown in Fig. 2, is to estimate at each frequency band the direction of arrival of sound, together with an estimate if the sound is arriving from one or multiple directions simultaneously. In principle, this can be performed using several techniques, however, the energetic analysis of sound field has been found to be suitable, as shown in Fig. 2. The energetic analysis can be performed when the pressure signal and velocity signals in 1–3 dimensions are captured from a single position. In first-order B-format signals, the omnidirectional signal is called the W-signal, which has been scaled down by $\sqrt{2}$. The sound pressure can be estimated as $P = \sqrt{2}W$, expressed in STFT domain. The X-, Y- and Z-channels have the directional pattern of a dipole directed along the Cartesian axis, which together form a vector $\mathbf{U} = [X, Y, Z]$. The vector estimates the sound field velocity vector; it is also expressed in the STFT domain. The energy E of the sound field can be computed as

$$E = \frac{\rho_0}{4}||\mathbf{U}||^2 + \frac{1}{4\rho_0 c^2}|P|^2, \tag{1}$$

where ρ_0 stands for the mean density of air, and c signifies the speed of sound. The capturing of B-format signals can be achieved with either coincident positioning of directional microphones, or with closely spaced set of omnidirectional microphones. In some applications, the microphone signals might be formed in the computational domain, i.e. simulated. The analysis is repeated as frequently as is needed for the application, typically with the update frequency of 100–1000 Hz.

The intensity vector $\mathbf{I}$ expresses the net flow of sound energy as a 3D vector. It can be computed as

$$\mathbf{I} = \overline{P}\mathbf{U}, \tag{2}$$

where $\overline{(\cdot)}$ denotes complex conjugation. The direction of sound is defined as the opposite direction of the intensity vector at each frequency band. The direction is denoted as corresponding to angular azimuth and elevation values in the transmitted metadata. The diffuseness of sound field is

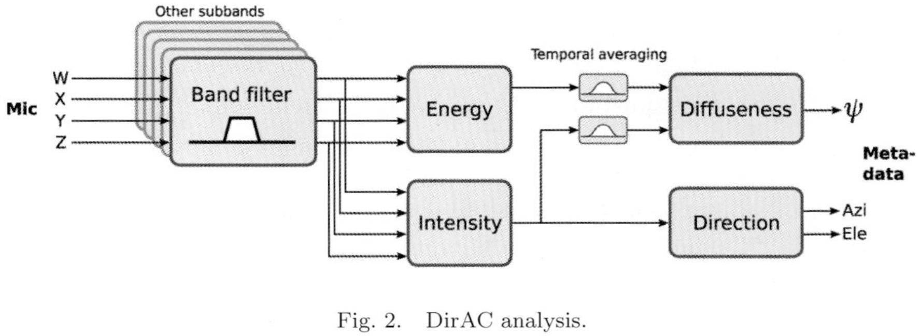

Fig. 2. DirAC analysis.

computed as

$$\psi = 1 - \frac{||\mathrm{E}\{\mathbf{I}\}||}{c\mathrm{E}\{E\}}, \tag{3}$$

where E is the expectation operator. The outcome of this equation is a real-valued number between zero and one, characterizing whether the sound energy is arriving from a single direction, or from all directions. This equation is appropriate in the case in which the full 3D velocity information is available. If the microphone setup delivers velocity only in 1D or 2D, then the equation

$$\psi_{cv} = \sqrt{1 - \frac{||\mathrm{E}\{\mathbf{I}\}||}{\mathrm{E}\{||\mathbf{I}||\}}} \tag{4}$$

yields estimates that are closer to the actual diffuseness of sound field than Eq. (1) is.[5]

2.3. *DirAC transmission*

In many applications, spatial sound must be transmitted from a location to another. In DirAC, this can be performed using different approaches. A straightforward technique is to transmit all signals of B-format. In such a case, no metadata are needed, and analysis can be performed on the receiving end. However, in the low-bit-rate version, only one channel of audio is transmitted, which is designated as a mono DirAC stream. This provides a large reduction in the data rate, the disadvantage of which is a slight decrease in timbral quality of reverberant sound and a decrease in the directional accuracy in multi-source scenarios.

In some cases, it is beneficial to merge multiple mono DirAC streams together. This is not a trivial task because there is no simple way to merge directional metadata. However, two methods have been proposed to provide artifact-free and efficient merging.[6]

2.4. *DirAC synthesis with loudspeakers*

The high-quality version of DirAC synthesis, shown in Fig. 3, receives all B-format signals, from which a virtual microphone signal is computed for each loudspeaker direction. The used directional pattern is typically a dipole. The virtual microphone signals are then modified in nonlinear fashion, depending on the metadata. The low-bit-rate version of DirAC is not shown in the figure, however; in it only one channel of audio is transmitted. The difference in processing is that all virtual microphone signals would be replaced by the single channel of audio received. The virtual microphone signals are divided into two streams (the diffuse and the non-diffuse streams), which are processed separately.

The non-diffuse stream is reproduced as point sources using vector base amplitude panning (VBAP).[7] In panning, a monophonic sound signal is applied to a subset of loudspeakers after multiplication with loudspeaker-specific gain factors. The gain factors are computed using the information of the loudspeaker setup and the specified panning direction. In the low-bit-rate version, the input signal is simply panned to the directions implied by the metadata. In the high-quality version, each virtual microphone signal is multiplied with the corresponding gain factor.

In many cases, the direction in metadata is subject to abrupt temporal changes. To avoid artifacts, the gain factors for loudspeakers computed with VBAP are smoothed by energy-weighted temporal integration with frequency-dependent time constant equaling about 50 cycle periods at each band, which effectively removes the artifacts. However, the changes in direction are not perceived to be slower than without averaging in most cases.

The aim of the synthesis of **the diffuse stream** is to create perception of sound that surrounds the listener. In the low-bit-rate version, the diffuse stream is reproduced by decorrelating the input signal and reproducing it from every loudspeaker. In the high-quality version, the virtual microphone signals of diffuse streams are already incoherent to some degree. They must be decorrelated only mildly. This approach provides better spatial quality for surrounding reverberation and ambient sound than the low-bit-rate version does.

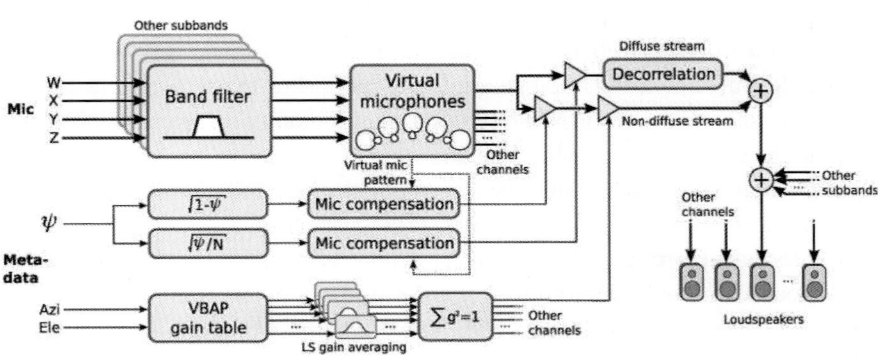

Fig. 3. DirAC synthesis.

2.5. *DirAC synthesis with headphones*

In,[8] different approaches to reproduce spatial audio over headphones with and without head tracking were investigated in the context of DirAC. Testing of different versions revealed that the best quality was obtained when the DirAC was formulated with about 40 virtual loudspeakers around the listener for non-diffuse stream, and 16 loudspeakers for the diffuse stream. The virtual loudspeakers are implemented as convolution of input signal with measured HRTFs.

A common problem in headphone reproduction is that the reproduced auditory space moves with the head of the listener, which causes internalized sound events. To prevent this, a method to use head tracking information in DirAC was also developed.[8] A simple and effective method was found to be to update the metadata, and to transform the velocity signals according to the head tracking information with about a 50 Hz update rate.[8]

2.6. *Similarity to channel-based coding*

DirAC shares many processing principles and challenges with existing spatial audio technologies in coding of multi-channel audio.[9,10] The most distinctive difference is the metadata. In DirAC, the metadata consist of direction and diffuseness values, which can be derived from sound field quantities directly, although the metadata in other approaches consist of quantities measured from multichannel audio files for loudspeaker playback. DirAC is useful similarly in processing of multi-channel audio files. However, DirAC is also applicable for recording real spatial sound environments.

3. DirAC Applications

Convolving reverberators. The first implementation of DirAC was reproduction of measured B-format impulse responses over arbitrary loudspeaker setups, Spatial impulse response rendering (SIRR).[11] The application was in convolving reverberators. An acoustically dry monophonic recording can be processed for multichannel listening to sound as though it were performed in the hall where the B-format impulse response was measured.

Teleconferencing. In a basic telecommunication application of DirAC, two groups of people want to have a meeting with each other. Both groups gather in the vicinity of typically 1D or 2D B-format microphone. The low-bit-rate version of DirAC is used, where directional metadata are encoded from the microphone signals, and the transmitted signal is mono DirAC stream. At the receiving end the mono DirAC stream can be rendered to whatever loudspeaker layout, which effectively spatializes the talkers in reproduction. Different microphone setups have been tested, revealing that the metadata can be encoded already from closely spaced pair of low-end omnidirectional or directional microphones.[12] Results also showed that a basic spatial segregation between talkers can be obtained from a stereo microphone, although the directional patterns are not known in DirAC encoding.

High-quality reproduction. As described above, the initial target for DirAC development was to reproduce a recorded sound scenario as realistically as possible. This is interesting at least at an academic level, and in some cases also in the audio industry.

The DirAC metadata seems also to be a good starting point for a **generic audio format**, which would tolerate different loudspeaker setups, and headphone listening. Methods to transform existing audio content into DirAC formats are under construction, and the audio synthesis in DirAC must be tested in various conditions.

Spatial filtering. In spatial filtering, a signal is formed which is sensitive only to a direction defined by the user. The DirAC method is useful also for this, in simplest form by listening to only one loudspeaker of a multi-loudspeaker setup, which already greatly emphasizes the sound arriving from the direction of the loudspeaker. A fine-tuned method to use directional analysis parameters to suppress diffuse sound and sound originating from arbitrary directions has already been suggested,[13] where it was found to outperform some traditional beam-forming methods in some aspects.

Source localization. Another application of directional analysis data is in localization of sound sources, where the usage of diffuseness and direction parameters provides an efficient method. For example, in a teleconferencing scenario, where energetic analysis is already performed from the microphone inputs, they can also be used to steer a camera to active talkers. In[14] it is shown that the localization method based on directional analysis data provides reliable estimates even in reverberant environments and with multiple concurrent talkers. The approach also allows for a tradeoff between localization accuracy and tracking performance of moving sound sources.

Applications in mixing, game sound, and in virtual realities. Prior applications concentrated on cases in which the directional properties of sound are captured with real microphones from a live situation. It is also possible to use DirAC in virtual or mixed realities.[15] In these applications, the directional metadata connected to a DirAC stream are defined by the user. For example, a single channel of audio is spatialized as a point-like virtual source with DirAC when the same direction for all frequency channels is added as metadata to the signal. In some cases, it would be beneficial to control the perceived extent or width of the sound source. A simple and effective method for this is to use a different direction value for each frequency band, where the values are distributed inside the desired directional extent of the virtual source.

Realization of this point is presented in Fig. 4. The virtual source signals are multiplied with distance attenuation gain. A distance delay can be added also in this point. The signals are then turned into mono DirAC streams consisting of an intact input audio channel and inserted directional metadata. The mono DirAC stream is then turned into B-format stream with a method presented in.[15] The B-format streams of virtual sources are then added together.

A common task in virtual worlds is the generation of a room effect or reverberation. In[15] it is shown that two single-output reverberators are sufficient for any horizontal loudspeaker setup; three is sufficient for any three-dimensional loudspeaker setup. The method is shown in Fig. 4. According to informal testing, this generates diffuse reverberation with good quality. If distinct localizable reflections are needed, feasible results are obtained if the reflections are reproduced as individual virtual sources with mono DirAC streams. Additionally, the created point-like or extended virtual sources with or without reverberation can be superposed to spatial sound in recorded B-format files, as shown in the figure.

4. Subjective evaluation

Loudspeaker versions of DirAC and SIRR in reproduction of actual sound scenes were researched by comparing reference scenarios to their reproductions. The reference scenarios were produced using more than 20 loudspeakers, which generate the direct sounds from sources, and also reflections and reverberation.[3,16] An ideal or real B-format microphone was used to capture the signal in the center of the setup. Then the recorded signal was reproduced using a high-quality implementation of DirAC or SIRR. The perceptual quality of reproduction of room impulse responses with SIRR was tested in anechoic listening. Results show that the difference between SIRR reproduction and the reference scenario was very small, rated as "perceptible, difference not annoying" in the worst case.[16]

To measure the perceptual audio quality of DirAC, two distinct listening tests were conducted, one with 3D reproduction in anechoic environment, and one with horizontal-only reproduction in a multi-channel listening room. Different loudspeaker layouts with 5–16 loudspeakers were used in both conditions for DirAC reproduction. In direct comparison to the reference, DirAC reproduction was rated in almost all conditions either excellent or good on an ITU scale. Only in off-sweet-spot 5.1 listening was the quality scored as fair. The results from the test in the listening room are shown in Fig. 5. Results also show clearly that DirAC produces better perceptual quality in loudspeaker listening than other available techniques using the same microphone input.

The binaural reproduction of DirAC was investigated in[8] with and without head-tracking in a test that did not use reference scenarios. The results with binaural head-tracked reproduction were very good. In an ITU scale, the overall quality was rated "excellent"; the spatial impression was rated "truly believable". The results of overall audio quality are presented in Fig. 6. Listeners reported that all sound objects were fully externalized; some found it difficult to believe that the sound came from the headphones, and not from the loudspeakers present in the same room.

For the teleconferencing application, the speech reception level with two concurrent talkers was measured with DirAC reproduction with 1-D and 2-D microphone setups.[17] Results show that the reception of speech was almost as good with DirAC with mono DirAC stream as it was with

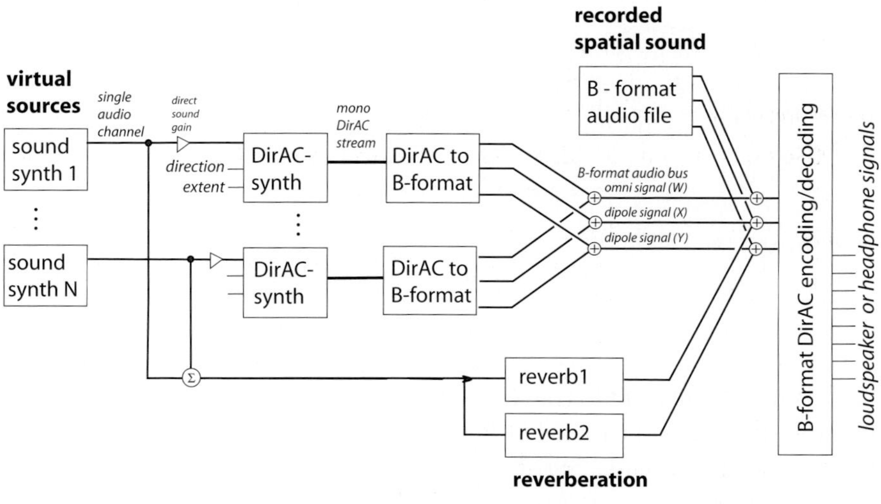

Fig. 4. Implementation of DirAC for virtual reality applications.

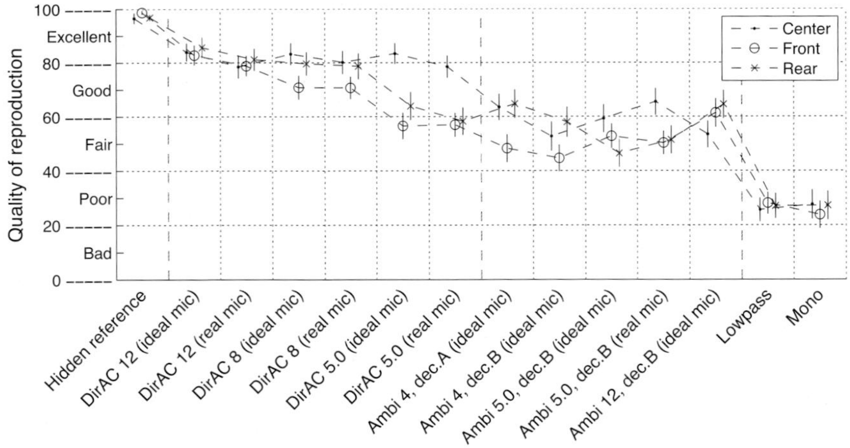

Fig. 5. Mean opinion scores for different reproduction methods in the listening room with 95% confidence intervals. The reference scenarios were created using 24 loudspeakers, with four different simulated acoustical conditions, and with three program materials. Scenarios were compared to the DirAC reproduction of them in a MUSHRA-type listening test. The B-format microphone signals were either simulated (ideal mic) or measured using an equalized SoundField ST350 microphone (real mic). An ITU-R BS.1116 listening room was used in the test, with three different listening positions (Center, Front-side, and Rear-side). The high-quality filter-bank version of DirAC was used with the stated number of loudspeakers. First-order Ambisonics playback was used as reference with two commonly available decoders, together with a 2-kHz-lowpassed original signal and a monophonic full-band signal. The number of listeners was 12. Details are reported in.[3]

two-channel transmission of dipole signals, which served as a reference. In[18] the data rate of directional metadata was studied in the context of teleconferencing application. Results show that the rate can be as low as 2–3 kBit/s, and that the directions of talkers are perceived correctly.

5. Summary

DirAC is a perceptually motivated signal-dependent spatial sound reproduction method based on frequency-band processing and directional sound field analysis. The time-dependent data originating from the analysis is used in the synthesis of sound. The method uses existing first-order microphones for recording and any surround loudspeaker setup, or headphones, for reproducing the spatial sound. In this paper, a review of different variants of the method was made. Applications of the method were discussed. The results from subjective tests conducted to date were also summarized.

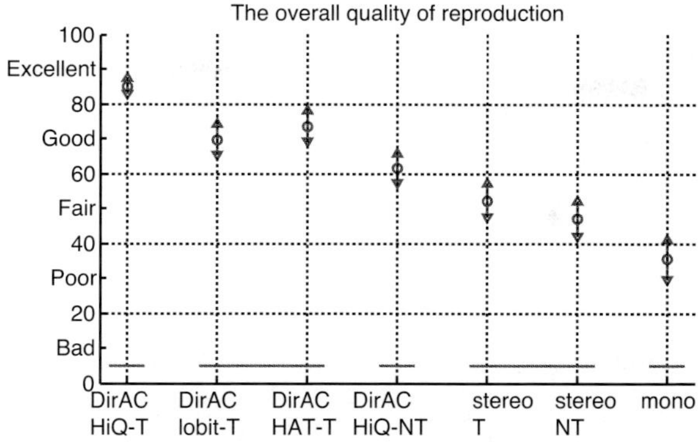

Fig. 6. Overall audio quality of DirAC with binaural listening as means and the 95% confidence intervals of data measured from eight subjects. Four B-format recordings were reproduced over headphones with or without head tracking with different versions of DirAC in anechoic chamber with visible loudspeakers. Data groups with non-significant differences in means are marked with a horizontal line. Acronyms: DirAC HiQ, high-quality version of DirAC with nonindividual HRTFs; DirAC lobit, low-bit-rate version of DirAC with nonindividual HRTFs; HAT, high-quality version of DirAC with HRTFs computed with simple head-and-torso model; T, head tracking on; NT, head tracking off; stereo, virtual cardioid microphones computed to directions ±60°: mono, the W channel of B-format signal presented diotically.

6. Acknowledgements

The Academy of Finland (Projects #105780 and #119092) and Fraunhofer Institute for Integrated Circuits IIS supported this work. The research leading to these results has received funding from the European Research Council under the European Community's Seventh Framework Programme (FP7/2007-2013) / ERC grant agreements n^o [240453] and n^o [203636].

References

1. J. Blauert, *Spatial Hearing* (The MIT Press, 1983).
2. V. Pulkki, *J. Audio Eng. Soc.* **55**, 503(June 2007).
3. J. Vilkamo, T. Lokki and V.Pulkki, *J. Audio Eng. Soc.* **57** (2009).
4. T. Pihlajamäki, Multi-resolution short-time fourier transform implementation of directional audio coding, Master's thesis, Helsinki Univ. Tech. (2009).
5. J. Ahonen and V. Pulkki, Diffuseness estimation using temporal variation of intensity vectors, in *Workshop on Applications of Signal Processing to Audio and Acoustics WASPAA*, (Mohonk Mountain House, New Paltz, 2009).
6. G. D. Galdo, V. Pulkki, F. Kuech, M.-V. Laitinen, R. Schultz-Amling and M. Kallinger, Efficient methods for high quality merging of spatial audio streams in directional audio coding, in *AES 126th Convention*, (Munich, Germany, 2009). Paper 7733.

7. V. Pulkki, *J. Audio Eng. Soc.* **45**, 456(June 1997).
8. M.-V. Laitinen and V. Pulkki, Binaural reproduction for directional audio coding, in *Workshop on Applications of Signal Processing to Audio and Acoustics WASPAA*, (IEEE, New Paltz, NY, 2009).
9. J. Herre, K. Kjörling, J. Breebaart, C. Faller, S. Disch, H. Purnhagen, J. Koppens, J. Hilpert, J. Rödén and W. Oomen, *J. Audio Eng. Soc.* **56**, 932 (2008).
10. M. M. Goodwin and J.-M. Jot, A frequency-domain framework for spatial audio coding based on universal spatial cues, in *120th AES Convention*, Paris, May 2006. Paper # 6751.
11. J. Merimaa and V. Pulkki, *J. Audio Eng. Soc.* **53**, 1115(December 2005).
12. J. Ahonen, V. Pulkki, F. Kuech, G. D. Galdo, M. Kallinger and R. Schultz-Amling, Directional audio coding with stereo microphone input, in *AES 126th Convention*, (Munich, Germany, 2009). Paper 7708.
13. M. Kallinger, G. D. Galdo, F. Kuech, D. Mahne and R. Schultz-Amling, Spatial filtering using directional audio coding parameters, in *Proceedings of the 2009 IEEE International Conference on Acoustics, Speech and Signal Processing*, 2009.
14. O. Thiergart, R. Schultz-Amling, G. D. Galdo, D. Mahne and F. Kuech, Localization of sound sources in reverberant environments based on directional audio coding parameters, in *127th AES Convention*, (New York, 2009). Paper # 7853.
15. V. Pulkki, M.-V. Laitinen and C. Erkut, Efficient spatial sound synthesis for virtual worlds, in *AES 35th Conf. Audio for Games*, (London, UK, 2009).
16. V. Pulkki and J. Merimaa, *J. Audio Eng. Soc.* **54**, 3(January/February 2006).
17. J. Ahonen, V. Pulkki, F. Kuech, M. Kallinger and R. Schultz-Amling, Directional analysis of sound field with linear microphone array and applications in sound reproduction, in *the 124th AES Convention*, (Amsterdam, Netherlands, 2008). Paper 7329.
18. T. Hirvonen, J. Ahonen and V. Pulkki, Perceptual compression methods for metadata in directional audio coding applied to audiovisual teleconference, in *the 126th AES Convention*, (Munich, Germany, 2009). Paper 7706.

CAPTURING AND RECREATING AUDITORY VIRTUAL REALITY

R. DURAISWAMI, D. N. ZOTKIN, N. A. GUMEROV and A. E. O'DONOVAN

Perceptual Interfaces & Reality Lab., Computer Science & UMIACS
Univ. of Maryland, College Park
E-mail: {ramani,dz,gumerov}@umiacs.umd.edu, adamod@gmail.com

Reproduction of auditory scenes is important for many applications. We describe several contributions to the capture and recreation of spatial audio that have been made over the past few years by the authors.

Keywords: Surrounding loudspeaker array; Auralization; Higher-order Ambisonics; Computational room acoustics

1. Introduction

The recreation of an auditory environment in virtual reality, augmented reality, or in remote telepresence is a problem of interest in many application domains. It involves first the capture of sound, and second its subsequent reproduction, in a manner that can fool the perceptual system to believe that the rendered sound is actually where the application requires it to be and consequently to make the user feel as if they are present in the real audio scene. Our natural sound processing abilities, such as acoustic source localization, selective attention to one stream out of many, and event detection, are often taken for granted; however, the percept of the spatial location of a source arises from many cues, including those that arise due to the scattering of the sound off the listeners. Accurate simulation of those scattering cues is necessary for the spatial audio reproduction task. We review the principles for virtual auditory space synthesis, describe problems that have to be solved in order for it to be convincing, and outline solutions available, including some developed in our lab. Numerous applications are possible, including audio user interfaces, remote collaboration, scene analysis, remote education and training, entertainment, surveillance, and others.

2. 3D Audio Reproduction

The goal of audio reproduction is to create realistic three-dimensional audio so that users perceive acoustic sources as external to them and located at the correct places.[12] One obvious solution is to place arrays of loudspeakers and just play the sounds from wherever they are supposed to be, panning as appropriate to permit sound placement in-between, e.g., as in.[1] Such a setup is obviously cumbersome, expensive, non-portable, and noisy to other people in the same environment, and is only used in special environments where these issues are not of concern. The following discussion assumes that the synthesis is done over headphones. Three building blocks of the virtual auditory reproduction system[2] are: head-related transfer function based filtering, reverberation simulation, and user motion tracking.

Head-related transfer function: Sound propagating through space interacts with the head and body of the listener, causing the wave reaching the ear to be modified from that was emitted at the source. Furthermore, due to the geometry of human bodies and in particular to the complex shape of the pinna, changes in sound spectrum depend greatly on the direction from which the sound arrive, and to a lesser extent on range. These spectral cues are linked to the perception of sound direction and are characterized as the head-related transfer function (HRTF) – the ratio between the sound spectrum at the ear and that at the source.[7] These are the filters to be applied to transform the original sound source to the perceived one.

Inter-personal differences in body geometry make the HRTF substantially different among individuals.[9] For accurate scene reproduction the HRTF can be measured for each participant, which may be time-consuming. Various methods of selecting the HRTF from a pre-existing database that "best-fits" individual in some sense were tried by researchers with matching either physical parameters (e.g. from a ear picture[2]) or perceptual experience (e.g., by asking which audio piece sounds closest to being overhead[5]). Successful attempts of computing HRTF using numerical methods on head/ear mesh have also been reported, though they usually require very significant computation time[15].[17]

Reverberation Simulation: In addition to the direct sound from the source, we hear multiple reflections from boundaries of the environment and objects in it, termed reverberation. From these we are able to infer the room size, wall properties, and perhaps our position in the room. Each reflection is not heard separately, but rather they are perceptually joined in one auditory stream. The perception of reverberation is complicated; roughly speaking, the energy decay rate in the reverberation tail implies room size, and relative magnitude at various frequencies is influenced by materials. The ratio of direct to reverberant sound energy is used to judge distance to the source. Hence, reverberation is a cue that is very important to perception of sound as being external to the listener. Also, the reflections that constitute the reverberation are directional (i.e., arrive to the listener from specific directions just like the direct sound) and should be presented as directional in an auditory display at least for the first few reflections.

User Motion Tracking: A key feature of any acoustic scene that is external to the listener is that it is stationary with respect to the environment. That is, when the listener rotates, auditory objects move in the listener-bound frame so as to "undo" the listener's rotation. If this is not done, listeners subconsciously assume that the sound's origin is in their head (as with ordinary headphone listening), and externalization becomes very hard. Active tracking of the user's head and adjustment of the auditory stream on the fly to compensate for the head motion/rotation is necessary. A tracker providing at least the head orientation in space is needed for virtual audio rendering. The stream adjustment must be made with sufficiently low latency so as not to create dissonance between the expected and perceived streams.[8]

Note that all these three problems do not exist if the auditory scene is rendered using loudspeaker array. Indeed, HRTF filtering for the correct direction and with correct HRTF is done by the mere act of listening; room reverberation is added naturally, given that the array is placed in some environment; and motion tracking is also unnecessary because in this case the sources are in fact external to the user and obviously shift when the user rotates. In essence, the loudspeaker array setup attempts to represent the "original" scene – with sources physically surrounding the listener. However, such a setup is expensive/non-portable and can be used only in large facilities, and may still find it difficult to reproduce specific scenes different from the environment in which the reproduction is being done.

2.1. *Signal Processing Pipeline*

We describe here the typical signal processing pipeline involved in virtual auditory scene rendering. The goal is to render several acoustic sources at certain positions in a simulated environment with given parameters (room dimensions, wall materials, etc.) Assume that the clean (reverberation-free) signal to be rendered for each source is given and that the HRTF of the listener is known. For each source, the following sequence of steps is done and the results are summed up together to produce auditory stream.

- Using source location and room dimensions, compute reflections (up to a certain order) using simple geometry.
- Obtain current user head position/rotation from the headtracker.
- Compute locations of the acoustic source and of its reflections in listener-bound coordinate system.
- Compose impulse responses (IR) for left and right ears by combining the HRIRs for the source and reflection directions.
- Convolve the source signal with left/right ear IR using frequency-domain convolution.
- Render the resulting audio frame to the listener and repeat.

In practice, due to limited computational resources available, reflections up to a certain order are computed in real time; the remaining reverberation tail is computed in advance and is appended to the IR unchanged. Also, artifacts such as clicks can arise at the frame boundaries due to the IR filter being changed significantly between frames. Usual techniques such as frame overlapping with smooth fade-in/fade-out windows can be used to eliminate artifacts.

3. Audio System Personalization

As mentioned before, the accuracy of the spatial audio rendering depends heavily on the knowledge of the HRTF for the particular user of the rendering system. The rendering quality and the localization accuracy degrade greatly when "generic" HRTF is used.[9] Therefore, it is necessary to acquire the HRTF for each user of the headphone spatial audio rendering system. Such acquisition can be done in several different ways, including direct HRTF measurement, selection of "best-fitting" HRTF from a database based on anthropometry, or HRTF computation using numerical methods on the pinna-head-torso mesh. Some relevant work done in the author's lab is described below.

3.1. *Fast HRTF Measurement*

We seek to measure the HRTF $H_l(k; s)$ defined as

$$H_l(k; s) = \frac{\Psi_l(k; s)}{\Psi_c(k)}, \qquad H_r(k; s) = \frac{\Psi_r(k; s)}{\Psi_c(k)},$$

where the signal spectrum at the ear is $\Psi_l(k; s)$ and at the center of the head in the listener's absence is $\Psi_c(k)$; indices l and r signify left/right ear. Here, s is the direction to the source and we use wavenumber k in place of the frequency f for notational convenience; note that $k = 2\pi f/c$. The traditional method of measuring the HRTF (e.g.,[10]) is to position a sound source (e.g., a loudspeaker) in the direction s and play the test signal $x(t)$. Microphones in the ears record the received versions of the test signal $y(t)$, which in the ideal case is $x(t)$ filtered with the head-related impulse response (HRIR) – the inverse Fourier transform of $H_{l,r}(k;s)$. Measurements are done on a grid of directions to sample the continuous HRTF at a sufficient number of locations. The procedure is slow because the test sounds have to be produced sequentially from many positions, with sufficient interval. This necessitates use of equipment to actually move the loudspeaker(s) between positions. As a result, measurement at an acceptable sampling density (about a thousand locations) requires an hour or more.

An alternative fast method was developed recently by us.[3] The method is based on Helmholtz' reciprocity principle.[11] Assume that in a linear time-invariant scene the positions of the source and the receiver are swapped. By reciprocity, the recording made would be the same in both cases. The fast HRTF measurement method thus places the source in the listener's ear and measures the received signal simultaneously at many microphones at a regular grid of directions. The time necessary for the measurement is thus reduced from hours to seconds.

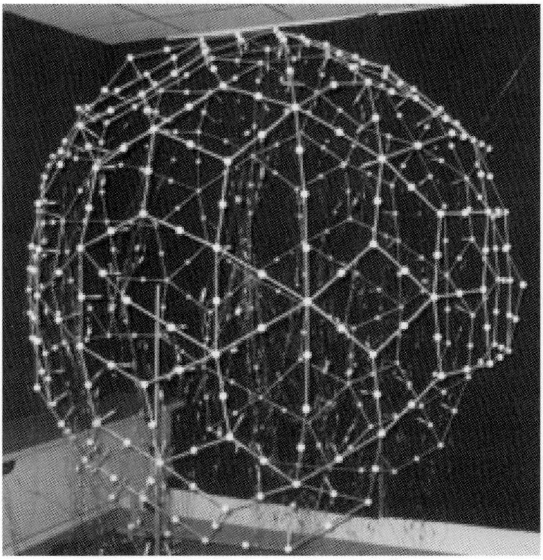

Fig. 1. A prototype fast HRTF measurement system based on the acoustic reciprocity principle. The diameter of the structure is about 1.4 meters.

The test signal $x(t)$ is selected to be wideband (to provide enough energy in frequencies of interest) and short (to be able to window out equipment/wall reflections). In our case, the test signal is about 96 samples (2.45 ms long) upsweep chirp. The sampling rate is 39062.5 Hz. The test signal is fed to the Knowles ED-9689 microspeaker, which is wrapped in a silicone plug so as to provide acoustic isolation and very safe sound levels, and is inserted into the subject's ear. 128 Knowles FG-3629 microphones are mounted at the nodes of an open sphere-like structure (Figure 1).

To measure the HRTF, a calibration signal is first obtained to compensate for individual differences in channels of the system. The speaker is placed at the center of the recording mesh, and a test signal recorded at all microphones as $y_i^c(t)$; i is the microphone index. The recorded signal was windowed so as to exclude reflections from the room boundaries. The test signal $x(t)$ is played through the microspeaker and recorded at ith microphone at s_i as $y_i(t)$. This is repeated 48 times, and the resulting signal is time-averaged to improve SNR. The same thresholding and windowing is used on $y_i(t)$. The estimated HRTF $H_l(k; s_i)$ is then determined as $Y_i(k)/Y_i^c(k)$, where capitals signify Fourier transform. The original signal $x(t)$ is not used in calculations but is present implicitly in $y_i^c(t)$ modified by channel responses.

The computed HRTF is windowed in the frequency domain with a trapezoidal window in order to dampen it to zero at the frequencies where measurements cannot be made reliably. In particular, the microspeaker is inefficient at low frequencies ($\lesssim$700 Hz), and there is no energy in the test signal above 14 kHz. The HRTF in the very low / very high frequency ranges is gradually tapered to zero. Inverse Fourier transform is used to generate HRIR, which is made minimum phase with appropriate time delays obtained from thresholding $y_i(t)$.

3.2. *HRTF Approximation Using Anthropometric Measurements*

The HRTF is a transfer function describing filtering of the sound source by the listener's anatomy, and HRTF features (such as ridges and valleys in the spectrum) are conceivably created by sound scattering in the outer ear and by the head and torso. It is therefore reasonable to assume that listeners having anatomical features "similar" to the certain extent would have similar HRTFs as well. Some authors (e.g.,[18]) have looked into this problem by building an HRTF model as a function of certain morphological measurements. However, the analysis was done on a limited number of subjects. Later, several HRTF databases became available. One of them is the CIPIC database,[10] which includes certain anthropometric measurements in addition to the HRTF data for 45 subjects. We explored a simple way of spatial audio rendering system customization by finding the "best-matching" database subject in terms of ear parameters and then further amending the HRTF in accordance with head and torso parameters.[19]

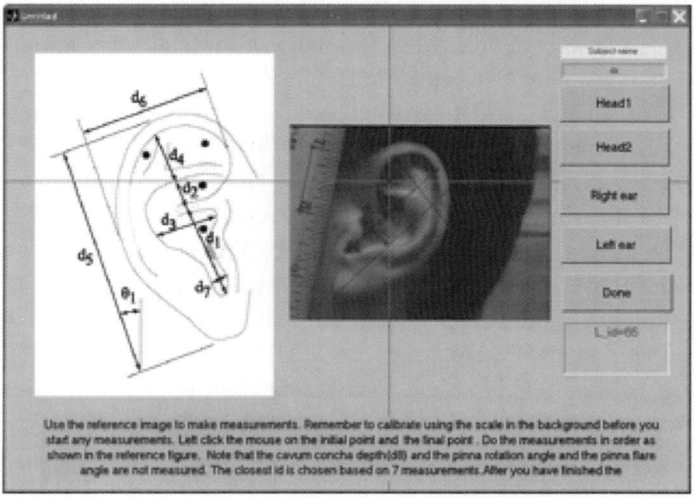

Fig. 2. A screenshot of our HRTF customization software.

Figure 2 shows the main interface of the system. On the left side, the reference image is shown with the seven ear dimensions $d_1, ..., d_7$ identified (they are, in order, cavum concha height, cymba concha height, cavum concha width, fossa height, pinna height, pinna width, and intertragal incisure width). The reference image is used to guide the operator in process of measuring these dimensions in the picture of real ear. On the right side, the image of the user's ear is acquired using a digital camera, along with a ruler to provide a scale. The operator identifies and marks feature points on the ear as shown. If $\hat{d}_i$ is the value of the ith parameter in the image, and d_i^k is the value of the same parameter for the kth subject in the database, then the matching is performed by minimizing the error measure E^k:

$$E^k = \sum_{i=1}^{7} \frac{(\hat{d}_i - d_i^k)^2}{\sigma_i^2}.$$

Here σ_i^2 is the variance of the ith parameter across all subjects in the database. Subject k, $k = \arg\min_k E^k$, is chosen to be the best match to the user. In the case shown, it is the subject with ID 45. Note that matching is done separately for left and right ears, which sometimes results in different best-matching subjects for left and right ears because of asymmetries in individual anatomy.

The HRTF of the best-matching subject is further refined using the head-and-torso (HAT) HRTF model.[20] This is a three-parameter (head radius, torso radius, and neck height) analytical model for computation of HRTF at low frequencies (below approximately 3-4 kHz). The HAT algorithm models head and torso as two separated spheres and simulates the wave propagation path(s) for various source positions. Note that contributions to the HRTF caused by the torso, the head, and the pinnae are more or less separated on the frequency axis; in particular, pinna-induced features are generally located above 4-6 kHz. Therefore, it can be assumed that the database matching method would produce reasonably matching HRTF at higher frequencies only, as only the pinna features are used for matching. To compensate for that, we take another frontal photograph of the subject, measure three HAT parameters from it, compute HAT model HRTF, and blend it with the database HRTF, using HAT model HRTF exclusively below 0.5 kHz, progressively blending in database HRTF and blending out HAT model HRTF between 0.5 kHz and 3 kHz, and using database HRTF exclusively above 3 kHz.

The localization accuracy was tested on eight subjects in two different conditions (with continuously repeating 1-second white noise bursts and with just one noise burst and silence afterwards). In both tasks, subjects were asked to point to the perceived location of the sound source with their nose. A head tracking unit (Polhemus Fastrack) was used to measure the pointing direction, and error was computed as an angle between the selected direction and the true one. It was generally found that incorporation of the HAT model consistently improves the localization accuracy for all subjects (by about 25% on average). Subjectively the HAT-processed HRTF also improves the quality of the scene (subjects describe the rendered sound as having "depth", being more "focused", and being more "stable"). The results obtained with ear-matching personalization were mixed, with some subjects showing limited improvement and others showing no changes. Based on the experiments, the HAT model is routinely incorporated into HRTF-based experiments in our lab. More detailed description of the algorithms, protocols, experiments, and results is available in.[19]

3.3. *HRTF Analysis and Feature Extraction*

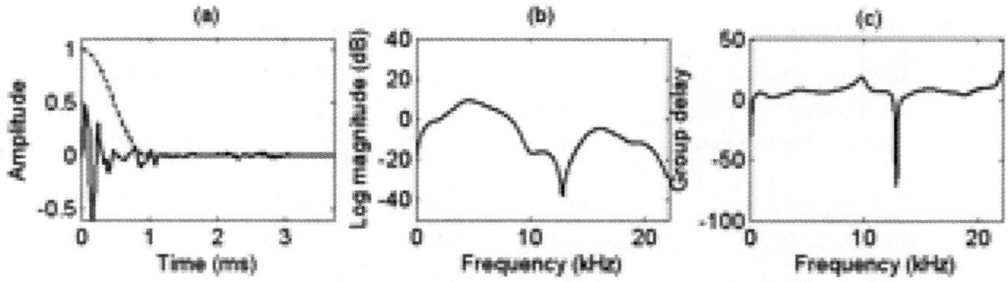

Fig. 3. An illustration to the HRTF feature extraction algorithm. a) HRIR (solid line) and 1.0 ms half-Hann window (dashed line). b) The spectrum of the windowed signal. c) The corresponding group-delay function.

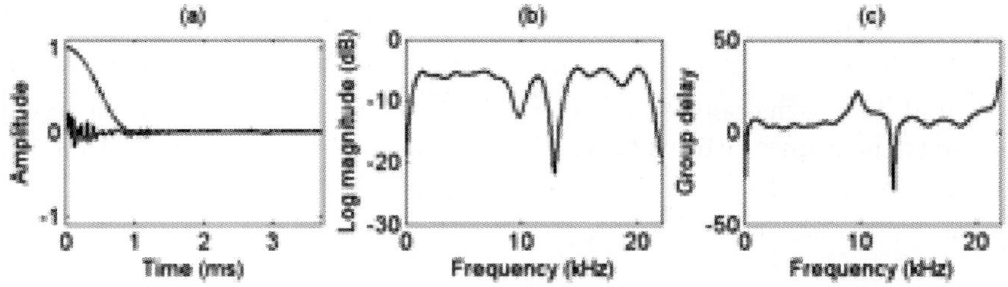

Fig. 4. Same as Figure 3 but for HRIR LP residual as an input instead of original HRIR.

To further advance towards reliable HRTF personalization method that does not require performing HRTF measurement, we have developed a method for extracting prominent features from HRTF and relating them to ear geometry.[21] A motivation for our work was that while many authors (e.g.,[22]) provide models for composing HRTF from known anthropometry, little work has been done in the opposite direction – i.e., to analyze the measured HRTF of a subject and decompose it into components.

The HRTF analysis method is a combination of several signal processing algorithms designed to extract HRTF notches caused by pinna and distinguish them from features caused by other body parts (e.g., shoulder reflection). In order to reject effects caused by body parts other than pinna, the HRTF is first converted to HRIR using an inverse Fourier transform

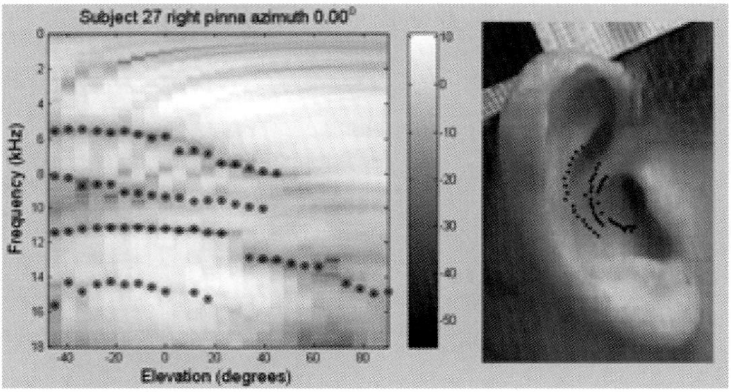

Fig. 5. The spectral notch frequencies extracted from HRTF measurements of CIPIC database subject 27 right pinna (left) are converted to ear dimensions and are marked on the pinna image of the same subject (right).

and then windowed using a half Hann window with the window length of 1.0 ms starting at the signal onset as shown in Figure 3(a). This particular window length was used to eliminate torso reflection (seen at around 1.6 ms) and knee reflection (at 3.2 ms). The spectrum of the windowed signal is shown in Figure 3(b). The notch is defined here more clearly than on the spectrum of the unwindowed signal (plot not shown). For further resolution enhancement, we extract the spectral notches from the group-delay function rather than from the magnitude spectrum, which was shown to be beneficial even for short data frames.[23] The group delay function is the negative of the derivate of the signal phase spectrum and can be computed directly via a combination of two Fourier transforms. It is shown in Figure 3(c) and clearly has a better resolution than a simple spectrum.

To reduce artifacts caused by windowing operation and remove broad HRTF features caused by resonances in the pinna cavities (and thus to better isolate narrow HRTF notches), we apply the 12th order linear prediction (LP) analysis to the original HRIR and then use the LP residual instead of the original HRIR as an input to the steps described above. Figure 4 shows the results of the processing of HRIR LP residual. It can be seen that the spectrum is flat now, allowing for higher accuracy in identifying the positions of the notches.

To verify the method, the notch frequencies were extracted for HRTFs of several subjects of the CIPIC HRTF database. A representative example is shown in Figure 5. On the left, the positions of prominent notches in the spectrum are marked. Note that each notch position traces a certain contour as sound source elevation changes. On the right, the extracted notch frequencies are projected onto the pinna photograph assuming that the notch frequency corresponds to the first anti-resonance between incoming wave and reflected wave. The contours obtained on the pinna photograph clearly trace the outlines of anatomical features of the ear, showing that the results are meaningful and that the shape of spectral contours is indeed related to fine details of the pinna shape. The method could potentially be used for HRTF personalization, either by HRTF synthesis "from scratch" based on pinna measurements or by modifying HRTF of one subject to match another subject in accordance with differences in ear geometry. A much more detailed description of the algorithm and a large number of additional examples are available in.[21]

3.4. *Numerical HRTF Computation*

Yet another alternative to measuring HRTF directly is to compute it using numerical methods. This requires one to have reasonably accurate representation of the surface of the subject's body, including head, torso, and fine pinna structure. The representation considered is usually a triangular surface mesh. The necessary mesh resolution is determined by the highest frequencies for which the HRTF is to be computed; the rule of thumb for most numerical methods is to have at least 5-6 mesh elements per wavelength. A fine discretization with very large number of elements is therefore necessary for upper frequency hearing limit of 20 kHz (wavelength of 1.7 cm). For faster convergence, it is also desirable that the mesh consists of mostly close-to-equilateral triangle patches (as opposed to narrow, elongated triangles).

The acoustic potential $\Phi(k;r)$ at a wavenumber k in any volume that does not enclose the acoustic sources must conform to the Helmholtz (wave) equation with the Sommerfeld radiation condition

$$\nabla^2\Phi(k;r) + k^2\Phi(k;r) = 0, \qquad \lim_{r\to\infty}\left[r(\frac{\partial\Phi(k;r)}{\partial r} - ik\Phi(k;r)\right] = 0,$$

where r is a radius-vector of an arbitrary point within the volume. Assume that an object (a head mesh, or head-and-torso mesh) is located within the volume and is irradiated by a plane wave $e^{ikr\cdot s}$ propagating in a direction s. Denote by S the boundary of the object. For simplicity, the sound-hard boundary condition

$$\left.\frac{\partial\Phi(k;r)}{\partial n}\right|_S = 0$$

is usually assumed, although impedance boundary condition can be stipulated as well. Any boundary element method can be used to iteratively solve the wave equation in presence of the irradiating field and to obtain the set of potentials $\Phi_m(k; r)$ at all surface elements S_m (e.g.,[24]). Note that by definition the HRTF is nothing but a value of the potential at the surface element corresponding to the ear canal entrance and is immediately available once the wave equation is solved.

However, such computation method is extremely wasteful. Indeed, the result of computation is the set of potentials for all S_m while only one value is ultimately needed (however, all the values are necessary for iterative solution procedure). Moreover, all computations have to be re-done from scratch for another direction s.

Similarly to how the physical HRTF measurement method can be sped up, the numerical computations can also be made several orders of magnitude faster using reciprocity principle. The same object mesh is used; however, the monopole source is placed directly on the boundary of the object at the place corresponding to the ear canal entrance, just like the loudspeaker is placed in the reciprocal HRTF measurement method. The BEM computations are then done to compute the acoustic field around the object, and the computed field is sampled at points corresponding to the locations of microphones in the reciprocal HRTF measurement method. An example paper using this technique is.[25]

However, the processing power required for direct implementation of BEM is extremely high. For reference, in[25] a head mesh with about 22000 elements is used, and the HRTF computations are done only up to the frequency of 5.4 kHz; still, it took about 28 hours to perform computations for one frequency only. Admittedly,[25] is eight years old; however, even with the present level of technology it is impossible to get up to 20 kHz because much finer mesh should be used at higher frequencies, and the BEM complexity grows as $O(N^6)$..

Based on our work in,[26] we have developed a method for the numerical HRTF computation to be done several orders of magnitude faster than other existing work. The key point in our work is to use fast multipole methods (FMM) to speed up the iterative process of computing the acoustic potential on the mesh.[17] Usually, the field computations are done pairwise for all mesh patches at each iteration, resulting in $O(N^2)$ cost per iteration. In FMM, the patches are grouped into sets according to certain rules, and for each set the field produced by all patches is computed at $O(N)$ cost. The computed fields are then applied to all patches also at $O(N)$ cost.

The total running time for the FMM BEM solver scales approximately as $O(N^{1+\alpha})$, for some $\alpha < 0.5$.

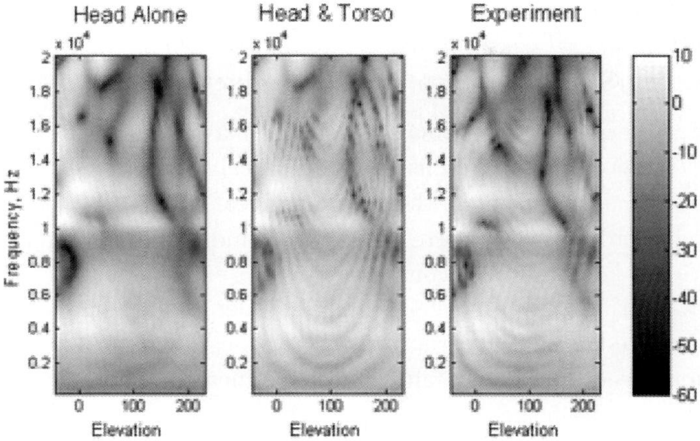

Fig. 6. BEM-obtained HRTF (in dB) for the KEMAR head mesh alone. Middle: BEM-obtained HRTF for the KEMAR head and torso mesh. Right: Experimentally measured KEMAR HRTF data. In all plots, azimuth is zero and elevation is varying from -45 to 225 degrees. "Large" KEMAR pinna is used.

We have computed the HRTF of the KEMAR manikin for which a lot of experimental measurements are available from different researchers. Our KEMAR mesh includes the torso and has approximately 445000 elements in total, and on a high-end desktop workstation (QX6700 2.66 GHz Intel CPU, 8 Gb RAM) we are able to compute the HRTF for one frequency in 35 minutes on average (total computation time for 117 frequencies spanning from 172 Hz to 20.155 kHz was 70 hours). We have performed comparison of our results with experimentally-measured KEMAR data and found very good agreement. One example is shown in Figure 6, where the contours of the HRTF features observed in the experimental data match those obtained in numerical simulations very well. The pinna mesh for this particular computation was obtained using a CT scan. We had also attempted to numerically compute HRTF using head/pinna meshes obtained from a 3-D laser scanner device and generally found that the agreement between computed and experimental HRTF is significantly worse, which is due to the fact that the laser scan is not able to reproduce the surface with required fidelity due to loss of features in concave areas. Obviously CT scan is not a good choice for pinna shape acquisition with human subjects because of associated time, cost, and radiation exposure concerns; perhaps a laser-scanned mesh can be made acceptable for BEM HRTF computation by stitching several meshes scanned at different angles. The research on obtaining quality meshes is ongoing.

4. Auditory Scene Capture and Reproduction

We wish to perform a multi-channel audio recording in a way so that the spatially/temporally removed listener can perceive the acoustic scene to be fully the same as if present at the recording location, so that sound sources are perceived to be in the correct directions and distances, environmental information is kept, and the acoustic scene stays external with respect to the listener when the listener moves/rotates. In order to do that, such recording should capture the spatial structure of the acoustic field. Indeed, a single-microphone recording could faithfully reproduce the acoustic signal at a given point; but from such a recording it is obviously impossible to tell which parts of audio scene arrive from which directions. Also, the information about spatial structure of the acoustic field is necessary to allow for listener motion. A microphone array that is able to sample such spatial structure is therefore necessary.

An important distinction from the VAS rendering system described above here is that in VAS rendering the synthesis is done "from scratch" by placing sound sources for which clean signals are available at known places in known environment and then artificially adding environmental effects. In scene capture and rendering application, the scene structure information is unknown. A tempting approach is to localize all sound sources, isolate them somehow, and then render them using VAS; however, localization attempts with multiple active sources (such as at a concert) are futile; moreover, such approaches would remove all but the most prominent components from the scene, which contradicts the goal of keeping the environmental information intact. Therefore, scene capture and reproduction system attempt to blindly re-create the scene without trying to analyze it. More specifically, it operates by decomposing the acoustic scene into pieces that come from various directions and then renders those pieces to the listener to appear to come from these directions.[4] The grid of these directions is fixed in advance and is data-independent. The placement of sources would be reproduced reasonably well with such a model.

Spherical microphone arrays constitute an ideal platform for such system due to their inherent omnidirectionality. A 64-microphone array is shown in Figure 7. Assume that the array has L_q microphones located

Fig. 7. A prototype 64-microphone array with embedded video camera. The array connects to the PC via a USB 2.0 cable streaming digital data in real-time.

at positions s'_q on the surface of a hard sphere of radius a. Assume that the time-domain signal measured at each microphone is $x(t; s'_q)$. Perform Fourier transform of those signals to obtain the potentials $\Psi(k; s'_q)$ at all microphones. The goal is to decompose an acoustic scene into waves $y(t; s_j)$ arriving from various directions s_j (the total number of directions is L_j) and then render them for the listener from directions $\tilde{s}_j$ (which are the directions s_j converted to the listener-bound coordinate system) using his/her HRTF $H_l(k; s)$ and $H_r(k; s)$. It is easier to work in the frequency domain. Let us denote the Fourier transform of $y(t; s_j)$ by $\lambda(k; s_j)$. Further, denote by Ψ the $L_q \times 1$ vector of $\Psi(k; s'_q)$, and denote by Λ the $L_j \times 1$ vector of $\lambda(k; s_j)$. We want to build a simple linear mechanism for decomposition of the scene into the directional components so that it can be expressed by equation $\Lambda = W\Psi$, where W is the $L_j \times L_q$ matrix of decomposition weights constructed in some way.

One way to proceed is to note that the operation that allows for selection of a part of audio scene that arrives from a specified direction is nothing but a beamforming operation (e.g.,[13]). A beamforming operation for direction s_j in frequency domain can be written as

$$\lambda(k; s_j) = -i\frac{(ka)^2}{4\pi}\sum_{n=0}^{p-1}(2n+1)i^{-n}h'_n(ka)\sum_{q=1}^{L_q}\Psi(k; s'_q)P_n(s_j \cdot s'_q),$$

where $h'_n(ka)$ is the derivative of the spherical Hankel function of order n

and $P_n(...)$ is an associate Legendre polynomial of order n as well. p here is the truncation number and should be kept about ka; increasing it improves the selectivity but also greatly decreases robustness, so a balance must be kept. This decomposition way can be expressed in the desired matrix-vector multiplication framework as

$$w(k; s_j, s'_q) = -i\frac{(ka)}{4\pi}\sum_{n=0}^{p-1}(2n+1)i^{-n}h'_n(ka)P_n(s_j \cdot s'_q).$$

Another method of decomposition[16] is to note that a relationship inverse to the desired one can be established as $\Psi = F\Lambda$, where F is a $L_q \times L_j$ matrix with elements $f(k; \acute{s}'_q, s_j)$ equal to the potential caused at microphone at s'_q by the wave arriving from the direction s_j. The equation for $f(k; \acute{s}'_q, s_j)$ is

$$f(k; s'_q, s_j) = \frac{i}{(ka)^2}\sum_{n=0}^{p-1}\frac{i^n(2n+1)P_n(s_j \cdot s'_q)}{h'_n(ka)}.$$

Then, a linear system $\Psi = F\Lambda$ is solved for Λ, exactly if $L_q = L_j$ or in least squares sense otherwise. For this method, the weight matrix $W_L = F^{-1}$ (generalized inverse for non-square F).

The decomposed acoustic scene is stored for later rendering and/or transmitted over network to the remote listening location. When rendering the scene for the specific user, his/her HRTF are assumed to be known. During the rendering, the directions s_j with respect to the original microphone array position are converted to directions $\tilde{s}_j$ with respect to the listener-bound coordinate frame, and the audio streams $y_l(t)$, $y_r(t)$ for left/right ears are obtained as[14]

$$Y_{l,r}(k) = \sum_{j=1}^{L_j} H_{l,r}(k; \tilde{s}_j)\lambda(k; s_j), \qquad y_{l,r}(t) = IFT[Y_{l,r}(k)](t),$$

where IFT is the inverse Fourier transform. Essentially the scene components arriving from different directions are filtered with their corresponding HRTF taking into account listener head position/orientation and are summed up. The above description is a brief outline of auditory capture and reproduction principles, and many practical details are omitted for brevity and are available in e.g.[27] or.[16]

5. Sample Application: Audio Camera

One particularly promising application of spherical microphone array signal processing is the real-time visualization of the acoustic field energy. The microphone array immersed in the acoustic field has the ability to analyze the directional properties of the field. The same spherical microphone array is used in this application as well, but the goal and the processing means are different. The goal is to generate an image that can be used by a human operator to visualize the spatial distribution of the acoustic energy in the space. Such image can further be spatially registered with and overlaid on the photograph of the environment so that the acoustically active areas can be instantly identified.[28]

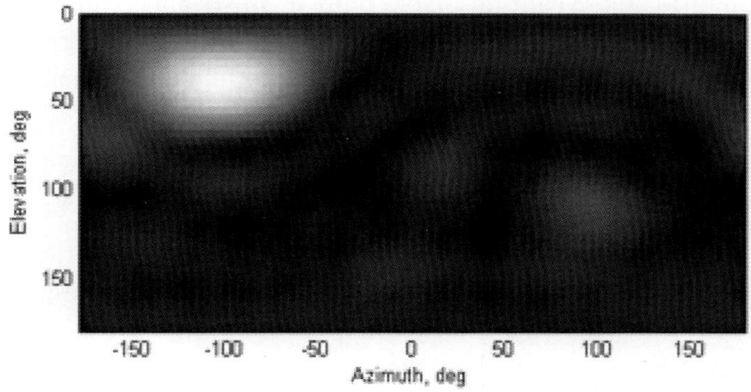

Fig. 8. Sample acoustic energy map. Map dimensions are 128 by 64 pixels. Pixel intensity is proportional to the energy in the signal beamformed to the direction indicated on axes.

To create the acoustic energy map, the beamforming operation is performed on a relatively dense grid of directions. The energy map is just a regular, two-dimensional image with the azimuth corresponding to X axis and the elevation corresponding to the Y axis. In this way, the array's full omnidirectional field of view is unwrapped and is mapped to the planar image. The energy in the signal obtained when beamforming to the certain direction is plotted as the pixel intensity for that direction. A sample map obtained with the array of Figure 7 is shown in Figure 8. It can be seen in the image that an active acoustic source exists in the scene and is located approximately in the direction $(-100, 40)$ (azimuth, elevation). In the same way, multiple acoustic sources can be visualized, or a reflection coming from the particular direction in space can be seen. Furthermore, if the original acoustic signal is short (like a pulse), an evolution of the reverberant sound field can be followed by recording it and then playing it back in the slow motion. In such a playback, the arrivals of the original sound and then of its reflections from various surfaces and objects in the room are literally visible painted over the places where the sound/reflections are coming from.[30] Such information is usually obtained indirectly in architectural acoustics for the purposes of adjusting the environment for better attendee experience, and a real-time visualization tool could prove very useful in this area. Furthermore, it has been shown that the audio energy visualization map is created according to the central projection equations – the same equations that regular (video) imaging devices conform to,[29] so the device has been tagged "audio camera", and the vast body of work derived for video imaging (e.g., multi-camera calibration or epipolar constraints) can be applied to audio or multimodal (auditory plus traditional visual) imaging. The reader is encouraged to refer to the references mentioned herein for further "audio camera" algorithm details and examples.

6. Conclusion

Possible applications of systems that capture and recreate spatial audio are numerous; one can imagine, for example, such a system being placed in the performance space and transmitting the audio stream in real time to the interested listeners, enabling them to be there without actually being there. Similarly, a person might want to capture some auditory experience – e.g., a celebration – to store for later or to share with family and friends. Another application from a different area could involve a robot being sent to environments that are dangerous to or unreachable by humans and relaying back the spatial auditory experience to the operator as if he/she is being there. The auditory sensor can also be augmented with video stream to open further possibilities, and the research in spatial audio capture and synthesis is ongoing.

References

1. V. Pulkki (2002). "Compensating displacement of amplitude-panned virtual sources", Proc. 22th AES Conf., Espoo, Finland, 2002, pp. 186-195.
2. D. N. Zotkin, R. Duraiswami, and L. S. Davis (2004). "Rendering localized spatial audio in a virtual auditory space", IEEE Transactions on Multimedia, vol. 6, pp. 553-564.
3. D. N. Zotkin, R. Duraiswami, E. Grassi, and N. A. Gumerov (2006). "Fast head-related transfer function measurement via reciprocity", J. Acoust. Soc. Am., vol. 120, pp. 2202-2215.
4. A. E. O'Donovan, D. N. Zotkin, and R. Duraiswami (2008). "Spherical microphone array based immersive audio scene rendering", Proc. ICAD 2008, Paris, France.

5. P. Runkle, A. Yendiki, and G. Wakefield (2000). "Active sensory tuning for immersive spatialized audio", Proc. ICAD 2000, Atlanta, GA.
6. J. B. Allen and D. A. Berkeley (1979). "Image method for efficiently simulating small-room acoustics", J. Acoust. Soc. Am., vol. 65, pp. 943-950.
7. W. M. Hartmann (1999). "How we localize sound", Physics Today, November 1999, pp. 24-29.
8. N. F. Dixon and L. Spitz (1980). "The detection of auditory visual desynchrony", Perception, vol. 9, pp. 719-721.
9. E. M. Wenzel, M. Arruda, D. J. Kistler, and F. L. Wightman (1993). "Localization using non-individualized head-related transfer functions", J. Acoust. Soc. Am., vol. 94, pp. 111-123.
10. V. R. Algazi, R. O. Duda, D. P. Thompson, and C. Avendano (2001). "The CIPIC HRTF database", Proc. IEEE WASPAA 2001, New Paltz, NY, pp. 99-102.
11. P. M. Morse and K. U. Ingard (1968). "Theoretical Acoustics", Princeton Univ. Press, New Jersey.
12. C. Kyriakakis, P. Tsakalides, and T. Holman (1999). "Surrounded by sound: Immersive audio acquisition and rendering methods", IEEE Signal Processing Magazine, vol. 16, pp. 55-66.
13. J. Meyer and G. Elko (2002). "A highly scalable spherical microphone array based on an orthonormal decomposition of the soundfield", Proc. IEEE ICASSP 2002, Orlando, FL, vol. 2, pp. 1781-1784.
14. R. Duraiswami, D. N. Zotkin, Z. Li, E. Grassi, N. A. Gumerov, and L. S. Davis (2005). "High order spatial audio capture and its binaural head-tracked playback over headphones with HRTF cues", Proc. AES 119th convention, New York, NY, preprint #6540.
15. M. Otani and S. Ise (2006). "Fast calculation system specialized for head-related transfer function based on boundary element method", J. Acoust. Soc. Am., vol. 119, pp. 2589-2598.
16. D. N. Zotkin, R. Duraiswami, and N. A. Gumerov (2009). "Plane-wave decomposition of acoustical scenes via spherical and cylindrical microphone arrays", IEEE Transactions on Audio, Speech, and Language Processing, in press.
17. N. A. Gumerov, A. E. O'Donovan, R. Duraiswami, and D. N. Zotkin (2009). "Computation of the head-related transfer function via the fast multipole accelerated boundary element method and its spherical harmonic representation", J. Acoust. Soc. Am., in press.
18. C. Jin, P. Leong, J. Leung, A. Corderoy, and S. Carlile (2000). "Enabling individualized virtual auditory space using morphological measurements", Proc. First IEEE Pacific-Rim Conference on Multimedia, Sydney, Australia, pp. 235-238.
19. D. N. Zotkin, J. Hwang, R. Duraiswami, and L. S. Davis (2003). "HRTF personalization using anthropometric measurements", Proc. IEEE WASPAA 2003, New Paltz, NY, pp. 157-160.
20. V. R. Algazi, R. O. Duda, and D. M. Thompson (2002). "The use of head-and-torso models for improved spatial sound synthesis", Proc. AES 113th

convention, Los Angeles, CA, preprint #5712.
21. V. C. Raykar, R. Duraiswami, and B. Yegnanarayana (2005). "Extracting the frequencies of the pinna spectral notches in measured head related impulse responses", J. Acoust. Soc. Am., vol. 118, pp. 364-374.
22. V. R. Algazi, R. O. Duda, R. P. Morrison, and D. M. Thompson (2001). "Structural composition and decomposition of HRTFs", Proc. IEEE WASPAA 2001, New Paltz, NY, pp. 103-106.
23. B. Yegnanarayana, D. K. Saikia, and T. R. Krishnan (1984). "Significance of group delay functions in signal reconstruction from spectral magnitude or phase", IEEE Transactions on Acoustics, Speech, and Signal Processing, vol. 32, pp. 610-623.
24. T. Xiao and Q.-H. Liu (2003). "Finite difference computation of head-related transfer function for human hearing", J. Acoust. Soc. Am., vol. 113, pp. 2434-2441.
25. B. F. G. Katz (2001). "Boundary element method calculation of individual head-related transfer function. I. Rigid model calculation", J. Acoust. Soc. Am., vol. 110, pp. 2440-2448.
26. N. A. Gumerov and R. Duraiswami (2004). "Fast multipole methods for the Helmholtz equation in three dimensions", Elsevier Science, The Netherlands.
27. B. Rafaely (2005). "Analysis and design of spherical microphone arrays", IEEE Trans. Speech Audio Proc., vol. 13(1), pp. 135-143.
28. A. E. O'Donovan, R. Duraiswami, and N. A. Gumerov (2007). "Real time capture of audio images and their use with video", Proc. IEEE WASPAA 2007, New Paltz, NY, pp. 10-13.
29. A. E. O'Donovan, R. Duraiswami, and J. Neumann. "Microphone arrays as generalized cameras for integrated audio-visual processing", Proc. IEEE CVPR 2007, Minneapolis, MN.
30. A. E. O'Donovan, R. Duraiswami, and D. N. Zotkin (2008). "Imaging concert hall acoustics using visual and audio cameras", Proc. IEEE ICASSP 2008, Las Vegas, NV, April 2008, pp. 5284-5287.

RECONSTRUCTING SOUND SOURCE DIRECTIVITY IN VIRTUAL ACOUSTIC ENVIRONMENTS

M. NOISTERNIG*

Acoustic and Cognitive Spaces Research Group, IRCAM–CNRS UMR STMS,
1 place Igor-Stravinksy, Paris, 75004, France
**E-mail: markus.noisternig@ircam.fr*
www.ircam.fr

F. ZOTTER

Institute of Electronic Music and Acoustics, University of Music and Performing Arts,
Inffeldgasse 10/3, Graz, 8010, Austria
E-mail: zotter@iem.at
www.iem.at

B. F.G. KATZ

Audio and Acoustics Group, LIMSI–CNRS,
BP 133, Orsay, 91403, France
E-mail: brian.katz@limsi.fr
www.limsi.fr

This study considers the directionality of acoustic sources as an important aspect of spatial perception and investigates aspects of reconstructing source radiation patterns in interactive virtual or augmented reality applications. The measurement of directional characteristics of sound sources applying spherical microphone arrays will be addressed, particularly with emphasis on the discrete spherical harmonic transform (DSHT) and its practical limitations. Finally, several case studies of practical implementations including real-time auralization, wave field synthesis (WFS) and spherical loudspeaker arrays will be briefly discussed.

Keywords: Spherical acoustic holography, acoustic radiation patterns, array processing, virtual acoustic environments, auralization.

1. Introduction

Olson[1] has described four necessary conditions for the perceptual illusion of realism in sound reproduction which can be extrapolated to virtual acoustic environments: (1) full spectral component content, (2) noiseless and distortion free reproduction over appropriate sound levels, (3) continuity of reverberation, and (4) preservation of spatial sound distribution. Numerous studies have shown that radiation patterns of natural sound sources vary with frequency and time; *e.g.* Meyer[2] provides a good summary of directional characteristics for different musical instruments. Thus, when virtual environments only consider spectral and temporal characteristics of the source, without considering the spatially varying radiation pattern, the perception of timbre at the listener's position can be both inaccurate and unrealistic. This is especially true in reverberant environments where reflections should be taken into account.[3–6]

One area of large concern that is a topic of ongoing research is the measurement, reproduction, and compact description of sound radiation patterns. It is shown in the following work that the wave field expansion in spherical coordinates provides a general description format, which is independent of the sound field rendering technique, offering compatibility and scalability in terms of reproduction detail. Several case studies for various real source directivities are highlighted with a presentation of examples of rendering architectures including auralization, wave field synthesis (WFS), and spherical beamforming loudspeakers.

2. Wave Field Decomposition using Spherical Harmonics

The solution of the wave equation in spherical coordinates is briefly discussed here. For notational simplicity, the following equations are given in the frequency domain with respect to time; the dependency on the frequency variable ω is omitted in the notation. It will be clear from the context of the discussion if the quantity is in the frequency or in the time domain.

The spherical Fourier transform provides a decomposition of acoustic wave fields into their spherical wave components, which is also referred to as *spherical wave spectrum*.[7,8] Solving the Helmholtz wave equation in spherical coordinates (r, θ, ϕ) results in separate equations with respect to angles θ and ϕ, and with respect to radius r. The angular portions of this solution are conveniently combined into a single function called a *spherical harmonic*, $Y_n^m(\theta, \phi)$, which – with the usual mathematical normalization – is defined as

$$Y_n^m(\theta, \phi) \equiv (-1)^m \sqrt{\frac{(2n+1)}{4\pi} \frac{(n-m)!}{(n+m)!}} P_n^m(\cos\theta) e^{im\phi} \qquad (1)$$

where n denotes the order and m the degree of the spherical harmonics,

P_n^m the associated Legendre functions, and $i = \sqrt{(-1)}$. It should be noted that various other normalizations can be found in the literature. In Refs. 8 and 9 it is described that applying the normalization as given in Eq. (1) the spherical harmonics are orthonormal, such that

$$\int_{\mathbb{S}^2} Y_n^m(\theta,\phi) Y_{n'}^{m'}(\theta,\phi)^* d\Omega = \delta_{nn'}\delta_{mm'} = \begin{cases} 1 \text{ if } n = n' \text{ and } m = m' \\ 0 \text{ otherwise} \end{cases} \tag{2}$$

where δ_{ij} denotes the Kronecker delta and the superscript $(\cdot)^*$ denotes complex conjugation.

Let $L^2(\mathbb{S}^2)$ denote the Hilbert space of functions square integrable with respect to the standard rotation invariant measure $d\Omega = \sin\theta d\theta d\phi$ on the 2-sphere surface $\mathbb{S}^2$. The Fourier transform of such a function amounts to its L^2-projection onto spherical harmonics providing an orthogonal decomposition in the Hilbert space. Assuming that the sound pressure $p(r,\theta,\phi)$ is known on a sphere with radius $r = r_0$, the complex spherical harmonic coefficients ψ_{nm} can be determined using analysis by forward harmonic transform[10]

$$\psi_{nm}(r_0) = \int_{\mathbb{S}^2} p(r_0,\theta,\phi) Y_n^m(\theta,\phi)^* d\Omega. \tag{3}$$

Referring to Driscoll and Healy[10], the expansion in terms of spherical harmonics becomes

$$p(r_0,\theta,\phi) = \sum_{n=0}^{\infty} \sum_{m=-n}^{n} \psi_{nm}(r_0) Y_n^m(\theta,\phi). \tag{4}$$

Once the spherical harmonic coefficients ψ_{nm} as given in Eq. (3) are determined, the radiated pressure field is uniquely defined. To avoid reconstruction errors for discrete considerations with a limited total number of measurement points over the whole sphere, the sound field has to be suitably captured considering the spatial sampling theorem (angular band limitations on the sphere) and the free field assumption (direct sound field only). This will be discussed in more detail in the following sections.

The spherical harmonic transform (SHT) therefore provides a reasonable description format for sound radiation pattern analysis and synthesis to what is typically found in practice for virtual acoustic environments.

2.1. *Holographic extrapolation of radiation patterns*

Spherical acoustic holography provides an evaluation of the radiating sound field on any radial distance from the origin, given a measured sound pressure (or surface radial velocity) distribution on another concentric sphere.

The "exterior" domain problem is defined as the case, in which all radiating sources are contained within a spherical boundary surface with radius r_b. Reconstructions are obtained from this boundary surface out to infinity. The sound pressure distribution (hologram) for calculating the SHT is measured on a sphere of greater radius $r_0 \geq r_b$. To extrapolate the spherical wave spectrum to concentric spherical surfaces of radius $r > r_b$, the radial solutions of the wave equation are taken into account, such that

$$\psi_{nm}(r) = \frac{h_n^{(2)}(kr)}{h_n^{(2)}(kr_0)} \psi_{nm}(r_0), \tag{5}$$

where $k = \omega/c$ denotes the wavenumber, c the speed of sound, and $h_n^{(2)}(kr)$ the spherical Hankel function of the second kind of order n. Eq. (5) determines the frequency dependent attenuation of the spherical wave spectrum components of the sound pressure field at a given radius r. This observation is very important from a practical point of view, as accurate sound radiation pattern synthesis requires one to appropriately equalize the wave propagation terms corresponding to the radius of the reproduction device.[11]

3. Discrete Spherical Harmonic Transform, Interpolation, and Approximation

The spherical harmonic coefficients ψ_{nm} can be determined from a given discrete sound pressure distribution on a spherical surface with radius r_0. In vector notation the discrete sound pressure distribution can be written as

$$\boldsymbol{p} = [p(\boldsymbol{\theta}_1) \quad p(\boldsymbol{\theta}_2) \quad \cdots \quad p(\boldsymbol{\theta}_{\mathrm{L}})]^{\mathrm{T}},$$

where $\boldsymbol{\theta}_j$ denotes the angular positions of sampling points $j = 1, \ldots, \mathrm{L}$ on the spherical surface, and $[\cdot]^{\mathrm{T}}$ the transpose of a matrix or a vector.

Real sources will exhibit a spherical wave spectrum which is essentially band limited, that is, only components below the order $n \leq \mathrm{N}$ exist. Suppose that $\boldsymbol{p}_{\mathrm{N}}$ results from such a finite spherical harmonic expansion. In this case, the weighted sum in Eq. (4) can be reformulated as the product of the matrix $\boldsymbol{Y}_{\mathrm{N}}$ consisting of sampled spherical harmonics $\boldsymbol{y}_{\mathrm{N}}$

$$\boldsymbol{Y}_{\mathrm{N}} = [\boldsymbol{y}_{\mathrm{N}}(\boldsymbol{\theta}_1) \quad \boldsymbol{y}_{\mathrm{N}}(\boldsymbol{\theta}_2) \quad \ldots \quad \boldsymbol{y}_{\mathrm{N}}(\boldsymbol{\theta}_{\mathrm{L}})]^{\mathrm{T}}$$
$$\boldsymbol{y}_{\mathrm{N}}(\boldsymbol{\theta}_j) = [y_{0,0}(\boldsymbol{\theta}_j) \quad y_{-1,1}(\boldsymbol{\theta}_j) \quad \cdots \quad y_{\mathrm{N,N}}(\boldsymbol{\theta}_j)]^{\mathrm{T}}$$

with the vector of corresponding spherical harmonic coefficients $\boldsymbol{\psi}_{\mathrm{N}}$

$$\boldsymbol{\psi}_{\mathrm{N}} = [\psi_{0,0} \quad \psi_{-1,1} \quad \cdots \quad \psi_{\mathrm{N,N}}]^{\mathrm{T}}$$

such that it writes as

$$\boldsymbol{p}_{\mathrm{N}} = \boldsymbol{Y}_{\mathrm{N}}\, \boldsymbol{\psi}_{\mathrm{N}}. \tag{6}$$

It is clear from Eq. (6) that determining the discrete spherical harmonic coefficients $\boldsymbol{\psi}_{\mathrm{N}}$ from $\boldsymbol{p}_{\mathrm{N}}$ requires one to invert the matrix $\boldsymbol{Y}_{\mathrm{N}}$. In many cases, the matrix $\boldsymbol{Y}_{\mathrm{N}}$ is badly conditioned and therefore direct inversion might fail. A well established solution method for this ill-posed inverse problem is based on the singular value decomposition (SVD), which computes the *generalized inverse* of a matrix.

The SVD decomposes the matrix $\boldsymbol{Y}_{\mathrm{N}} = \boldsymbol{U}\,\boldsymbol{S}\,\boldsymbol{V}^{\mathrm{T}}$ into a diagonal matrix $\boldsymbol{S} = \mathrm{diag}\{\boldsymbol{s}\}$ with the *singular values* of $\boldsymbol{Y}_{\mathrm{N}}$ on its main diagonal, and two orthogonal matrices $\boldsymbol{U}$ and $\boldsymbol{V}$ containing the left and right *singular vectors* of $\boldsymbol{Y}_{\mathrm{N}}$, respectively. In this case, the diagonal entries of $\boldsymbol{S}$ can be arranged to be nonnegative and in order of decreasing magnitude. Keeping only the K non-vanishing singular values in $\tilde{\boldsymbol{S}}$ and cropping the orthogonal matrices accordingly, the SVD provides a regularized inverse $\boldsymbol{Y}_{\mathrm{N}}^{\dagger} = \tilde{\boldsymbol{V}}\,\tilde{\boldsymbol{S}}^{-1}\,\tilde{\boldsymbol{U}}^{\mathrm{T}}$. More details can be found in Ref. 12.

Applying the SVD, the solution of the linear system of equations in Eq. (6) becomes

$$\tilde{\boldsymbol{\psi}}_{\mathrm{N}} = \boldsymbol{Y}_{\mathrm{N}}^{\dagger}\, \boldsymbol{p}. \tag{7}$$

Depending on the dimensions of $\boldsymbol{Y}_{\mathrm{N}}$ and the number of non-vanishing singular values, K, Eq. (7) has the following properties:

(1) $\mathrm{K} = (\mathrm{N}+1)^2 \leq \mathrm{L}$: *Discrete spherical harmonic transform (DSHT)*, assuming a spherical harmonics order limited by N, the pseudo-inverse provides an exact spherical harmonic analysis.
(2) $\mathrm{K} = \mathrm{L} \leq (\mathrm{N}+1)^2$: *Discrete spherical harmonic interpolation (DSHI)*, the pseudo-inverse behaves as an interpolation that achieves exact representation of the sampling nodes.
(3) $\mathrm{K} < \min[(\mathrm{N}+1)^2, \mathrm{L}]$: *Discrete spherical harmonic approximation (DSHA)*, the inversion is neither an exact spherical harmonics transform, nor an exact interpolation.

In the first DSHT case, applying the pseudo-inverse to strictly band-limited radiation patterns $\boldsymbol{p} = \boldsymbol{p}_{\mathrm{N}}$ provides the exact spherical harmonics transform $\tilde{\boldsymbol{\psi}}_{\mathrm{N}} = \boldsymbol{\psi}_{\mathrm{N}}$. The pseudo-inverse becomes $(\boldsymbol{Y}_{\mathrm{N}}^{\mathrm{T}}\boldsymbol{Y}_{\mathrm{N}})^{-1}\boldsymbol{Y}_{\mathrm{N}}^{\mathrm{T}}$ and inverts the spherical harmonics expansion in Eq. (6) from the left. In the second DSHI case, the spherical harmonic coefficients determined by the pseudo-inverse exactly reconstruct arbitrary patterns $\boldsymbol{p}_{\mathrm{N}} = \boldsymbol{p}$ by band-limited spherical harmonics interpolation. It becomes $\boldsymbol{Y}_{\mathrm{N}}^{\mathrm{T}}(\boldsymbol{Y}_{\mathrm{N}}\boldsymbol{Y}_{\mathrm{N}}^{\mathrm{T}})^{-1}$ and inverts the spherical harmonics expansion from the right. For the third DSHA case, neither exact transform, nor exact interpolation is feasible.

Fig. 1 summarizes different fundamental discretization schemes on the sphere, and compares them providing a classification into the above mentioned three part scheme for discrete spherical harmonics of order $\mathrm{N} = 9$ and a varying number of sampling points L. The following spherical discretization schemes are considered: Extremal points for hyperinterpolation[13] (*hi*), spiral points[14] (*sp*), equal-area partitions[15] (*eqa*), HEALPix[16] (*healpix*), Gauss-Legendre grid[17] (*gl*), equi-distant cylindrical grid[17] (*ecp*), and equiangle grid[10] (*equiangle*).

The condition number of the matrix $\boldsymbol{Y}$, *i.e.* the ratio between its maximum and minimum singular value, measures the sensibility and stability of the solution of the inverse problem. If the condition number is too large, regularization by SVD, as above mentioned, is required. In Fig. 1(b), condition numbers less than $\mathrm{cond}\{\boldsymbol{Y}_9\} < 20\mathrm{dB}$ are considered to provide a pseudo-inverse without regularization, *i.e.* the case of DSHT or DSHI. Greater condition numbers are considered to require limitation by regularization, hence DSHA.

It can be easily seen from Fig. 1(b) that near the critical number of sampling nodes $\mathrm{L} = (\mathrm{N}+1)^2$, most discretization schemes only allow for DSHA. In the context of critical sampling, the extremal points designed for *hyperinterpolation* provide an exact and well-conditioned inverse, which is both left and right inverse to the spherical harmonics expansion. This inversion is then called *hyperinterpolation* and provides both DSHT and DSHI.

3.1. *Spatial aliasing of low-order radiators: the acoustic centering problem*

In principle, the spherical harmonic coefficients of a low-order acoustic source can be determined by DSHT, using just a few angular sampling points of the radiation pattern. However, acoustic sources that are dislocated from the coordinate origin $r = 0$ cannot be fully represented at low orders anymore as translation produces higher order patterns.[9] If such radiation patterns are being discretized, spatial aliases might occur. On the other hand, if the order is truncated to N, substantial information might be lost. Fig 2 gives an example of the truncation error of a monopole source that is shifted out of the center by a radial distance r_d.

As a practical design example, consider a measurement facility system

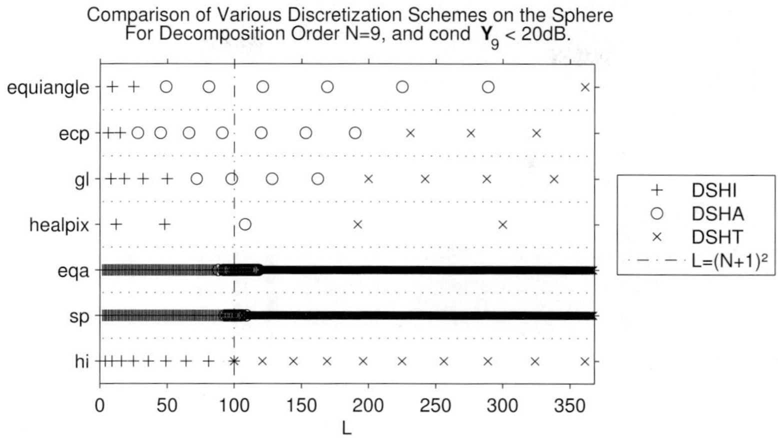

(a)

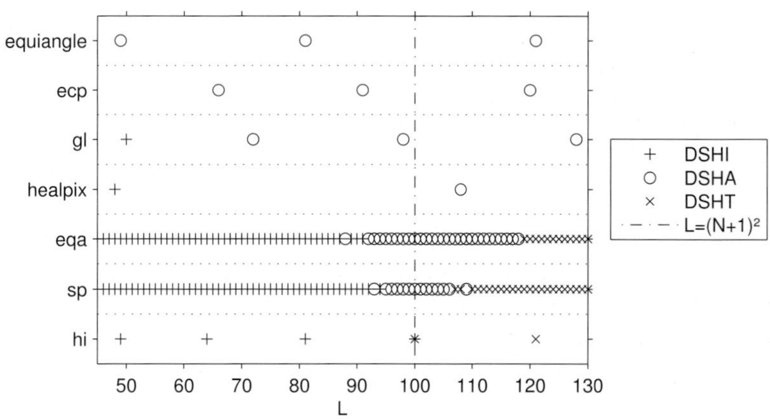

(b)

Fig. 1. The above figures indicate the existence and kind of spherical harmonic decompositions for different spherical discretization schemes; the discrete spherical harmonics interpolation (or pseudo-spectral analysis[18]) is labeled as DSHI, the discrete spherical approximation as DSHA, and the discrete spherical transform (or spectral analysis[18]) as DSHT, respectively. The condition number has been limited to $\mathrm{cond}\{\boldsymbol{Y}_9\} < 20\mathrm{dB}$ for the purpose of regularization. The vertical dash-dot line marks the critical number of sampling points $\mathrm{L} = (\mathrm{N}+1)^2$. To better illustrate the unique behavior of the *hyperinterpolation* subfigure (b) highlights the region around the critical sampling points.

that observes one single monopole source located inside a surrounding measurement array. One might choose the radius of the facility to be bigger than twice the possible displacements of the source from $r = 0$. As a rule of thumb for preventing inaccuracies because of source displacement, the spherical harmonic order should be at least six times the displacement in wavelengths, *i.e.* $\mathrm{N} > 6\, r_d/\lambda$.

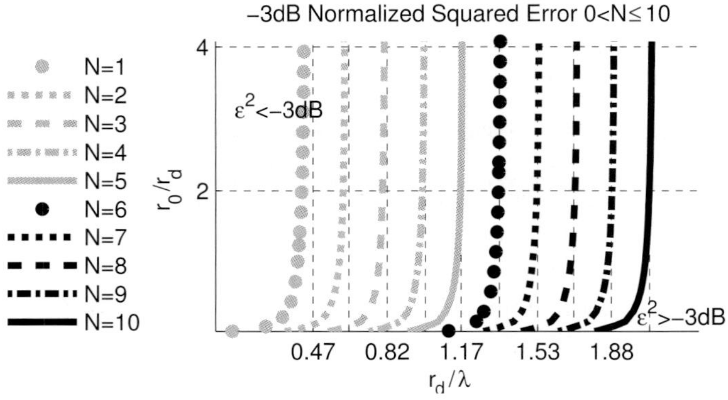

Fig. 2. This figure shows the -3dB contours of the normalized squared truncation error ε^2 of a monopole source that is shifted out of the center by r_d (see x-axis). The truncation error is the normalized average error of the sound pressure of the source evaluated at r_0 (see y-axis). It can be easily seen that the truncation error depends on the order of decomposition and remains small for small shifts r_d.

3.2. *Angular interpolation of far-field radiation patterns*

If DSHT becomes infeasible, for instance due to incomplete sets of data (as has been discussed in Refs. 19 and 20), or irregular distributions of points, holographic extrapolation will also become ambiguous. In fact, the result of the holographically extrapolated sound field will depend on the kind of regularization for DSHA, or DSHI, which may even be altered by weighted optimization criteria.[21] However, in the acoustic far-field, where different spherical harmonic orders share the same extrapolation term

$$\lim_{\substack{kr \gg \frac{n(n+1)}{2} \\ kr_0 \gg \frac{n(n+1)}{2}}} \frac{h_n^{(2)}(kr)}{h_n^{(2)}(kr_0)} = \frac{h_0^{(2)}(kr)}{h_0^{(2)}(kr_0)} = \frac{r_0}{r}\, e^{-ik(r-r_0)}, \tag{8}$$

interpolation and approximation (DSHI, DSHA) seem suitable to interpolate radiation patterns smoothly. Care has to be taken when extrapolating towards the acoustic near-field as ambiguous aliases of DSHI or DSHA yield sound pressures deviating by several orders of magnitude.

4. Measurement and Analysis of Sound Radiation Patterns

Weinreich and Arnold[22] have shown that the expansion of the complex sound pressure measured on two concentric spheres in spherical harmonics allows one to determine the forward and backward propagation of the sound waves independently, resolving both the interior and exterior solutions. This is applied to derive a *spatiotemporal fingerprint* of the radiating sound source even under semi-anechoic conditions. As has been discussed in Sec. 3, for practical reasons the spherical wave spectrum is often determined through discrete observations on the sphere, *e.g.* from synchronous recordings with a spherical microphone array.[23]

Generally, the inclusion of higher order coefficients corresponds to finer spatial resolutions of the sound radiation pattern. In most practical applications, the functions given on the surface of the sphere $\mathbb{S}^2$ are band-limited with a *band-limit* or *bandwidth* $N \geq 0$ in the sense that all $\psi_{nm} \equiv 0 \ \forall \ n \geq N$, *i.e.* only a finite number of coefficients are nonzero. For band-limited functions various quadrature schemes are well known, *e.g.* equiangular, equiareal or other discretizations, which reduce the integrals in Eq. 3 to finite weighted sums of a sampled data vector with quadrature weights q_u. See Refs. 17, 24, and 25 for more details.

Assuming a regular grid of angles $\theta_u = u\pi/2N$ and $\phi_v = v\pi/N$ with $u, v = 0, \ldots, 2N-1$ the spherical harmonic analysis can be written as

$$\psi_{nm}(k, r_0) = \sum_{u=0}^{2N-1} \sum_{v=0}^{2N-1} q_u \, p(k, r_0, \theta_u, \phi_v) Y_n^m(\theta_u, \phi_v)^* \qquad (9)$$

and the spherical harmonic synthesis can be formulated as

$$p(k, r_0, \theta, \phi) = \sum_{n=0}^{N-1} \sum_{m=-n}^{n} \psi_{nm}(k, r_0) Y_n^m(\theta, \phi) \qquad (10)$$

The spherical harmonic expansion will not necessarily result in a perfect reconstruction of the function on the 2-sphere surface, but will provide a least-squares approximation. The different quadrature schemes can vary substantially in their computational efficiency and accuracy, especially for high degrees and orders.[26] For most acoustic radiation problems relatively low spatial resolutions can be assumed, *e.g.* compared to geopotential models, which apply SHT of orders and degrees greater ≥ 2000.

4.1. *Measurement setup*

As an example of the utilization of the use of spherical harmonics for sound radiation, we present preliminary results of a recent measurement study regarding the directivity patterns of the saxophone. All measurements were made in the anechoic chamber at IRCAM. The chamber is equipped with a remote-controlled mechanical arm that supports a half-circle of 24 microphones which can be rotated in elevation $-40^\circ \leq \phi \leq +90^\circ$. The microphone capsules (Panasonic MCE2000) were arranged with constant angular spacing at a distance $r = 1.48\,m$ from the origin. Prior to measurements all microphones were individually calibrated, and gain and phase correction filters were later applied. The instrument was mounted on a turntable and is centered at the origin of the measurement system using three coincident laser beams, see Fig. 3.

(a) Measurement setup with 24 microphones. (b) Driver.

Fig. 3. Measurement setup in the anechoic chamber for a saxophone (a) and the coupled acoustic driver for excitation (b).

The current measurement setup at IRCAM does not allow for simultaneous recordings on the entire sphere. Hence, to measure the sound pressure on the 2-sphere surface the acoustic signal must be reproducible. A coupled acoustic driver was developed for wind instruments which includes a probe microphone to monitor the excitation signal at the entrance of the instrument, see Fig. 3(b). The driver chassis was acoustically damped to minimize sound radiation, which would bias the measurement results.

4.2. *Measurement results*

Caussé *et al.*[27] have presented a simple and efficient physical model for calculating the directional pattern of woodwind instruments with curved tubes. This model calculates the far-field sound pressure on a sphere surrounding the instrument, also taking into account the directivity of the openings (holes and bells). Fig. 4 compares the measured radiation pattern of a saxophone, reconstructed from its relevant spherical harmonic coefficients up to order 3, with the simulation results.

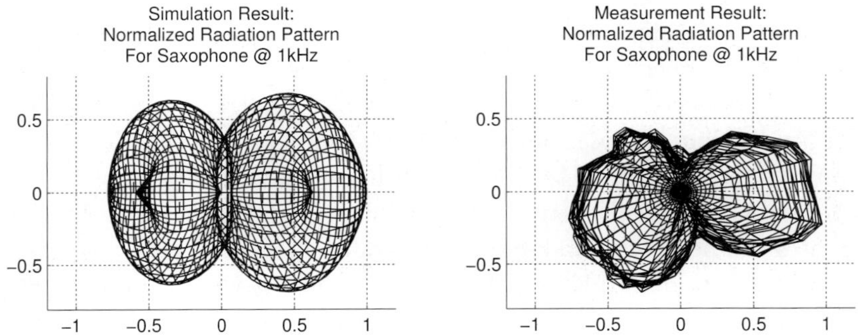

(a) Radiation pattern at $f_0 = 1$ kHz.

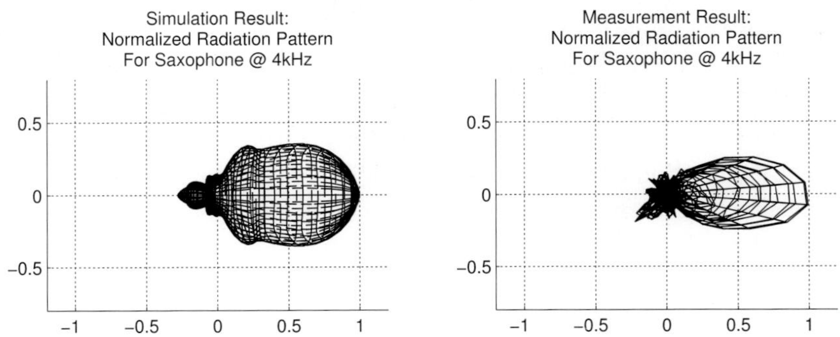

(b) Radiation pattern at $f_0 = 4$ kHz.

Fig. 4. Simulated (left) and measured (right) radiation pattern of a saxophone at frequencies $f_0 = 1$ kHz (a) and $f_0 = 4$ kHz (b). The radiation patterns are normalized to the maximum sound pressure.

5. Source Directivity Reconstruction Examples

Referring to Eq. 4 the function value, *i.e.* the sound pressure $p(k, r, \theta, \phi)$, at any given point on the surface of the sphere $\mathbb{S}^2$ can be evaluated from the expansion coefficients by weighted summation of the corresponding spherical harmonics. Several case studies of practical implementations are given in the following sections.

5.1. *Auralization*

A software environment for real-time auralization of complex geometries comprising novel algorithms for accelerated beam tracing is described in Ref. 28. Room acoustic modeling is performed using a beam tracing technique, and the audio signal processing for 3D sound field reproduction is realized using spherical harmonic decomposition of the sound field arriving at the listeners position via a mixed order implementation of higher order Ambisonics (HOA). One of the real-time optimizations of this environment relates to the use of spherical harmonic encoding of the spatial room response before rendering over headphones or loudspeakers. This step separates the positional updates of the room impulse response and the rotational updates due to listener orientation when using headphones which do not affect image source visibility checks.

In addition to the beam tracing algorithm providing the direction of arrival of each reflection (or image source) to the listener, the communication protocol of the proposed framework also provides the first reflection point relative to the sound source; hence the direction of radiation can be easily determined. Considering Eq. 4, the sound pressure for each reflection can be weighted to account for source directivity through a computationally efficient summation of the spherical harmonics weighted by the expansion coefficients that correspond to the sound source's radiation pattern. As such, spatial filters can be easily applied to incorporate the directionality of sound sources, which can be modified dynamically to represent changes to source orientation or variable directionality of complex sources.

5.2. *Wave field synthesis*

Wave field synthesis (WFS) is a spatial sound field reproduction technique that is principally based on the Huygens-Fresnel principle, which was reformulated for use in acoustics by Snow[29] and Berkhout *et al.*[30] WFS aims to authentically reproduce any given sound field over an extended listening area. Most of the current system designs use linear or circular loudspeaker arrays and are therefore restricted to the sound field in a planar listening area. Practical implementations can be shown to introduce audible artifacts, which result mainly from the finite length of the array, irregularities and constraints in loudspeaker spacing, and the spectral and temporal response of loudspeakers. The WFS approach was originally limited to omnidirectional sound sources. Corteel[31] has proposed the use of a subset of circular harmonics to manipulate the radiation characteristics of virtual sound sources in WFS systems; hence spherical harmonics as general description format for sound source directivity are directly applicable to this approach.

5.3. *Spherical beamforming loudspeakers*

Multichannel spherical loudspeaker arrays have been proposed for the use in sound radiation pattern synthesis. They typically consist of individually controllable drivers mounted on the surface of a sphere or of a convex polyhedron. See Refs. 11, 32, 33, and 34 for more details.

Warusfel *et al.*[32] have initially presented a basic method, derived from a minimum error criterion, for reproducing the radiation characteristics of musical instruments. A similar approach was taken by Kassakian *et al.*[35] for studying the limitations and error bounds of different array geometries. Modeling the loudspeakers as vibrating caps on a sphere allows deriving computationally efficient MIMO filters to reduce crosstalk due to acoustic coupling of the chassis.[36] In Ref. 11 it is shown that accurate radiation pattern synthesis requires appropriate equalization of the wave propagation terms corresponding to the radius of the reproduction device.

Spherical loudspeaker beamforming is a suitable method to efficiently control the sound radiation pattern at a variable distance. It directly applies spherical harmonics expansion to the array achievable subspace. Practical limitations, such as angular aliasing at high frequencies and dynamic range bounds at low frequencies, usually constrain the achievable resolution.

6. Perception of Directivity Rendering

It is important to consider the perceptual effects of integrating directivity into a rendering system. While there have been many studies on the directivity patterns of musical instruments (see Refs. 2 and 37 for summaries), there have been few studies on the perception of controlled source directivity. Misdariis *et al.*[38] have recently conducted perceptual studies employing both planar (2D) and spherical (3D) loudspeaker arrays. This preliminary study used synthesized radiation patterns for vibrating plates as well as measured phoneme directivity patterns for spoken[39] and sung voice[40]. Through the use of binaural recordings, the reproduced directivity patterns were individually rendered and then evaluated over headphones. Subjective impression ratings of source width and to a larger extent source distance varied significantly with respect to the rendered directivity and number of rendered dimensions. Perceived source width and distance increased with dimensional reproduction, implying a perceptual relationship with respect to the interaction of source directivity and room acoustics. This is interestingly similar to studies on perception of reproduced 1st order Ambisonic sound fields using 2D and 3D arrays, where perceived distance increased with increasing dimensional rendering.[41]

7. Conclusions

This paper has been concerned with various aspects of acoustical source directivity. The fundamental use of spherical harmonic decomposition has been presented as a generic method of representation for complex directivities. The effect of a limited spherical harmonic order and discretization on the measurement precision and results has also been discussed. A variety of examples incorporating controllable source directivity have also been presented, focusing on the measurement, reproduction, and perceptual evaluation of the directivity patterns of real sources using spherical loudspeaker arrays or directivity synthesis in a Wave Field Synthesis reproduction context.

8. Acknowledgements

This research was supported in part by French ANR RIAM 004 02 "EarToy" (first author); the funding of Austrian Research Promotion Agencies (FFG, SFG), COMET program (second author); and LIMSI-CNRS project ASP "Tete Parlante" (third author). The authors are grateful to Eric Boyer, Alexandre Lang, Gaëtan Parseihan, Hannes Pomberger, and Joseph Sanson for help with experimental procedures, and to René Caussé, Nicolas Misdariis, and Olivier Warusfel for fruitful discussions of various aspects and contributions to this work.

References

1. H. F. Olson, *Modern Sound Reproduction* (Van Nostrand Reinhold, New York, 1972).

2. J. Meyer, *Acoustics and the Performance of Music*, 5th edn. (Springer, New York, 2009).
3. J. Meyer, Directivity of the bowed string instrument and its effects on orchestral sound in concert halls, *J. Acoust. Soc. Am.* **51**, 1994 (1972).
4. J. Meyer, The influence of directivity on sound heard by an audience during an orchestral performance, *J. Acoust. Soc. Am.* **99**, 2526 (1996).
5. A. H. Marshall and J. Meyer, The directivity and auditory impressions of singers, *Acustica* **58**, 130 (1985).
6. D. Takahashi and S. Kuroki, Effects of directivity of a sound source on speech intelligibility in the sound field, in *Proc. Int. Congress on Acoustics, ICA*, (Madrid, Spain, 2007).
7. P. M. Morse and K. U. Ingard, *Theoretical Acoustics.* (McGraw-Hill, Inc., New York, 1968).
8. E. G. Williams, *Fourier Acoustics* (Academic Press, London, 1999).
9. N. A. Gumerov and R. Duraiswami, *Fast Multipole Methods for the Helmholtz Equation in Three Dimensions* (Elsevier Science, Amsterdam, 2004).
10. J. R. Driscoll and D. M. Healy, Computing Fourier transforms and convolution on the 2-sphere, *Adv. Appl. Math.* **15**, 202 (1994).
11. F. Zotter and M. Noisternig, Near- and farfield beamforming using spherical loudspeaker arrays, in *Proc. 3rd Congr. Alps Adria Acoust. Assoc.*, (Graz, Austria, 2007).
12. G. Golub and W. Kahan, Calculating the singular values and pseudo-inverse of a matrix, *SIAM Numer. Anal.* **2**, 205 (1965).
13. I. H. Sloan and R. S. Womersley, Extremal system of points and numerical integration on the sphere, *Adv. in Comp. Math.* **21**, 107 (2004).
14. E. A. Rakhmanov, E. B. Saff and Y. M. Zhou, Minimal discrete energy on the sphere, *Mathematical Research Letters* **1**, 647 (1994).
15. E. B. Saff and A. B. J. Kuijlaars, Distributing many points on a sphere, *The Mathematical Intelligencer* **19**, 5 (1997).
16. K. M. Górski, E. Hivon, A. J. Banday, B. D. Wandelt, F. K. Hansen, M. Reinecke and M. Bartelmann, HEALPix: A framework for high-resolution discretization and fast analysis of data distributed on the sphere, *The Astrophysical Journal* **622**, 759 (2005).
17. N. Sneeuw, Global spherical harmonic analysis by least-squares and numerical quadrature methods in historical perspective, *Geoph. J. Int.* **118**, 707 (1994).
18. J. P. Boyd, *Chebyshev and Fourier Spectral Methods* (DOVER Publications, 2000).
19. R. Pail, G. Plank and W.-D. Schuh, Spatially restricted data distributions on the sphere: the method of orthonormalized functions and applications, *Journal of Geodesy* **75**, 44 (2001).
20. H. Pomberger and F. Zotter, An ambisonics format for flexible playback layouts, in *Proc. Int. Ambisonics Symposium*, (Graz, Austria, 2009).
21. F. Zotter, Sampling strategies for acoustic holography / holophony on the sphere, in *Proc. NAGA / DAGA*, (Rotterdam, Netherlands, 2009).
22. G. Weinreich and E. B. Arnold, Method for measuring acoustic radiation

fields, *J. Acoust. Soc. Am.* **68**, 404 (1980).
23. F. Giron, *Investigations about the Directivity of Sound Sources* (EAA Fenestra, Shaker Publishing, Bochum, 1996).
24. O. L. Colombo, *Numerical methods for harmonic analysis on the sphere*, Tech. Rep. 310, Dept. of Geodetic Science and Surveying, Ohio State University (Columbus, Ohio, 1981).
25. P. N. Swarztrauber and W. F. Spotz, Generalized discrete spherical harmonic transform, *J. Comp. Phys.* **159**, 213 (2000).
26. J. A. R. Blais, Discrete spherical harmonic transforms: Numerical preconditioning and optimization, *Springer Lect. Notes Comp. Sc.* **5102**, 638 (2008).
27. R. Caussé and C. L'Heureux, Modeling in 3D of directional radiation of curved woodwind instruments (A), *J. Acoust. Soc. Am.* **103**, p. 2874 (1998).
28. M. Noisternig, B. F. G. Katz, S. Siltanen and L. Savioja, Framework for real-time auralization in architectural acoustics, *Acta Acustica United with Acustica* **94**, 1000 (2008).
29. W. B. Snow, Basic principles of stereophonic sound, *IRE Trans. on Audio* **3**, 42 (1955).
30. A. J. Berkhout, D. de Vries and P. Vogel, Acoustic control by wave field synthesis, *J. Acoust. Soc. Am.* **93**, 2764 (1993).
31. E. Corteel, Synthesis of directional sound sources using wave field synthesis, possibilities, and limitations, *EURASIP J. Applied Sig. Proc.* **11**, 188 (2007).
32. O. Warusfel, P. Dérogis and R. Caussé, Radiation synthesis with digitally controlled loudspeakers, in *Proc. 103rd AES Conv.*, (New York, 1997).
33. R. Avizienis, A. Freed, P. Kassakian and D. Wessel, A compact 120 independent element spherical loudspeaker array with programmable radiation patterns, in *Proc. 120th AES Conv.*, (Paris, France, 2006).
34. B. Rafaely, Spherical loudspeaker array for local active control of sound, *J. Acoust. Soc. Am.* **125**, 3006 (2009).
35. P. Kassakian and D. Wessel, Characterization of spherical loudspeaker arrays, in *Proc. 117th AES Conv.*, (San Francisco, CA, 2004).
36. F. Zotter, A. Schmeder and M. Noisternig, Crosstalk cancellation for spherical loudspeaker arrays, in *Proc. DAGA*, (Dresden, Germany, 2008).
37. N. H. Fletcher and T. D. Rossing, *The Physics of Musical Instruments*, 2nd edn. (Springer, New York, 2008).
38. N. Misdariis, A. Lang, B. F. G. Katz and P. Susini, Perceptual effects of radiation control with a multi-loudspeaker device (A), *J. Acoust. Soc. Am.* **123**, p. 3665 (2008).
39. B. F. G. Katz, F. Prezat and C. d'Alessandro, Human voice phoneme directiviy pattern measurements (A), *J. Acoust. Soc. Am.* **120**, p. 3359 (2006).
40. B. F. G. Katz and C. d'Alessandro, Directivity measurements of the singing voice, in *Proc. Int. Congress on Acoustics, ICA*, (Madrid, Spain, 2007).
41. C. Guastavino and B. F. G. Katz, Perceptual evaluation of multi-dimensional spatial audio reproduction, *J. Acoust. Soc. Am.* **116**, 1105 (2004).

IMPLEMENTATION OF REAL-TIME ROOM AURALIZATION USING A SURROUNDING 157 LOUDSPEAKER ARRAY

T. OKAMOTO[1,2], B. FG KATZ[3], M. NOISTERNIG[4],
Y. IWAYA[1,5] and Y. SUZUKI[1,5]

[1] *Research Institute of Electrical Communication, Tohoku University,*
[2] *Graduate School of Engineering, Tohoku University,*
[5] *Graduate School of Information Sciences, Tohoku University,*
2-1-1 Katahira, Aoba-ku, Sendai, 980-8577, Japan
E-mail: {okamoto@ais., iwaya@, yoh@}riec.tohoku.ac.jp

[3] *LIMSI-CNRS,*
BP 133, F91403 Orsay, France
E-mail: brian.katz@limsi.fr

[4] *IRCAM - UMR CNRS*
1, place Igor Stravinsky 75004 Paris, France
markus.noisternig@ircam.fr

This chapter presents the implementation of a real-time room acoustic auralization system using a 157-loudspeaker array. The room acoustic model combines an iterative image-source model and feedback delay networks to create early reflections and late reverberation. Higher-order Ambisonics (HOA) is used to generate spatial room impulse responses. A distributed network system is then used to generate the auralization output.

Keywords: Surrounding loudspeaker array; Auralization; Higher-order Ambisonics; Computational room acoustics

1. Introduction

Auralization[1] has become a useful tool for the acoustic design of three-dimensional architectural environments; it enables rendering processes that can simulate the sound field from the source to receiver and therefore the auditory perception at the listener's position. Most conventional auralization techniques are computationally intensive, providing detailed results, but they are poorly suited to real-time applications. Consequently, they are not applicable to interactive virtual environments. Although pre-calculation of impulse responses and real-time panning is possible, this approach has some limitations, which include a predefined static source and receiver positions. For a truly interactive virtual acoustic environment it is important to develop room acoustic auralizations that allow for dynamic source and listener movements in real time.

Noisternig *et al.*[2] proposed a modular processing environment for rendering of complex geometry room acoustics within interactive update times. Although the proposed system was presented using headphone rendering, the system is flexible and adaptable to playback via loudspeakers or headphones. The proposed system uses a combination of an iterative beam tracing approach[3] optimized for a moving listener for creating early reflections and feedback delay networks (FDN)[4] for creating late reverberation. In a beam tracing approach, early reflections are calculated using an image method with a source and receiving position, and room geometry data.[3] In addition, FDN can estimate and calculate the late reverberation using the slope of impulse responses of early reflections. These calculated impulse responses are encoded to Higher order Ambisonics (HOA)[5,6] format. HOA is a sound field decomposition and reproduction method based on spherical harmonics.[7] Finally, each sound signal at each loudspeaker is calculated by HOA decoding using the HOA encoded format. This combination produces a satisfactory compromise between precise calculation and rendering of early geometrical reflections and a statistically based less-precise late reverberation. All software components of this system are published as open source.

In this chapter, we discuss the implementation of this system using a three-dimensional loudspeaker array surrounding the listener. The installation consists of a dense grid of 157 speakers located on the walls and ceiling of an acoustically damped rectangular room.[8] A distributed network system has been used to create the auralization.

The system as presented in Noisternig *et al.*[2] has been initially developed around the binaural rendering of spatial room impulses for auralization. Although the system architecture allows for the different modules to be distributed over a network, each module was designed to exist in a single instance, which is well suited for playback over a small number of audio channels (headphones, stereo, etc.). In contrast, the use of 157 audio channels requires the distribution of the audio output over a cluster of several PCs to drive the loudspeaker array. The proposed system uses five PCs (one master and four parallel slaves), which communicate via network using User Datagram Protocol (UDP). In addition to a presentation of the system cluster architecture, the overall system latency and the synchronization of the different audio streams will be discussed in detail.

2. Surrounding 157 loudspeaker array system

A total of 157 loudspeakers (FE38E; Fostex Co.) were installed on a regular grid with 0.5 m spacing and at a distance of 0.3 cm from the wall surface. Figure 1 shows the surrounding loudspeaker array; Fig. 2 shows the arrangements of the loudspeakers. The mid-frequency reverberation time RT30 of the reproduction room was approximately 0.2 s, as measured using a real time octave band analyzer (SR-5300; Ono Sokki Co. Ltd.) with pink noise and calculated based on Schroder integration method. Therefore, from Sabine's reverberation equation,[9] the average sound absorption coefficient $\overline{\alpha}$ is about 0.45. The audio rendering cluster consists of 14 units of the digital-to-analog converter (D/A) (HD192; Motu Inc.) connected to four PCs (MacPro; Apple Computer Inc.). Clock synchronization was achieved using a global clock generator (Nanosyncs HD; Rosendahl Studiotechnik GmbH) connected to each D/A module.

3. Real-time room auralization system using the surrounding loudspeaker array

The surrounding loudspeaker auralization system requires (i) modeling the acoustic paths from the source to the listener according to the given room model, (ii) generation of the impulse responses at the listener's position, (iii) convolution of these impulse responses with audio source material, and (iv) conversion of the resulting audio stream to a format applicable to the multi-channel reproduction system which is rendered synchronously.

To realize this system, we have developed the system architecture shown in Fig. 3. In this architecture, the Master machine contains three modules: the scene modeler module, VirChor (http://virchor.wiki.sourceforge.net/), which is the real-time 3D graphics rendering engine that maintains the source and receiver positions as well as the room geometry model; the acoustic modeler module, EVERTims,[10] which receives the various scene components and calculates the early reflection paths; and a distribution/control module. The distribution/control module, Pure Data (Pd) (http://puredata.info/) an open source and multi-OS software environment for real time audio processing, collects the results from the modeler and forwards this information to the different rendering modules on the four slave machines using network broadcasting. There are also general parameter controls which are provided through an interface in this module. The slave machines receive the reflection paths and other acoustical data, which are transformed into HOA-encoded spatial impulse responses in the audio ren-

Fig. 1. Appearance of the surrounding loudspeaker array.

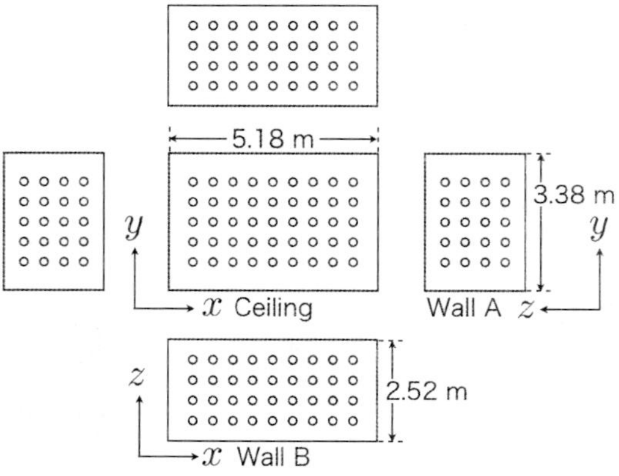

Fig. 2. Arrangement of loudspeakers.

dering module. These responses are used to render audio material which is then decoded from the HOA stream to the 157 loudspeaker channels.

Separate from the acoustical rendering, the early reflection path results are also redistributed back to the scene modeler, allowing for their visualization. Figure 4 shows visualization results for different reflection orders.

3.1. *Spatial audio rendering*

The audio rendering module is implemented on four parallel slave processing units running Pure Data. Each processing unit drives a subset of the 157 loudspeakers. All inter-machine communication was managed via Pure Data using the Open Sound Control (OSC) Protocol (http://opensoundcontrol.org/) and User Datagram Protocol (UDP). The sound source position, the listener's position and the image source information are broadcast from the Master PC to the four audio control PCs over UDP. In the image source or beam tracing method, the computation time increases with the reflection order. The EVERTims module functions in an iterative manner such that results are sent upon the completion of each subsequent order, up to a configurable highest order. Statistically, the number of reflections increases concomitantly with increasing reflection order. In the rendering module, each reflection is treated individually to provide correct spatial positioning. Therefore, the CPU load increases directly with the number of HOA encoded reflections. The additional CPU load is attributed to the HOA to loudspeaker decoding. With the current system configuration, the maximum number of individual reflection paths (image sources) able to be encoded before audible artifacts was set to 100. For the test geometry used, this was comparable to limiting the maximum reflection order to 3.

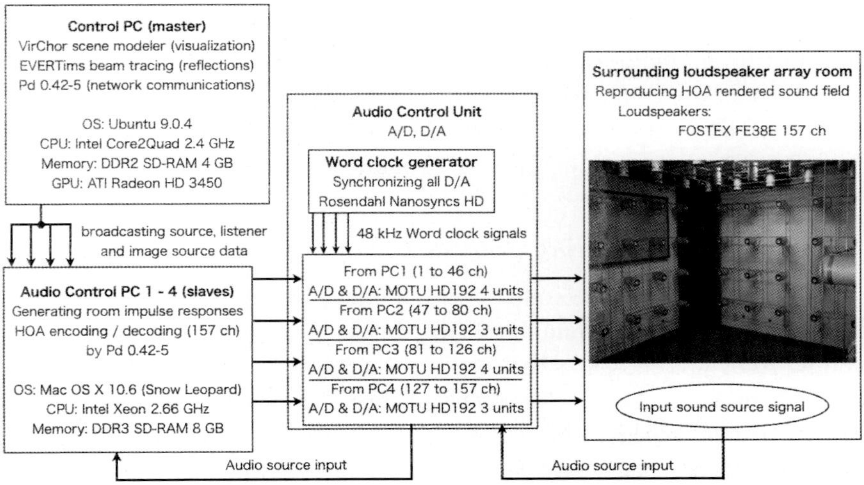

Fig. 3. Real-time auralization system using the surrounding 157 ch loudspeaker array.

Fig. 4. Visualization of a test room model: left shows no reflection, middle shows first order reflection paths, and right shows third order reflection paths.

3.2. *Auralization based on High-order Ambisonics encoding*

Three-dimensional sound rendering at the listener's position from the simulated room impulse responses was implemented using a method based on HOA.[11] It is a characteristic of HOA that the encoding system and the decoding system are mutually independent completely. Moreover, the higher the order in encoding and decoding, the more spatially precise is the rendered sound field.

In the current architecture, the sound pressure of the direct sound path and each of the image sources calculated by the EVERTims were encoded using fourth order HOA (25 channels). These elements of the impulse response were created using a tap delay line method in which the input signal was delayed in time, and modified in level (for different frequency bands), depending on the distance of the calculated image source and the accumulated acoustic absorption. The late reflections generated by the FDN of the image sources, in the current implementation, were assigned arbitrary directional information and encoded using first-order HOA. The results of each encoder are combined on a common bus to create a single HOA audio stream. Each of the slave audio rendering modules performs an identical function, generating an identical HOA audio stream.

3.3. *Ambisonic decoding over surround loudspeaker array*

The HOA stream was decoded corresponding to the arrangement of the loudspeakers of the surrounding loudspeaker array. The sweet spot for decoding HOA was chosen as the center of the room (coordinate: $x = 2.59$ m, $y = 1.69$ m, $z = 1.26$ m). The HOA decoding matrix was generated according to the loudspeaker positions relative to the center position. In the current architecture, each slave rendering module performed the HOA decode for the entire speaker array, and only those channels which were connected to each respective slave were then further processed and rendered over the loudspeakers. Subsequent processing consisted primarily of correcting the level and delay of each loudspeaker to create a virtual spherical speaker array. A simple delay was included for each speaker to place it radially at the distance of the farthest loudspeaker from the center, thereby creating a virtual partial sphere (because of a lack of floor speakers) surrounding the listener. Simple propagation attenuation following the spherical law was then applied.

3.4. *Audio source*

For real-time auralization, we can imagine audio source signals of two different types: prerecorded audio files and live audio streams. The live audio stream is obtainable using microphones in the reproduction room, thereby allowing not only for position and orientation interactivity, but also acoustic interactivity within the auralization system.

In the case of prerecorded audio content, the audio files are duplicated on each slave machine and the playback is controlled from the master controller module. The case of a live audio input presents two alternatives. The first is to input the microphone signal to the analog-to-digital converter (A/D) on one machine and then to distribute this signal over the network to the different rendering modules. The second alternative is to distribute the audio signal to the A/D of each slave PC. This second option should provide a reduced result as compared to the first option.

4. Latency performance evaluation

The difference between the previous architecture[2] and our proposed system is the distribution of the audio rendering module over a number of the machines to perform the HOA encoding and decoding for the large loudspeaker array. Because of this modification, it is important to quantify the degree of synchronicity between the audio outputs of each Slave PC.

In this section, we examine two basic synchronous reproduction estimations using a simplified Pure Data architecture, without the room impulse HOA encoding and decoding or any other process.

The latency was measured using a time stretched pulse (TSP)[12] signal, a variant of the swept-sine. The sampling frequency was 48 kHz, the quantization bit was 16 bit, and the format was linear PCM. Using Pure Data the audio buffer size was minimized through several listening comparisons. The output sounds became clipped under high CPU load conditions if the buffer size was set less than 10 ms. Therefore, the minimum audio buffer size for each machine was set to 11 ms (= 528 samples).

A deconvolution of the TSP signal from the recorded signals at each channel provides an impulse whose temporal position is equivalent to the delay. Using this characteristic, we examined the temporal synchronization.

Signal acquisition was performed using an 8 channel audio interface (ProFire Lightbridge; M-Audio), A/D (ADA8000; Behringer) with the recording software (Pro Tools 8 M-Powered; Desidesign). Using only 8 input channels, channel combinations were tested in batches to test all 157 channels. The measurements were repeated five times.

4.1. *File playback synchronicity*

For file playback latency, a control signal was sent from one slave (PC1) to the three other slave machines (PC2–PC4). All communications without file playback command were realized within Pure Data using the OSC protocol over UDP because the file playback command should be set at all PCs with synchronous. Therefore, we introduced the audio splitter (MX882 Ultralink Pro; Behringer) to realize a set file playback command at all four PCs using the trigger signal. The latency measurement system is shown in Fig. 5. The inter-channel latency between D/A channels was measured relative to the first output channel of the control machine: PC1.

Results show that the output signals from all D/A channels from the same PC were completely synchronous. For example, the signals from channels 1 to 46 were all synchronous at the 1 sample level. In contrast, the output signals between different PCs were not synchronous at the 1 sample level. The unsigned average latency of PC2–PC4 relative to PC1 (control signal) was 28 samples (maximum 51 samples (= 1.1 ms)).

4.2. *Audio streaming synchronicity using a signal splitter*

For audio streaming latency, the TSP signal was input to channel #160, which was the A/D connected to PC4. This signal was then routed to the three other PCs using the audio signal splitter (MX882 Ultralink Pro; Behringer). The signal was received at all four PCs and then distributed over all available loudspeaker channels, resulting in a measured latency for the 157 output channels. The latency measurement system is shown in Fig. 6. The results show that all 157 output signals were completely synchronous and that the overall I/O system latency is constant 981 taps (= 20 ms).

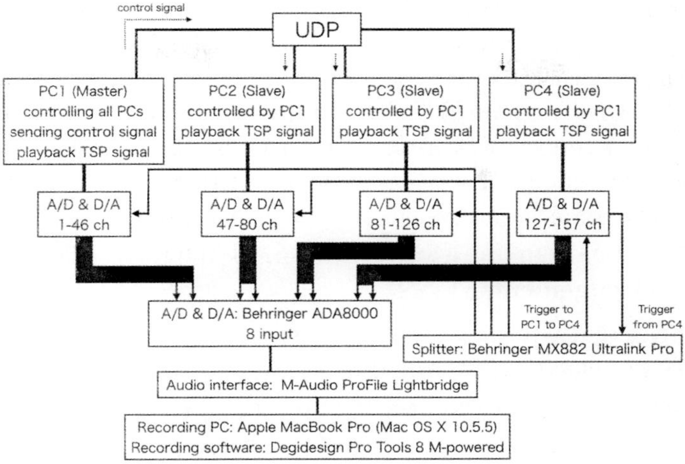

Fig. 5. Measurement system for file playback synchronicity.

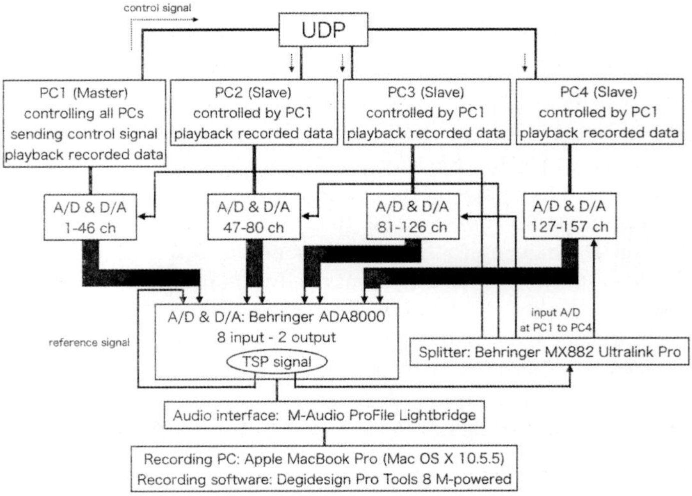

Fig. 6. Measurement system for audio streaming.

5. Conclusions

In this chapter, we introduced an implementation of a real-time room auralization system using a 157 channel surrounding loudspeaker array. Various latency issues were investigated because they pertain to audio source material being either prerecorded or live streaming. Results showed latencies to be less than the audio buffer size, which is currently limited to 1.1 ms; this result was nearly synchronous.

In future works, we would like to improve our system to realize 157 channel synchronous reproduction together with the trigger signals such as AES/EBU. Additionally, we would like to estimate the delay evaluation through total auralization system and the degree of reproducibility.

6. Acknowledgements

This study was partly supported by the GCOE program (CERIES) of the Graduate School of Engineering, Tohoku University and a Grant-in-Aid for Specially Promoted Research No. 19001004 from MEXT, Japan.

References

1. M. Kleiner, B. I. Dalenback and P. Svensson, Auralization: an overview, *J. Audio Eng. Soc.* **41** 861–875 (1993)
2. M. Noisternig M, B. FG Katz, S. Siltanen and L. Savioja, Framework for real-time auralization in architectural acoustics, *Acta Acustica United with Acustica* **94** 1000–1015 (2008)
3. J. B. Allen and D. A. Berkley, Image method for efficiently simulating small-room acoustics, *J. Acoust. Soc. Am.* **65** 943–950 (1979)
4. J.-M. Jot and A. Chaigne A, Digital delay networks for designing artificial reverberators *in Proc. AES 90th Int. Convention* 3030 (1991)
5. M. A. Poletti, Three-dimensional surround sound systems based on spherical harmonics, *J. Audio Eng. Soc.* **53** 1004–1025 (2005)
6. S. Moreau, J. Daniel and S. Bertet, 3D sound field recording with high order ambisonics – Objective measurements and validation of a 4th order spherical microphone, *in Proc. AES 120th Int. Convention* 6857 (2006)
7. E. Williams, *Fourier Acoustics: Sound Radiation and Nearfield Acoustic Holography*, London, UK: Academic Pres (1999)
8. T. Okamoto, R. Nishimura and Y. Iwaya, Estimation of sound source positions using a surrounding microphone array, *Acoust. Sci. & Tech.* **28** 181–189 (2007)
9. R. W. Young, Sabine reverberation equation and sound power calculations, *J. Acoust. Soc. Am.* **31** 912–921 (1959)
10. S. Laine, S. Siltanen, T. Lokki and L. Savioja, Accelerated beam tracing algorithm, *Applied Acoustics* **70** 172–181 (2009)
11. M. Noisternig, T. Musil, A. Sontacchi, and R. Höldrich, 3D binaural sound reproduction using a virtual ambisonic approach, *in Proc. IEEE Int. Symp. VECIMS* 174–178, (2003)
12. Y. Suzuki, F. Asano, H.-Y. Kim and T. Sone, An optimum computer-generated pulse signal suitable for the measurement of very long impulse responses, *J. Acoust. Soc. Am.* **97** 1119–1123 (1995)

SPATIALISATION IN AUDIO AUGMENTED REALITY USING FINGER SNAPS

H. GAMPER and T. LOKKI*

Department of Media Technology, Aalto University,
P.O.Box 15400, FI-00076 Aalto, FINLAND
**E-mail: [Hannes.Gamper,ktlokki]@tml.hut.fi*

In audio augmented reality (AAR) information is embedded into the user's surroundings by enhancing the real audio scene with virtual auditory events. To maximize their embeddedness and naturalness they can be processed with the user's head-related impulse responses (HRIRs). The HRIRs including early (room) reflections can be obtained from transients in the signals of ear-plugged microphones worn by the user, referred to as instant binaural room impulse responses (BRIRs). Those can be applied on-the-fly to virtual sounds played back through the earphones. With the presented method, clapping or finger snapping allows for instant capturing of BRIR, thus for intuitive positioning and reasonable externalisation of virtual sounds in enclosed spaces, at low hardware and computational costs.

Keywords: Audio Augmented Reality; Finger snap detection; Binaural Room Impulse Response; Head Related Transfer Functions.

1. Introduction

Augmented reality (AR) describes the process of overlaying computer generated content onto the real world, to enhance the perception thereof and to guide, assist, or entertain the user.[1,2] In early AR research the focus was primarily on purely visual augmentation of reality,[3] at the expense of other sensory stimuli such as touch and sound. This imbalance seems unfortunate, given the fact that sound is a key element for conveying information, attracting attention and creating ambience and emotion.[4] Audio augmented reality (AAR) makes use of these properties to enhance the user's environment with virtual acoustic stimuli. Examples of AAR applications range from navigation scenarios,[5] social networking[6] and gaming[7] to virtual acoustic diaries[8] and binaural audio over IP.[9,10]

The augmentation is accomplished by mixing binaural virtual sounds into the ear input signals of the AAR user, thus overlaying virtual auditory events onto the surrounding physical space. The position of a real or a virtual sound source is determined by the human hearing based on localisation cues.[11] Encoding them into the binaural signals determines the perceived position of the virtual sounds. In the case of a real sound source, these localisation cues stem from the filtering behaviour of the human head and torso, as well as room reflections. A Binaural Room Impulse Response (BRIR) is the time domain representation of this filtering behaviour of the room and the listener, for given source and listener positions. It contains the localisation cues that an impulse emitted by a source at the given position in the room would carry when reaching the ear drums of the listener. Convolving an appropriate BRIR (for left and right ear) with a monaural virtual sound recreates the listening experience of the same sound as emitted from a real source at the position defined by the BRIR. The BRIR can thus be used to position a virtual source in the acoustic environment.

The chapter is organised as follows: section 2 describes the real-time acquisition of BRIRs. In section 3 a real-time implementation of the proposed algorithm for spatialisation in audio augmented reality (AAR) is presented. Results from informal listening tests of the real-time implementation are discussed in section 4. Section 5 concludes the chapter.

2. Instant BRIR acquisition

We present a simple and cost-effective way to acquire BRIRs on-the-fly and their application to intuitively position virtual sound sources, using finger snaps and/or hand claps. The BRIRs are obtained in the actual listening space, thus the filtering behaviour of the actual room is contained in them, as well as the filtering behaviour of the actual listener. Applying the BRIRs obtained with the presented method in the actual listening space to virtual auditory sources yields a natural and authentic spatial impression. In a telecommunication scenario with multiple remote talkers, the spatial separation achieved by processing each talker with a separate BRIR can improve the speech intelligibility and speaker segregation.[12,13]

2.1. *Hardware*

Unlike virtual reality (VR) systems, AAR aims at augmenting, rather than replacing, reality. This implies that the transducer setup used to reproduce virtual sounds for AAR must allow for the perception of the real acoustic environment. At the same time precise control over the ear input signals

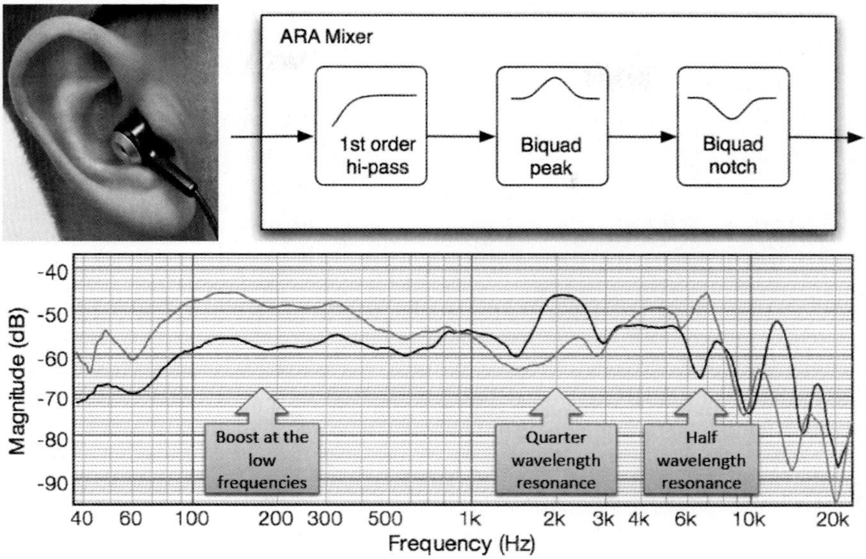

Fig. 1. The MARA headset and the basic principle of the analogue equalisation. Microphones embedded into insert-earphones record the acoustic surroundings at the ears of the MARA user (top figure, left). The bottom graph shows HRTF measurements at the ear drum with earphone (grey line) and without earphone (black line). To compensate for the impact of the earphones on the HRTF, the microphone signals are filtered in the ARA mixer before being played back to the user via the earphones, to ensure acoustic transparency of the transducer setup.[16]

must be ensured for correct playback of the binaural virtual sounds. Using earphones as transducers provides the advantages of excellent channel separation, easily invertible transmission paths and portability.

The transducer setup used in this work is a MARA (mobile augmented reality audio) headset, as proposed by Härmä et al.[14] It consists of a pair of earphones with integrated microphones and an external mixer (see Fig. 1). The microphones record the real acoustic environment, which is mixed with virtual audio content and played back through the earphones. Analogue equalisation filters in the mixer correct the blocked ear canal response to correspond to the open ear canal response, thus they ensure acoustic transparency of the earphones.[15] This allows for an almost unaltered perception of the real acoustic environment and the augmentation thereof with virtual audio content.

2.2. *Algorithm description*

If a transient in the microphone signals of the MARA headset is detected, the signals are buffered and the transient is extracted in each channel. These transients are taken as an approximation of the BRIR. A monaural input signal is filtered with this BRIR. The resulting binaural signals carry the same localisation cues as the recorded transient and the reverberation tail contains the information of the surrounding environment. Thus the monaural input signal is enhanced with the localisation cues of an external sound event at a certain position in the actual listening space. By generating a transient in the immediate surroundings of the user, for example by snapping fingers or by clapping, a user can therefore intuitively position a virtual sound source in his or her acoustic environment.

2.2.1. *Detection of transients*

Room impulse responses are usually measured with a deterministic signal, e.g. with a maximum length sequence (MLS) or a sweep.[17] By deconvolving the known input signal out of the recorded room response, the impulse response of the room can be derived. If an impulse is used as the excitation signal, the recorded response corresponds to the room impulse response.

In the presented algorithm, a finger snap is taken as the excitation signal to estimate a BRIR on-the-fly. As the spectrum of the finger snap is however not flat, the measured frequency response is in fact "coloured" by the snap spectrum. The BRIR derived from a finger snap excitation is thus only the coloured approximation of the real BRIR. The implications of this in the presented usage scenario are discussed in section 3.

To facilitate the detection of the snap, the microphone signals are preprocessed: The energy of finger snaps is mainly contained between 1500 and 3500 Hz.[18] A bandpass filter with a centre frequency of 2100 Hz and the mentioned bandwidth is applied to the microphone signals to remove frequency components above and below this band. This improves the detection performance in the presence of background noise considerably (see Fig. 2). To detect transients in the bandpass-filtered microphone signal, a method presented by Duxbury et al.[19] is employed. The energy of the signal is calculated in time frames of 256 samples each with 50 % overlap. Transients in the time domain are characterised by an abrupt rise in the short-time energy estimate. The derivative of the energy estimate is a measure for the abruptness of this rise in energy. If the derivative exceeds the detection threshold, the peak of the derivative is determined and the mi-

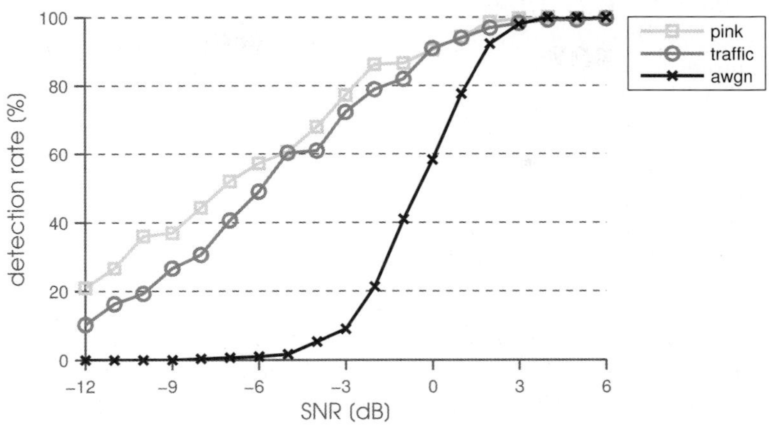

Fig. 2. Finger snap detection. The detection rate of finger snaps in noisy signals is given as a function of the signal-to-noise ratio (SNR), for various noise signals (pink noise, traffic noise, and additive white Gaussian noise). Pink and traffic noise yield higher detection rates, as their power spectral density decreases with frequency, thus less noise energy is present around 2000 Hz, where most of the finger snap energy is concentrated.

crophone signals of the MARA headset are buffered. Due to its simplicity the computational cost of the algorithm is very low. The algorithm proved to be quite robust also in the presence of background noise, which is an important criterion especially for mobile AAR applications. The performance of the transient detection in the presence of noise is depicted in Fig. 2.

2.2.2. *Extraction and application of the BRIRs*

The BRIR is extracted by windowing the buffered raw microphone signals around the detected snap. Thus, the BRIR is approximated by the unprocessed finger snap detected in the MARA signals. A flat top hanning window is applied to the buffers, starting 15 to 100 samples (i.e. 0.3–2.3 ms at 44100 Hz sampling rate, depending on the total window length) before the position of the finger snap, to ensure the onset of the transient is preserved. The length of the window is variable. For a short window (128 to 256 samples) only the early part of the impulse response is captured. It contains the direct signal and signal components that arrive 3–6 ms after the direct signal due to traveling an additional path length of up to 1–2 m, e.g. reflections from the shoulders and pinnae. Thus with a short window the room influence is eliminated, and only a coloured HRIR is extracted. Longer windows also include signal components that arrive after 3–6 ms, i.e. reflections from walls and objects inside the room. It is known that inclusion of this room reverberation improves the externalisation of virtual sounds.[4,20] With impulse response lengths of 200 to 400 ms (i.e. window lengths of 8192–16384 samples) reasonable externalisation could be achieved.

The BRIR estimated in this way can directly be applied to a monaural input signal, thus enhancing the signal with the localisation cues of the recorded snap. This allows the user to position a virtual source intuitively in his/her environment by snapping a finger. To reduce the colouration of the BRIR with the finger snap spectrum, inverse filtering could be considered to whiten the BRIR. However, for virtual speech sources the colouration was not found to be disturbing, and postprocessing of the BRIR was thus omitted in the present implementation. A possible application scenario to study the usability of the presented method was implemented in the programming environment Pure Data.[21]

3. Real-time implementation

A real-time implementation of the proposed algorithm for spatialisation in audio augmented reality was presented at the IWPASH 2009 (International Workshop on the Principles and Applications of Spatial Hearing) conference in Japan.[22] The Pure Data implementation of the described algorithm simulates a multiple-talker condition in a teleconference. In the simulated teleconference three participants (two remotes and one local) are discussing. The local participant is wearing the MARA headset. The remote end speakers are simulated by monaural recordings of a male and a female speaker and played back to the local participant over the earphones of the MARA headset. As the simulated remote end speakers are talking simultaneously, a multiple-talker condition arises. The unprocessed monaural speech signals are perceived inside the head, with no spatial separation. When the local participant snaps his or her fingers, the snap is recorded via the microphones of the MARA headset and convolved with the monaural speech signals. Snapping in two different positions, one for each of the remote speakers, allows the local participant to position the speakers in his or her auditory environment. The remote speakers are externalised and spatially separated, which improves intelligibility and listening comfort. The structure of the algorithm is depicted in Fig. 3.

As the excitation signal, i.e. the finger snap, does not have a flat spectrum, the input signal will be coloured with the snap spectrum after convolution. The colouration can be controlled by the user by varying the spectrum of the transient, e.g. by clapping instead of finger snapping. This was

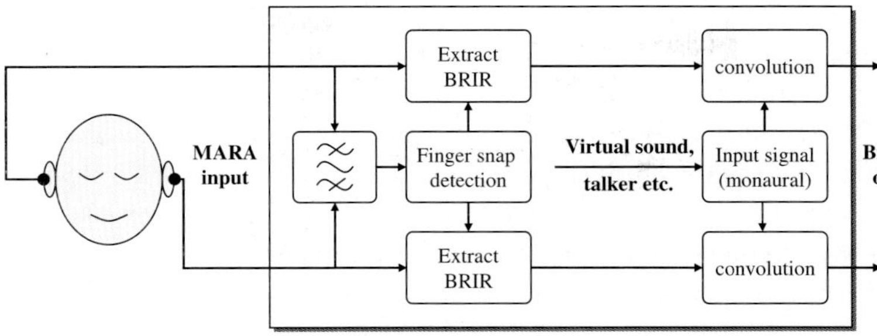

Fig. 3. Structure of the algorithm. If a finger snap is detected, a BRIR is extracted from each microphone channel and convolved with the input signal, i.e. a monaural speech signal of a virtual remote teleconference participant. Convolving each speaker with a separate snap, the participants can be spatially separated.

found to be an interesting effect in informal listening tests. Furthermore, as the finger snap energy is mostly contained in a frequency band in particular important for speech perception and intelligibility, the colouration was not found to deteriorate the communication performance.

4. Discussion

It has been shown that the spatial separation of simultaneously talking speakers improves their intelligibility. This is a phenomenon known as the "cocktail party effect".[23] In addition to the implications on speech intelligibility, the externalisation is also considered to "add a pleasing quality" to virtual sounds.[24] In the present work the spatialisation is performed by applying a separate BRIR to each signal. The BRIRs are acquired in the actual environment of the listener and recorded at the ear canal entrances of the listener.

Informal listening tests suggest that the use of a locally acquired individual BRIRs allows for reasonable externalisation of virtual sources, given a sufficient filter length. We believe that there are two main reasons for this result. Firstly, it has been shown that individual HRTFs are in general superior than generic non-individual HRTFs in terms of the localisation performance.[25,26] As the BRIRs are recorded at the user's own ears with the presented method, the filtering behaviour of the user's own head, torso and pinnae is captured. Applying the BRIRs to virtual sounds simulates the listening experience of that very user when exposed to a real source, leading to localisation cues in the binaural virtual sounds similar to normal listening.

Secondly, the influence of the listening environment on the sound field in the form of reflections and (early) reverberation is preserved in the tail of the BRIR. We assume that spatialising a virtual sound source with a room resembling the actual listening room leads to a more natural and physically coherent binaural reproduction. This is especially beneficial in the context of AAR, where embeddedness and immersion of virtual content in the real surroundings is required. To perceive the virtual and real environment as one, the characteristics of the virtual world have to resemble the ones of the real world.

5. Conclusions and Future Work

Instant individual BRIRs acquired with the described method and applied to monaural speech signals provide reasonable externalisation of virtual talkers. This can be a considerable improvement of intelligibility and listening comfort in multiple-talker conditions in telecommunication. The colouration of speech signals with the non-white input spectrum of a finger snap was not found to be disturbing, and could in fact be seen as an entertaining side effect.

A real-time implementation of the system was presented at the IWPASH 2009 (International Workshop on the Principles and Applications of Spatial Hearing) conference in Japan.[22] During the demo session, it was found that the described method of BRIR acquisition using finger snaps or clapping provides a very intuitive and straightforward way of defining the positions of virtual auditory events.

A major improvement of the presented system would be to include head-tracking, to allow for stable externalised sources by dynamically panning them according to the head movements of the user. Another potential enhancement might be to whiten the transient spectrum, thus minimising the colouration, if high fidelity or reproduction of signals other than speech is required. Matched filtering could be applied as an efficient alternative to the proposed transient detection.

Acknowledgements

The research leading to these results has received funding from Nokia Research Center [kamara2009], the Academy of Finland, project no. [119092] and the European Research Council under the European Community's Seventh Framework Programme (FP7/2007-2013) / ERC grant agreement no. [203636].

References

1. R. Azuma. A survey of augmented reality. *Presence: Teleoperators and Virtual Environments*, pp. 355–385 (1997).
2. R. Azuma, Y. Baillot, R. Behringer, S. Feiner, S. Julier, and B. Macintyre. Recent advances in augmented reality. *IEEE Computer Graphics and Applications* **21**, pp. 34–47 (2001).
3. M. Cohen and E.M. Wenzel. The design of multidimensional sound interfaces. In W. Barfield and T.A. Furness, editors, *Virtual environments and advanced interface design*, pp. 291–346 (Oxford University Press, Inc., New York, NY, USA, 1995).
4. R. Shilling and B. Shinn-Cunningham. Virtual auditory displays. In K. Stanney, editor, *Handbook of Virtual Environments*, pp. 65–92 (Lawrence Erlbaum Associates, Mahwah NJ, 2002).
5. B.B Bederson. Audio augmented reality: a prototype automated tour guide. In *ACM Conference on Human Factors in Computing Systems (CHI)*, pp. 210–211 (New York, NY, USA, 1995).
6. J. Rozier, K. Karahalios, and J. Donath. Hear&there: An augmented reality system of linked audio. In *Proceedings of the International Conference on Auditory Display (ICAD)*, pp. 63–67 (Atlanta, Georgia, USA, 2000).
7. K. Lyons, M. Gandy, and T. Starner. Guided by voices: An audio augmented reality system. In *Proceedings of the International Conference on Auditory Display (ICAD)*, pp. 57–62 (Atlanta, Georgia, USA, 2000).
8. A. Walker, S.A. Brewster, D. McGookin, and A. Ng. Diary in the sky: A spatial audio display for a mobile calendar. In *Proceedings of the 15th Annual Conference of the British HCI Group*, pp. 531–540 (Lille, France, 2001. Springer).
9. T. Lokki, H. Nironen, S. Vesa, L. Savioja, and A. Härmä. Problem of far-end user's voice in binaural telephony. In *the 18th International Congress on Acoustics (ICA'2004)*, volume II, pp. 1001–1004 (Kyoto, Japan, April 4-9 2004).
10. T. Lokki, H. Nironen, S. Vesa, L. Savioja, A. Härmä, and M. Karjalainen. Application scenarios of wearable and mobile augmented reality audio. In *the 116th Audio Engineering Society (AES) Convention* (Berlin, Germany, May 8-11 2004). paper no. 6026.
11. J. Blauert. *Spatial Hearing. The psychophysics of human sound localization*, pp. 36–200. (MIT Press, Cambridge, MA, 2nd edition, 1997).
12. R. Drullman and A. W. Bronkhorst. Multichannel speech intelligibility and talker recognition using monaural, binaural, and three-dimensional auditory presentation. *Journal of the Acoustical Society of America* **107**, pp. 2224–2235 (2000).
13. H. Gamper and T. Lokki. Audio augmented reality in telecommunication through virtual auditory display. In *Proceedings of the 16th International*

Conference on Auditory Display (ICAD), pp. 63–70 (Washington, DC, USA, 2010).
14. A. Härmä, J. Jakka, M. Tikander, M. Karjalainen, T. Lokki, J. Hiipakka, and G. Lorho. Augmented reality audio for mobile and wearable appliances. *Journal of the Audio Engineering Society* **52**, pp. 618–639 (June 2004).
15. M. Tikander. Usability issues in listening to natural sounds with an augmented reality audio headset. *Journal of the Audio Engineering Society* **57**, pp. 430–441 (June 2009).
16. M. Tikander, M. Karjalainen, and V. Riikonen. An augmented reality audio headset. In *Proceedings of the 11th International Conference on Digital Audio Effects (DAFx-08)*, pp. 181–184 (Espoo, Finland, 2008).
17. S. Müller and P. Massarani. Transfer function measurement with sweeps. *Journal of the Audio Engineering Society* **49**, pp. 443–471 (June 2001).
18. S. Vesa and T. Lokki. An eyes-free user interface controlled by finger snaps. In *Proceedings of the 8th International Conference on Digital Audio Effects (DAFx-05)*, pp. 262–265 (Madrid, Spain, 2005).
19. C. Duxbury, M. Davies, and M. Sandler. Improved time-scaling of musical audio using phase locking at transients. In *the 112th Audio Engineering Society (AES) Convention* (Munich, Germany, May 10-13 2002). preprint no. 5530.
20. U. Zölzer, editor. *DAFX: Digital Audio Effects*, pp. 151–153. (John Wiley & Sons, May 2002).
21. M. Puckette. Pure data: another integrated computer music environment. In *Proceedings of the International Computer Music Conference (ICMC)*, pp. 37–41 (Hong Kong, 1996).
22. IWPASH Organizing Committee. IWPASH 2009 International Workshop on the Principles and Applications of Spatial Hearing. `http://www.riec.tohoku.ac.jp/IWPASH/` (November 2009).
23. A.W. Bronkhorst. The cocktail party phenomenon: A review of research on speech intelligibility in multiple-talker conditions. *Acta Acustica united with Acustica* **86**, pp. 117–128 (January 2000).
24. B. Kapralos, M. R. Jenkin, and E. Milios. Virtual audio systems. *Presence: Teleoperators and Virtual Environments* **17**, pp. 527–549 (2008).
25. H. Møller, M.F. Sørensen, C.B. Jensen, and D. Hammershøi. Binaural technique: Do we need individual recordings? *Journal of the Audio Engineering Society* **44**, pp. 451–469 (1996).
26. H. Møller, C.B. Jensen, D. Hammershøi, and M.F. Sørensen. Evaluation of artificial heads in listening tests. *Journal of the Audio Engineering Society* **47**, pp. 83–100 (1999).

GENERATION OF SOUND BALL: ITS THEORY AND IMPLEMENTATION

YANG-HANN KIM[†]

Department of Mechanical Engineering, Center for Noise and Vibration Control, Korea Advanced Institute of Science and Technology, Science Town Daejeon, 305-701, Korea

MIN-HO SONG

Graduate School of Culture Technology, Korea Advanced Institute of Science and Technology, Science Town Daejeon, 305-701, Korea

JI-HO CHANG, JIN-YOUNG PARK

Department of Mechanical Engineering, Center for Noise and Vibration Control, Korea Advanced Institute of Science and Technology, Science Town Daejeon, 305-701, Korea

It is well known that the problem of generating sound in the region of interest, by using finite number of speakers, is mathematically ill-posed. With additional constraints and suitable object function given, this problem can be regarded to be well-posed. In other words, the way to drive the speakers to make desired sound field in the prescribed zone can be directly determined. This problem is called the sound manipulation problem. It is noteworthy, however, that the arrangement and radiation characteristics of speakers have to be assumed to be known or predetermined. With this assumption, we can manipulate a desired sound field in a selected zone. A novel way to generate the desired sound field in space will be introduced. Theoretical formulation of sound manipulation theory is given and a simple example – sound ball - will be introduced. The sound ball is defined with acoustic contrast control method, and it is implemented using 32 channel system.

1. Introduction

Sound manipulation is to generate a desired sound field in space using multiple sound sources. The "desired sound field" can be, in theory, any sound field in space. It can be a spatial shape of sound, which is reproduced in a zone that we want to put sound energy, or zones that can have different shape of wave

[†] E-mail : yanghann@kaist.ac.kr

field or frequency contents. For example, if we want to generate a zone with a high acoustic energy level compare to other region, it is a problem of manipulating multiple speakers to make 'acoustically bright or dark zones' (Figure 1) [1]. We can also try to manipulate sound intensity of the sound field in a selected region with desired level and direction; it is a sound intensity manipulation problem [2]. If we need to control the direction of propagating sound in some area, it is a wave front manipulation problem [3, 4]. A desired sound field is defined differently with respect to its application and the theory is still open for a new sound field satisfying the needs. However, it is noteworthy that the theory associate with these manipulation problem stems from identical mathematical formulation.

The theory starts to define the object functions. These can be, for example, acoustic energy in the zone that we defined. Then it tries to find the optimal solution that determines the magnitudes and phases of the sound sources that are arranged to be manipulated. These often are expressed in vector form, expressing the magnitude and phase relations between sound sources in rather compact and physically realizable manner.

As we mentioned earlier, the theory does not limit to generate sound field in space and time. However, in this paper, we select a "sound ball" as an example of applying sound manipulation theory. The ball (sphere shape) is the simplest example of the sound shape or zone/zones, but it represents everything related with the theory.

The "sound ball" is an "acoustically bright" sphere shaped region; a zone with high acoustic potential energy density. The sound ball can be used to give a listener in the zone to have a private listening experience. Again, the theory does not limit to generate sound field in space. It means that the location of desired sound ball is not restricted with loudspeaker characteristic and location - the ball can be moved in space. This will lead us not only just making certain sound zone/zones in the desired position and space, but having new opportunities to use the ball in an artistic performance or even play with the ball by moving it freely in space.

We have attempted to make a sound ball by using *acoustic contrast control* algorithm [1]. 32 loudspeakers were used in reverberant condition [5].

(a) (b)

Figure 1. Examples of sound manipulation : (a) Brightness Control, (b) Contrast control

2. Mathematical Formulation

2.1. *Kirchhoff-Helmholtz Integral Equation (Ideal Case)*

Figure 2 shows a system with arbitrary boundary condition. Let $P(\vec{r};f)$ be a complex valued pressure at $\vec{r} \in S_0$ with frequency f and $G(\vec{r}\,|\,\vec{r}_0;f)$ is a Green's function. Then we can predict sound pressure at $\vec{r}$ ($P(\vec{r};f)$) by using the Kirchhoff-Helmholtz integral equation with respect to frequency f, that is,

$$P(\vec{r};f) = \int_{\partial S_0} \left[G(\vec{r}\,|\,\vec{r}_0;f) \frac{\partial P(\vec{r})}{\partial\ n}\bigg|_{(\vec{r}=\vec{r}_0;f)} - P(\vec{r}_0) \frac{\partial G(\vec{r}\,|\,\vec{r}_0;f)}{\partial\ n} \right] dS \,. \quad (1)$$

Eq. (1) shows that the sound pressure $P(\vec{r};f)$ at any point in S_0 is determined if we know the sound pressure $P(\vec{r}_0;f)$ and velocity $\left.\partial P(\vec{r})/\partial n\right|_{(\vec{r}=\vec{r}_0;f)}$ on boundary ∂S_0. The vector $\vec{r}_0$ denotes a position vector on the boundary. With Neumann boundary condition ($\partial G(\vec{r}\,|\,\vec{r}_0;f)/\partial n = 0$), the equation can be simplified as

$$P(\vec{r};f) = \int_{\partial S_0} \left[G(\vec{r}\,|\,\vec{r}_0;f) \frac{\partial P(\vec{r})}{\partial\ n}\bigg|_{(\vec{r}=\vec{r}_0;f)} \right] dS \,. \quad (2)$$

The derivative $\partial P(\vec{r}) / \partial n$ represents a surface normal velocity on the boundary. It means that the sound pressure in S_0 is a function of source velocity on the boundary. Eq. (2) essentially says that any desired sound field can be manipulated in the region if we can control the velocity values at all point on the boundary.

In sound manipulation theory, the contour connecting the volume velocity sources (loudspeakers) act as a boundary ∂S_0. Velocity values at ∂S_0 – the magnitudes and phases information - are controlled in order to generate desired sound field in the region S_0. We can select any region $V \subset S_0$ to be controlled. In generation of sound ball, region is ball shape, centered at $\vec{r}_c$ with radius R (Figure 2, (b)). Therefore, the domain of Eq. (2) is V_{ball}, that is,

$$P(\vec{r};f) = \int_{\partial S_0} \left[G(\vec{r} \mid \vec{r}_0;f) \left. \frac{\partial P(\vec{r})}{\partial \ n} \right|_{(\vec{r}=\vec{r}_0;f)} \right] dS, \quad |\vec{r}-\vec{r}_c| \leq \mathrm{R}. \quad (3)$$

But, in practice, finite number of sources has to be used to generate a desired sound field and finite number of point in control space will be selected to sample spatial information of the control zone. Therefore, Eq. (2) and Eq. (3) have to be discretized in space.

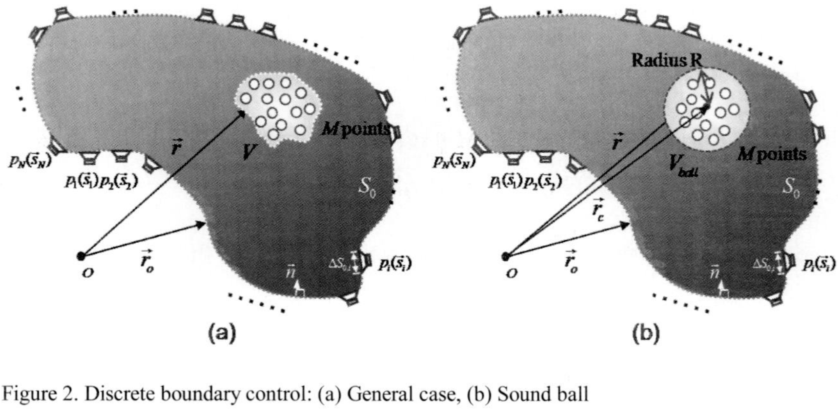

Figure 2. Discrete boundary control: (a) General case, (b) Sound ball

2.2. *Pressure Field Representation in Discrete Case*

Let's assume that we can only control velocities at finite points on the boundary. And also let $\mathbf{s} = \{\vec{s}_i\}_1^N$ be a location set for N control sources. Sound field are sampled at M control points, which we call as a measurement points and they are denoted as $\mathbf{r} = \{\vec{r}_i\}_1^M$.

In Eq. (2), position information of $\vec{r}$ and $\vec{r}_0$ are needed to calculate the Green's function. In practice, we measure the Green's function, which is a transfer function between each source and M measurement points. Let $h(r_i;s_j;f)$ denote the single transfer function value between ith measurement point and jth control source with respect to frequency f. The vector $\mathbf{h_j} = \{h(r_i;s_j;f)\}_{i=1}^M$ represents a transfer function sequence or set between all points in the desired region and jth control source. Then we can define a transfer function matrix with respect to control region, that is

$$\mathbf{H} = \begin{bmatrix} | & | & & | \\ \mathbf{h_1} & \mathbf{h_2} & \cdots & \mathbf{h_N} \\ | & | & & | \end{bmatrix} \tag{4}$$

The matrix $\mathbf{H}$ is an M x N linear operator which maps velocity value set of sound sources with sound pressure field at M measurement points. Each column vector represents the relation between the zone and single loudspeakers. Note that the matrix $\mathbf{H}$ constructs the relation between the target zone and the multiple sources (Figure 3).

The pressure field vector $\mathbf{p}$ of M points in control region is represented by multiplication of $\mathbf{H}$ with control vector $\mathbf{q} = (q_1, q_2, \cdots, q_N)^T$, which denotes the velocity value vector of control sources (Eq. (4))

$$\mathbf{p} = \mathbf{Hq} = \begin{bmatrix} | & | & & | \\ \mathbf{h_1} & \mathbf{h_2} & \cdots & \mathbf{h_N} \\ | & | & & | \end{bmatrix} \begin{bmatrix} q_1 \\ q_2 \\ \vdots \\ q_N \end{bmatrix} = \sum_{i=1}^{N} q_i \mathbf{h_i} \tag{5}$$

We can see that Eq. (4) is a discrete representation of Kirchhoff-Helmholtz integral equation with Neumann boundary condition for selected zone.

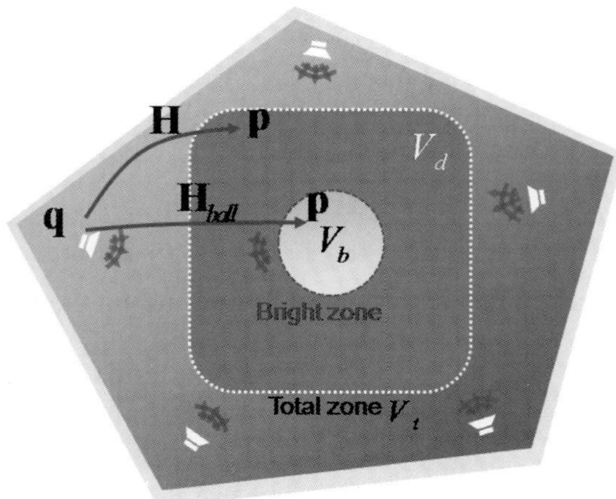

Figure 3. The transfer function is function of zone

In the case of sound ball generation, everything holds the same. But, the transfer function matrix $\mathbf{H}$ for sound ball changes. If we denote $\mathbf{H}_{ball}$ as a transfer function matrix for generating the sound ball, that is,

$$\mathbf{H}_{ball} = \begin{bmatrix} | & | & & | \\ \mathbf{h_1} & \mathbf{h_2} & \cdots & \mathbf{h_N} \\ | & | & & | \end{bmatrix}, \mathbf{h_j} = \{h(\vec{r}_i;\vec{s}_j;f), \vec{r}_i \in V_{ball}\}_{j=1}^{N} \quad .(6)$$

Note that the dimension of $\mathbf{H}_{ball}$ is determined by the number of points in V_{ball}. If the number of measure points in the ball is M', the matrix has M' x N dimension.

2.3. *Energy Density Control: Brightness and Contrast Control [1]*

We have now expressed the cause and effect relation between sources and sound pressure at the position of interest, in terms of selected acoustic and geometrical variables. The next is to express or define a measure; for example, ''acoustical brightness''. The first choice to express the overall acoustical brightness of a zone will be space averaged potential energy density, as a first approximation [1]. The acoustic potential energy density is proportional to square of complex valued pressure magnitude. Note that actually, Eq. (4) is a matrix equation denoting M equations for each measurement point. For arbitrary single measurement point x, Eq. (4) is simplified as,

$$p(x) = \sum_{i=1}^{N} q_i h(x; s_i; f) \tag{7}$$

The acoustic potential energy density in desired zone (a bright zone V_b) is defined as a space averaged integral of square of pressure magnitude over V_b that is, [1]

$$e_b = \frac{1}{V_b} \int_{V_b} | p(\vec{r}) |^2 \, dV \tag{8}$$

This is a representation of acoustic brightness for continuous measurement point case. If we assume that the number of measurement point is proportional to the control region volume, space averaged potential energy density for discrete representation is

$$\begin{aligned} e_b &\equiv \frac{1}{M} \|\mathbf{p}\|^2 \\ &= \frac{1}{M} (\mathbf{Hq})^{\mathbf{H}} \mathbf{Hq} \\ &= \mathbf{q}^{\mathbf{H}} (\frac{1}{M} \mathbf{H}^{\mathbf{H}} \mathbf{H}) \mathbf{q} \\ &= \mathbf{q}^{\mathbf{H}} \mathbf{R_b} \mathbf{q} \end{aligned} \tag{9}$$

where M is the number of measurement point within the selected zone. The brightness of V_b is represented with a vector form. The superscript $*^{\mathbf{H}}$ denotes the Hermitian operator. Note that $\mathbf{p}$ is a vector with M element, $\|\mathbf{p}\|^2$ corresponds to the integral defined in Eq. (6). For simplicity, $\mathbf{R_b} = \frac{1}{M} \mathbf{H}^{\mathbf{H}} \mathbf{H}$

is called “spatial correlation matrix” for the control zone and $\mathbf{R_b}$ is an N×N Hermitian matrix which represents the spatial correlation of the sound pressure field in the control zone and control sources.

Now, we can define the acoustic brightness. Acoustic brightness α is defined by the ratio between the potential energy density of the bright zone with the input power $J_0 = \mathbf{q}^{\mathbf{H}}\mathbf{q}$, that is, [1]

$$\alpha = \frac{e_b}{J_0} = \frac{\mathbf{q}^{\mathbf{H}}\mathbf{R_b}\mathbf{q}}{\mathbf{q}^{\mathbf{H}}\mathbf{q}} \tag{10}$$

Acoustic contrast β is defined by ratio of the potential energy density between the bright and the dark zone as follows: [1]

$$\beta = \frac{e_b}{e_d} = \frac{\mathbf{q}^{\mathbf{H}}\mathbf{R_b}\mathbf{q}}{\mathbf{q}^{\mathbf{H}}\mathbf{R_d}\mathbf{q}} \tag{11}$$

$\mathbf{R_d}$ is a spatial correlation matrix for the dark zone. Note that Eq. (7) stands for arbitrary region, $\mathbf{R_d}$ can be calculated in same manner with $\mathbf{R_b}$.

Since this is a generalized eigenvalue problem, the solution maximizing the acoustic contrast β is an eigenvector corresponding to maximum eigenvalue of the matrix $\mathbf{R_d^{-1}}\mathbf{R_b}$.

It is noteworthy that Eq. (7) can be used for any type of shape or zone/zones, which have different correlation matrix. For specific case of generating sound ball, we can select bright zone V_b as V_{ball},

$$V_b = \{\vec{r} \mid |\vec{r} - \vec{r}_c| \le R\} \tag{12}$$

From Eq. (6), the potential energy density of the ball is,

$$e_b = \frac{1}{4\pi R^3/3} \int_{|\vec{r}-\vec{r}_c| \le R} | p(\vec{r}) |^2 \, dV \tag{13}$$

For discrete representation, we can construct correlation matrix with respect to discrete points in V_{ball}. Assume $\mathbf{H}_{ball}$ is the transfer matrix for sound ball V_{ball} and its dimension is M' x N , the correlation matrix for sound ball is,

$$R_b = \frac{1}{M'} H_{ball}^H H_{ball} \tag{14}$$

Now we are ready to generate sound ball with *acoustic brightness control* method [1]. Also, with the constraint of the dark zone, we can generate sound ball with *acoustic contrast control* method [1]. Note that the correlation matrix is a function of $\vec{r}_c$ and R. For moving sound ball, these matrices will be function of time and trajectory that we want to move.

3. Implementing Sound Ball

3.1. *The Sound ball with Acoustic Contrast Control*

As we mentioned in the introduction section, we attempted to implement the sound ball as a practical example of the sound manipulation theory.
The sound ball is a sphere shaped sound zone with high energy density compare to the other region. With the sound ball, we can generate a personal listening zone (Figure 4, (a)). Furthermore, by moving the sound ball one can experience a virtual sound source (Figure 4, (b)).

To generate the sound ball, we decided to use the *acoustic contrast control method.* The *acoustic contrast control method* is known to be effective method in focusing acoustic energy on designated region [6, 7]. We aim to generate a sound ball with different in location, size and frequency.

For simplicity, instead of 3-D sound ball, 2-D sound "circle" is implemented due to huge measurement point problem. Although the measurement region is limited as a plane but from the linearity, we can easily predict that 3-D sound ball is just an extension of 2-D case.

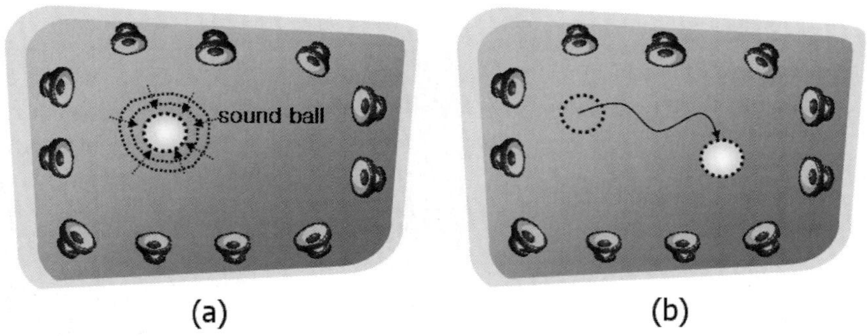

Figure 4. The sound ball

3.2. *Experimental Setup*

The zone of interest is surrounded by 32 loudspeakers (Figure 5). Transfer functions are measured in 1m x 1m region of the system. Total 729(27x27) measurement points are equally distributed in square shaped region. The number of measurement points is selected in order to avoid spatial aliasing effect. Transfer functions are measured step by step using chirp-signal.

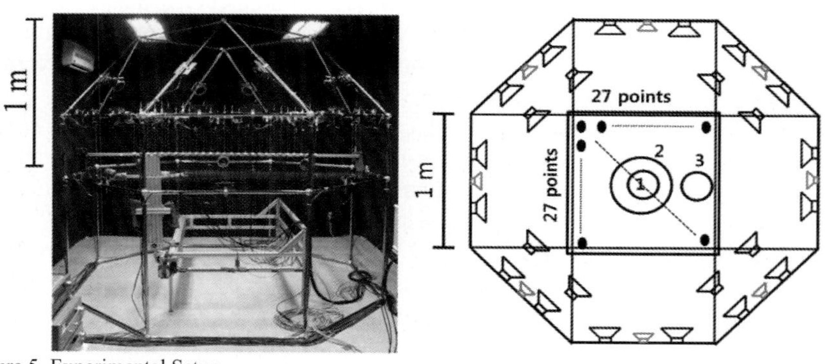

Figure 5. Experimental Setup

3.3. *Implementation of the Sound Ball*

We have generated three kind of sound ball. The first one is a ball located in the center with radius 10cm (Figure 5, “1”) and the second one has same location but with radius is 20cm (Figure 5, “2”). The last one is located in 50cm from the center (Figure 5. “3”) with radius 10cm. Total 33 sound ball is generated from frequency 500Hz to 1500Hz with 100Hz interval for case 1, 2 and 3. For frequency 500Hz, 1000Hz and 1500Hz, the results are shown below.

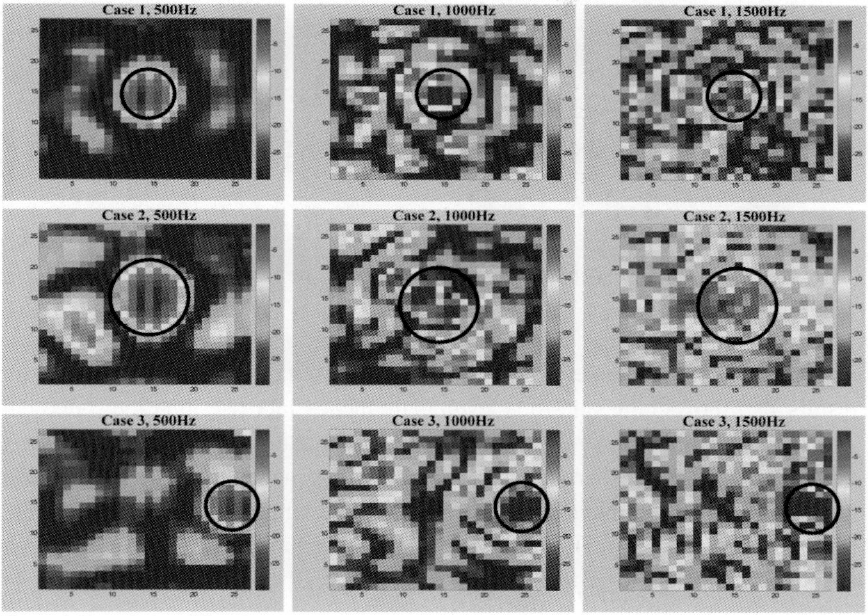

Figure 6. Experimental Results

Figure 6 shows that as frequency increases, the sound ball has no longer preserves its shape. This is because the sound ball is defined to maximize the space averaged potential energy density instead of keeping the ball shape. Although the solution is an optimal solution for maximizing contrast, we can see that at high frequency, the dark zone is brightened and overall contrast is low. To analyze these results, comparison with the computer-simulated results would be helpful.

3.4. *Comparison with Computer-Simulation Results*

Figure 7 shows computer-simulated results for case 1 for 500Hz, 1000Hz and 1500Hz. The sources are assumed to be 32 monopoles with same location as in Figure 5 and free-field green's functions are used.

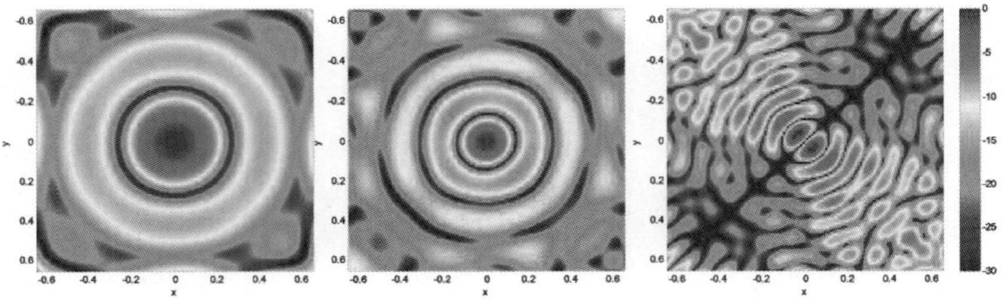

(a) 500Hz (b) 1000Hz (c) 1500Hz

Figure 7. Computer Simulation Results

Figure 7 shows that with a free field condition, different sound field is reproduced from the experimental results. But we can observe that simulation results also shows high side-lobe level as frequency increases and the sound ball has no longer preserves its shape. This phenomenon is assumed to be related with transfer function characteristic, such that if frequency increases, normalized radius of bright zone increases with respect to the length of the wavelength and transfer functions are likely to be orthogonal to each other in the region, so that the sound focusing effect decreases (See Appendix).

4. Summary

As an example of sound manipulation theory, we introduced a way to manipulate a "sound ball" in space. Acoustic contrast control was used to manipulate sound ball in space. A sound ball was implemented using KAIST-NOVIC multichannel system. From the results, we could observed that in the low frequency range (500~800Hz), a ball shaped bright zone was generated. But, in the high frequency range (above 800Hz), we noticed high side-lobe energy level and the ball no longer preserves its shape. Since this phenomenon observed in both numerical, experimental results, we predict that this is related with the transfer function set characteristic. To generate a sound ball in the high frequency region, the linear dependency relation between transfer functions should be changed. How can we change the transfer function characteristics? Note that the transfer function characteristic is related with source and zone. This leads us to an array design problem and control zone aperture problem. These problems are open problems for contrast control and leaves further work to do.

Appendix

From Eq. (5), following inequality holds [8].

$$\frac{1}{\left\|(H^*H)^{-1}\right\|}\cdot\sum_{i=1}^{N}|q_i|^2 \le \|\mathbf{p}\|^2 \le \left\|H^*H\right\| \cdot\sum_{i=1}^{N}|q_i|^2 \qquad \text{(A.1)}$$

The inequality (A.1) represents that the potential energy density we can generate in the desired region is determined by the constant $A = 1/\left\|(H^*H)^{-1}\right\|$ and the constant $B = \left\|H^*H\right\|$ and input power $\sum_{i=1}^{N}|q_i|^2$. The constant $B = \left\|H^*H\right\|$ is called the upper bound of basis constant for sequence $\{\mathbf{h_k}\}_{k=1}^{N}$ and $A = 1/\left\|(H^*H)^{-1}\right\|$ is called the lower bound. These constant A, B is known to be related with linear dependency, or correlation of the vector set [8].

For example, if the vector set $\{\mathbf{h_k}\}_{k=1}^{N}$ is an orthogonal set, the constant A=B=1 holds (Parseval's Identity). It means that the brightness is not controllable when the basis set $\{\mathbf{h_k}\}_{k=1}^{N}$ is orthogonal set. On the contrary, if vectors are correlated and linear dependency increases, we can generate more brightness in the zone. It is known that if the zone aperture increases, transfer function set $\{\mathbf{h_k}\}_{k=1}^{N}$ are likely to be orthogonal and therefore the possible potential energy density we can generate decreases.

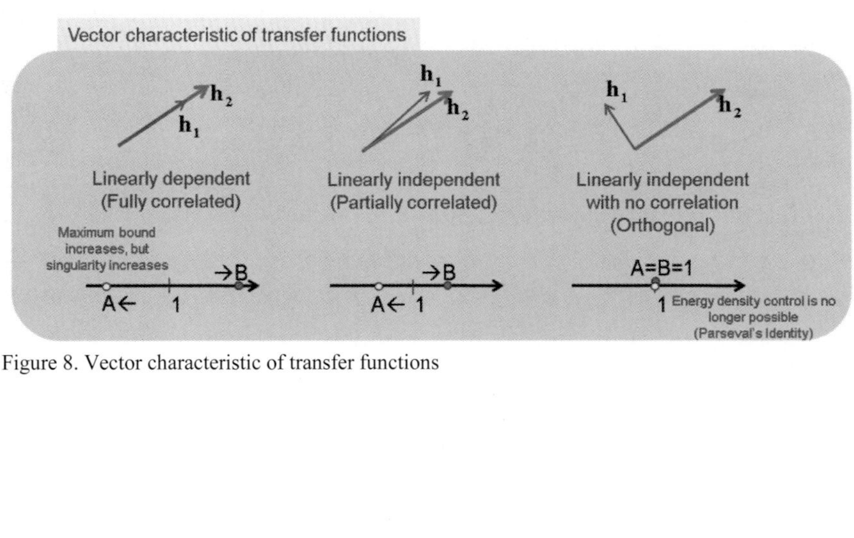

Figure 8. Vector characteristic of transfer functions

References

1. J.-W. Choi and Y.-H. Kim, "Generation of an acoustically bright zone with an illuminated region using multiple sources," *J. Acoust. Soc. Am.* **111**, 1695–1700 , 2002.
2. J.-W. Choi and Y.-H. Kim, "Manipulation of sound intensity within a selected region using multiple sources," *J. Acoust. Soc. Am.* **116(2),** 843-852, 2004
3. J.-W. Choi, Spatial Manipulation and Implementation of Sound, (Ph.D. Thesis, Korea Advanced Institute of Science and Technology, Daejeon, Korea, 2005)
4. J.-H. Chang and Y.-H. Kim, "A planewave generation method by wavenumber domain point focusing," submitted, *J. Acoust. Soc. Am.* 2009
5. http://soundmasters.kaist.ac.kr, 2009
6. J.-H. Chang, C.-H. Lee, J.-Y. Park, and Y.-H. Kim, "A realization of sound focused personal audio system using acoustic contrast control," *J. Acoust. Soc. Am.* **125(4)**, 2091-2097, 2009.
7. S. J. Elliott and M. Jones, "An active headrest for personal audio," *J. Acoust. Soc. Am.* **119**, 2702-2709, 2006
8. Ole Christensen, An Introduction to Frames and Riesz Bases (Birkhauser, Boston, 2003)

ESTIMATION OF HIGH-RESOLUTION SOUND PROPERTIES FOR REALIZING AN EDITABLE SOUND-SPACE SYSTEM

T. OKAMOTO[1,2], Y. IWAYA[1,3] and Y. SUZUKI[1,3]

[1] *Research Institute of Electrical Communication, Tohoku University,*
[2] *Graduate School of Engineering, Tohoku University,*
[3] *Graduate School of Information Sciences, Tohoku University,*
2-1-1 Katahira, Aoba-ku, Sendai, 980-8577, Japan
E-mail: {okamoto@ais., iwaya@, yoh@}riec.tohoku.ac.jp

Using signal processing and recording with numerous microphones, a sound field can be decomposed into its attributes such as original sound source signals, sound source positions, directivity of sound sources, early reflections, and late reverberation. Sound field editing would be highly versatile after such decomposition. Moreover, the original sound field and a modified sound field can be synthesized flexibly by modifying and exchanging attributes. We designate such a system as an editable sound-space system. To realize such systems, we are developing signal processing techniques to decompose such properties from recorded sound based on a surrounding array of 157 microphones. After introducing this system, this report describes the estimation of source positions, original sound source signals, and directivity of sound sources.

Keywords: Surrounding microphone array; editable sound-space system; sound localization; dereverberation; directivity of sound source.

1. Introduction

A sound field in an actual environment comprises various sound objects. Each has its own properties such as original sound source signals, sound position, and directivity. Moreover, each sound object is affected by a room's acoustic features (reflections and reverberation). If signal processing techniques to decompose received sound signals into these properties and to recompose them after versatile editing of these properties were realized, then not only the original sound field[1] but also a modified sound field could be synthesized. Moreover, such a sound field could be modified into a form that would be irreproducible using conventional sound field reproduction techniques. We designate such a system as an editable sound-space system.

Figure 1 portrays a conceptual diagram of an editable sound-space system. To realize such a system, we have been striving to develop methods to decompose such properties from recorded sound.

Previous studies have often treated sound sources as ideal point sources that radiate spherical sound waves to all directions equally, i.e. an omnidirectional characteristic. In an actual environment, sound sources are not point sources: their radiation has directivity. Musical instruments have directivity.[2] Therefore, techniques to estimate directivity from recorded sounds and reproduction techniques reflecting estimated and edited directivity are strongly demanded. Such techniques seem promising for use, especially in combination with 3D image display arrangements in which viewers can move around displayed 3D objects. This type of 3D image display, for which holography might be a key technology, is considered to be an important type of medium for use in future 3D displays with screen images. Using a combination of visual and auditory displays, people can move around a virtual object just as they are able to do in a real environment.

Therefore, in this paper, we propose a system to record sound information including the directivity of a sound source, in addition to estimation methods of sound source properties such as sound source positions, original sound source signals, and sound source directivity.

2. Construction of the surrounding microphone array system

To record sound information including the sound source directivity and to realize an editable sound-space system as described above, we constructed a test-bed room for sound acquisition in which a microphone array consisting of 157 microphones (Type 4951; Bruel and Kjaer) is installed on all

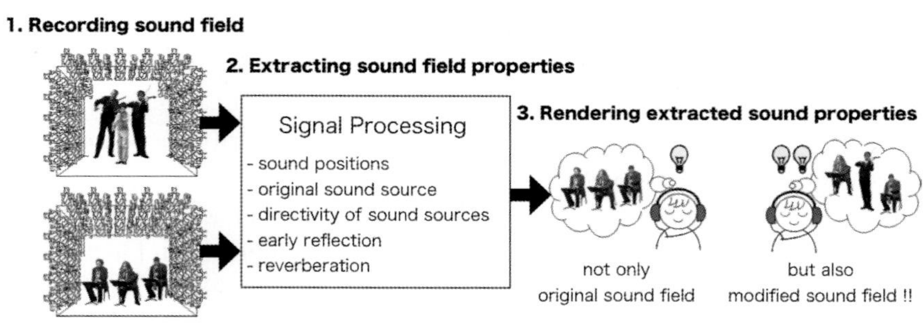

Fig. 1. Conceptual diagram of an editable sound space system.

four walls and the ceiling of a room. We designate this as a 'surrounding microphone array'. All microphones are installed 30 cm inside from all four walls and the ceiling using pipes. They are separated from each other by 50 cm. The microphone arrangement is portrayed in Fig. 2. We introduced a recording system for this microphone array to enable synchronous recording of 157 channels at the sampling frequency of 48 kHz with the linear PCM audio format.

3. Estimation of sound source properties

3.1. *Directivity model of a sound source in a room*

We consider the directivity of a direction θ_0 from the sound source position to a microphone of the surrounding microphone array. The original sound signal $s(n)$ is radiated with directivity $d(\theta_k, n)$ for a direction θ_k, where n represents time. The wave component radiated to direction θ_k can be described as $s(n) * d(\theta_k, n)$, where '$*$' signifies the convolution. Each component related to θ_k, $s(n) * d(\theta_k, n)$ is convolved with room impulse responses $h(\theta_k, n)$ and arrives at the microphone. Consequently, the output signal of the microphone $x(n)$ can be described as a summation of each component, as

$$x(n) = \sum_{k=0}^{\infty} s(n) * d(\theta_k, n) * h(\theta_k, n) = s(n) * \sum_{k=0}^{\infty} \{d(\theta_k, n) * h(\theta_k, n)\}. \quad (1)$$

Here, it is necessary to estimate $d(\theta_0, n)$. The equation can be described as

$$x(n) = s(n) * \{d(\theta_0, n) * h(\theta_0, n) + \sum_{k=1}^{\infty} d(\theta_k, n) * h(\theta_k, n)\}. \quad (2)$$

If the distance between the sound source and the microphone is $r(\theta_0)$, then Eq. 2 can be described as

$$x(n) = s(n) * \left\{ d(\theta_0, n) \cdot \frac{1}{r(\theta_0)} + d(\theta_0, n) * h'(\theta_0, n) + \sum_{k=1}^{\infty} d(\theta_k, n) * h(\theta_k, n) \right\} \quad (3)$$

$$= s(n) * \{h_D(n) + h_R(n)\} \quad (4)$$

$$= s(n) * h(n), \quad (5)$$

where $h_D(n) = d(\theta_0, n)/r(\theta_0)$ expresses the direct wave component, $h'(\theta_0, n)$ represents the reflection component of $h(\theta_0, n)$, $h_R(n)$ denotes the sum of the second and third term in Eq. 3, and $h(n)$ in Eq. 5 is the abbreviation of the component in Eq. 4.

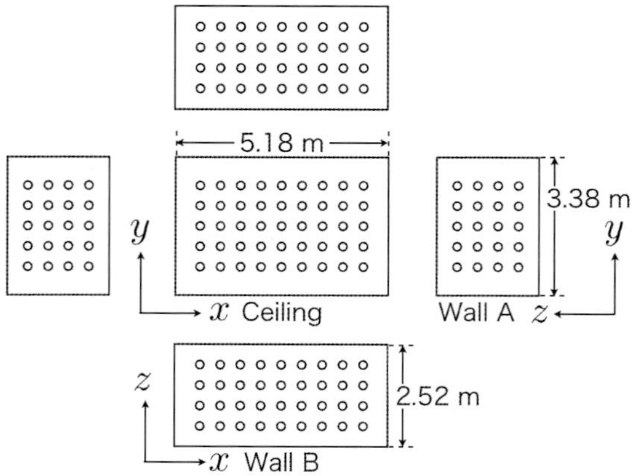

Fig. 2. Appearance of the surrounding microphone array.

3.2. *Estimation of sound source positions*

During estimation of the sound source positions in a room, reflected sounds reduce the accuracy of estimation results. To solve this problem, the use of a spatial smoothing technique has been proposed.[3] However, this technique is only useful to estimate the directions of arrival of sound sources: not their positions. Therefore, we developed a new method through integration of the multiple signal classification (MUSIC) algorithm[4] and delay-and-sum[5] beam-forming. We named it "rearrangement and presmoothing for MUSIC" (RAP-MUSIC). Measurement in an actual room showed that, using this method, source positions in a reverberant room can be detected more accurately than when using the conventional MUSIC method.[6]

3.3. *Estimation of a sound source signal*

To estimate an original sound source signal in a reverberant environment, dereverberation algorithms are an important tool to estimate sound field properties correctly. Although many methods have been proposed, the linear-predictive multichannel equalization (LIME) algorithm[7] is among the most promising. Using the original LIME algorithm, however, the dereverberation performance is inadequate for sampling frequencies higher than or equal to 16 kHz if the source signal is colored, as are speech signals and musical signals. An algorithm that is useful at high sampling frequencies is necessary to process sound that richly includes high-frequency components such as musical sounds. Therefore, to solve this problem, we propose to improve LIME with preprocessing of all input signals $x_i(n)$ with filters to whiten their spectra. The pre-whitening filter was calculated as an averaged AR polynomial[8,9] of input signals $x_i(n)$. Each linear prediction coefficient $b_i(n)$ is calculated from each input signal by solving the Yule–Walker equation using the Levinson–Durbin algorithm.[10]

$$x_i(n) = \sum_{k=1}^{N} b_i(k)x_i(n-k) + x_i'(n), \tag{6}$$

$$c(z) = 1 - \{E\{b_i(1)\}z^{-1} + \cdots + E\{b_i(n)\}z^{-N}\} \tag{7}$$

Thereby, this algorithm can function for colored signals at high sampling frequencies. We named this algorithm White-LIME.[11]

3.4. *Estimation of sound source directivity*

Several studies have been conducted to measure the directivity of the sound source in an anechoic environment.[12,13] However, Nakadai *et al.* proposed a method to estimate the sound source position and the front direction of the sound source simultaneously;[14] other directions were not considered. No previous report has described estimation of the all-around directivity in a reverberant environment. Therefore, we propose a simple and novel method to estimate the directivity of a sound source in a room environment from recorded signals using information of the estimated source position and the original sound signal provided respectively by RAP-MUSIC and White-LIME.

3.4.1. *Method of estimation of the directivity component from impulse response*

If the original signal $s(n)$ can be estimated, then $h(n)$, which is the impulse response between source and a microphone, can also be estimated by deconvolving the microphone output by $s(n)$. The early response of $h(n)$ would indicate $h_{Di}(n)$ if the length of $h_{Di}(n)$ were shorter than the time at which other reflective waves including $H_{Ri}(n)$ come. With the surrounding microphone array, because the microphone is installed from the wall at distance d, the minimum reflection path is not oblique-incidence reflection but rather head-on incidence reflection. Therefore, the minimum arrival time interval between the direct sound and the reflected sound is $t = 2d/c$, where c stands for the acoustic wave velocity. Consequently, $h_{Di}(n)$ can be extracted as a section of duration of $t = 2d/c$ from the first response of the early response. As inferred from Eqs. 3 and 4, the amplitude of each $h_{Di}(n)$ must be corrected using distance r_i, thereby yielding the estimated directivity as $d_i(n) = r_i h_{Di}(n)$. The distance between the source and each microphone can be estimated using RAP-MUSIC,[6] as described above.

The impulse responses $\hat{h}_i(n)$ are estimated from the observed signals $x(n)$ using the estimated sound signal $\hat{s}(n)$ obtained using White-LIME.[11] Each impulse response $\hat{h}_i(n)$ is estimated by deconvolving the microphone output signal $x_i(n)$ by this estimated source signal $\hat{s}(n)$. In practice, it is necessary to insert zeros in the first L taps of $\hat{s}(n)$ and to transpose each $x_i(n)$ and $\hat{s}(n)$ to each signal in the frequency domain $X_i(f)$ and $\hat{S}(f)$ using the discrete Fourier transpose (DFT). In calculating the DFT, the FFT orders of $x_i(n)$ and $\hat{s}(n)$ are expected to be the same. From $X_i(f)$ and $\hat{S}(f)$, we obtained the transfer function in the frequency domain $\hat{H}_i(f) = X_i(f)/\hat{S}(f)$. Then, each impulse response $\hat{h}_i(n)$ in the time domain is obtained using the Inverse DFT (IDFT). Figure 3 presents a schematic diagram that is useful to estimate impulse responses from the estimated source signal.

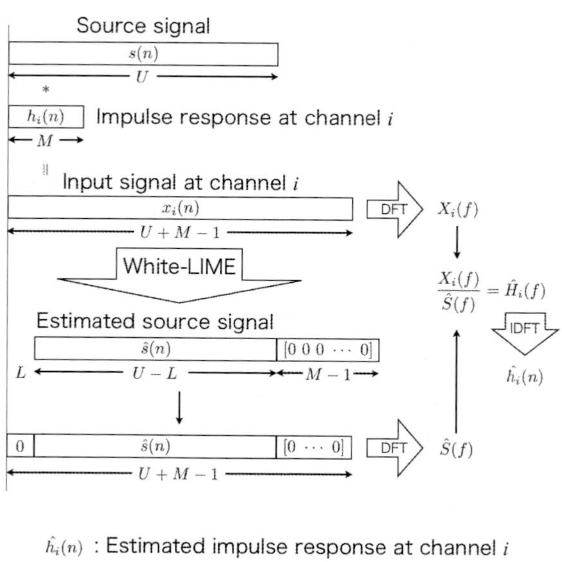

Fig. 3. Extracting impulse response $\hat{h}_i(n)$ from the deconvolving estimated source signal $\hat{s}(n)$ to the observed signal $x_i(n)$.

3.4.2. *Simulation*

To evaluate the effectiveness of the proposed method, a computer simulation was performed using impulse responses measured in the actual environments. The impulse responses including both the directivity of the sound source and the reflected properties were measured using recorded signals from the surrounding microphone array system. The sampling frequency of all signals was set to 44.1 kHz. As a reference, we measured the impulse responses from the loudspeaker to all directions to obtain the directivity pattern of this loudspeaker in an anechoic room in advance. In this measurement, the front direction is 0 deg: measurements were taken from 0 deg to 180 deg with a clockwise rotation with 15 deg-steps; 13 directional measurements were taken. Examples of the amplitude–frequency response measured in the anechoic room are portrayed in Fig. 4 (a). Figure 5 portrays the arrangement of the loudspeaker and 28 microphones in the room. The reverberation time of this room was 0.15 s. Consequently, the length of the room impulse responses was 6615 points. The observed signal at each microphone $x_i(n)$ was obtained by convolving the source signal $s(n)$ by each measured impulse response $h_i(n)$. The source signal was a fragment of a musical piece[15] (duration: 2.7 s, classical music). The estimated source signal $\hat{s}(n)$ was obtained using White-LIME. The score of the signal to distortion ratio (SDR) of the original signal $s(n)$ to the estimated signal $\hat{s}(n)$ was 57.3 dB. Each estimated impulse response $\hat{h}_i(n)$ was calculated from each observed signal $x_i(n)$ and the estimated source signal $\hat{s}(n)$. The average of the score of the SDR of each original response $h_i(n)$ to each estimated response $\hat{s}(n)$ was 62.6 dB. These results show that the estimated responses can be inferred accurately. The distance between each microphone of the surrounding microphone array system and the wall was 30 cm. Therefore, the clipping length of each estimated response $h_D(n)$ was approximately $44100 \times 2 \times 0.6/340 \approx 78$ taps. The direction–frequency responses measured in the anechoic room and in the reverberant room, and those estimated by clipping 78 taps are portrayed, respectively, in Figs. 4 (a–c). Moreover, the directivity patterns corresponding to these conditions at 2000 Hz in terms of 1/3 octave-band analysis at 2000 Hz are shown respectively in Figs. 4 (d–f). Here, panels 4 (c) and 4 (f) show the estimated directivity of the sound source using the proposed method. The results presented in Fig. 4 demonstrate that the proposed method was capable of estimating the directivity of the sound source. To confirm the effectiveness of the proposed method, similarity in terms of the nearest-neighbor method[16] was calculated with 1/3 octave-band analysis. Results presented in Fig. 6 show that the directivity patterns obtained using the proposed method exhibit high scores (similarities) at all frequency bands, whereas the patterns simply derived from the originally received signals, in which reverberation is included, exhibit low similarities at high-frequency bands.

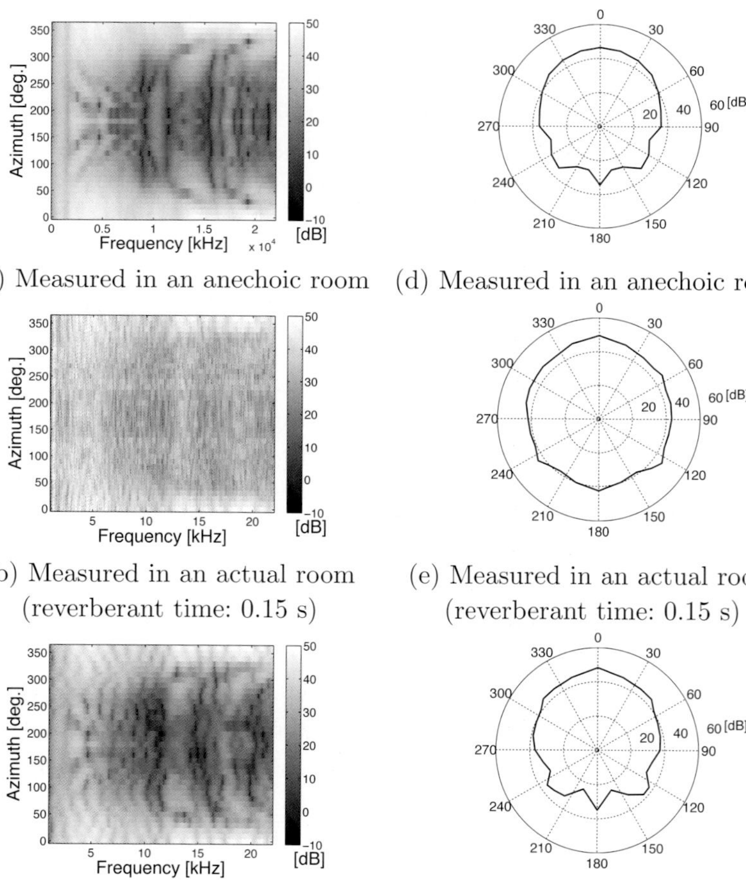

(a) Measured in an anechoic room (d) Measured in an anechoic room

(b) Measured in an actual room (reverberant time: 0.15 s) (e) Measured in an actual room (reverberant time: 0.15 s)

(c) Proposed method (f) Proposed method

Fig. 4. Amplitude–frequency characteristics: (a–c), and measured/estimated directivity at 2000 Hz in the 1/3 octave-band: (d–f).

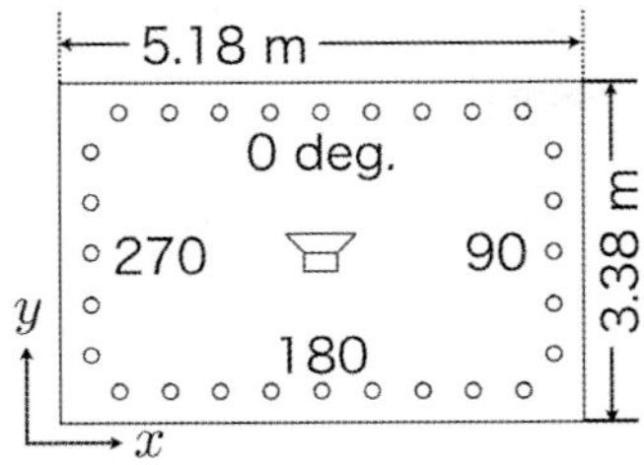

Fig. 5. Arrangement of the loudspeaker (z=1.1) and 28 microphones (z=1.0).

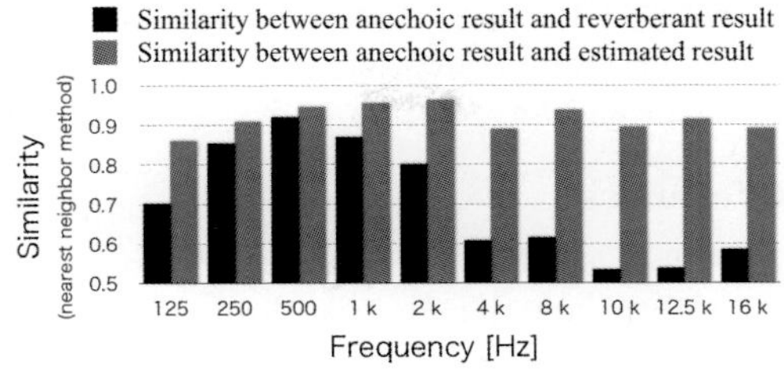

Fig. 6. Result of similarity analysis (nearest neighbor method) of the reverberant response and the response extracted using the proposed method.

4. Concluding remarks

This chapter introduces our proposal for an editable sound-space system. If a sound field could be decomposed into attributes such as original sound source signals, sound source positions, directivity of sound sources, early reflections, and late reverberation, then sound field editing would be highly versatile after such decomposition. Moreover, the original sound field and a modified sound field could be synthesized flexibly by modifying and exchanging these attributes. This is an editable sound-space system. For the realization of such systems, we first introduced a surrounding microphone array system consisting of 157 microphones installed in a room. Then we introduced signal processing techniques developed along with this microphone array to estimate sound source properties such as sound source positions, an original sound source signal and sound source directivity. Our proposed methods show high performance, even at the sampling frequency of 44.1 kHz.

As future work, a high-performance estimation method for directivity of a sound source in a noisy environment is necessary to make this method a useful one in a practical sense. In this study, the algorithms included the presumption that the sound sources were only one in the sound field. Several sound sources usually exist in actual environments. Therefore, development of a good sound source separation algorithm that is suitable for an editable sound-space system should also be addressed in future studies.

Acknowledgements

We wish to thank Prof. Masato MIYOSHI, Dr. Tomohiro NAKATANI, Mr. Keisuke KINOSHITA, Mr. Takuya YOSHIOKA, and Dr. Ryouichi NISHIMURA for their intensive discussion related to this study. This study was supported by the GCOE program (CERIES) of the Graduate School of Engineering, Tohoku University.

References

1. S. Ise, A principle of sound field control based on the Kirchhoff–Helmholtz integral equation and the theory of inverse systems, *Acustica – Acta Acustica* **85**, 78–87 (1999)
2. F. Saito, M. Kasuya, T. Harima and Y. Suzuki, Sound power levels of musical instruments and the estimation of the influence on players' hearing ability, *in Proc. Inter-Noise 2003* 2259–2266 (2003)
3. T. Shan, M. Wax and T. Kailath, On spatial smoothing for direction-of-arrival estimation of coherent signals, *IEEE Trans. Acoust. Speech Signal Process.* **33** 806–811 (1985)
4. R. O. Schmidt, Multiple emitter location and signal parameter estimation, *IEEE Trans. Antennas, Propag.* **34** 276–280 (1986)
5. J. L. Flanagan, J. D. Johnson, R. Zahn and G. W. Elko, Computer-steered microphone arrays for sound transduction in a large room, *J. Acoust. Soc. Am.* **78** 1508–1518 (1985)
6. T. Okamoto, R. Nishimura and Y. Iwaya, Estimation of sound source positions using a surrounding microphone array, *Acoust. Sci. & Tech.* **28** 181–189 (2007)
7. M. Delcroix, T. Hikichi and M. Miyoshi, Precise dereverberation using multichannel linear prediction, *IEEE Trans. Audio Speech Lang. Process.* **15** 430–440 (2007)
8. N. D. Gaubitch, P. A. Naylor and D. B. Ward, On the use of linear prediction for dereverberation of speech, *Proc. Int. Workshop Acoust. Echo Noise Control*, **1** 99–102 (2003)
9. K. Kinoshita, M. Delcroix, T. Nakatani and M. Miyoshi, Suppression of late reverberation effect on speech signal using long-term multiple-step linear prediction, *IEEE Trans. Audio Speech Lang. Process.*, **17**, 534–545 (2009)
10. L. Ljung, *System Identification: Theory for the User*, Prentice Hall, (1987)
11. T. Okamoto, Y. Iwaya and Y. Suzuki, New blind dereverberation method based on multichannel linear prediction using pre-whitening filter, *in Proc. 2009 Spring Meet. Acoust. Soc. Jpn.* 675–676 (2009) (in Japanese)
12. B. F. G. Katz and C. d'Alessandro, Directivity measurements of the singing voice, *Proc. ICA 2007*, (2007)
13. D. Devoy and F. Zotter, Acoustic center and orientation analysis of sound-radiation recording with a surrounding spherical microphone array, *Proc. the 2nd Int. Symp. Ambisonics and Spherical Acoust.*, (2010)
14. K. Nakadai, H. Nakajima, K. Yamada, Y. Hasegawa, T. Nakamura and H. Tsujino, Sound source tracking with directivity pattern estimation using a 64 ch microphone array, *IROS 2005* 1690–1696 (2005)
15. http://staff.aist.go.jp/m.goto/RWC-MDB/index.html
16. P. J. Clark and F. C. Evans, Distance to nearest neighbor as a measure of spatial relationships in populations, *Ecology* **35** 445–453 (1954)

Section 4

Applying Virtual Sound Techniques in the Real World

BINAURAL HEARING ASSISTANCE SYSTEM BASED ON FREQUENCY DOMAIN BINAURAL MODEL

T. USAGAWA* and Y. CHISAKI**

Graduate School of Science and Technology, Kumamoto University,
Kumamoto, 860–8555, Japan
**E-mail: tuie@cs.kumamoto-u.ac.jp*
***E-mail: chisaki@cs.kumamoto-u.ac.jp*

Based on the spatial selectivity using binaural hearing, a binaural hearing assistance system has been proposed using a frequency domain binaural model. This system has a simple but very stable howling canceller. This chapter presents a comprehensive discussion of the design of this system and an evaluation of its performance.

Keywords: binaural hearing assistance system, frequency domain binaural model, binaural howling canceller, spatial selectivity

1. Introduction

Binaural hearing assistance systems have become increasingly popular year after year using up-to-date technologies such as a digital signal processing, modern high-energy density batteries, and sophisticated power management. Although those systems use two microphone inputs to provide directional selectivity based on binaural hearing models of various types, some problems remain to be solved, especially in relation to robust howling cancellation.

Various binaural hearing models have been proposed since the 1940s, such as a coincidence-based model by Jeffress.[1] Blauert[2] and his colleagues made continuous contributions[3],[4] and as a complete model, Bodden proposed the cocktail party processor[5] in 1993. Aside from this series of activities, many others have been undertaken related to the binaural model. Some of those were reviewed in the recently published book edited by DeLiang Wang and Guy J. Brown.[6] Among those approaches, the frequency domain binaural model (FDBM)[7] has advantages of signal segregation based on two-dimensional direction, i.e. azimuth and elevation, without a heavy computational load. In addition, for the application of hearing aids, the FDBM has a very efficient and robust howling cancellation method based on the maximum of realistic interaural level differences for each frequency bin.[8]

In this paper, a binaural hearing assistance system is proposed using the frequency domain binaural model, which has a simple but very robust howling cancellation mechanism. The performance of this system is evaluated using various measures including PESQ[9] because PESQ considers human perceptual and psychoacoustic models; moreover, PESQ provides a similar score to the mean opinion score (MOS) given by a human listener.

2. Hearing assistance system based on FDBM

Figure 1 shows a block diagram of FDBM for a hearing assistance system including a howling canceller. This model consists of four sub-blocks. Although the directional information is obtainable in the higher frequency range such as greater than 5 kHz, especially as the spectral peaks and dips,[10] 16 kHz sampling is used because the major target of hearing assistance systems is a frequency range corresponding to that of a conventional telephone.

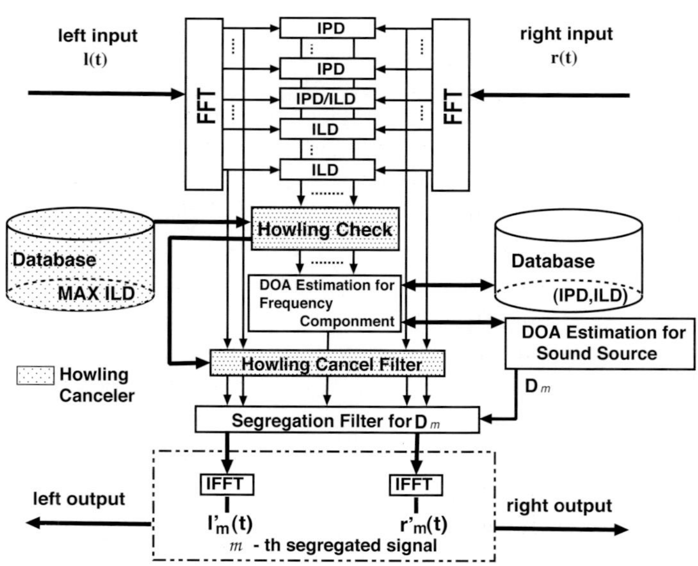

Fig. 1. Block diagram of frequency domain binaural model with a howling canceller.

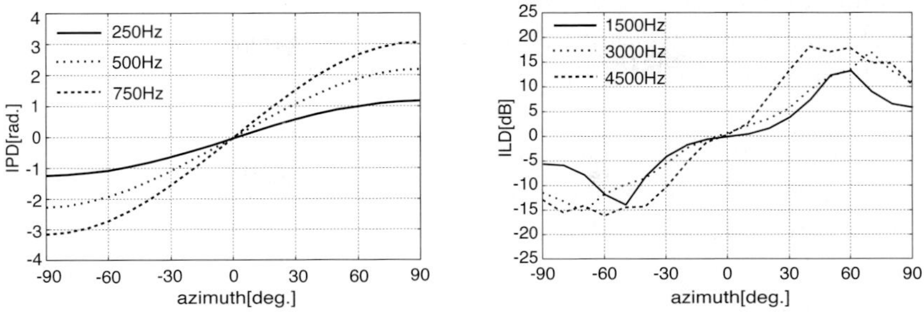

(a) Interaural phase difference (b) Interaural level difference

Fig. 2. IPD and ILD at elevation $\psi = 0°$ of a KEMAR dummy-head HRTF database provided by the MIT Media Laboratory.

2.1. *FFT Analysis Sub-block*

Assuming that the input signals, $l(n)$ and $r(n)$, observed by microphones at ear positions of a human being or a dummy-head are transformed into spectra, $L(k)$ and $R(k)$, using fast Fourier transformation(FFT) for each input channel, as shown in Fig. 1.

2.2. *Sub-block for estimating sound source directions based on interaural phase and level differences*

The process of estimation of sound source direction is conducted using an interaural phase difference (IPD) and an interaural level difference (ILD). Figure 2 presents an example of IPD and ILD derived from HRTFs of KEMAR Dummy Head microphone, as provided by the MIT Media Laboratory.[11] In the low-frequency range, ILD is quite small because the low-frequency components are well diffracted by the head and torso, but the IPD changes according to the direction of sound source. Therefore, mainly in the lower frequency range, IPD carries the cue. On the other hand, at higher frequencies greater than 1500 Hz, ILD becomes large, and IPD has an ambiguity of direction because of a wavelength against the dimension of head. Consequently, ILD takes a major part of directional cues. However, at frequencies higher than 3000 Hz, ILD provides many candidates of sound source direction for various reasons including the resonances of pinna. It means that ILD has no deterministic information to estimate a sound source direction in a high frequency range.

For the lower frequency range, the sound source direction of each frequency bin is obtained by IPD. Actually, IPD is calculated through a cross spectrum $C_{lr}(k)$, which is defined as

$$C_{lr}(k) = L(k)R(k)^*, \tag{1}$$

where * denotes a complex conjugate. Consequently, IPD, $\theta_{lr}(k)$, at a frequency bin k is obtained as

$$\theta_{lr}(k) = \tan^{-1}\left\{\frac{\mathrm{Im}[C_{lr}(k)]}{\mathrm{Re}[C_{lr}(k)]}\right\} \tag{2}$$

by the cross spectrum, $C_{lr}(k)$. The ILD, $\xi_{lr}(k)$, for each frequency bin is obtained as

$$\xi_{lr}(k) = 20\log\left|\frac{C_{lr}(k)}{C_{ll}(k)}\right|, \tag{3}$$

where $C_{ll}(k)$ represents the power spectrum of $L(k)$. In addition, $\theta_{lr}(k)$ and $\xi_{lr}(k)$ are used, respectively, to determine the sound source direction by comparing it with the IPD map and ILD map.

The sound source direction obtained from the IPD and ILD for each frequency bin is combined based on the frequency bin.

2.3. *Howling Canceller Sub-block*

As is well known, the howling control for hearing aids is very important, but it is still under discussion. To produce an effective howling canceller, it is most important to know how to distinguish whether howling occurs or whether only a large input signal is fed. A howling canceller equipped with FDBM[8] is extremely stable and robust against the change of operational conditions. Basic concepts of this howling canceller can be described as follows.

If we can assume that the howling will not occur for both left and right channels at exactly simultaneous timing, we can say that howling occurs if and only if ILD is larger than the maximum ILD in the database. Therefore, if the howling occurs, then the signal level of the specific frequency bin will exceed the maximum, so that we can easily find out the occurrence of the howling. Furthermore, we can control the howling cancel filter according to the detected information. Figure 3 shows the maximum ILD threshold $\xi_{max}(k)$ constructed from HRTFs of 96 subjects,[12] to distinguish whether howling occurs or not.

Once howling is detected, the proposed howling cancel filter functions as illustrated in Fig. 4. It is assumed that a power spectrum of the observed signal is shown in Fig. 4(a) when the howling occurs. The howling cancel filter is set as shown in Fig. 4(b) according to the detected information; Fig. 4(c) shows the controlled power spectrum after the howling cancel filter is applied.

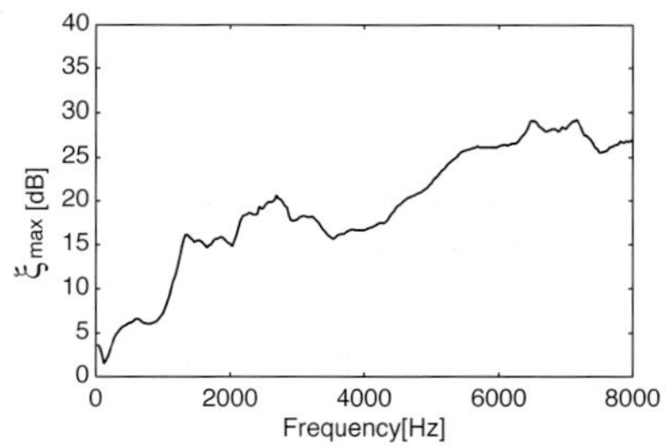

Fig. 3. Maximum ILD used as a threshold for howling cancellation. The mean ILD is obtained from 96 sets of HRTFs.

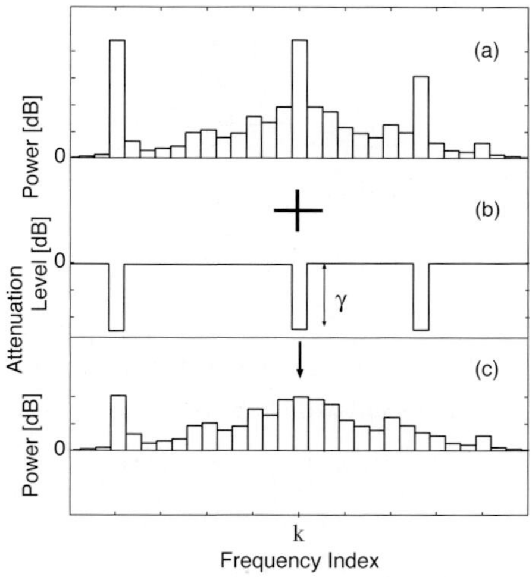

Fig. 4. Schematic diagram of the howling canceller. (a) presents an example of input spectrum under a howling canceller, (b) shows a configured cancellation filter based on the detection of howling, and (c) shows the spectra obtained with the howling canceller applied.

2.4. *Segregation Filter Sub-block*

The segregation filter is set based on the estimated direction in each frequency bin and is applied to both channels' spectra to obtain filter spectra so that spectral components of each frame only from the specified direction are kept and others are diminished for reproduction for both channel signals. A pair of segregated signals is obtained using inverse FFT of filtered spectra. The binaural information in the segregated signals is preserved.

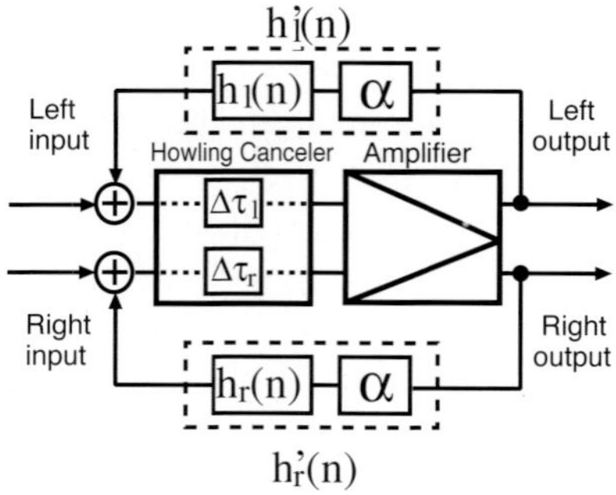

Fig. 5. Block diagram of a feedback model for the simulation for performance evaluation of the howling canceller.

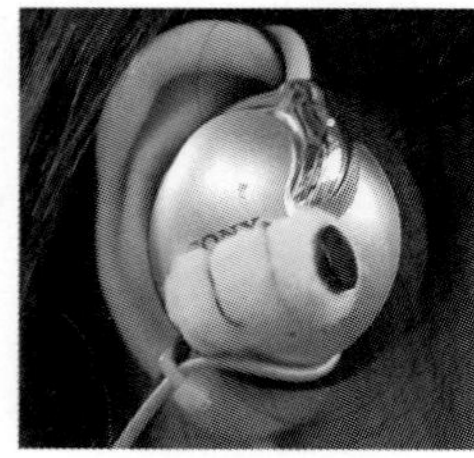

Fig. 6. Prototype of the headset for the hearing assistance system.

3. Performance evaluation of the proposed hearing assistance system

3.1. *Howling canceller*

To evaluate performance of the howling canceller, simulations using a feedback model represented in Fig. 5 are performed. The aim of the model is to simulate amplification and signal leakage from the loudspeaker into the microphone. In the feedback model, $h_l^{'}(n)$ and $h_r^{'}(n)$ are calculated by multiplying impulse responses $h_l(n)$ and $h_r(n)$; feedback loop attenuation α is set to 15 dB. Actually, $h_l(n)$ and $h_r(n)$ are impulse responses from headphones to microphones at each ear when the headset is put on a head. In addition, $\Delta\tau_l$ and $\Delta\tau_r$ in Fig. 5 are delays to compensate the delay because of frame based processing attributable to FFT. The amplifier provides a sufficient gain for a listener including hearing-impaired people. Gain and phase characteristics of the amplifier are, respectively, set to flat and linear.

Simulation is designed assuming a prototype hearing assistance system depicted as Fig. 6, in which the microphone is attached to the consumer headphone.

In simulation I, the howling canceller algorithm is confirmed using pink noise with a simulated feedback path whose characteristics are designed based on one of the results in preliminary experiments. Simulation II presents results of the howling canceling using Japanese vowel /a/ uttered by a male with a measured feedback path of the prototype of the headset shown in Fig. 6. The delay of the feedback path is measured using a preliminary experiment. The gain of feedback path for simulation I is also designed based on the results of the preliminary experiment. Figure 7 shows that, based on the preliminary measurement of frequency characteristics of acoustic coupling of the feedback path are designed. To confirm basic performance in simulation I and II, input signals are generated by convolving the measured head-related impulse response with arriving signals. The example signals for simulation I and II are, respectively, pink noise and the Japanese vowel /a/. In simulation III, performance is evaluated by howling margin using the observed feedback path and pink noise as the sound source. Pink noise is used as an input signal to confirm performance of the howling canceller against a wide band. Sampling frequency and quantization are respectively 16 kHz and 16 bits. Attenuation levels γ_l and γ_r are set accordingly to reduce spectral components as low as an average of the spectrum level of the specified frame.

3.2. *Simulation I*

The proposed algorithm uses interaural level difference to distinguish whether howling occurs or not. Consequently, the feedback frequency response affects judgment. Although a feedback frequency response depends on a device and the wearing position, Nakao[13] showed a measured frequency response which had multiple local peaks in the frequency domain. Feedback paths for $h_l(n)$ and $h_r(n)$ are designed as shown in Fig. 7 to verify multiple oscillating frequency component suppression in simulation.

Figure 8 portrays the output waveforms. Figure 9 depicts the frequency characteristics. A frame length in FFT is 32 points at 16 kHz sampling. The amplifier gain for the hearing assistance system is set to 5.8 dB. In Fig. 8, (a) is the input signal without a feedback signal, (b) indicates the results under a howling condition for the left channel, and (c) presents results with howling canceller for the left channel. In Fig. 9, (a) shows power spectrum of the input signal without a feedback signal, (b) and (c) indicate the results under howling condition for left and right channels, and (d) and (e) present results with howling canceller for left and right channels. Oscillating frequencies of 2 kHz up to 7 kHz with 1 kHz interval, as shown in Fig. 9(b), are observed.

In Fig. 9(c), oscillation occurs simultaneously at 3 kHz, 4 kHz and 5 kHz for the right channel. Howling is suppressed at all oscillating frequencies simultaneously in both left and right channels, as shown in Fig. 9(d) and Fig. 9(e).

Consequently, results confirm that the proposed howling canceller can control multiple frequencies simultaneously.

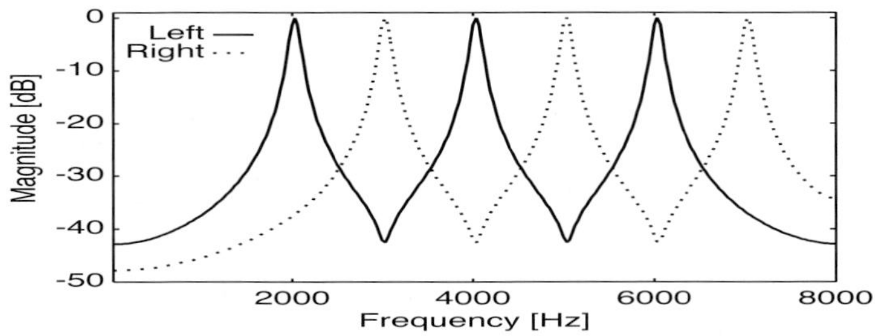

Fig. 7. Frequency characteristics of simulated feedback path of a hearing assistant system.

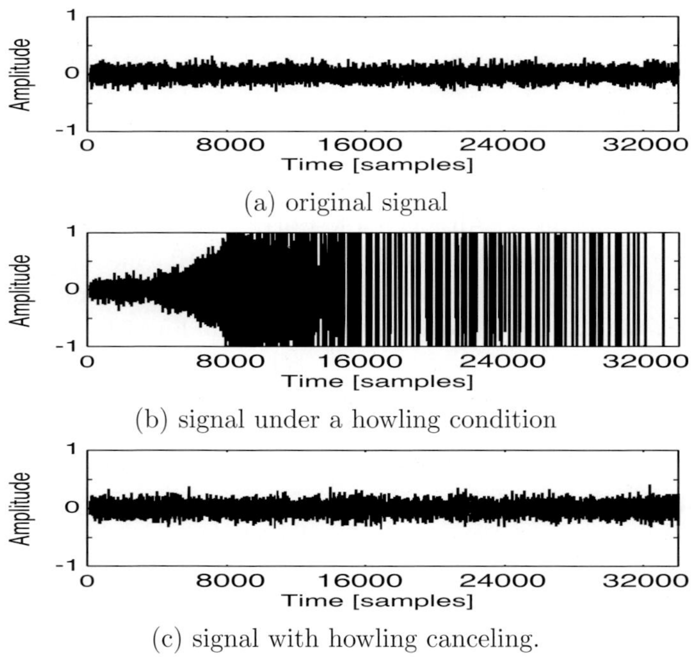

Fig. 8. Waveforms: (a) input signal without a feedback signal, (b) left channel signal under a howling condition, (c) signal processed using the proposed method.

3.3. *Simulation II*

Figure 6 shows that simulation using a measured transfer function for a prototype of a hearing assistant system is performed. The transfer functions $h_l(n)$ and $h_r(n)$ from a headset to a microphone are measured in an anechoic room using the setup shown in Fig. 10. The headset is attached to

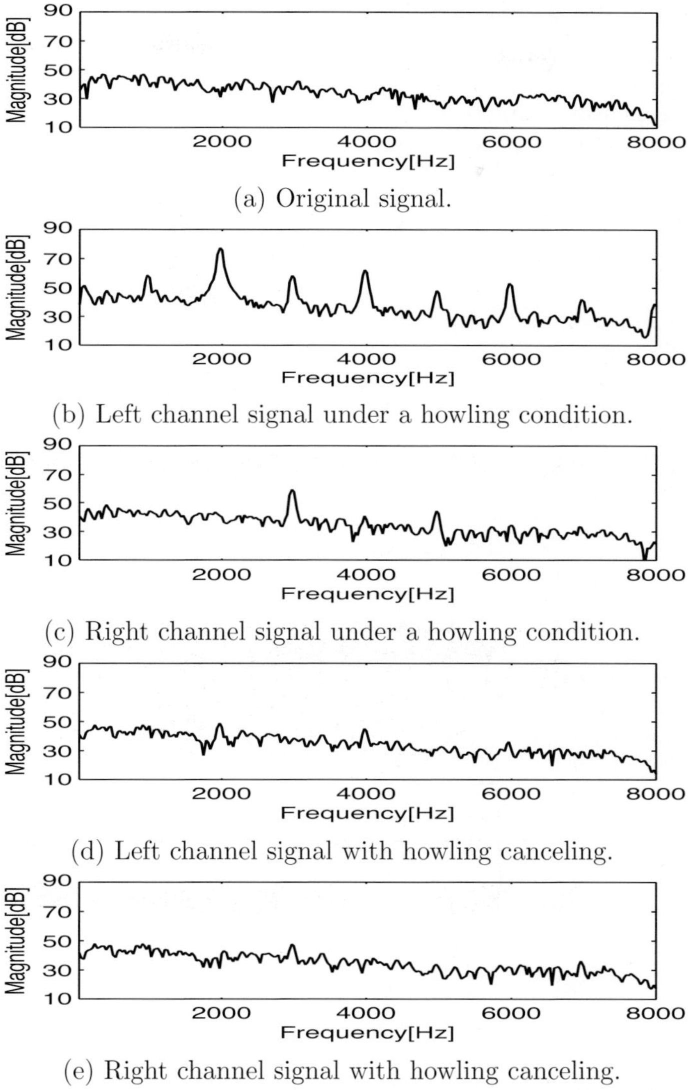

(a) Original signal.

(b) Left channel signal under a howling condition.

(c) Right channel signal under a howling condition.

(d) Left channel signal with howling canceling.

(e) Right channel signal with howling canceling.

Fig. 9. Frequency characteristics: (a) ideally amplified signal, (b) and (c) are amplified signals for the left and right channel signal with howling, (d) and (e) are results of the howling canceller obtained using the proposed method for the left and right channel.

a dummy head (Type 4128; Bruel and Kjaer). A microphone (ECM-44B; Sony Corp.) and headphones (MDR-Q22SL; Sony Corp.) were used. An amplifier was used for the microphone (MA8; Tascam), along with DAT for playback (ZA5ES; Sony Corp) and a DAT recorder (PC216Ax; Sony Corp). A time-stretched pulse (TSP) is radiated 10 times from an actuator of the headset. The feedback impulse response is measured as the average of the measure impulse responses. The frequency characteristics of the feedback transfer function are shown in Fig. 11. Figure 12 shows the output waveforms. Figure 13 portrays the frequency characteristics. A frame length in FFT is set to 32 for 16 kHz sampling. The amplifier gain for a hearing assistance system is set to 7.2 dB. In both Fig. 12 and Fig. 13, (a) is the input signal without a feedback signal: (b) indicates the results of the howling condition; and (c) presents results obtained under howling canceling. Oscillating frequencies around 2 kHz, 4 kHz, and 6 kHz in Fig. 12(b) are observed. In Fig. 12(c) and 13(c), it is apparent that those oscillating frequency components are suppressed. However, residue of the oscillating frequency component remains. In addition, small peeps of the howling occur during listening.

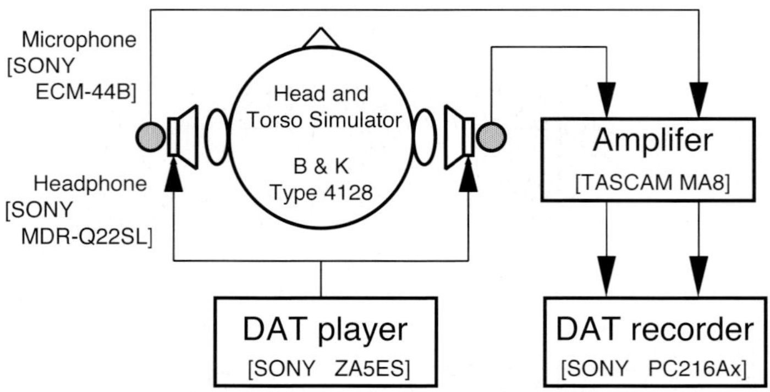

Fig. 10. Measurement environment.

3.4. *Quality of the enhanced speech signal*

In this subsection, the quality of the enhanced speech signal obtained using the hearing assistance system is examined using SNR, coherence, and PESQ. The observed signals at ear positions are generated based on the transfer functions measured by the microphones of the hearing assistance system instead of ordinary HRTF. The SNR and coherence are measured between the original signals and the segregated ones for each channel.

In addition, the observed signals at both ear positions are defined as

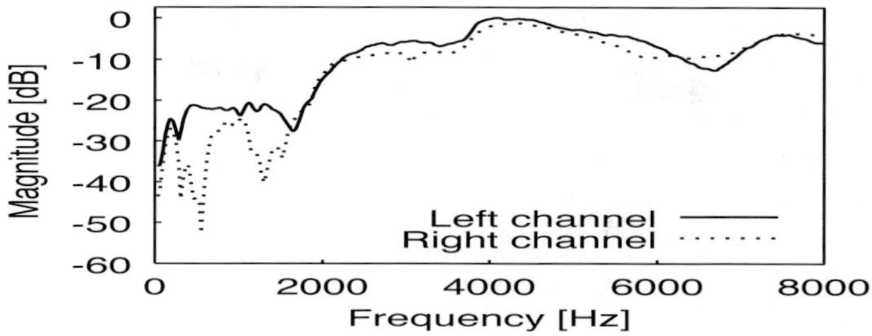

Fig. 11. Frequency characteristics of feedback paths.

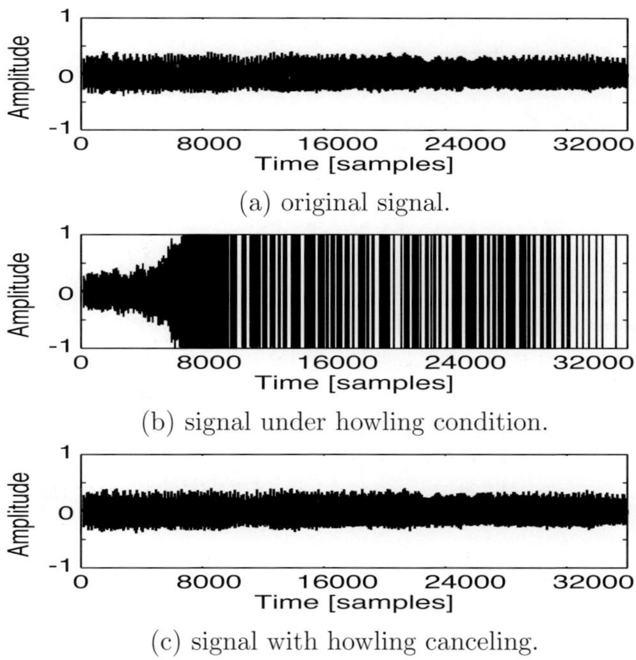

Fig. 12. Waveforms: (a) ideal amplified signal, (b) amplified signal, and (c) signal processed using the proposed method.

$l_t(n)$, $r_t(n)$ for the target and $l_i(n)$, $r_i(n)$ for interference. Target and interference signals are expressed as $s_t(n)$ and $s_i(n)$. The observed signals at the ear position are generated based on the transfer functions measured using the microphones of the hearing assistance system instead of ordinary HRTFs.

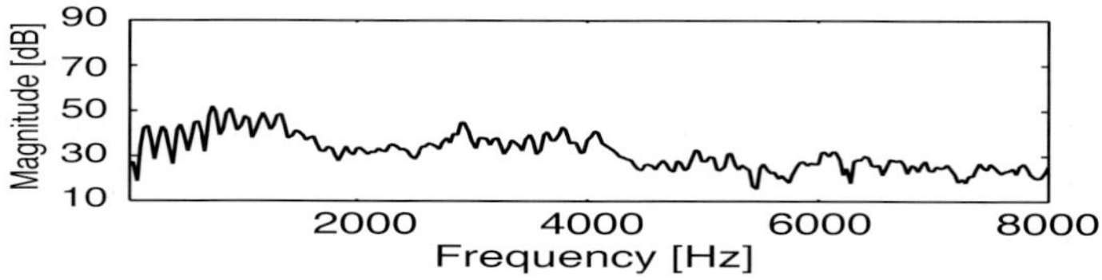

(a) original signal.

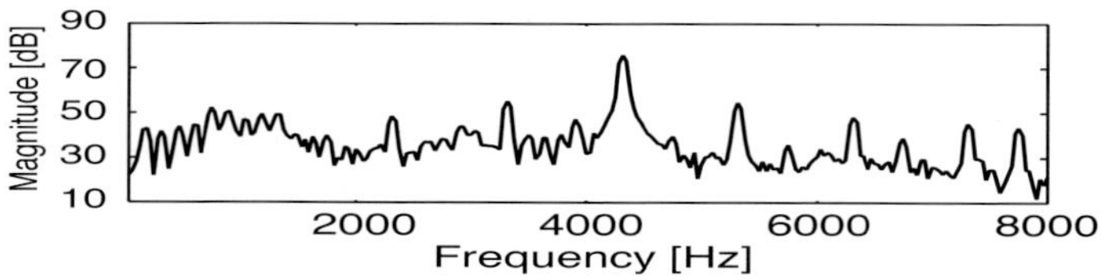

(b) signal under howling condition.

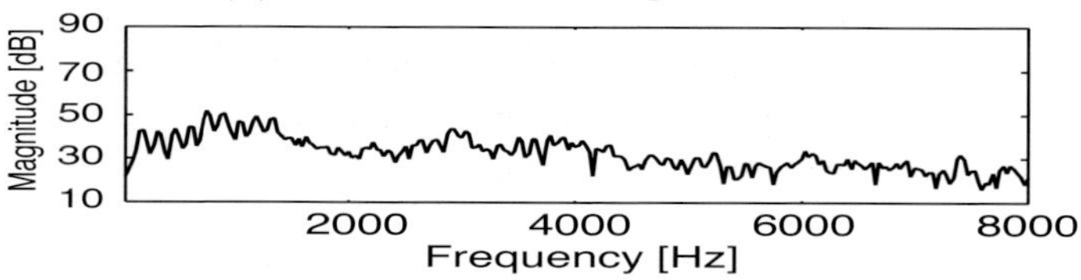

(c) signal with howling canceling.

Fig. 13. Frequency characteristics: (a) ideal amplified signal, (b) amplified signal, and (c) signal processed using the proposed method.

- **SNR**
  The definition of SNR at left input channel is the following:

$$SNR_{in\_l} = 10log\left|\frac{E[l_t(n)^2]}{E[(l_t(n) - l(n))^2]}\right|, \tag{4}$$

  and SNR_{in_r} is defined as Eq. (4), but all $l(n)$ and $l_t(n)$ are replaced by $r(n)$ and $r_t(n)$, respectively. Additionally, the SNR at left output channel is also defined as

$$SNR_{out\_l} = 10log\left|\frac{E[l_t(n)^2]}{E[(l_t(n) - \hat{l}_t(n))^2]}\right|, \tag{5}$$

  where $E[\cdot]$ denotes the expected value. Furthermore, SNR_{out_r} is defined as Eq. (5), but all $l(n)$ and $\hat{l}_t(n)$ are replaced respectively by $r(n)$ and $\hat{r}_t(n)$.
- **Coherence**
  Let $L_t(k)$ and $R_t(k)$ respectively represent spectra of the target signal obtained at left and right ears. And let $L(k)$ and $R(k)$ respec-

tively denote the spectra of the input signals obtained at left and right ears. Actually, $\hat{L}_t(k,\phi,\psi)$ and $\hat{R}_t(k,\phi,\psi)$ are the estimated spectra of the target signals when the interference is arranged at (ϕ,ψ). $\hat{L}_t(k,\phi,\psi)$ and $\hat{R}_t(k,\phi,\psi)$ are obtained using the FDBM. Based on the spectra, the coherence for the left input signal, $C(h)_{in_l}$ is defined as

$$C(k)_{in\_l} = C^2(L_t(k), L(k)), \tag{6}$$

where $C^2(X_1, X_2)$ is defined as

$$C^2(X_1, X_2) = \frac{(\overline{X_1 X_2^*})^2}{\overline{X_1}^2\,\overline{X_2}^2}, \tag{7}$$

and X^* denotes a complex conjugate of X, and $\overline{X}$ signifies an average of X. Actually, $C(k)_{in\_r}$ is defined as Eq. (6) but all $L(k)$ and $L_t(k)$ are replaced respectively by $R(k)$ and $R_t(k)$. Coherence of the left output channel is defined as

$$C(k)_{out\_l} = C^2(L_t(k), \hat{L}_t(k,\phi,\psi)), \tag{8}$$

and $C(k)_{out\_r}$ is defined as Eq. (8), but all $L_t(k)$, $\hat{L}_t(k)$ are replaced respectively by $R_t(k)$ and $\hat{R}_t(k)$. In Eq. (6) and (8), coherence is defined as the average of $C(k)$ across the frequency bin.

- **PESQ**

  Figure 14 presents an overview of PESQ. First, the original signal and processed signal are transformed to a representation based on perceptual frequency (Bark) and loudness (Sone) using the perceptual model. Then, the cognitive model gives the estimated subjective Means Opinion Score (MOS) by evaluating the difference between the original and processed signals. In this simulation, the reference signal, i.e. the original signal shown in Fig. 14, is always the target signal. The signals used for evaluation are defined in Table 1. The range of the PESQ score is defined as -0.5 to $+4.5$ and $+4.5$ means that the observed signal is the same as the original signal from a perceptual perspective.

original signal	processed signal	result
$l_t(t)$, $r_t(t)$	$l(t)$, $r(t)$	$PESQ_{in}$
$l_t(t)$, $r_t(t)$	$\hat{l}_t(t)$, $\hat{r}_t(t)$	$PESQ_{out}$

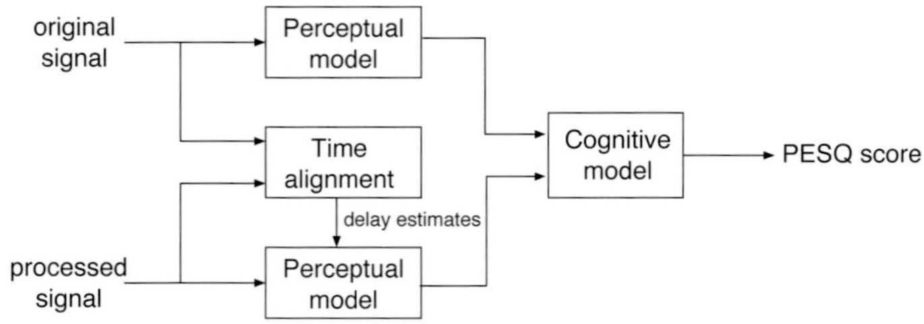

Fig. 14. Overview of PESQ.

The target sound source is set at (0°, 0°) in the azimuth–elevation coordinates, whereas the interference is set at (60°, 0°). Male speech is used as a target signal; white noise or female speech is used as interference. The SNR corresponding to the target of the interference "input SNR" is varied from −30dB to +30dB.

Figure 15 show the results of quality evaluation when the target is a male speech signal at (0°, 0°) and the interference is a female speech signal at (60°, 0°). The abscissa shows overall SNR defined as the power level ratio of target signal against interference, labeled as input SNR. The ordinate of Fig. 15(a) is the overall SNR. Figure 15(a) shows that the SNR of segregated left and right signals become similar. Furthermore, they improved when the input SNR rose to 10 dB. Although the segregated left signal's SNR is improved when input SNR is lower than 10 dB, it is degraded even when input SNR is greater than it is. Because 10 dB is the threshold to distinguish whether the proposed howling canceller is effective or not, and because it is defined as the cross of input and output measurement, it is designated as the "crossing point" of the performance in this chapter.

Figure 15(b) shows results of evaluation using coherence. Although the crossing point for SNR is 10 dB for both channels, the crossing points in the coherence is different between left and right channels, +5 dB for left channel and +20 dB for right channel. This is mainly attributable to the high coherence score for the input signal of the left channel.

Figure 15(c) presents the PESQ score. The PESQ scores show different performance from both SNR and coherence cases. For all input SNR condition, the PESQ score is improved, even if the input SNR is higher than 20 dB. In addition, like SNR, PESQ scores of segregated left and right channels are almost identical whereas those of inputs differ depending on the input SNR.

Figure 15 shows that the degradations of SNR and coherence at high

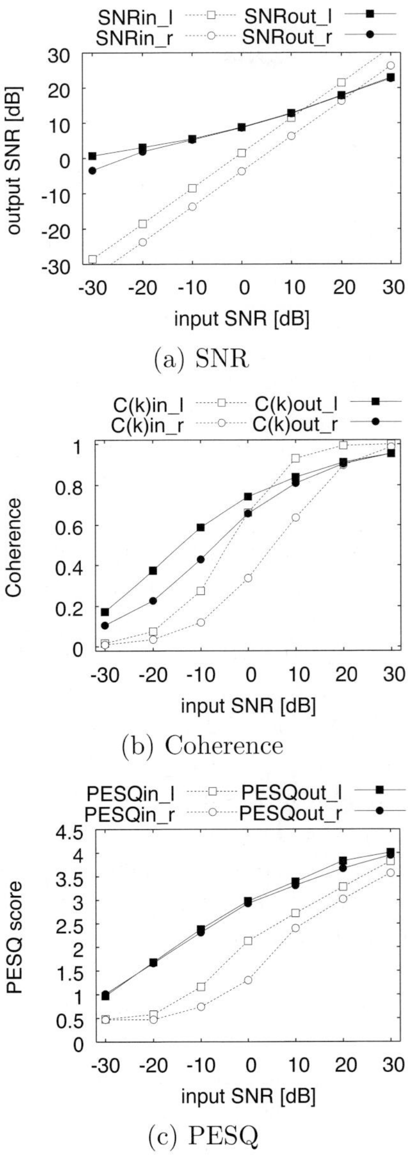

(a) SNR

(b) Coherence

(c) PESQ

Fig. 15. Evaluation results for left and right channel signals, when the respective target and the interference are male and female speech.

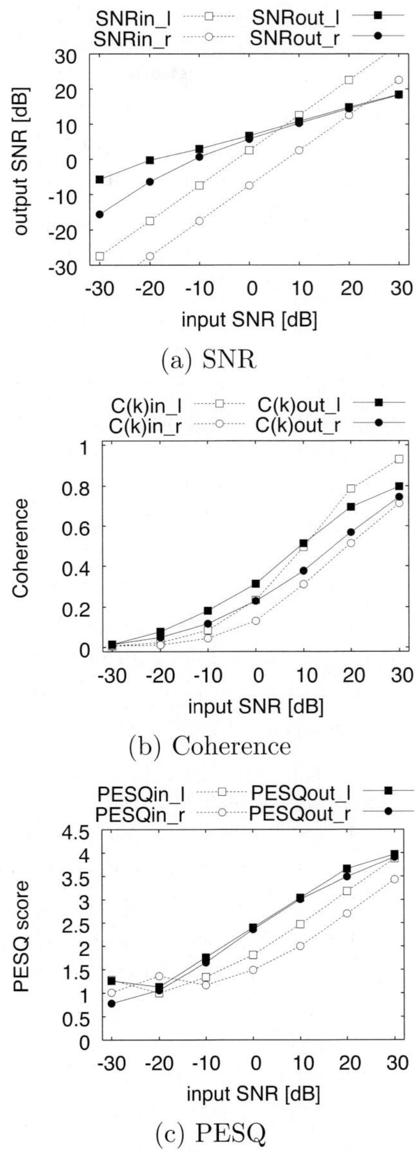

(a) SNR

(b) Coherence

(c) PESQ

Fig. 16. Evaluation results for left and right channel signals, when the target and the interference are male speech and white noise, respectively.

input SNR suggest that the segregated signal is distorted by segregation processes. On the other hand, the PESQ score at the same input SNR is improved, meaning that the distortion attributable to the segregation process can be negligible.

The noise source is located at right-hand-side of the front in this experiment. Therefore, all three evaluation scores of the observed right channel signal are lower than those of left channel signal. However, all scores of segregated right channel signal are improved and approached to those of the segregated left channel signal.

Figure 16 shows the evaluation results when interference is changed to white noise, whereas the target is the same as male speech. Although the improvements of evaluation scores shown in Fig. 16(a) and 16(b) are smaller than those in Fig. 15(a) and 15(b), the tendency of segregated signals is similar.

The highest coherence of segregated signal remains as 0.8 in Fig. 16(b) although it reached 0.9 in Fig. 15(b). As a result, the distortion in segregated signal provided by the FDBM is larger for the white noise case, Fig. 16, than for the female case, Fig. 15. This difference reflects that FDBM can segregate signals if the signal and interference have a sparseness in their spectra.

Figure 16(c) shows that PESQ scores are improved when the input SNR is −20 dB to 30 dB. When the input SNR is −30 dB, the PESQ score is higher than that at input SNR: −20 dB. That result might arise from error of evaluation attributable to a very low SNR condition.

Figure 16(a) shows that the cross point of input and segregated scores is shifted from that in Fig. 15(a); however no cross point exists in the coherence of the right channel shown in Fig. 16(b). The PESQ score in Fig. 15(c) is improved except that the input SNR is below −20 dB.

4. Conclusions

As described in this paper, a binaural hearing assistance system is proposed using the frequency domain binaural model, which has a simple but very robust howling cancellation mechanism. Additionally, the system performance was assessed both for howling cancellation and segregation quality. Results show sufficient performance for howling cancellation. However, the quality of segregation must be improved.

Acknowledgments

Part of this work was conducted for Cooperative Research Projects of the Research Institute of Electrical Communications, Tohoku University (H19/A10).

References

1. L. A. Jeffress, *J. Comparative Physiology and Psychology* **41**, 35 (1948).
2. J. Blauert, *Spatial Hearing – The Psychophysics of Human Sound Localization*, 1st edn. (MIT Press, Boston, MA, 1983).
3. W. Lindemann, *J. Acoust. Soc. Am.* **80**, 1608 (1986).
4. W. Gaik, *J. Acoust. Soc. Am.* **94**, 98 (1993).
5. M. Bodden, *Acta Acoustica* **1**, 43 (1993).
6. D. Wang and G. J. Brown, *Computational Auditory Scene Analysis*, 1st edn. (IEEE Wiley Inter-science, 2006).
7. H. Nakashima, Y. Chisaki, T. Usagawa and M. Ebata, *Acoust. Sci. & Tech.* **24**, 172 (2003).
8. Y. Chisaki, K. Matsuo and T. Usagawa, *Acoust. Sci. & Tech.* **28**, 90 (2007).
9. ITU-T, *Perceptual evaluation of speech quality (PESQ): An objective method for end-to-end speech quality assessment of narrow-band telephone networks and speech codecs* (ITU-T, 2001).
10. K. Iida, M. Itoh, A. Itagaki and M. Morimoto, *Applied Acoustics* **68**, 835 (2007).
11. B. Gardner and K. Martin, *HRTF measurements of a KEMAR dummy head microphone*, MIT Media Lab Perceptual Computing Technical Report 280, MIT (1994).
12. N. U. Itakura laboratory, *Head related transfer function for 96 subjects* (Itakura Lab, Nagaoya University. `http://www.sp.m.is.nagoya-u.ac.jp/HRTF/`.
13. K. Nakao, R. Nishimura and Y. Suzuki, Estimation of the feedback transfer function of a hearing-aid using a microphone in the ear, in *Proc. Spring Meet. Acoust. Soc. Jpn. (in Japanese)*, (Tokyo, Japan, 2002).

A SPATIAL AUDITORY DISPLAY FOR TELEMATIC MUSIC PERFORMANCES

J. BRAASCH

School of Architecture, Rensselaer Polytechnic Institute,
Troy, NY 12180, USA, www.rpi.edu, E-mail: braasj@rpi.edu

N. PETERS

CIRMMT, Schulich School of Music, McGill University,
Montreal, Quebec H3A 1E3, Canada

P. OLIVEROS

Arts Department, Rensselaer Polytechnic Institute,
Troy, NY 12180, USA

D. VAN NORT

School of Architecture & Arts Department, Rensselaer Polytechnic Institute,
Troy, NY 12180, USA

C. CHAFE

CCRMA, Department of Music, Stanford University,
Stanford, CA 94305, USA

This paper describes a system which is used to project musicians from two or more co-located venues into a shared virtual acoustic space. The sound of the musicians is captured using near-field microphones and a microphone array to localize the sounds. Afterwards, the near-field microphone signals are projected at the remote ends using spatialization software based on Virtual Microphone Control (ViMiC) and an array of loudspeakers. In order to simulate the same virtual room at all co-located sites, the ViMiC systems communicate using the Open Sound Control (OSC) protocol to exchange room parameters and the room coordinates of the musicians. Using OSC they also receive localization data from the microphone arrays.

Keywords: Virtual Auditory Display, Telepresence, Telematic Music.

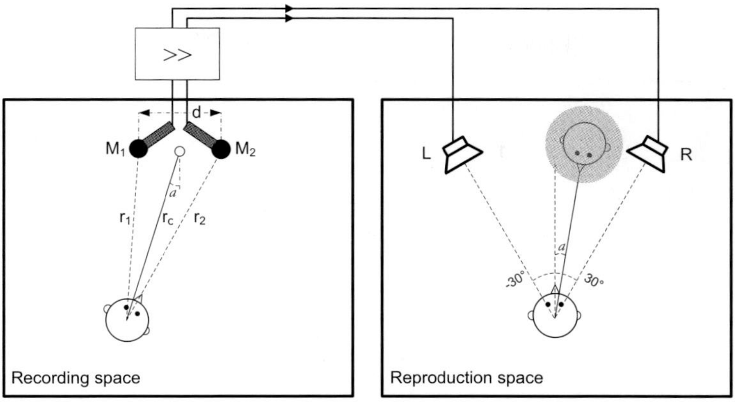

Fig. 1. Sketch of a microphone-based recording and reproduction set-up.

1. Introduction

Live networked music performances have gained in popularity over the last few years. In these concerts, musicians are distributed over at least two remote venues and connected via the internet. Some of the challenging technical requirements that these projects have imposed on the underlying research have been addressed in previous work.[1–6] One of the problems that has not been solved to the full extent is the accurate spatial reproduction of the broadcasted sound field at the remote end. Especially with the introduction of High-Definition (HD) video, the need for accurate spatial sound reproduction has become more pressing. In this chapter, a system for accurate spatial sound reproduction in telematic music performances is described.

The idea of this research goes back to the 1930s, when Steinberg and Snow described a system that enabled the world-renowned conductor Leopold Stokovski and the Philadelphia Orchestra to broadcast music live from Philadelphia to Washington, D.C. The authors used the then newly invented main-microphone techniques to produce a stereophonic image from the recorded sound. Figure 1 shows a general diagram of how microphones and loudspeakers have to be set up and the signals have to be routed for stereophonic imagery. The spatial positions of sound sources in the recording space are encoded by placing and orienting two or more microphones—the main microphone array—strategically, capturing spatial information by utilizing time and level differences between the different mi-

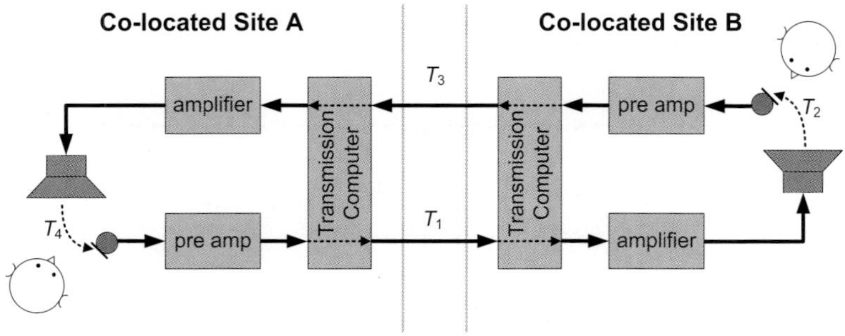

Fig. 2. Feedback loop in a telematic transmission.

crophone channels. Each channel is then transmitted separately, amplified and fed to the matching loudspeaker of an array of at least two speakers, for example the classic stereo set-up shown in Fig. 1. The box in this figure that connects the microphones to the loudspeakers can either be an amplifier, a broadcasting unit or a sound-recording/reproduction system. Steinberg and Snow used two to three parallel telephone lines to transmit the spatially encoded sound from Philadelphia to Washington, D.C.[7]

While we now experience music broadcasts via radio, satellite, and the internet in our daily life, music collaborations in which ensemble members are distributed over long distances are still in the experimental stage, due to technical difficulties associated with two-way or multicast connections. A major challenge is the susceptibility of bidirectional set-ups to feedback loops, which can easily lead to audible colorations and echoes. Figure 2 demonstrates the general problem: the microphone signal recorded at Site A is broadcast through a loudspeaker at Site B, where it is picked up by a second microphone. This microphone signal is then broadcast back to the original Site A, where it is re-captured by the first microphone. Due to the transmission latency, the feedback becomes audible as echo at much lower gains when compared to the feedback situation known from local public address systems. Many popular audio/videoconferencing systems such as *iChat* or *Skype* use echo-cancellation systems to suppress feedback. In speech communication echo-cancellation systems work well, since the back-and-forth nature of spoken dialogue usually allows for suppressing the transmission channel temporarily in one direction. In simultaneous music communication, however, this procedure tends to cut off part of the performance. Spectral alterations are a common side effect if the echo-cancellation system operates with a filter bank.

For the given reasons, the authors suggest to avoid using echo-cancellation systems completely. Instead, it is proposed to capture all instruments from a close distance (e.g., lavalier microphones) to minimize the gain and therefore the risk of feedback loops. Unfortunately, the exclusive use of near-field microphones contradicts the original idea of Steinberg and Snow, since the main microphones have to be placed at a further distance to capture the sound field stereophonically. To resolve this conflict, this paper describes an alternative approach to simulate main microphone signals from closely captured microphone signals and geometric data. The system—called Virtual Microphone Control (ViMiC)—includes a room simulation software to (re-)construct a multichannel audio signal from a dry recording as if it had been recorded in a particular room.[8–10] The position data of the sound sources, which is needed to compute the main microphone signals, are estimated using a microphone array. The array, which is optimized to locate multiple sound sources, is installed at each co-located venue to track the positions of the sound sources. The recorded position data is transmitted to the remote venue(s) along with the acoustic signals that were recorded in the near-field of the instruments. At the remote end, the sound can then be projected with correct spatial image using the ViMiC system. A sketch of the transmission system, which also includes video broadcast, is shown in Fig. 3. The low-latency audio transmission software *Jacktrip*[11,12] and the *Ultravideo Conferencing*[4] system are used for telecommunication.

2. Sound Spatialization using Virtual Microphone Control

2.1. *Basic Concept*

The following section deals with the fundamental principles of ViMiC. The system basically simulates a multichannel main microphone signal from the near-field recordings, using descriptors about the room size, wall-absorption coefficients, and the sound-source positioning data. The latter is provided by a sound-localization microphone array as described in Section 3. To auralize the signals, each virtual microphone signal is then fed to a separate (real) loudspeaker.

The core concept of ViMiC involves an array of virtual microphones with simulated directivity patterns. The axial orientation of these patterns can be freely adjusted in 3D space, and the directivity patterns can be varied between the classic patterns that are found in real microphones: omnidirectional, cardioid, hyper-cardioid, sub-cardioid, or figure-eight characteristics. The transfer function between the sound source (e.g., a musical instrument

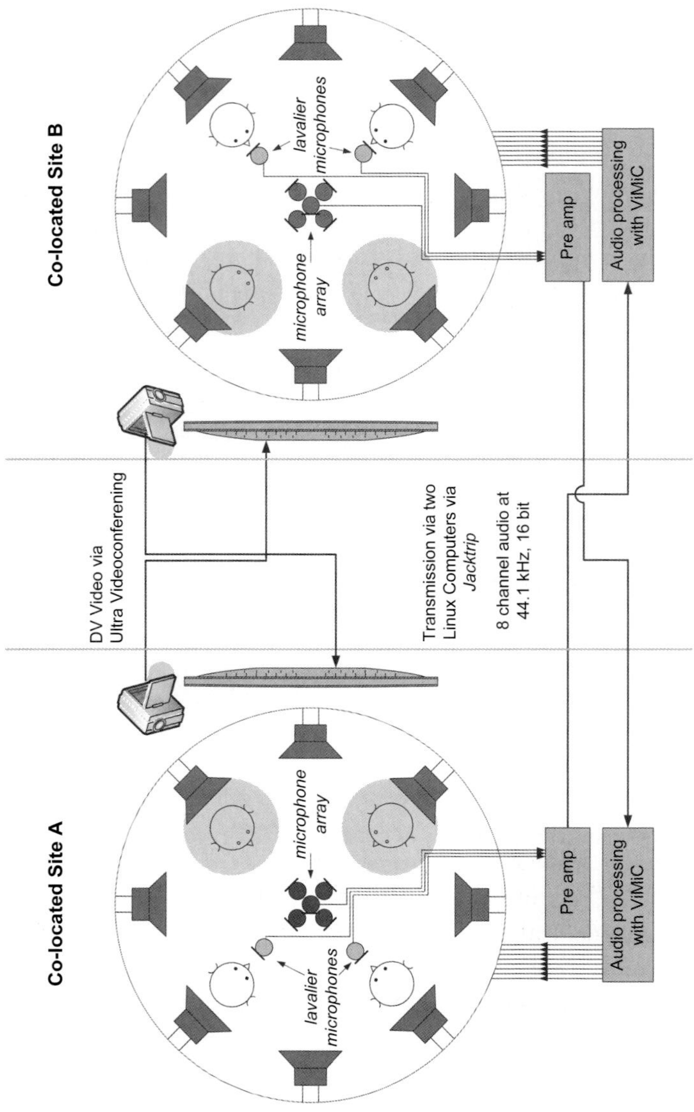

Fig. 3. Sketch of the internet-based telematic music system used by the authors.

which can be treated as a one-dimensional signal in time $x(t)$) and a virtual microphone is then determined by the distance and the orientation between the microphone's directivity pattern and the recorded sound source (e.g., musical instrument). The distance determines the delay τ between the radiated sound at the sound-source position and the microphone signal:

$$\tau(r) = \frac{r}{c_s}, \tag{1}$$

with the distance r in meters and the speed of sound c_s. The latter can be approximated as 344 m/s at room temperature (20° C). According to the $1/r$ law, the local sound-pressure radiated by a sound source will decrease by 6 dB with each doubling of the distance r:

$$p(r) = \frac{p_0 \cdot r_0}{r}, \tag{2}$$

with the sound pressure p_0 of the sound source at a reference distance r_0. In addition, the system considers that the sensitivity of a microphone varies with the angle of incidence according to its directivity pattern. In theory, only omni-directional microphones are equally sensitive towards all directions, and in practice even this type of microphone is more sensitive toward the front for high frequencies. The circumstance that real microphones generally have rotational directivity patterns simplifies their implementation in ViMiC, since these types of directivity patterns $\Gamma(\alpha)$ can be written in a simple general form:

$$\Gamma(\alpha) = a + b\cos(\alpha). \tag{3}$$

The variable α is the incoming angle of the sound source in relation to the microphone axis. Typically, the maximum sensitivity $a+b$ is normalized to one ($b = 1 - a$), and the different available microphones can be classified using different combinations of a and b, with omnidirectional: a=1, b=0; cardioid: a=0.5, b=0.5, and figure-eight: a=0, b=1.

The overall gain g between the sound source and the virtual microphone can be determined as follows:

$$g = g_d \cdot \Gamma(\alpha) \cdot \Gamma(\beta), \tag{4}$$

with the distance-dependent gain $g_d = r_0/r$, and the sound source's rotational radiation pattern $\Gamma(\beta)$.

The transfer function between the sound source and the microphone signal can now be derived from two parameters only, the gain g and the delay τ, if the microphone and source directivity patterns are considered to

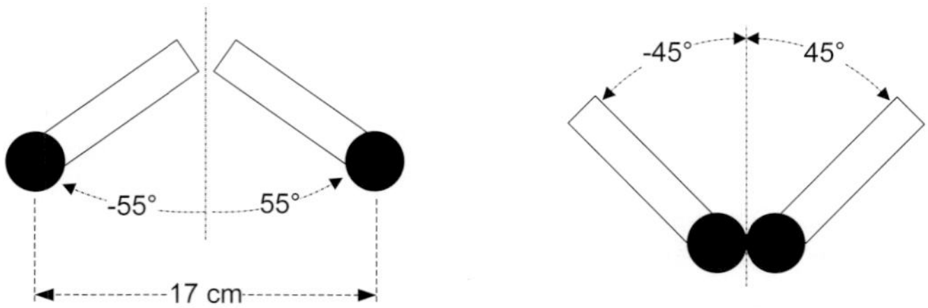

Fig. 4. Microphone placements for the ORTF technique with two cardioid microphones (left) and the Blumlein technique with two figure-eight microphones (right).

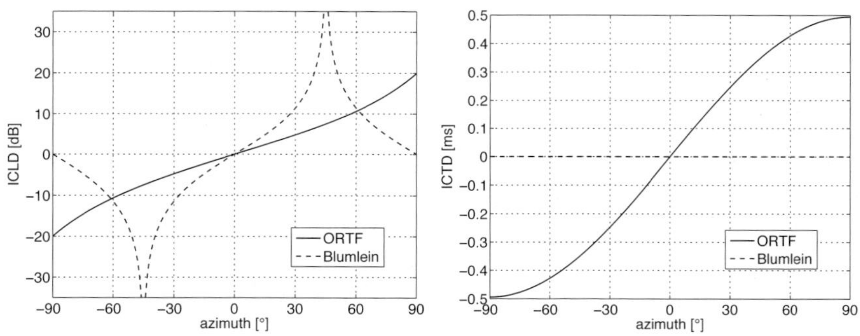

Fig. 5. The left graph shows calculated inter-channel level differences (ICLDs) for the ORTF technique as a function of azimuth in comparison to the Blumlein/XY technique. The Blumlein/XY technique is based on two figure-eight microphones with coincident placement at an angle of 90°. The right graph shows the results for inter-channel time differences (ICTDs).

be independent of frequency. It is noteworthy that the directivity patterns of most real microphones are not fully independent of frequency, although this is often a design goal. The relationship between the sound radiated from a point source $x(t)$ and the microphone signal $y(t)$ is found to be:

$$y(t, r, \alpha) = g \cdot x(t - \tau) = g_d(r) \cdot \Gamma(\alpha) \cdot \Gamma(\beta) \cdot x(t - \frac{r}{c_s}). \tag{5}$$

By simulating several of the virtual microphones as outlined above, the sound sources can be panned in virtual space according to standard sound recording practices.

2.2. *ORTF-technique implementation*

A good example to demonstrate ViMiC is the classic ORTF microphone technique, which is named after the French national broadcasting agency *Office de Radiodiffusion et de Télévision Française* where it was first introduced. The ORTF microphone placement is shown in Fig. 4. Due to the relative broad width of the directivity lobe of the cardioid pattern, the angle between both microphones is adjusted to 110°. The ratio between the signal amplitude at the sound source x and microphone signal amplitudes for the left and right channels, y_1 and y_2, varies with the angle of incidence according to Eq. 5:

$$y_1(t) = g_{d_1} \cdot 0.5 \cdot (1 + \cos(\alpha + 55°)) \cdot x(t - \tau), \qquad (6)$$

$$y_2(t) = g_{d_2} \cdot 0.5 \cdot (1 + \cos(\alpha - 55°)) \cdot x(t - \tau). \qquad (7)$$

In general, both amplitude and time differences between the microphone channels determine the position of the spatial image that a listener will perceive when both microphone signals are amplified and played through two loudspeakers in standard stereo configuration (see Fig. 1). When a virtual sound source is encircling the microphone set-up in the frontal horizontal plane at a distance of 3 m (α=−90° to 90°), the inter-channel level difference (ICLD) ρ as shown in Fig. 5 can be calculated as follows:

$$\rho(\alpha) = 20 \cdot \log_{10}\left(\frac{y_2(t)}{y_1(t)}\right) = 20 \cdot \log_{10}\left(\frac{g_{d_2} \cdot (1 + \cos(\alpha - 55°))}{g_{d_1} \cdot (1 + \cos(\alpha + 55°))}\right). \qquad (8)$$

In the far-field—when the distance between the sound source and the center of the recording set-up r is much larger than the distance between both microphone diaphragms d ($r \gg d$), the $1/r$ term can be neglected in the ICLD calculation. Further, the occurring ICLDs are almost solely generated by the different orientations of the cardioid patterns of both microphones. Figure 5 shows the ICLDs as a function of the angle of incidence α. Apparently, the level difference between the microphones remains rather low for all angles when compared to coincidence techniques like the Blumlein/XY technique. However, increasing the angle between the microphones is rather problematic, as this would result in very high sensitivity toward the sides. Instead, the diaphragms of both microphones are spaced 17 cm apart in the ORTF configuration (compare Fig. 4. This way ICTDs τ_Δ are generated in addition to the ICLDs. The ICTDs, which are also shown in Fig. 5, can be easily determined from the geometry of the set-up (compare Fig. 1):

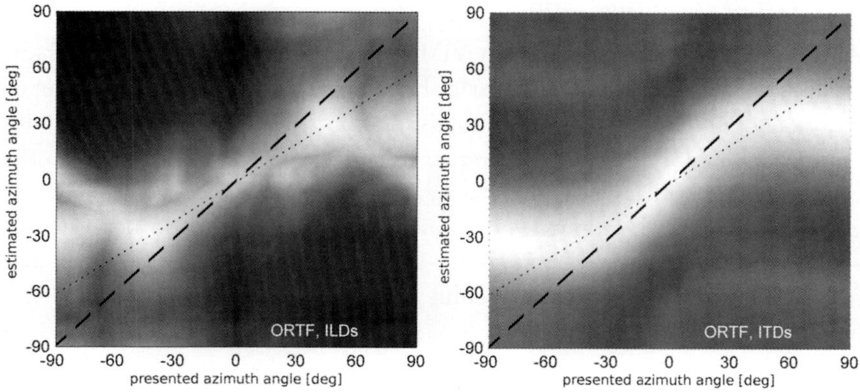

Fig. 6. Results for a binaural model to localize an ORTF en- and decoded sound source from various positions in the horizontal plane. The left graph shows the results of the ILD analysis, the one in the right the ITD-analysis results.

$$\tau_\Delta(\alpha) = \frac{(r_1 - r_2)}{c_s}, \tag{9}$$

with the speed of sound c_s and the far-field approximation:

$$\tau_\Delta(\alpha) = \frac{d}{c_s}\sin(\alpha). \tag{10}$$

One of the core ideas of ViMiC is to be able to play with the spatial imagery of sound in a similar fashion to microphone-based sound recording practice. To illustrate this approach, the output of a binaural model is shown in Fig. 6. For this graph, the model was used to analyze the reproduced sound field of an ORTF recording via a dummy head. The figures show the estimated localization curves, the relationship between the azimuth of the original source position and the azimuth of the auditory event when listening to the reproduced signal. The left graph shows the analysis of interaural level differences (ILDs), and the estimated position of the auditory events are highlighted in white or light gray. The right graphs shows the same context but for interaural time differences. The figure shows that within the range of interest ($-45°$ to $+45°$), the ILD cues project the sound source at a narrower angle compared to the natural condition, while the ITD cues suggest a wider angle. The mismatch between both cues leads to the perceptual widening of the auditory objects, which is often preferred

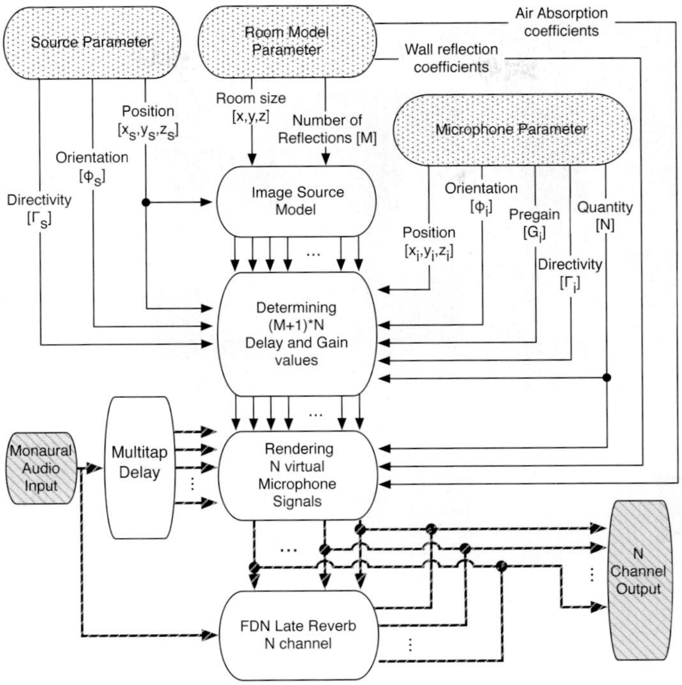

Fig. 7. Architecture of the Virtual Microphone Control (ViMiC) auditory virtual environment.

and which makes the use of classic microphone techniques so interesting. Further details about the model analysis can be found in Braasch (2005)[13] and Blauert and Braasch (2008).[14]

2.3. *Software Implementation*

Figure 7 shows the system architecture of the current ViMiC implementation, which is part of the Jamoma package.[15,16] The system has three larger signal processing components: an Image Source Model, a Multitap Delay Unit, and a Reverberation Unit. The Image Source Model determines the gains and delays between the sound source positions and the receiving virtual microphones. The algorithm considers the positions and orientations of both sources and receivers including their directivity characteristics. The model uses the mirror image technique[17] to calculate the positions and strengths of early room reflections for a rectangular enclosure with adjustable dimension and wall-absorption characteristics.

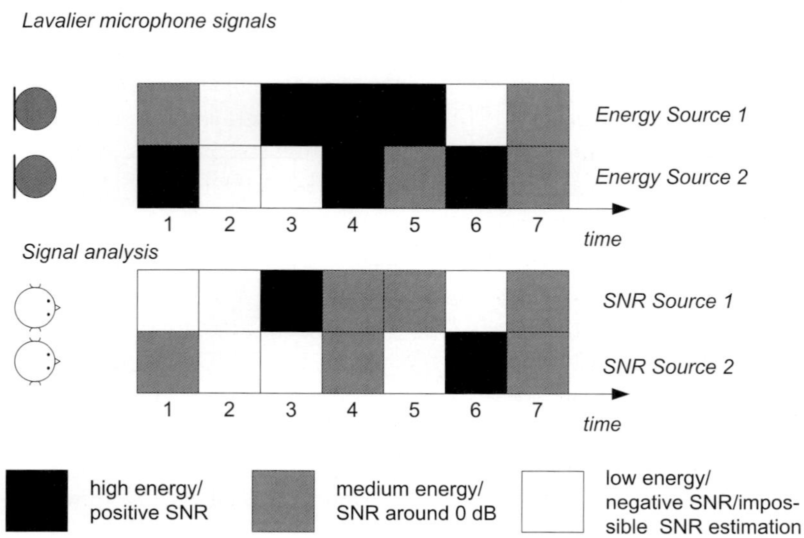

Fig. 8. Estimation of the signal-to-noise ratios for each sound source.

Using the gain and delay data provided by the Image Source Model, the dry sound is processed using a multi-tap delay network for spatialization. Typically a high number of delays have to be computed – for example, 42 delays have to be processed per primary sound source in a 6 channel surround system, if first-order reflections are considered (1 direct source plus 6 first-order reflections $\times$ 6 output channels). This number increases to 114 delays if second-order reflections are simulated as well. Several measures have been taken to reduce the computational load. One of them is the automated shift between 4-point fractional delays for moving sound sources and non-fractional delays, which are activated once the sound source remains stationary. The late reverberant field is considered to be diffuse and simulated through a feedback delay network[18] with 16 modulated delay lines, which are diffused by a Hadamard mixing matrix. By feeding the outputs of the room model into the late reverb unit a diffuse reverb tail is synthesized (see Fig. 7), for which timbral and temporal character can be modified. This late reverb can be efficiently shared across several rendered sound sources.

3. Sound Source Tracking System

So far, we have described the spatial decoding method using the ViMiC system, but we have not discussed how the spatial positions can be captured at the remote site. For fixed instrument positions, as is often the case in classical music, a manual adjustment of the sound source positions is a viable option. However, this procedure can be cumbersome if the positions of the sound source vary over time.

The solution that was integrated into our telematic music system is based on a pyramidal five-microphone array, which has been described earlier.[10,19] The five omni-directional microphones are arranged in a square-based pyramid with 14-cm base side and 14-cm triangular side dimensions. Traditional microphone-array based systems work well to localize an isolated sound source by utilizing arrival time differences or amplitude differences of the sound source between the individual array microphones.[20,21] In multiple-sound-source scenarios (e.g., a music ensemble), however, determining the sound-source positions from the mixed signal and assigning them to the corresponding source is still a real challenge.

A solution for this problem is to use the near-field microphone signals in conjunction with a traditional microphone-array based localization system. The near-field microphone signals are then used to determine the signal-to-noise ratios (SNRs) between several sound sources, for example concurrent musicians, while still serving the main purpose of capturing the audio signals. The running SNR is calculated frequency-wise from the acoustic energy recorded in a certain time interval:

$$\mathrm{SNR}_{i,m} = 10\log_{10}\left(\frac{1}{a}\int\limits_{t_m}^{t_m+\Delta t} p_i^2 \cdot dt\right) \tag{11}$$

with:

$$a = \sum_{n=1}^{i-1}\int\limits_{t_m}^{t_m+\Delta t} p_i^2 \cdot dt + \sum_{n=i+1}^{N}\int\limits_{t_m}^{t_m+\Delta t} p_i^2 \cdot dt \tag{12}$$

and p_i the sound pressure captured with the ith near-field microphone, t_m the beginning of the measured time interval m, Δt its duration and N, the number of near-field microphones.

Basically, the SNRs are measured for each time interval between each observed sound source and the remaining sound sources. The data can then be used to select and weight those time slots in which the sound source dominates the scene, assuming that in this case the SNR is high enough for the microphone array to provide stable localization cues. Figure 8 depicts the core idea. In this example, a good time slot is found for the third time frame for Sound Source 1, which has a large amount of energy in this frame, because the recorded energy for Sound Source 2 is very low. Time Slot 6 depicts an example where a high SNR is found for the second sound source.

To improve the quality of the algorithm, all data are analyzed frequency-wise. For this purpose the signals are sent through an octave-band filter bank before the SNR is determined. Basically, the SNR is now a function of frequency f, time interval t, and the index of the sound source.

The sound source position is determined for each time/frequency slot by analyzing the time delays between the microphone signals of the microphone array. The position of the sound source is estimated using the cross-correlation technique, which is used to determine the direction of arrival (DOA) from the measured internal delay (peak position of the maximum of the cross-correlation function) via this equation as shown by Würfel[22] among others:

$$\alpha = \arcsin\left(\frac{c}{f_s} \cdot \frac{\tau}{d}\right), \tag{13}$$

with the speed of sound c, the sampling frequency f_s, the internal delay τ, and the distance between both microphones d.

Since this technique cannot resolve two sound sources within one time-frequency bin, the estimated position is assigned to the sound source with the highest SNR. Alternatively, the information in each band can be weighted with the SNR in this band. To save computational cost, a minimum SNR threshold can be determined, below which the localization algorithm will not be activated for the corresponding time/frequency slot.

4. Integrated system

Figure 9 depicts the whole transmission chain which includes the sonification system. At the recording site, the raw sound signals are captured through the near-field microphones which also feed the localization algorithm with information to calculate the instantaneous SNR. Both the audio data and the control data—which contains information on the estimated sound source position—is transmitted live to the co-located site(s). Here, the sound field is resynthesized from the near-field audio signals and the control data using rendering techniques such as ViMiC.

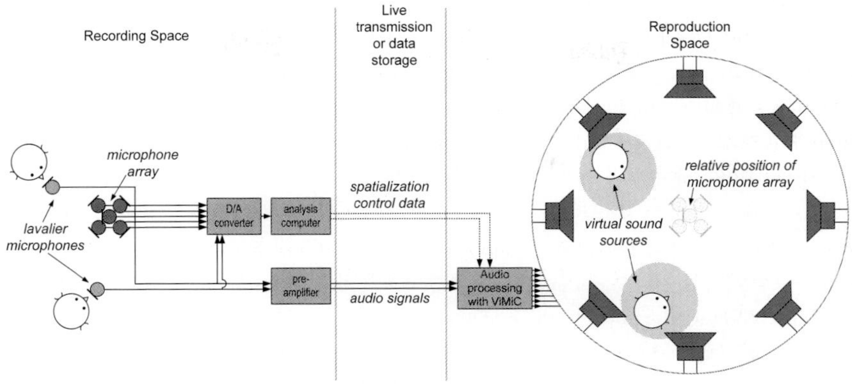

Fig. 9. Sketch of the spatial sound recording and reproduction set-up.

The sound source tracking unit is currently implemented in Matlab, which allows easier prototyping than an implementation in Max/MSP. The Matlab module runs in real-time using the *Data Acquisition Toolbox*. The module receives multichannel audio input and returns the calculated results (positions of individual sound sources) via the Open Sound Control (OSC) protocol.[23] Currently, we are also experimenting with an ambisonics-based microphone array (1st-order, B-Format) for sound localization.[24,25] Since the spatial positions can be derived from amplitude differences, this requires less computational resources than the current pyramidal array, which localizes sounds through time delay analysis. The expected decrease in localization accuracy is acceptable for the given application and the described algorithm to analyze multiple sound sources can be applied equally well.

The ViMiC system has been used in several projects to spatialize telematically transmitted sound. The first commercial album using ViMiC in a telepresence scenario has been released with the Deep Listening Record Label in Kingston, New York.[26] The 5-channel Quicktime video is a recording of the Tintinnabulate and Soundwire ensembles performing live at the ICAD 2007 conference in Montreal, Canada (McGill University), RPI, Stanford University and KAIST, Seoul, South Korea. For a telematic concert at SIGGRAPH 2007, *Dynamic Spaces*,[27] we used ViMiC to create a dynamically changing acoustical space. In this piece, the room acoustics were altered in realtime using a handheld controller. The system was used to vary the acoustics in San Diego during a remote clarinet solo that was played by Bobby Gibbs at Rensselaer Polytechnic Institute. Reverberation time, room size, sound pressure level of early reflections, and frequency response were among the parameters that were controlled. The project was a milestone in our current paradigm to explore the possibility of changing the acoustics of the concert space during the performance. This new possibility adds substantially to the way we perform and listen to music—creating a new awareness for the space surrounding us.

The project reported here has received support from the National Science Foundation (#0757454), the Canadian Natural Sciences and Engineering Research Council (NSERC, New Media Initiative), and a seed grant from Rensselaer Polytechnic Institute and the Experimental Media and Performing Arts Centers (EMPAC). We would also like to thank Johannes Goebel and Todd Vos from EMPAC for their support.

References

1. P. Oliveros, J. Watanabe and B. Lonsway, *A collaborative Internet2 performance*, tech. rep., Offering Research In Music and Art, Orima Inc. Oakland, CA (2003).
2. E. Chew, A. Sawchuk, R. Zimmerman, V. Stoyanova, I. Tosheff, C. C. Kyriakakis, C. Papadopoulos, A. Franois and A. Volk, Distributed immersive performance, in *Proceedings of the 2004 Annual National Association of the Schools of Music (NASM) Meeting*, (San Diego, CA, 2004).
3. R. Rowe and N. Rolnick, The technophobe and the madman: an internet2 distributed musical, in *Proc. of the Int. Computer Music Conf. Miami*, (Florida, 2004).
4. J. Cooperstock, J. Roston and W. Woszczyk, Broadband networked audio: Entering the era of multisensory data distribution, in *18th International Congress on Acoustics*, (Kyoto, 2004).
5. F. Schroeder, A. Renaud, P. Rebelo and F. Gualdas, Addressing the network: Performative strategies for playing apart, in *Proc. of the 2007 International Computer Music Conference (ICMC 07)*, (Copenhagen, Denmark, 2007).
6. P. Oliveros, S. Weaver, M. Dresser, J. Pitcher, J. Braasch and C. Chafe, *Leonardo Music Journal* **19**, 95 (2009).
7. J. C. Steinberg and W. B. Snow, *Electrical Engineering* , 12(Jan 1934).
8. J. Braasch, A loudspeaker-based 3D sound projection using virtual microphone control (ViMiC), in *Proc. of the 118th Convention of the Audio Eng. Soc.*, (Barcelona, Spain, 2005). Paper Number 6430.
9. J. Braasch, T. Ryan and W. Woszczyk, An immersive audio environment with source positioning based on virtual microphone control (ViMiC), in *Proc. of the 119th Convention of the Audio Eng. Soc.*, (New York, NY, 2005). Paper Number 6546.
10. J. Braasch, N. Peters and D. Valente, *Computer Music Journal* **32**, 55 (2008).
11. J. Cáceres, R. Hamilton, D. Iyer, C. Chafe and G. Wang, To the edge with China: Explorations in network performance, in *ARTECH 2008: Proceedings*

of the 4th International Conference on Digital Arts, (Porto, Portugal, 2008).
12. J. Cáceres and C. Chafe, JackTrip: Under the hood of an engine for network audio, in *Proceedings of International Computer Music Conference*, (Montreal, QC, Canada, 2009).
13. J. Braasch, A binaural model to predict position and extension of spatial images created with standard sound recording techniques, in *Proc. of the 119th Convention of the Audio Eng. Soc.*, (New York, NY, 2005). Paper Number 6610.
14. J. Blauert and J. Braasch, Räumliches Hören [Spatial hearing], in *Applications of digital signal processing to audio and acoustics*, ed. S. Weinzierl (Springer Verlag, Berlin-Heidelberg-New York, 1998) pp. 75–108.
15. T. Place and T. Lossius, Jamoma: A modular standard for structuring patches in Max, in *Proc. of the 2006 International Computer Music Conference (ICMC 06)*, (New Orleans, LA, 2006).
16. N. Peters, T. Matthews, J. Braasch and S. McAdams, ViMiC – A novel toolbox for spatial sound processing in Max/MSP, in *Proceedings of International Computer Music Conference*, (Belfast, Northern Ireland, 2008).
17. J. B. Allen and D. A. Berkley, *J. Acoust. Soc. Am.* **65**, 943 (1979).
18. J. Jot and A. Chaigne, Digital delay networks for designing artificial reverberators, in *Proc. of the 90th Convention of the Audio Eng. Soc.*, (Paris, France, 1991). Paper Number 3030.
19. J. Braasch, D. Valente and N. Peters, Sharing acoustic spaces over telepresence using virtual microphone control *Proc. of the Convention of the Audio Eng. Soc. New York, NY* **123**2007. Paper Number 7209.
20. A. Quazi, *IEEE Transactions on Acoustics, Speech and Signal Processing* **29**, 527(June 1981).
21. R. Hickling, W. Wei and R. Raspet, *J. Acoust. Soc. Am.* **94**, 2408(Oct 1993).
22. W. Würfel, Passive akustische lokalisation [passive acoustical localization], Master's thesis, Technical University Graz (1997).
23. M. Wright, A. F. A. and A. Momeni, Opensound control: State of the art 2003, in *Proceedings of the 2003 Conference on New Interfaces for Musical Expression (NIME-03)*, (Montreal, Canada, 2003).
24. B. Gunel, Loudspeaker localization using b-format recordings, in *Proceedings of the IEEE Workshop on Applications of Signal Processing to Audio and Acoustics*, (New Paltz, NY, USA, 2003).
25. V. Pulkki, J. Merimaa and T. Lokki, Reproduction of reverberation with spatial impulse response rendering, in *Proc. of the 116th Convention of the Audio Eng. Soc.*, (Berlin, Germany, 2004). Paper Number 6057.
26. Tintinnabulate & Soundwire, J. Braasch, C. Chafe, P. Oliveros and B. Woodstrup, *Tele-Colonization* (Deep Listening Institute, Ltd., DL-TMS/DD-1, 2009).
27. P. Oliveros, C. Bahn, J. Braasch, C. Chafe, T. Hahn, Soundwire Ensemble, Tintinnabulate Ensemble, D. Valente and B. Woodstrup, Dynamic spaces(August 2007), SIGGRAPH 2007.

AUDITORY ORIENTATION TRAINING SYSTEM DEVELOPED FOR BLIND PEOPLE USING PC-BASED WIDE-RANGE 3-D SOUND TECHNOLOGY*

Y. SEKI

National Institute of Advanced Industrial Science and Technology (AIST), 1-1-1 Higashi Tsukuba, Ibaraki 305-8566, Japan

Y. IWAYA, T. CHIBA†, S. YAIRI

Tohoku University, 2-1-1 Katahira, Aoba-ku Sendai, Miyagi 980-8577, Japan

M. OTANI

Shinshu University, 4-17-1 Wakasato Nagano, Nagano 380-8553, Japan

M. OH-UCHI

Tohoku Fukushi University, 1-8-1 Kunimi, Aoba-ku Sendai, Miyagi 981-8522, Japan

T. MUNEKATA

National Institute of Special Needs Education, 5-1-1 Nobi Yokosuka, Kanagawa 239-0841, Japan

K. MITOBE

Akita University, 1-1 Tegata Gakuencho Akita, Akita 010-8502, Japan

A. HONDA

Iwaki Meisei University, 5-5-1 Chuodai, Iino Iwaki, Fukushima 970-8551, Japan

* This study was partially funded by the Research Grants from the Okawa Foundation for Information and Telecommunications, 2008, and the Cooperative Research Project Program of the Research Institute of Electrical Communication, Tohoku University, 2007–2009.

† Has worked for Hitachi, Ltd., since 2009.

We developed a new auditory orientation training system for the orientation and mobility instruction of blind people using PC-based wide-range 3-D sound technology. Our training system can conduct both sound localization and obstacle perception training by producing virtual objects such as cars, walls, and roads in a virtual environment. Our training system has the following features: (i) The HRTF simulations are conducted by CPU using a mobile PC. (ii) The listener's head position and orientation are measured by the gyro, terrestrial magnetism sensor, acceleration sensor, and global positioning system (GPS), which can be connected to a PC using a USB interface.

1. Introduction

People with blindness must be able to cognize their environment using acoustic information through their auditory sense when they are walking or conducting daily activities. This skill, known as "auditory orientation", includes sound localization and obstacle perception. Sound localization is the ability to identify a sound source location, such as a vehicle or pedestrian. Obstacle perception is the ability to detect a silent object, such as a wall or pole, using sound reflection and insulation. It is sometimes called "human echolocation." [1]

Training of auditory orientation is usually conducted for people with blindness as one lesson in orientation and mobility (O&M) instruction. Such O&M instruction is usually conducted in a real environment; the trainee is expected to acquire auditory orientation capability by listening to ambient sounds experientially [2]. However, training in a real environment where actual vehicles are present is sometimes dangerous and stressful for novice trainees. Furthermore, the trainee must spend a long time to acquire auditory orientation using this training method because it is very difficult for the novice trainee to discern and listen to important sounds selectively from many other environmental noises. To reduce the risk and stress, and to shorten the period of training, a new training method in an ideal sound field reproduced by acoustical simulation is considered very effective. Methods to pursue training in the simulated sound field have been investigated in previous studies (see section 2.1).

We reported previously that we developed an auditory orientation training system that can conduct both sound location and obstacle perception training [3]. Our training system reproduced 3-D sound through headphones by simulating head-related transfer functions (HRTFs); it can measure positions and directions of the head and knee using magnetic six degrees of freedom (6DOF) sensors. We also reported that results of evaluation experiments showed that our system is effective in reducing both stress and veering in novice trainees (see section 2.2).

However, two major problems were apparent in our system: our system was very expensive, and the range of the trainee's walking was restricted to about 1 m. Consequently, our training system has not been introduced to education or rehabilitation facilities yet.

As described in this paper, we report that we have developed a new training system that solved the previous problems encountered when using PC-based wide-range 3-D sound technology. The new system has the following features: (i) HRTF simulations are conducted by CPU in mobile PC. (ii) The listener's head position and orientation are measured by the gyro, terrestrial magnetism sensor, acceleration sensor, and global positioning system (GPS), which can be connected to a PC through a USB.

2. Previous Studies

2.1. *General survey*

Methods to train in the simulated sound field have been investigated in previous studies.

Previous studies [4–10] were undertaken mainly to make the blind trainee acquire the "spatial concept" using the acoustic VR. The authors reported that their systems are efficient to some degree, but their virtual fields represented actual objects metaphorically, and did not reproduce the actual sound world such as towns, roads, or environment of daily living. Consequently, their systems cannot be used directly for O&M instruction. The other purpose of previous studies [11, 12] was navigation for people with blindness who had already acquired O&M skills, and not for O&M instruction.

Some studies have specifically addressed O&M [1, 13]. However, these focused on either sound localization or obstacle perception, but not both.

2.2. *Auditory Orientation Training System ver. 1.0 (AOTS 1.0) [3]*

In 2005, we developed an auditory orientation training system, ver. 1.0 (AOTS 1.0) that was able to reproduce not only the sound sources but also sound reflection and insulation, so that a trainee became able to learn both sound location and obstacle perception skills. Actually, AOTS 1.0 can reproduce a virtual training field for O&M instruction; the trainee can walk through the virtual training field safely while listening to sounds such as those of vehicles, stores, and ambient noise in 3-D through headphones.

AOTS 1.0 comprises 10 3-D sound processors (RSS-10; Roland Corp.), 10 sound recorders/players (AR-3000; Roland Corp.), two sound mixers (RFM-186; Roland Corp.), a magnetic 6DOF position and direction sensor (3SPACE Fastrak; Polhemus), headphones and an amplifier (SRS-4040; Stax Ltd.), and a computer (iBook G4; Apple Computer Inc.). Software was developed (REALbasic; REAL Software Inc.) to function on Apple Mac OS X. The 3-D sound processors and sound recorders/players are controlled through MIDI; the magnetic sensor is controlled through RS-232C (Figures 1 and 2).

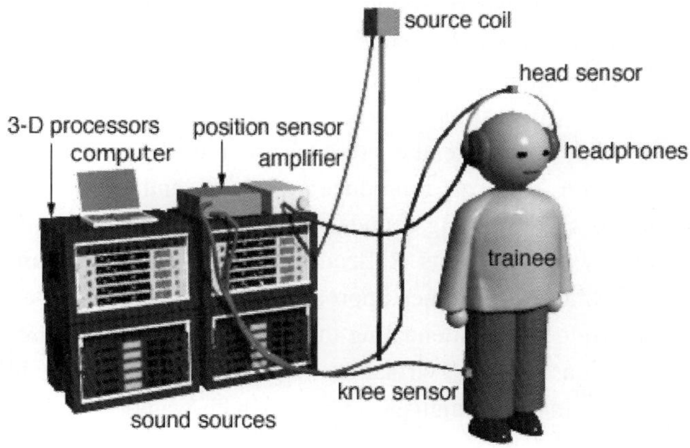

Fig. 1. Composition of AOTS 1.0.

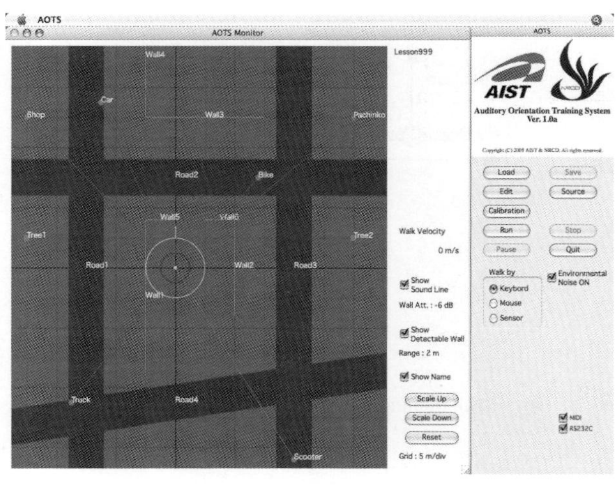

Fig. 2. Monitor display of AOTS 1.0.

A trainee can listen to sounds in the virtual training environment through headphones while changing the head direction. A trainee can also walk though the virtual training environment by moving their feet. The head and foot movements are measured by the magnetic 6DOF position and direction sensors.

The virtual training environment of AOTS 1.0 can include elements of four kinds: sound sources, walls, roads, and landmarks (Figure 3). The sound source can represent the sound of a vehicle, pedestrian, store, etc., and move in a constant speed and direction. The wall is used for training of the obstacle perception, and gives rise to sound reflection and insulation. The road and landmarks do not influence the sound propagation, but they are very helpful in the design of the virtual training environments. AOTS 1.0 can reproduce six sound sources and four ambient noises (from east, west, north, and south), simultaneously. To reproduce the presence of a wall for obstacle perception training, AOTS ver. 1.0 can reproduce reflection and insulation of ambient noise, and insulation of moving sounds. These reproductions enable the trainee to learn to detect walls and paths. Reflection and insulation of the ambient noises are reproduced when the listener approaches to within 2 m of a wall; sound insulation is reproduced by attenuating the sound by 6 dB. The O&M instructor can design and "construct" the virtual training environments easily by describing them in extensible markup language (XML), which was originally proposed for this system. This technology is now patent pending.

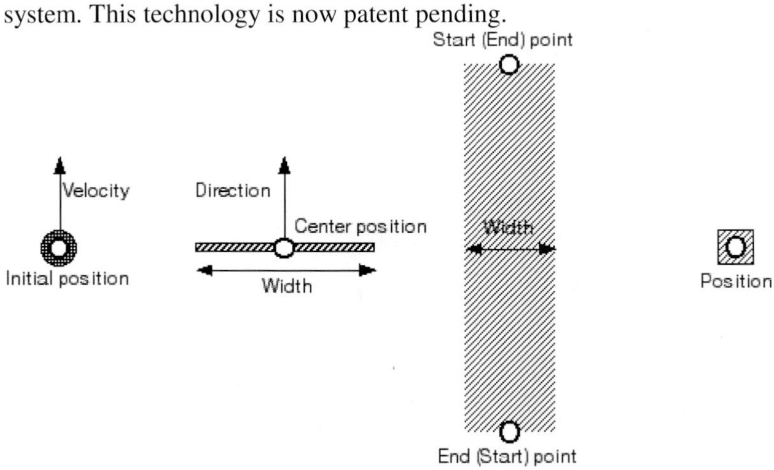

Fig. 3. Elements of the virtual training environment: (from left) sound source, wall, road, and landmark.

Some effectiveness assessments of AOTS 1.0 were conducted. Subjects were 30 sighted people who had been blindfolded. They were divided into three groups: Control, AOTS, and O&M. The Control group was not trained. The AOTS group was trained using AOTS. The O&M group was trained using a usual O&M program. The training course was a 50-m-long straight sidewalk.

The stress reduction effect of AOTS was measured using the stress pulse ratio (SPR) (Figure 4 left), which was calculated as

$$\mathrm{SPR} = 100\,(P - P0)/\,P0\ [\%], \qquad (1)$$

where P is the measured heart rate, and $P0$ is the heart rate when the trainee feels no stress during walking. If the trainee feels no stress during the walk, then P is equal to $P0$ and SPR = 0. As the stress increases, the heart rate also increases proportionally to the stress. For this discussion, DSPR is the difference between the SPRs of post-training and pre-training. Results show that actual O&M training is effective for reducing stress, although novice trainees feel great stress initially. AOTS was also effective, but slightly less so than O&M.

The veering reduction effect of AOTS was measured using a travel locus (Figure 4 right). The veering score (DVS) was calculated using the veering distance, where a smaller DVS represents smaller veering. Results show that AOTS is the most effective method for training auditory orientation skills. A possible reason is that no other factors (tactile, smell, etc.) were included in the virtual training space of AOTS. Therefore, the trainee was able to concentrate on learning the auditory orientation.

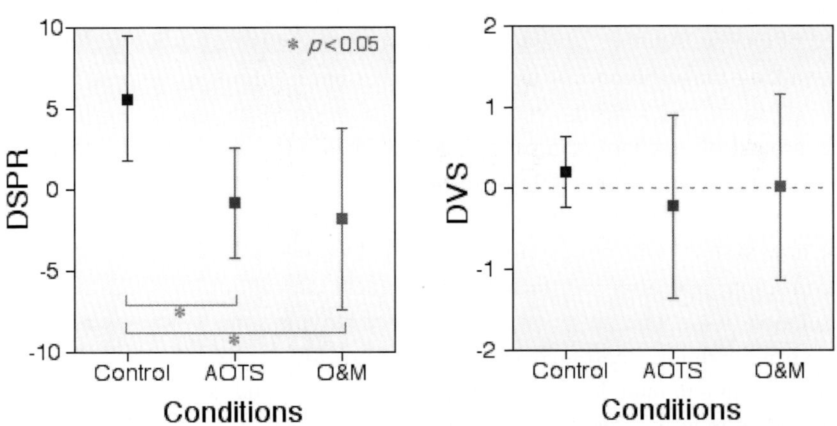

Fig. 4. Evaluation results of AOTS 1.0: stress reduction results (left) and veering reduction results (right).

Nevertheless, two major problems remain with AOTS 1.0: it is very expensive (about 5 million yen); and the trainee's walking range is restricted to about 1 m because the magnetic 6DOF sensor has a limited detection range. Consequently, our training system has not been introduced to education or rehabilitation facilities yet.

3. Wide-Range Auditory Orientation Training System (WR-AOTS)

We developed a new training system that solves the problems presented above. Our new training system has the following features: (i) All HRTF simulations are conducted by CPU in a mobile PC, whereas the previous system simulated them using expensive digital signal processors (DSPs) in external devices. The Simulative environment for 3-D Acoustic Software (SiFASo) technology developed by Oh-uchi et al. [9] was used for the PC-based HRTF simulation. The SiFASo can reproduce at least eight 3-D sounds simultaneously using a Pentium IV 2 GHz or better processor. (ii) The listener's head position and orientation are measured using the gyro, terrestrial magnetism sensor, acceleration sensor, and global positioning system (GPS), which can be connected to the PC through a USB whereas the previous system used an expensive magnetic 6DOF sensor. These two features reduce the price of our system. The education/rehabilitation facilities with PCs available are expected to pay only a few tens of thousands of yen for GPS and other sensor equipment. The software of our new system can be distributed for a low price, possibly even free of charge. Another advantage of our new system is that it has no limitation in walking range because it uses sensors that have unlimited range to measure the position and orientation. Therefore, a wide open space, such as the playground of a school, can be used as a wide virtual space in which a trainee can actually walk around while performing auditory orientation training.

Moreover, the important advantages of the previous system are retained in the new system. Our new system can reproduce not only the factors of sound localization but also the factors of obstacle perception by reflection and insulation of ambient noise, and insulation of moving sounds. Representation of the virtual training environment is described in XML format. The virtual training environment can include sound sources, walls, roads, and landmarks. The sound source can present vehicle, pedestrian, and store sounds, and present simulated movement with a constant speed and direction. The wall is used for the training of obstacle perception. It gives rise to the sound reflection and insulation. The road and landmark do not influence sound propagation, but they are helpful to design virtual training environments. The O&M instructor can design and "construct" virtual training environments easily by describing them in XML, as originally proposed for this system.

The simulation of sound reflection and insulation in the previous system was qualitative. The reflection and insulation of the ambient noise were reproduced when the listener approached within 2 m of a wall. The reflection of sound sources other than ambient noise was not reproducible because of a lack of 3-D sound channels. The sound insulation was reproduced by attenuating the sound by 6 dB. These simulations were not correct quantitatively, but they sounded plausible. We are attempting to improve the sound rendering fidelity. The improved rendering algorithm will be included in the system software. It can be provided to users as a software update.

The prototype of our new system consists of a mobile PC (Toughbook, Core 2 Duo™ 1.06 GHz; Panasonic Inc.), a GPS (GM-158-USB, 5 Hz sampling in NMEA GGA format; San Jose Navigation Inc.), a 3-D motion sensor containing a ceramic gyro, a terrestrial magnetism sensor, and an acceleration sensor (MDP-A3U9S; NEC Tokin Corp,), headphones (HD 280 pro; Sennheiser Electronic GmbH and Co.), and a audio stream input output (ASIO) sound adapter (Transit USB; M-Audio) (Figures 5–7).

The hardware of the prototype comprises a single mobile PC and three small USB peripherals. A trainee can carry it easily. The source code of the prototype is written in Microsoft Visual C++, a Windows application on Win32API (Microsoft Corp.). The application can run on either Windows XP or Vista (Microsoft Corp.).

We have verified that this new system can reproduce at least 10 3-D sounds simultaneously.

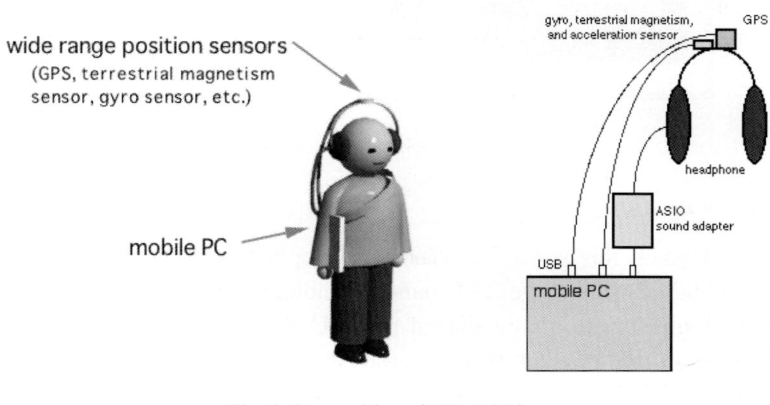

Fig. 5. Composition of WR-AOTS.

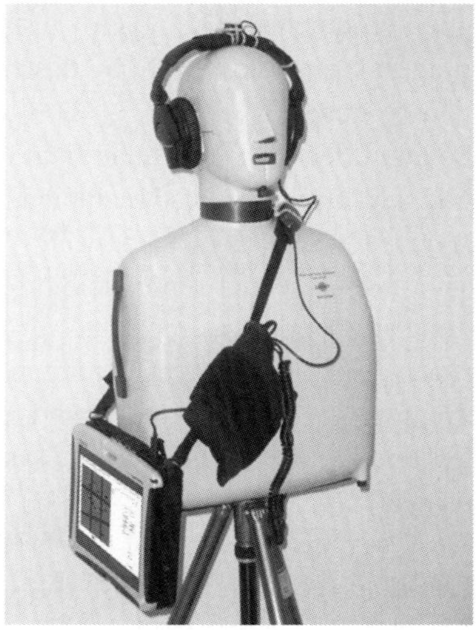

Fig. 6. Photo of WR-AOTS.

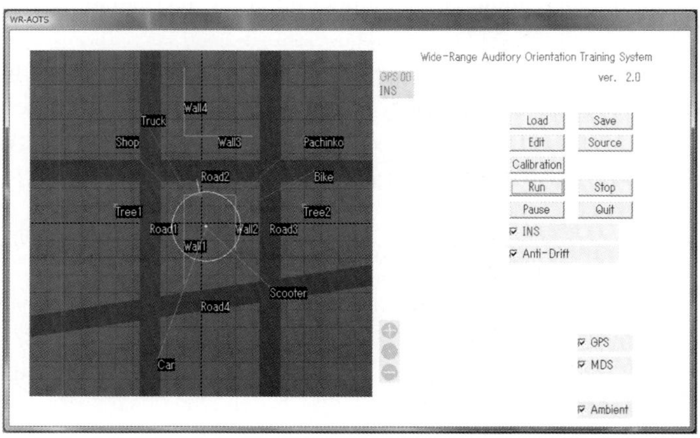

Fig. 7. Monitor display of WR-AOTS.

4. Summary

As described in this paper, we report development of a new training system using PC-based wide-range 3-D sound technology with the following features: (i) HRTF simulations are conducted by CPU in mobile PC. (ii) The listener's head position and orientation are measured by the gyro, terrestrial magnetism sensor, acceleration sensor, and global positioning system (GPS), which can be connected to the PC through a USB interface.

Our new system can resolve two important problems posed by the previous system: high cost, and restriction of the trainee's walk range. We believe that our new system can be introduced easily to education and rehabilitation facilities.

References

1. Y. Seki and K. Ito, Obstacle perception training system and CD for the blind, *Proc. CVHI* **2004**, CD-ROM (2004).
2. B. B. Blasch, W. R. Wiener and R. L. Welsh, *Foundations of Orientation and Mobility* **2nd Ed.**, 1977.
3. Y. Seki and T. Sato, Development of auditory orientation training system for the blind by using 3-D sound, *Proc. CVHI* **2006**, CD-ROM (2006).
4. M. Shimizu, K. Itoh and T. Kanazawa, Pattern representation system using movement sense of localized sound, *Proc. HCI Int.* **2**, 990 (1999).
5. M. L. Max and J. R. Gonzalez, Blind persons navigate in virtual reality (VR); hearing and feeling communicates "reality", *Medicine Meets Virtual Reality, Global Healthcare Grid*, (1977).
6. J. L. González-Mora, A. Rodríguez-Hernández, L. F. Rodríguez-Ramos, L. Díaz-Saco and N. Sosa, Development of a new space perception system for blind people, based on the creation of a virtual acoustic space, *Lecture Notes in Computer Science* **1607**, 321 (1999).
7. M.W. Krueger and D. Gilden, Going places with "KnowWare": virtual reality maps for blind people, *Lecture Notes in Computer Science* **2398**, 565 (2002).
8. J. Sánchez, User-centered technologies for blind children, *Human Technology* **4**, 96 (2008).
9. M. Oh-uchi, Y. Iwaya, Y. Suzuki and T. Munekata, Cognitive-map forming of the blind in virtual sound environment, *Proc. 12th Int. Conf. Auditory Display*, 1 (2006).
10. A. Honda, H. Shibata, J. Gyoba, K. Saitou, Y. Iwaya and Y. Suzuki, Transfer effects on sound localization performance from playing a virtual three-dimensional auditory game, *Applied Acoustics* **68**, 885 (2007).
11. S. Hollan and D. R. Morse, Audio GPS: spatial audio in a minimal attention interface, *Proc. Human Computer Interaction with Mobile Devices*, **28** (2007).
12. J. M. Loomis, J. R. Marston, R. G. Golledge and R. L. Klatzky, Personal guidance system for people with visual impairment: a comparison of spatial displays for route guidance, *J. Visual Impairment & Blindness*, **219** (2005).

13. D. P. Inman and M. S. Ken Loge, Teaching orientation and mobility skills to blind children using simulated acoustical environments, *Proc. HCI Int.* **2**, 1090 (1999).

MAPPING MUSICAL SCALES ONTO VIRTUAL 3D SPACES

J. VILLEGAS and M. COHEN*

Spatial Media Group, University of Aizu
Aizu-Wakamatsu, Fukushima-ken 965-8580; Japan
**E-mail: julovi@yahoo.com, mcohen@u-aizu.ac.jp*
`www.u-aizu.ac.jp/~mcohen/spatial-media`

We introduce an enhancement to the Helical Keyboard, an interactive installation displaying three-dimensional musical scales aurally and visually. The Helical Keyboard features include tuning stretching mechanisms, spatial sound, and stereographic display. The improvement in the audio display is intended to facilitate pedagogic purposes by enhancing user immersion in a virtual environment. The newly developed system allows spatialization of audio sources controlling the elevation and azimuth angles at a fixed range. In this fashion, we could overcome previous limitations on the auditory display of the Helical Keyboard, for which we heretofore usually displayed only azimuth.

Keywords: Music 3D Visualization, Music 3D Auralization, Multimodal Musical Interface, Immersive Environment, Musical Scales Topology, Pedagogy of Music, Visual Music

1. Introduction

Most cultures have developed mechanisms to represent, store, and preserve musical content, partly because of the ephemeral nature of sound. Circa the 11th century, Arezzo proposed a musical notation (staff notation) that can be considered the most successful of such attempts, judging by its enduring usage and current ubiquity. In his system, height on a written staff corresponds intuitively to aural pitch height. Arezzo's set of seven notes and their respective chromatic alterations (flats—♭ and sharps—♯) constitute the most widely used discretization of the musical octave (twelve tones per octave). These twelve tones are commonly evenly distributed— i.e., they are equal tempered.

Staff notation adequately captures linear dimensions of musical scales (pitch height and time), but other scale properties are difficult to represent. For example, staff notation lacks an intuitive way to visualize pitch chroma, the fact that tones separated by an integral number of octaves (comprising pitch classes) are judged as more similar than other intervals. Shepard[1] proposed a geometrical representation of scales expressing pitch height and chroma as well as the circle of fifths (tones separated by perfect fifths, which are regarded as harmonically closer than other intervals besides the unison and octave). This structure is shown in Figure 1.

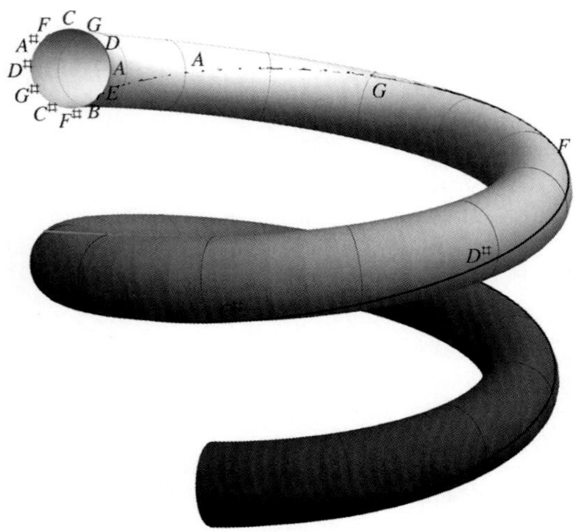

Fig. 1: Multidimensional scale model proposed by Shepard. The circle of fifths corresponds to the minor axis cross-section, chroma to the major axis cross-section, and pitch to the height.

The completeness of Shepard's model makes it impractical for many applications, so projections into lower dimensional spaces have been preferred. Chew & Chen[2] use one such projection to visualize and track tonal patterns in real-time, and Ueda & Ohgushi[19] confirmed with subjective experiments the suitability of helices to represent tone height and chroma (as had been theoretically proposed by Shepard).

2. Helical Keyboard

The Helical Keyboard[3] is a permanent installation at the University of Aizu University-Business Innovation Center (UBIC).[a] This Java application allows visualization, auralization, and manipulation of equal tempered scales. Pitch height and chroma are mapped into a left-handed helix with low pitches at the bottom. The standard (unstretched) helix has one revolution per octave, as shown in Fig-

[a]A Java webstart version of this application can be launched from `http://julovi.net/webstart/hkb/hkb.jnlp`

ure 2a. Coördinated visual and aural display is featured by the Helical Keyboard

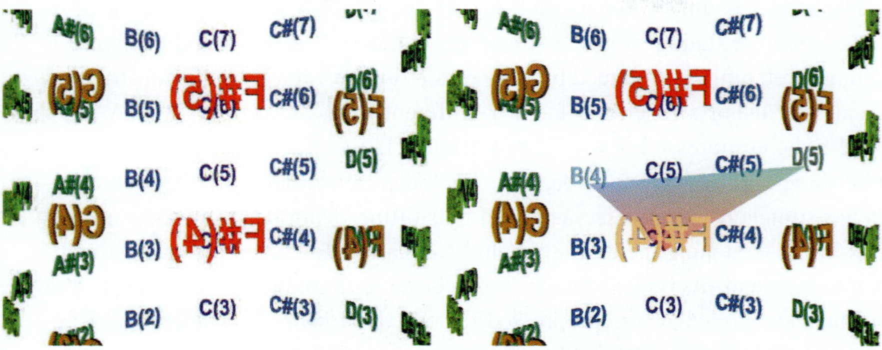

(a) standard representation

(b) a chord-kite for Bm (2nd inversion) interpolating a triangle between $F^{\sharp}_4$, B_4, and D_5

Fig. 2: The nominal (unstretched) Helical Keyboard with chromastereoscopic rendering.

in different ways. Realtime chord recognition is visualized by "chord kites," polygons connecting constituent chord notes (as shown in Figure 2b). When several keys are activated (by pressing the corresponding real keys on a MIDI controller, for example), the Helical Keyboard displays a polygon interconnecting them if their compound forms a simple chord (triads or tetrads) in any inversion.

(a) Stretched (2.1:1)

(b) Compressed (1.1:1)

(c) Inverted (1:2)

Fig. 3: The Helical Keyboard with different stretching/compressing ratios.

The Helical Keyboard can receive and transmit MIDI events from various controllers (including piano-style keyboards and the notes of the GUI 3D helix model) to user-specified synthesizers. Using a joystick or computer keyboard, visitors can fly, rotate and translate the helix, and invoke cinematographic rotational perspective effects such as roll, tilt, and pan (a.k.a. roll, pitch, and yaw), and translational effects such as dolly, boom, and track (a.k.a. surge, heave, and sway).

The model is unique for its multimodal display characteristics and the ability to stretch the tuning of a MIDI stream in realtime,[5,6] as illustrated in Figure 3. 'Tuning Stretching,' introduced by Mathews & Pierce,[7] refers to the construction of scales by using a repetition factor ρ different from the octave (2:1). A scale is said to be compressed if $\rho < 2$, or stretched if $\rho > 2$. Mathews & Pierce describe three kinds of tuning stretching: 'melodic,' for which only the tone intervals are stretched; 'harmonic,' achieved by shifting overtones from their normal ratios by the stretching factor; and 'melodic & harmonic,' simultaneously stretching intervals and overtones. Music is perceived as harmonious if intervals and overtones are stretched (or compressed) by the same factor.[8,9] The Helical Keyboard can melodically stretch any MIDI stream if the deployed synthesizer implements the MIDI commands `pitch-bend` and `pitch-bend sensitivity`. Since realtime harmonic stretching it is not a common feature in modern synthesizers, we created a JavaSound additive synthesizer capable of such expression, available via the Helical Keyboard application menu.

Stereographic rendering is achieved by a combination of dual projectors with orthogonal polarizing filters on their beams, silver (polarity preserving) screen, and passive filter eyewear. Alternatively, visual 3D cues can also be generated chromastereoscopically[4] and viewed with ChromaDepth eyewear.[b] The latter technique is simpler (and more limited), backwards compatible (so no ghosting when viewing without eyewear), and used when printing (as in Figures 2 and 3) or off-site demonstrations are performed. Audio spatialization is discussed in the next section.

3. Previous Spatial Audio Display Solutions

Originally, sound spatialization for the Helical Keyboard was achieved with the Pioneer Sound Field Control System (PSFC) using a loudspeaker array in hemispherical configuration[10] (shown in Figure 4). Later, the PSFC system was eclipsed by a system integrating four Roland Sound Space Processors (RSS-10s).[11] The most recent spatialization solution (illustrated in Figure 5) is performed by directly manipulating via MIDI messages an audio crossbar mixer (Roland VM-7200[13]) connected to one of the loudspeaker arrays.[12]

The PSFC system can directionalize only two channels, limiting possibilities to display chords spatially polyphonically, and this legacy system is difficult to maintain and debug. Spatialization based on the RSS-10 processors can manage only a

[b]`www.chromatek.com`

Fig. 4: Ubic 3d Theater, with two suspended loudspeaker arrays

single dynamic audio channel, and the internal protocol used to communicate with the Rss-10s does not allow distance control.

Communication between locally authored Java applications and control/display systems is via the Collaborative Virtual Environment protocol (cve), a simple but robust synchronization mechanism developed in our laboratory used for sharing information regarding position and attributes of virtual objects among multiple applications.[14] Despite these improvements, the latest implementation has some of the same restrictions as its predecessors: the virtual space is collapsed into a plane where only azimuth and distance can be displayed, and the loudspeaker array is meters above the listeners, hindering cross-talk cancelation and impairing localization.

The discussed difficulties do not prevent listeners from enjoying a partially immersive experience when the Helical Keyboard is presented at the Ubic 3d Theater. But, when the Helical Keyboard is presented in other venues, auralization had been limited to a projection of the notes over the left–right axis of the listener's head, degrading significantly the immersive experience. We develop an extension of the Helical Keyboard using Head-Related Impulse Response (hrif) filters to ameliorate this experience, as explained in the following section.

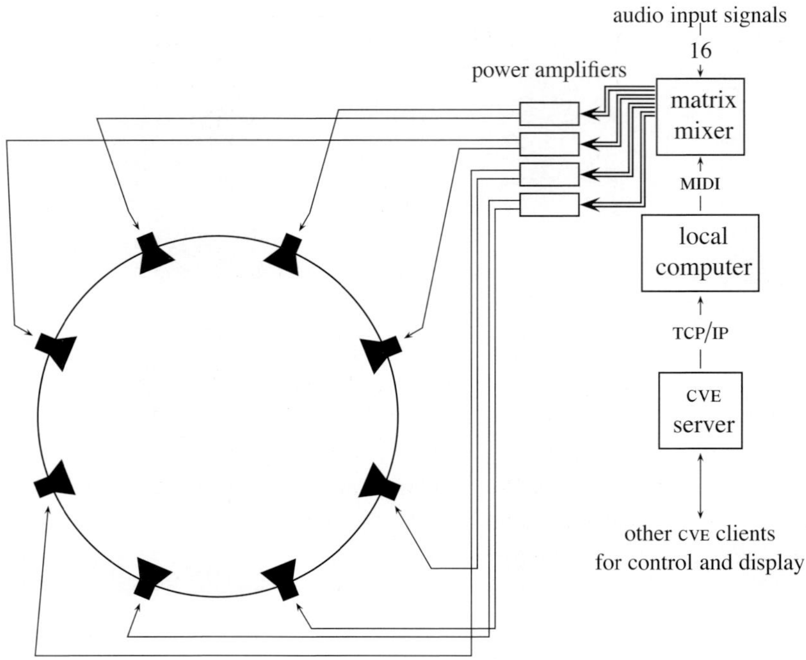

Fig. 5: The UBIC 3D Theater and newest speaker array spatialization installation

4. Implementation

Java3D originally performed audio spatialization through the JavaSoundMixer library. This library had some disabled functions, and was subsequently replaced with JOAL (Java Bindings for OpenAL),[c] a cross-platform 3D audio API. We envision a system in which MIDI notes are independently directionalized according to the relative direction between the sources and sink (i.e., the listener). Such spatial polyphonic richness is possible with JOAL, but requires MIDI synthesizers to be Java software. Alternatively, MIDI channels can be associated with independent audio channels which are spatialized. We followed the latter approach with a combination of Java (not JOAL) and `Pure-data` (`Pd`), as described in the following paragraphs. `Pd` is a realtime graphical programming environment for audio, video, and graphical processing.[d]

All musical notes in our virtual environment are modeled as spherical sound sources, radiating energy isotropically (with equal intensity in all directions in a

[c]`https://joal.dev.java.net`
[d]`http://puredata.info`

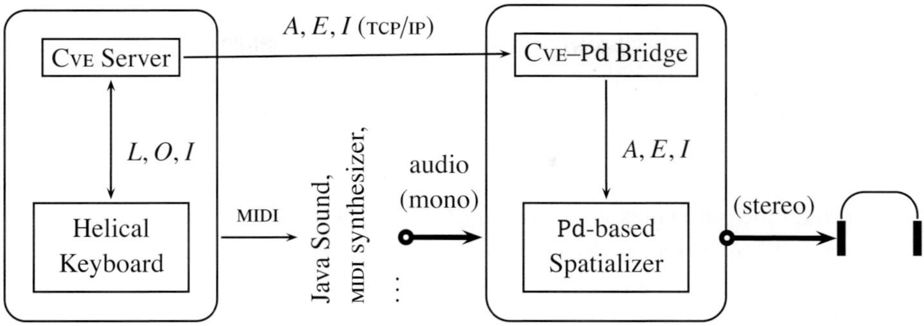

Fig. 6: Diagram of the system. Location L, Orientation O, and sound Intensity I are sent from the Helical Keyboard to a CVE server TCP/IP sockets. This server relays such information—as Azimuth A, Elevation E, and Intensity I— to connected clients, including the CVE–Pd Bridge. Communication between the CVE–Pd Bridge and the Pd-based spatializer is also via TCP/IP sockets. The monaural output of the designated synthesizer in the Helical Keyboard is connected to the audio input of the machine hosting the spatializer. The Pd program spatializes this audio stream.

homogeneous medium). The virtual head of a listener represents the origin of a local coördinate system into which the sources are projected. We consider only the locations of the sound sources and the position of the virtual listener to create the 3D soundscape. Position P comprises location L (x, y, z, in cartesian or rectangular coördinates) and orientation O (roll ϕ, pitch θ, and yaw ψ angles). Projection into the virtual head frame-of-reference is performed by a double transformation: from the sources' local coördinates to virtual world coördinates, and thence to the sink's local system. Note that Java3D coördinates observe a right-hand convention, with the positive z axis pointing out of the screen, and positive x axis to the left. Azimuths for auralization are measured counterclockwise from the front sagittal plane.

The distance between source and sink is used to modulate sound amplitude. Basically (and unrealistically), a distance gain attenuator monotonically decreases the amplitude of the sound as the distance between the source and the sink increases.

We developed a three-part spatialization solution: The Helical Keyboard tracks object positions and sends them to a CVE server whenever there is a change; a CVE–Pd bridge receives azimuths, elevations, and ranges from the CVE server and relays the source→sink vector, via TCP sockets, to a spatializer Pd program.

The spatializer instantiates a `earplug~` object, a binaural filter[15] which allows spatialization of sources with bearing angle $0 \leq A < 2\pi$ and elevation angle

$-2\pi/9 \leq E \leq \pi/2$. A mono audio source is connected to the input of the computer running the spatializer. This mechanism allows switching among different MIDI synthesizers. Figure 6 illustrates the system.

`Earplug~` uses the KEMAR compact set of impulse response binaural measurements provided by Bill Gardner and Keith Martin at the MIT Media Lab (1994),[e] who claim angular error in their measurements of about ±0.5°. The number of azimuth measurements varied with elevation. Elevation was measured every 10° from −40° to +90° (i.e., above the vertical axis of the dummy head). Whereas only one measurement was taken for the zenith, 72 measurements (every 5°) were taken around the equator. Equalization was performed to compensate for the spectral coloring of the loudspeaker used in the measurements. The 'compact set' comprises equalized filters with only "128 point symmetrical HRTFS derived from the left ear KEMAR responses." Up to four of the closest measurements are linearly interpolated to convolve with the audio signal in the time domain.

The `Pd` application was tested on a MacBook with 2 GB of RAM running Mac OS X v. 10.6.1; `Pd` v. 0.41.4-extended was connected to Jack OS X audio server v. 0.82. The sampling rate and frame buffer in the audio server are 44.1 kHz and 512 samples (about 11 ms). The frame size is 64 samples with no overlap (the default `Pd` setup).

5. Discussion and Future Work

Interesting musical experiences can result when users stretch tuning *ad libitum* (freely) while rendering MIDI streams. One may collapse a scale to a single tone, or invert it (as in Figure 3c); reproduce any equal-tempered scale; and create complex patterns of beatings, compressing the tuning harmonically to close to unity. Understanding of such musical curiosities is reinforced by their visual display. For instance, extreme compression ratios cause notes to tend to fuse in the multiple display modalities: visualization, pitch height, pitch chroma, and auditory direction.

Our group has explored different display modalities for virtual and augmented reality, including haptic interfaces such as the $^{S}_{c}$ha$_{i}$r^{e} rotary motion platform.[16] The $^{S}_{c}$ha$_{i}$r^{e} could be used in conjunction with the Helical Keyboard to reproduce and experiment with interesting psychoacoustic phenomena, such as the tritone paradox reported by Diana Deutsch.[17] In her experiments, participants judged sequences of Shepard tones[18] as ascending or descending. Differences between judgements of the two groups are apparently related to their exposure to different languages or dialects. It would be interesting to explore how this illusion is

[e] `http://sound.media.mit.edu/resources/KEMAR.html`

affected by the inclusion of haptic and proprioceptive cues.

The extended Helical Keyboard is useful for demonstrating the benefits of HRTF filters for audio spatialization. However, this proof of concept should be improved in many aspects. For instance, the number of audio channels (polyphonic degree) should be increased to directionalize chords articulately regardless of the spatialization technique (i.e., performing directionalization on a different computer, or using JOAL).

In the future, UBIC visitors might wear wireless gyroscopically tracked headphones as well as stereoscopic eyewear to enjoy a more immersive virtual reality experience. Currently, we are exploring the implementation of such features using an Arduino FIO microcrontroller[f] connected via Xbee wireless radios[g] to the machine hosting the Helical Keyboard.

In collective experience, general HRTF filters offer adequate aural rendition. For individual use, however, it is desirable to improve the overall quality of the experience by personalizing such earprints.

6. Conclusions

The feasibility of implementing an audio spatializer capable of working in collaboration with other clients of our development suite (CVE) has been confirmed in the extension and modernization of the Helical Keyboard. Pd was used for development, adding a degree of freedom in the auditory display (i.e., elevation) and exposing some restrictions such as limited polyphonic degree and fixed range HRTFS measurements. Such limitations need to be circumvented in future implementations (probably using JOAL as backend). Observed results of our preliminary prototype encourage integration of this solution with the UBIC permanent exhibition and exploration of new research directions. The permanent exhibition of the Helical Keyboard features include multiple stereoscopic options, multiple spatial sound options, elementary chord recognition and display (chord-kites), and harmonic and melodic stretching.

7. Acknowledgments

We thank Prof. Yôiti Suzuki for his valuable observations.

References

1. Roger Shepard. *Structural Representation of Musical Pitch.* Academic-Press, New York, NY, USA, 1982.

[f] `http://arduino.cc/en/Main/ArduinoBoardFio`
[g] `www.digi.com/products/wireless/point-multipoint/xbee-series1-module.jsp`

2. E. Chew and Y-C Chen. Mapping MIDI to the Spiral Array: Disambiguating Pitch Spelling. In *Computational Modeling and Problem Solving in the Networked World–Proc. of the 8th INFORMS Computer Soc. Conf.*, pages 259–275. Computational Modeling and Problem Solving in the Networked World, Kluwer, 2003.
3. Jens Herder and Michael Cohen. The Helical Keyboard: Perspectives for Spatial Auditory Displays and Visual Music. *J. of New Music Research*, 31(3):269–281, 2002.
4. Richard A. Steenblik. Chromastereoscopy. In *Stereo Computer Graphics and Other True 3D Technologies*, David F. McAllister, editor, pages 183–195. Princeton University Press, 1993.
5. Julián Villegas and Michael Cohen. Melodic Stretching with the Helical Keyboard. In *Proc. Enactive: 2nd Int. Conf. on Enactive Interfaces*, Genoa, Italy, November 2005.
6. Julián Villegas, Yuuta Kawano, and Michael Cohen. Harmonic Stretching with the Helical Keyboard. *3D Forum: J. of Three-Dimensional Images*, 20(1):29–34, 2006.
7. M. V. Mathews and J. R. Pierce. Harmony and Nonharmonic Partials. *J. Acoust. Soc. Am.*, 68:1252–1257, 1980.
8. A. J. M. Houtsma, T. D. Rossing, and W. M. Wagenaars. Auditory Demonstrations, 1987. Philips compact disc No. 1126–061.
9. John Pierce. *Consonance and Scales*. MIT Press, Cambridge, MA; USA, 2001.
10. Katsumi Amano, Fumio Matsushita, Hirofumi Yanagawa, Michael Cohen, Jens Herder, William Martens, Yoshiharu Koba, and Mikio Tohyama. *TVRSJ: Trans. of the Virtual Reality Soc. of Japan*, 3(1):1–12, 1998.
11. Masahiro Sasaki. Dancing Music: Motion Capture Data Parameterizing Musical Synthesis and Spatialization via Speaker Array. Master's thesis, University of Aizu, 2005.
12. Yoshiyuki Yokomatsu. Primassa: Polyphonic Spatial Audio System with Matrix Mixer and Speaker Array Integrated with CVE. Master's thesis, University of Aizu, 2007.
13. Roland Corp. *V-Mixer VM-7200/VM-7100 Users Manual*, 2003. PDF version available on Roland's Website.
14. Takashi Mikuriya, Masataka Shimizu, and Michael Cohen. A Collaborative Virtual Environment Featuring Multimodal Information Controlled by a Dynamic Map. In *Proc. HC2000: Third Int. Conf. on Human and Computer*, pages 77–80, Aizu-Wakamatsu, Japan, 2000.
15. Pei Xiang, David Camargo, and Miller Puckette. Experiments on Spatial Gestures in Binaural Sound Display. In *Proc. of* ICAD*—Eleventh Meeting of the Int. Conf. on Auditory Display*. ICAD, 2005.
16. Uresh Chanaka Duminduwardena and Michael Cohen. Control System for the $^{S}_{C}$ha$_{i}$r^{e} Internet Chair. In *Proc. CIT-2004: Int. Conf. on Computer and Information Technology*, pages 215–220, 2004.
17. Diana Deutsch. The Tritone Paradox: Effects of Spectral Variables. *Perception & Psychophysics*, 41(6):563–75, 1987.
18. R. Shepard. Circularity in Judgments of Relative Pitch. *J. Acoust. Soc. Am.*, 36:2345–2353, 1964.
19. Kazuo Ueda and Kengo Ohgushi. 多次元尺度法による音の高さの二面性の空間的表. Perceptual components of pitch: Spatial representation using a multidimensional scaling technique. *J. Acoust. Soc. Am.*, 82(4):1193–1200, 1987. In Japanese.

SONIFYING HEAD-RELATED TRANSFER FUNCTIONS

D. CABRERA, and W. L. MARTENS
Faculty of Architecture, Design and Planning, The University of Sydney, Sydney, NSW 2006, Australia
E-mail: densil.cabrera@sydney.edu.au, william.martens@sydney.edu.au
sydney.edu.au

This chapter describes a set of techniques that can be used to make the spectral and temporal features of head-related transfer functions (HRTFs) explicitly audible, whilst preserving their spatial features. The purpose of sonifying HRTFs in such a way is to enhance the understanding of HRTFs though the listening experience, and this is especially applicable in acoustics education.

1. Introduction

Head-related transfer functions (HRTFs) play a major role in spatial hearing [1], and so gaining an understanding of them is an important part of an education in spatial hearing. Such understanding involves a combination of theoretical development and experience – and such experience might involve measuring HRTFs, visualizing HRTF data, listening to sound convolved with one's own and other people's HRTFs, synthesizing HRTFs, and so on. The purpose of this chapter is to demonstrate how a set of techniques can be combined to produce a sonification of HRTFs that conveys rich information to the listener. The sonifications are intended to make the spectral, temporal and spatial structures of HRTFs plainly audible, which is in stark contrast to the 'click' that one hears when listening directly to a HRTF. It is also different in purpose to the process of convolving a signal (such as speech) with HRTFs, in that we are aiming to provide an explicit experience of the HRTFs' features, rather than just an experience of the HRTFs' effect on a signal. The purpose of the sonification is to supplement other experiences and theoretical exposition in education on spatial hearing.

Sonification is analogous to visualization, presenting data for listening to, rather than looking at. According to Hermann [2], the sound of a sonification reflects objective data properties, through a systematic transformation that has reproducible results, and the sonification can be applied to different data. Perhaps the simplest form of sonification is 'audification', which involves playing data with little transformation as if the raw time series comprised an audio recording (for example, the computer program SONIFYER [3]). The techniques described in this chapter could be thought of as a sophisticated approach to audification, in that although the data are played as audio, the transformations that we apply are more complex than usual for audification, and we combine three interpretations of the data into a single sonification.

General principles of sonification are likely to be similar to general principles of data visualization. Visualizations should focus on data rather than introducing distractions [4]. Including the full dataset within a visualization (rather than merely averages, for example) can allow a user to appreciate the scale and form of the detail as well as gaining an immediate overview [5]. Visualizations should be attuned to perception, and pre-attentive displays (in which the display is understood to some extent without conscious effort or learning) are preferred [6]. However, time is almost inevitably used as the primary display domain in sonifications, whereas space dominates in visualizations, so exploration of a full dataset is likely to be approached in a different way.

HRTFs are sound phenomena (although they are system responses, not acoustic signals), and so are amenable to being played as an audio recording even without transformation (in this chapter, the term HRTF is used to include its time-domain equivalent head-related impulse response). The sonification of sound phenomena for acoustics education is a concept developed by Cabrera and Ferguson [7] based on the concept that listening to acoustic data can provide a rich experience of the relevant phenomena, to complement the reductive representation given by visualization and the profoundly abstract representation given by large arrays of numbers. The sheer relevance of sonification to sound phenomena makes a straightforward case for its use.

2. Sonification Techniques

The sonification techniques described in this paper are intended to facilitate an expanded perception of HRTF features – and could be thought of as ways of 'zooming in' to the various features. The features of HRTFs that we are concerned with are their temporal and spectral content, including interaural. Used in combination, the sonification techniques provide multiple perspectives on HRTF data. In defining these techniques, we have chosen to use simple operations on the waveform, which can be implemented in a simple function (given in the appendix).

2.1. *Appropriate Duration*

The brevity of a HRTF listened to directly makes it difficult to catch much information from it. Visualization in the form of charts, and listening through convolution with a signal such as white noise, both provide a steady state stimulus that a student can mentally explore. The latter, however, also provides a spatial perception (which may or may not correspond to the direction from which the HRTF was measured) and a direct experience of the HRTF timbre, and thus is a rich representation. The appropriate duration for mental exploration probably should be longer than the loudness integration time (*i.e.*, not less than 100 ms [8, 9]), and depends on the detail of the exploration. For example, a short duration is desirable if many HRTFs are being compared in succession, but durations of several seconds may be helpful if a small number of HRTFs are being examined. Interaural time cues are preserved by applying the same noise signal to the HRTFs of both ears.

In our implementation we generate a random phase spectrum for components other than 0 Hz and the Nyquist frequency, which is then combined with the magnitude spectra for each ear, before returning to the time domain (this is equivalent to convolution with white noise). The duration of the sonification is determined by the fast Fourier transform size (using zero-padded data).

2.2. *Scaling of the Magnitude Spectrum*

White noise convolved with a binaural pair of HRTFs conveys spatial and timbral features, but the features (such as peak frequencies) of the amplitude spectrum are difficult to discern. The sound is noise-like, and at best, the timbre could be described as 'colored noise'. One of the ways in which features can be brought to the fore is through a non-linear scaling of the amplitude spectrum, which can be achieved by raising the absolute value of the spectrum to a power (whilst preserving the original spectrum phases). This operation is similar to autocorrelation and autoconvolution of the waveform, except that the original time structure of the waveform is better preserved. A power greater than 1 increases the contrast of the amplitude spectrum, and a high power will transform a noise-like signal to almost a pure tone. An intermediate power produces results of more interest – where multiple peaks may be discerned by ear (when applied to steady-state rendered HRTF data). Less usefully, powers between 0 and 1 reduce spectral contrast (with 0 yielding a flat spectrum), and negative powers invert the magnitude spectrum.

While interaural time cues are preserved in this operation, this may not be helpful except for small powers, because when the magnitude spectrum contrast of each binaural channel is exaggerated they usually have little energy at common frequencies. As the scaling of amplitude is non-linear, it is also not possible to preserve the frequency-dependent interaural level differences. However, an alternative to this is to restore the broadband interaural level difference to the transformed data, which is the approach taken in our implementation.

Figure 1 shows the resulting waveforms, envelope functions and magnitude spectra from raising the spectrum magnitude to a power (without steady state rendering). The HRTF illustrated was measured from the second author, at 50° azimuth on the horizontal plane. In the illustration, the broadband interaural level difference is preserved. As can be observed, interaural time difference is preserved throughout, although the peak of the envelope broadens at high powers. However, the binaural spectrum for a power of 4 and above has almost no power at frequencies common to the two channels.

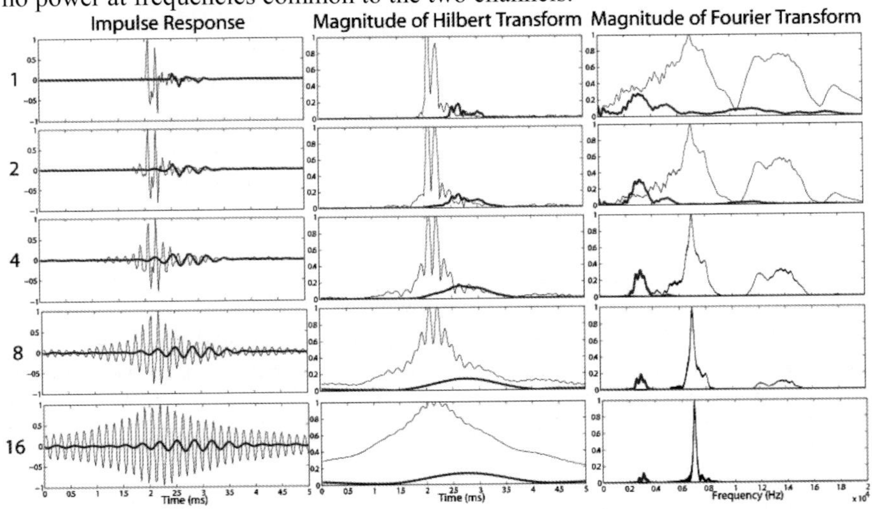

Figure 1. Effect of raising spectrum amplitude of a measured HRTF to a power of 1 (i.e., no change), 2, 4, 8, and 16, as indicated by the numbers on the left. The fine line represents the ipsilateral ear, and broad grey line represents the contralateral ear. All charts are normalized, and units are amplitude units (i.e., not squared amplitude nor decibels).

2.3. *Scaling of Frequency*

Arguably, much of the spectral information in HRTFs is between 2 kHz and 20 kHz, which is a part of the spectrum in which pitch sensation is not strong. Scaling frequency by a factor of 0.1 transposes this spectral content to a range of high pitch acuity and potentially strong pitch strength [8, 9, 10], thereby allowing the listener to hear the spectrum in a different way. Considering that the 200 Hz – 2 kHz decade does not overlap the untransposed decade, it is straightforward to create a sonification in which the original and transposed spectra are presented simultaneously. This is an important point, because transposition will distort interaural cues: interaural time differences are expanded by a factor of 10; and while interaural level differences are not changed, in their transformed version they do not map well to the listener's auditory spatial template (typically, they are also expanded relative to the interaural level difference that would be expected for a given frequency at a given source direction). Hence, the untransposed decade can provide spatial and timbral information to the listener, while the transposed decade provides a clearer representation of the spectral peak structure.

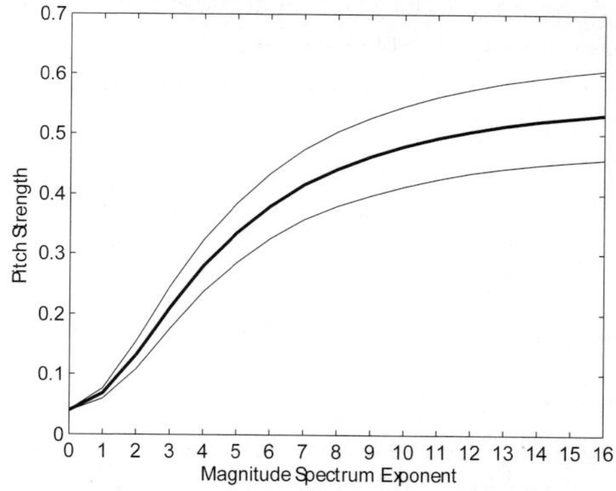

Figure 2. Mean pitch strength (±1 standard deviation) of the full set of monaural HRTFs of an individual (5° azimuth and 10° elevation intervals, data measured from the first author) transposed down by one decade, with the steady-state magnitude spectrum raised to powers from 0 to 16. Pitch strength is estimated from the SWIPE' algorithm [11], which yields a value from 0 (no pitch) to 1 (maximum strength).

When non-linear magnitude spectrum scaling is applied to the frequency-scaled decade, the result can be more useful than applying it to the untransposed decade for two reasons: firstly, the untransposed decade retains spatial cues, and it may be preferable not to distort these, or at least not by much; and secondly, the transposed decade is in the range of best pitch acuity, and so increasing the pitch strength of that decade takes advantage of this. Figure 2 shows how calculated pitch strength of transposed HRTFs is affected by raising the transposed magnitude spectrum to various powers. In interpreting the pitch strength scale, values of 0.1 and less are noise-like, and values around 0.5 are clearly tone-like. A magnitude spectrum exponent within the range of 3 to 8 appears to be most useful.

The maximum peak frequencies of an individual's HRTF set depends on the direction of sound incidence, as shown in Figure 3 (which is an analysis of the first author's HRTFs). These maximum peak frequencies are brought to prominence when the magnitude spectrum is raised to a power greater than one. Low frequency peaks occur at very low elevations and for azimuths around 270° (*i.e.,* where the ear is contralateral), and the highest peak frequencies occur close to 90° azimuth at elevations around -30°.

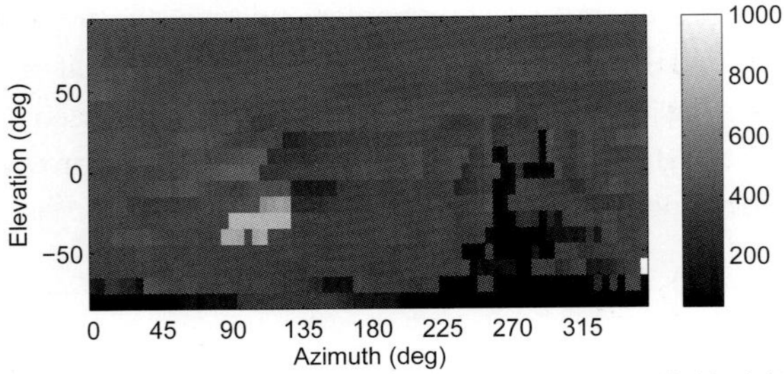

Figure. 3. Peak frequency of the full set of monaural HRTFs of an individual (5° azimuth and 10° elevation intervals, single-pole coordinates) transposed down by one decade (units are Hz). These are the frequencies that remain after raising the transposed magnitude spectrum to a high exponent.

2.4. *Scaling of Time*

Close examination of binaural impulse responses shows a pattern of 'echoes', for example, from structures in the pinna and the shoulder. This pattern of echoes changes with source direction, and also depends on individual physiological form. However, this temporal structure is much too fine to hear when listening to HRTFs directly. Time-stretching by a factor of 1000 transposes 20 kHz (the approximate upper limit of hearing) to 20 Hz (towards the upper limit of auditory fluctuation sensation) [8]. HRTFs often have interesting spectral features around 3-5 kHz, and transposing these to 3-5 Hz puts the in the range of maximum fluctuation sensitivity [8]. Alternatively, time-stretching by a factor of 5000 allows the listener to mentally track the 'rhythm' of the binaural room impulse response, rather than mainly experiencing the sensation of fluctuation.

Of course, simple transposition is of no use because the result would be inaudible 'infrasonifications'. Instead we extract the envelope of the binaural impulse response (by taking the absolute value of the Hilbert transform), and use this to amplitude-modulate a steady state carrier. The duration of a time-stretched sonification is the product of the original binaural impulse response and the time-stretching factor – for example, a 256-sample impulse response with a sampling rate of 44.1 kHz yields a sonification duration of 5.8 s for a factor of 1000, or 29 s for a factor of 5000. The latter is somewhat impractical if many HRTFs are being sonified for comparison.

The dynamic range of the envelope may be expanded or compressed by raising it to a positive power. A high power leaves little more than the highest peak of the envelope. A low power (less than 1) may make noise (including temporal smearing) from the HRTF measurement audible in the periods before and after the signal (if such noise exists, and has not been truncated). Figure 4 shows the calculated short-term loudness functions for a measured HRTF processed in this way with various envelope exponents. The time-varying loudness model of Glasberg and Moore [12] was used, assuming a listening level in the vicinity of 70 dB. In this instance (and in others that we have examined), an exponent of about 1.5 provides a good balance between envelope detail and noise suppression.

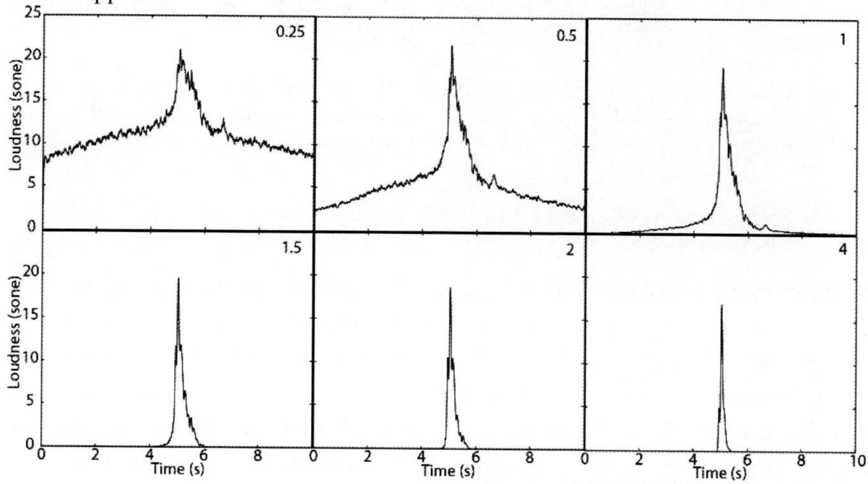

Figure 4. Calculated short-term loudness (following Glasberg and Moore [12]) of an HRTF envelope (0° azimuth and elevation) for a listening level peaking in the vicinity of 70 dB (using the temporal integration described by Glasberg and Moore). The carrier is steady noise, and the envelope function is raised to various exponents (shown in the top right-hand corner of each subplot).

3. Combining Techniques

The techniques described above have been combined in a sonification process that retains the spatial information of the original HRTF, provides an appropriate listening duration through steady-state rendering, improves audibility of the peak structure through transposition and non-linear amplitude scaling, and provides an experience of the temporal structure of the binaural impulse response. This is done by combining three elements: (i) the steady-state rendered HRTF, which is heard as a quiet hiss (an attenuation of 40 dB is applied relative to the following components); (ii) the transposed HRTF (by one decade), steady-state rendered, with its peak structure exaggerated (we tend to use an exponent of 3), recovering the broadband interaural level difference; and (iii) the envelope of the binaural room impulse response time-stretched by 1000, raised to the power of 1.5, which is used to modulate (ii). Figure 5 shows spectrograms of two measured HRTFs sonified in this way, both measured from the second author, on the same cone of confusion.

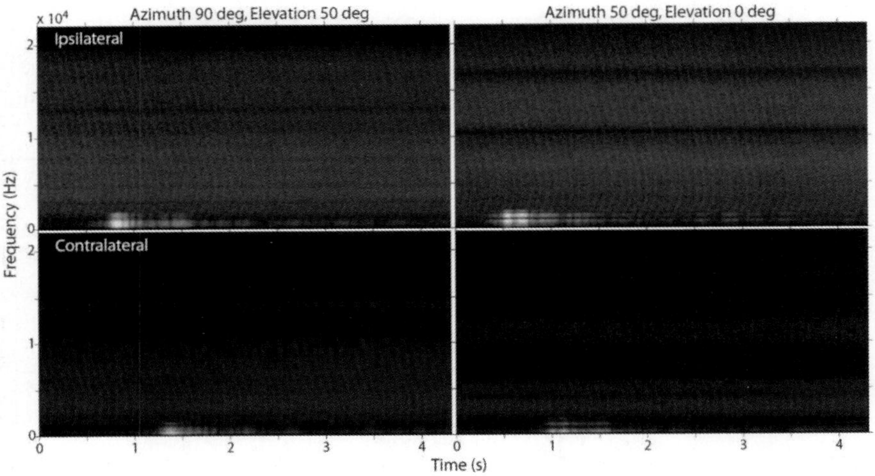

Figure 5. Spectrograms showing sonifications of two HRTFs on the same cone of confusion. There is a 90 dB range between white and black.

While the spectrograms of Figure 5 do not display the fine temporal structure, the time-stretched interaural time difference is clearly visible in the low frequency range envelope. There are clear differences between the sonifications at the two angles: seen in the different high frequency hiss spectra, the different 'rhythm' of the low frequency envelopes, and the different frequencies present in the low frequency range. We hope that the reader can grasp something of the sound from the visualization. Such sonifications are interesting and informative as stand-alone sounds, but become more informative when the sound is heard alongside graphical displays of the relevant HRTF features.

4. Conclusions

The effectiveness of a sonification depends on how well the information is conveyed (and understood) by the listener – and so depends on what information is being sought (or explored for), and on making a good match between the information representation and the perceptual and cognitive sensitivity of the listener. There are many ways in which HRTFs could be sonified, and this chapter has focussed on a simple approach to the problem. In a single sonification, one can clearly hear at once the spatial, spectral and temporal structure of an HRTF, and this approach can be used for comparing HRTFs from different directions, different distances, and different individuals. The approach meets Hermann's criteria for sonification [2], and is also well-attuned to audition. The sonification is entirely based on the data, and so does not introduce distractions analogous to "chart junk", and it presents three perspectives on the full dataset of an HRTF. The sonification is, to a reasonable extent, pre-attentive because sound is being represented by itself, with transformations that emphasize key features: it represents space with space, time with time, and spectrum with spectrum. Nevertheless, the interpretation of the sonification is facilitated by both knowledge of the sonification process, and a basic knowledge of the general characteristics of HRTFs (for example, interaural and spectral cues). Presenting the sonification as code (rather than as a black box computer program or as pre-processed sound recordings) is helpful in explaining the sonification process to students who normally do audio signal processing in Matlab. The purpose of this sonification is to allow students to hear key features of HRTFs, and in doing so, to add experience to the learning process. While the sonifications are not intended for ear training, students should become more aware of the auditory characteristics of HRTFs through exposure to such sonifications.

Appendix

Table 1 presents a Matlab function (requiring Matlab's Signal Processing Toolbox) that implements the sonification process described in this chapter. The inputs are a head-related impulse response (HRIR, which consists of two columns, one for each channel) and its audio sampling rate in Hz (fs). The level of the untransposed hiss, relative to the rms level of the transposed sonification is set by 'hisslevel'. Three exponents (e1, e2 and e3) control the spectrum magnitude contrasts for the hiss, the transposed spectrum magnitude, and the envelope respectively; and the time-stretching factors of the transposed spectrum and the envelope are controlled by s1 and s2 respectively.

The frequency domain processing in the function steadystate() from line 17, which requires an even number of samples (this is ensured in line 8, because s2 might not be even). This function raises the spectrum magnitude to an exponent, preserves the phase, and introduces a random phase offset for each spectrum component between DC and the Nyquist frequency, before returning to the time domain (real(...) in line 23 fixes small rounding errors). Line 24 adjusts the output to the same rms value for each channel as the original HRIR, and this principle continues through the main function in lines 14 and 15, so that normalization or further attenuation to avoid peak clipping is usually unnecessary if the original HRIR values are between -1 and 1. If an exponent other than 1 is used for the envelope, then only the peak levels of the transposed content match the rms levels of the original data.

Table 1. Matlab code for sonifying HRTFs.

```
1.  function y  = sonifyHRIR(HRIR,fs)
2.  hisslevel   = -40;                  % relative level of hiss in dB
3.  e1          = 1;                    % exponent for untransposed magnitude spectrum
4.  e2          = 3;                    % exponent for transposed magnitude spectrum
5.  e3          = 1.5;                  % exponent for envelope
6.  s1          = 10;                   % time-stretch for carrier
7.  s2          = 1000;                 % time-stretch for envelope
8.  HRIR        = HRIR(1:2*floor(end/2),1:2); % even length, 2 channels
9.  outlength   = length(HRIR) * s2;    % length of output wave
10. rms         = mean(HRIR.^2) .^ 0.5; % root-mean-square of each HRIR channel
11. hiss        = steadystate(HRIR, outlength, e1, rms) .* 10 .^ (hisslevel/20);
12. carrier     = steadystate(resample(HRIR,s1,1), outlength,e2,rms);
13. envelope    = abs(hilbert(resample(HRIR,s2,1))) .^ e3;
14. envelope    = envelope ./ repmat(max(abs(envelope)), length(envelope), 1);
15. y           = (hiss+carrier.*envelope) ./ (1+10^(hisslevel/20));
16. sound(y,fs)
17. function y  = steadystate(HRIR, outlength,exponent,rms)
18. spectrum    = fft(HRIR, outlength); % zero-padded fast Fourier transform
19. magnitude   = abs(spectrum).^exponent;
20. phase       = angle(spectrum);
21. randphase   = exp(1i*2*pi.*rand(outlength /2-1,1));
22. noise       = [0; randphase; 0; flipud(conj(randphase))];
23. y           = real(ifft(magnitude.*exp(1i*phase).*[noise,noise]));
24. y           = repmat(rms, outlength,1) .* y ./ repmat(mean(y.^2).^0.5, outlength,1);
```

References

1. J. Blauert, *Spatial Hearing* (The MIT Press, Cambridge, 1997).
2. T. Hermann, Taxonomy and definitions for sonification and auditory display, in *Proc. 14th Int. Conf. Auditory Display* (Paris, France, 2008).
3. F. Dombois, O. Brodwolf, O. Friedli, I. Rennert and T. Koenig, SONIFYER: a concept, a software, a platform, in *Proc. 14th Int. Conf. Auditory Display* (Paris, France, 2008).
4. E. R. Tufte, *The Visual Display of Quantitative Information* (Graphics Press, Cheshire Connecticut, 1983).
5. J. W. Tukey, *Exploratory Data Analysis* (Addison-Wesley, Reading Massachusetts, 1977).
6. C. Ware, *Information Visualization: Perception for Design* (Morgan Kaufman, San Fransisco, 2000).
7. D. Cabrera and S. Ferguson, Sonification of sound: tools for teaching acoustics and audio, in *Proc. 13th Int. Conf. Auditory Display* (Montreal, Canada, 2007).
8. E. Zwicker and H. Fastl, *Psychoacoustics: Facts and Models* (Springer-Verlag, Berlin, 1999).
9. B. C. J. Moore, *An Introduction to the Psychology of Hearing* (Academic Press, Boston, 2003).
10. E. Terhardt, G. Stoll and M. Seewan, Algorithm for extraction of pitch and pitch salience from complex tonal signals, *J. Acoust. Soc. Am.* **71**, 679-688 (1982).
11. A. Camacho and J. G. Harris, A sawtooth waveform inspired pitch estimator for speech and music, *J. Acoust. Soc. Am.* **124**, 1638-1652 (2008).
12. B. Glasberg and B. C. J. Moore, A model of loudness applicable to time-varying sounds, *J. Audio Eng. Soc.* **50**, 331-342 (2002).

EFFECTS OF SPATIAL CUES ON DETECTABILITY OF ALARM SIGNALS IN NOISY ENVIRONMENTS

N. KURODA[1], J. LI[1]*, Y. IWAYA[2], M. UNOKI[1], and M. AKAGI[1]

[1]*School of Information Science, Japan Advanced Institute of Science and Technology*
1-1 Asahidai, Nomi, Ishikawa 923–1292 Japan
{n-kuroda,junfeng,unoki,akagi}@jaist.ac.jp
[2]*Research Institute of Electrical Communication, Tohoku University*
2-1-1 Katahira, Aoba-ku, Sendai 980-8577 Japan
iwaya@riec.tohoku.ac.jp

It is crucial to correctly detect alarm signals in real environments. However, alarm signals are possibly masked by varieties of noise. Therefore, it is necessary to clarify the perceptual characteristics of alarm signals in noisy environments. The aim of this study was to investigate how spatial cues influenced the detectability of alarm signals in noisy environments. We measured the masked thresholds of alarm signals in the presence of car noise in virtual acoustic environments regenerated by using head-related transfer functions, which could control spatial cues. Our two main conclusions were: (a) when the frequency of the alarm signal was 1.0 kHz, the detectability was improved by using the interaural time difference (ITD) and interaural phase difference (IPD). (b) When the frequency of the alarm signal was 2.5 kHz, not only ITD and IPD but also the interaural level difference (ILD) played an important role in improving alarm-signal detection when the signal was fixed in front of listeners.

Keywords: Detectability, SRM, ITD, IPD, ILD, HRTFs

1. Introduction

Alarm signals are sounds that provide starting, ending, and alarm information to users.[1] It is important for all of us to interpret these correctly in many types of scenarios. There are, however, cases where alarm signals cannot be correctly perceived in real environments because they are masked or partially masked by background noise. For example, an accident may occur if a driver inside a car fails to hear important alarm signals. Therefore, it is necessary to clarify the perceptual characteristics of alarm signals in noisy

*Junfeng Li is now with Institute of Acoustics, Chinese Academy of Sciences.

environments. Moreover, a method of designing alarm signals to provide people with correct information and one of presenting alarm signals so that they are accurately perceived by drivers are required.

A policy of universal design has been incorporated in recent years as a method of designing alarm signals. For example, Mizunami *et al.* reported benefits from intermittent signal patterns for attentional and ending signals.[2] However, it is important to investigate how robust the detectability of alarm signals in noisy environments is, which has not yet been investigated in developing a method of presenting alarm signals, to safety convey information to drivers. We took the following into consideration in investigating how robust the detectability of alarm signals is. Ebata *et al.* reported that the detectability of signals could be improved by using directional information[3] to detect signals in the presence of noise. Moreover, Saberi *et al.* found that the detectability of signals was improved in a free sound field when the signal and masker were spatially separated.[4] This phenomenon is referred to as spatial release from masking (SRM). It is well known that the interaural time difference (ITD) and interaural level difference (ILD) are also used as significant spatial cues in SRM.[5] Saberi *et al.* carried out detection experiments with a pulse-train as a target signal and white noise as a masker. As a result, about 14 dB of SRM occurred at maximum.[4]

The aims of our work were to confirm whether SRM occurred for alarm signals in noisy environments and then to determine whether SRM could be accounted for by these spatial cues. However, ITD and ILD in these experiments had a complex effect on SRM because the stimuli were presented through loudspeakers in free sound as a function of the direction of either the signal or noise with respect to the subject. It was therefore difficult to investigate the separate influences of ITD and ILD in SRM.

As the first step toward investigating the detectability of binaural alarm signals in SRM, we first scaled down the experiments in a free field (loudspeaker presentation) to those in a sound-proof room (headphone presentation) in terms of the experimental design to cancel out the effects of surrounding background noise and to separately control the spatial cues (ITD and ILD). Nakanishi *et al.* investigated what effect ITDs had on SRM.[6] They carried out detection experiments using pulse trains masked by white noise (the same stimuli as Saberi *et al.*) with ITDs only via headphones in a sound-proof room. Their results indicated that ITD is used as a significant spatial cue in SRM. In addition, they carried out other detection experiments in which they replaced the target signals in the former experiments with alarm signals (1.5, 2.0, and 2.5 kHz) based on JIS S 0013.[1] Their results revealed that both the ITD and interaural phase difference (IPD) of alarm signals have an effect under noise as interfering noise.

In the second step, Uchiyama *et al.* carried out detection experiments on alarm signals (1.0, 1.5, 2.0, 2.5, 4.0 kHz) in the presence of realistic noise (using car noise instead of white noise) without ILDs. They obtained the same results as Nakanishi *et al.*[7] and clarified that ITDs and IPDs were used as significant spatial cues. Their results mainly revealed that SRM could be accounted for by a function of the relationship between ITD and IPD corresponding to the binaural masking level difference (BMLD) and the amount of masking release could similarly be accounted for by interpreting it as BMLD because these releases depended on the signal frequency. What effect ILDs have, however, has not yet been considered under these conditions so that we are not sure whether ILD is used as a significant spatial cue in SRM. We expect that the detectability of alarm signals in noisy environments can be further improved by introducing ILDs.

We therefore investigated what effect three spatial cues had on the detectability of alarm signals in noisy environments in the next step. We conducted experiments using Head-related transfer functions (HRTFs), which are acoustic transfer functions from a sound source to the ear drum of a listener. The advantage of using HRTFs is that it is possible to extract ITDs, IPDs, and ILDs and easily control the direction of presentation of a signal. Here, we first focused on the individualization of HRTFs. Second, we conducted experiments to detect different alarm signals masked by noise.

2. Individualization of HRTFs

Since HRTFs are highly dependent on people, they must be individualized before experiments are done on detecting signals in noise to ensure that the HRTFs used in the experiments for all listeners match their HRTFs as closely as possible. We accomplished this task in two steps: we individualized HRTFs for all listeners and then evaluated the individualized HRTFs.

To individualize the HRTFs for all listeners, we carried out the following experiments using the determination method of optimum impulse response by sound orientation (DOMISO).[8] We used 114 sets of HRTFs measured in an anechoic chamber at Tohoku University in our experiments. We first selected 32 sets of HRTFs from the 114 HRTFs using a technique of vector quantization, and further convolved them with pink noise to generate stimuli that provided rich cues for localizing sound. These stimuli were then presented to each listener who assessed the fitted set of HRTFs through subjective listening tests based on a tournament procedure (DOMISO).

Next, localization experiments for sound sources were carried out to objectively evaluate whether the individualized HRTFs obtained by DOMISO were suitable. The virtual sound sources were set from 0° to 350° in step sizes of 10°, and the stimuli were randomly presented to listeners. The evaluation criteria were set over 70% when the range of correct answers was ±10°, over 90% when the range of correct answers was ±20°, and under 10% in front-back confusion error. Only listeners who satisfied the evaluation criteria were allowed to participate in the next round of experiments.

3. Experiment I

3.1. *Purpose and Method*

This experiment had two purposes: (1) to confirm whether the virtual acoustic environment generated by using individualized HRTFs on a particular participant could simulate a presentation using loudspeakers in an anechoic chamber, and (2) to confirm to what degree ILDs affected SRM in such a simulated environment.

A 1-s long pulse-train, which was composed of 100 rectangular pulses that were 62.5 μs in duration, was used as the target signal, and white noise that was 2-s long was used as the masker. The sampling frequency was 48 kHz. The direction of arrival of sound in this experiment was controlled by varying its ITD, which was calculated by

$$\mathrm{ITD} = \frac{d}{c} = \frac{r(\theta + \sin\theta)}{c}, \tag{1}$$

$$d = r\theta + r\sin\theta, \tag{2}$$

where r in meters is the radius of the head, θ in radians is the direction of the sound source, c in meters per second is the sound velocity and d in meters is the path difference from the sound source to both ears. In this study, r was set to 0.09 m and c was set to 343.5 m/s. The directions of presentation of the target signal (or the masker) were varied from 0° to 90° in steps of 15°, and the direction in front of the listeners was 0°. The configurations for the target signal and the masker have been assumed to be $\mathrm{S}_m\mathrm{N}_0$ ($m = 0, 15, \cdots, 90$) in this paper when the masker was fixed at the front and $\mathrm{S}_0\mathrm{N}_m$ ($m = 0, 15, \cdots, 90$) when the target signal was fixed at the front. For example, $\mathrm{S}_{60}\mathrm{N}_0$ means that the directions of arrival of the target signal and masker corresponded to 60° and 0°. When ITDs and ILDs could be used as the spatial cues, the directions of arrival of the sound sources were varied by convolving the stimuli with the corresponding HRTF for each direction.

The experiment was carried out in a sound-proof room by using Tucker-Davis Technologies (TDT) System III. The TDT System III was controlled with a personal computer (HP dc5750 Microtower Base DT). The stimuli were presented to all listeners through inner-ear type earphones (SONY, MDR-EX90SL).

3.2. *Procedure*

We measured the masked thresholds by using a method of limits in this experiment which method included descending and ascending series. In the descending series, the sound pressure level (SPL) of the target signal in the stimuli at the beginning of the experiment was randomly chosen from a range where the listener could distinctly detect the target signal. Then, the SPL of the target signal was varied from high to low in steps of 1 dB. In the ascending series, the SPL of the target signal in the stimuli at the beginning of the experiment was randomly chosen from a range where the listener could not distinctly detect the target signal. Then, the SPL of the target signal was varied from low to high in steps of 1 dB. In addition, the starting position for the target signal in the stimuli was randomly chosen. The SPL of the masker was fixed at 65 dB SPL. Both descending and ascending series were each carried out for 10 trials. When the difference in the mean of each series was 2 dB or less, the masked threshold was determined as the mean for all measurements. Additional trials were done until the difference was within 2 dB.

Six graduate students aged from 23 to 26, five males and one female, participated in this experiment. All had normal hearing (15 dB HL from 0.125 to 8 kHz) and experience with participating in other experiments.

3.3. *Results and discussion*

Figure 1(a) plots the results for the mean masked thresholds for each azimuth to find the pulse-train signals in white noise. The vertical axis has the relative masked thresholds, which have been normalized by the masked thresholds at $\mathrm{S_0N_0}$. The horizontal axis indicates the azimuth of either the pulse-train signal or the white noise. The thin lines denote the results under ITD-only conditions, and the thick lines plot the results when individualized HRTFs were used. In addition, the solid lines specify the results for $\mathrm{S}_m\mathrm{N}_0$ ($m = 0, 15, \cdots, 90$), and the dotted line shows the results for $\mathrm{S}_0\mathrm{N}_m$ ($m = 0, 15, \cdots, 90$). The error bars represent the standard deviations for relative masked thresholds.

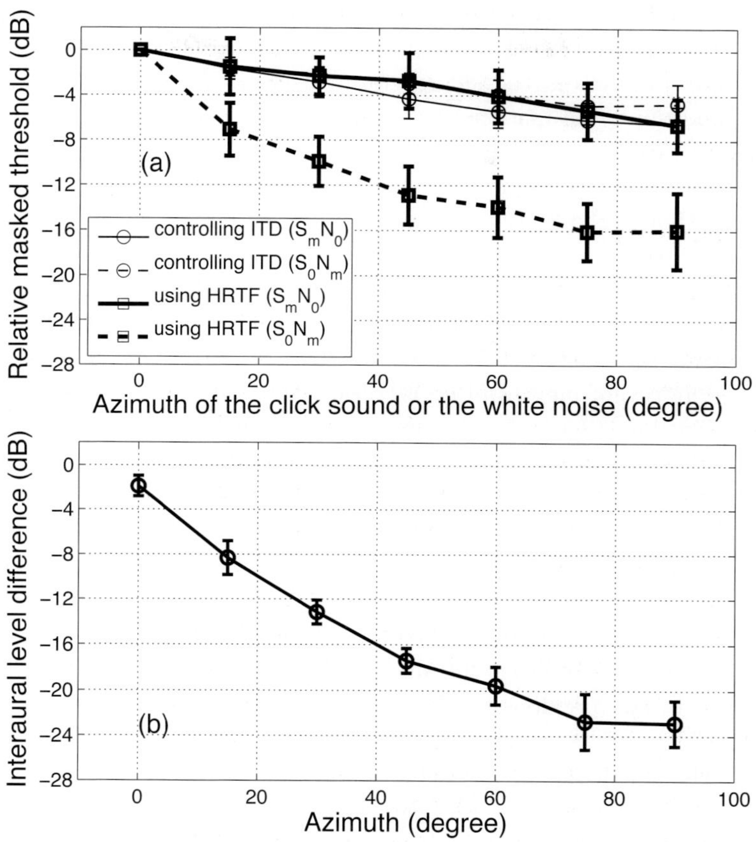

Fig. 1. Results of experiments: (a) mean masking thresholds for perception of pulse-train signal under white noise and (b) ILDs for each direction of presentation of signal.

The thin lines indicate that SRM occurred for all signals and that the detectability of alarm signals could be improved by utilizing ITD. This was the same tendency as that in the previous study.[6] Hence, we confirmed that ITDs were used as a significant spatial cue in SRM.

In S_mN_0, the thick lines have the same tendency as the thin ones. From these results, ITD was used as a significant spatial cue in SRM. However, in S_0N_m, the thick lines indicate a larger amount of masking release than that in the thin lines. Under these conditions, ITD and ILD could be used as significant spatial cues in SRM. In addition, the amount of SRM was 16 dB in this case and this was the same as that obtained by Saberi *et al.*[4]

Figure 1(b) plots the mean value and standard deviation of ILDs applied in the above HRTF-based stimuli as a function of the increasing azimuth direction of the target signal and the masker. The ILDs were calculated by subtracting the SPL at the left ear from the SPL at the right ear. Under $S_m N_0$ conditions, the masker was presented to both ears at about 65 dB and the SPL of the target signal at the left ear decreased as the direction of presentation for the signal increased and the SPL at the right ear was constant. Hence, the SNR at the left ear decreased and that at the right ear was constant as the direction of presentation for the signal increased. Therefore, we expected that SRM would not occur. However, the results revealed that SRM occurred by using ITDs (Fig. 1(a), thick solid line). We can explain these results from the above by assuming that listeners could use ITDs by using interaural correlation even if the SNR in the left ear was great reduced by the introduction of ILD. In contrast, in $S_0 N_m$, the SNR at the left ear increased and SNR at the right ear was constant as the direction of presentation of the masker increased. However, the effect of the better ear was also very important in spatial hearing. Although the SNR difference due to this effect can be used as a cue to detect the signal, the SNR difference in binaural hearing can also be related to spatial effects such as specific HRTF so that this difference results in the better-ear effect. Therefore, we expected that the amount of masking release would be larger than that in the previous study.[6] The results matched this expectation (Fig. 1(a), thick dotted line), and we confirmed that SRM occurred by using ITDs and ILDs. ILDs particular had a great effect on SRM.

Finally, for purpose (1), we confirmed from the above that the virtual acoustic environment regenerated by using individualized HRTF on a particular participant could simulate presentation using loudspeakers in an anechoic chamber. For purpose (2), we confirmed that the availability of ILD cues greatly improved SRM.

4. Experiment II

4.1. *Purpose, method, and procedure*

This experiment had two purposes: (1) to enable investigations into how ILDs affected the detectability of alarm signals in car noise, and (2) to confirm whether superior spatial cues existed according to different frequencies of alarm signals.

The target signals in this experiment were alarm signals and the masker was car noise. Alarm signals convey the highest degree of warning according to the Japanese Industrial Standards (JIS S 0013).[1] These signals had repeated patterns of ON and OFF (ON = 0.1 s and OFF = 0.05 s) for 1 s. The frequencies of the alarm signals were 1.0 and 2.5 kHz. The car noise was recorded via omnidirectional microphones at the ear canals of a driver inside a vehicle with the window open while the automobile was traveling at 60 km/h. The sampling frequency was 48 kHz.

The target and masker signals were spatialized either by ITDs alone or by using HRTFs, as described in Experiment 1. The procedure was the same as the one described in Experiment I.

4.2. *Results and discussion*

Figure 2 plots the results for the mean relative masked thresholds as a function of the target (or masker) direction for the alarm signals masked by car noise. The thin lines indicate SRM occurred for all signals and that the detectability of alarm signals could be improved by not only utilizing ITDs but also the IPDs of the signal. This was the same tendency as that in the results from a previous study.[7] We confirmed that ITD and IPD greatly affected the detectability of alarm signals in car noise.

In S_mN_0, regardless of the frequency of the alarm signals, the thick lines reveal the same tendency as that indicated by the thin lines. From these results, the amount of masking release in the thick lines was smaller than that in the thin lines. Let us reconsider why it was difficult to calculate interaural correlation due to the effect of ILDs. The interaural correlation, on the other hand, could be calculated in S_0N_m, because the SPL of the signals was equal at both the left and right ears. When the frequency of the alarm signals was 1.0 kHz, their detectability was not improved by using ILDs because the thick line indicates the same tendency as that in the thin lines. In contrast, when the frequency of the alarm signals was 2.5 kHz, their detectability was greatly improved by using ILDs because the thick line has lower masked thresholds than the thin line. Although this is not as surprising as what we expected, we found what effect the three spatial cues had on the detectability of alarm signals in noisy environments.

First, the same phenomenon (S_mN_0 versus S_0N_m) occurred as that in Experiment I for the positional relation between the signal and the masker. Therefore, the listeners detected the alarm signals by using the interaural correlation in S_mN_0, and they detected the alarm signals by using the large SNR at the left ear in S_0N_m. However, under the same conditions, when the frequency of the alarm signals was 1.0 kHz, we could not observe whether their detectability was improved by using ILDs. This means that ITDs greatly affect SRM for 1.0-kHz alarm signals and ILDs greatly affect SRM

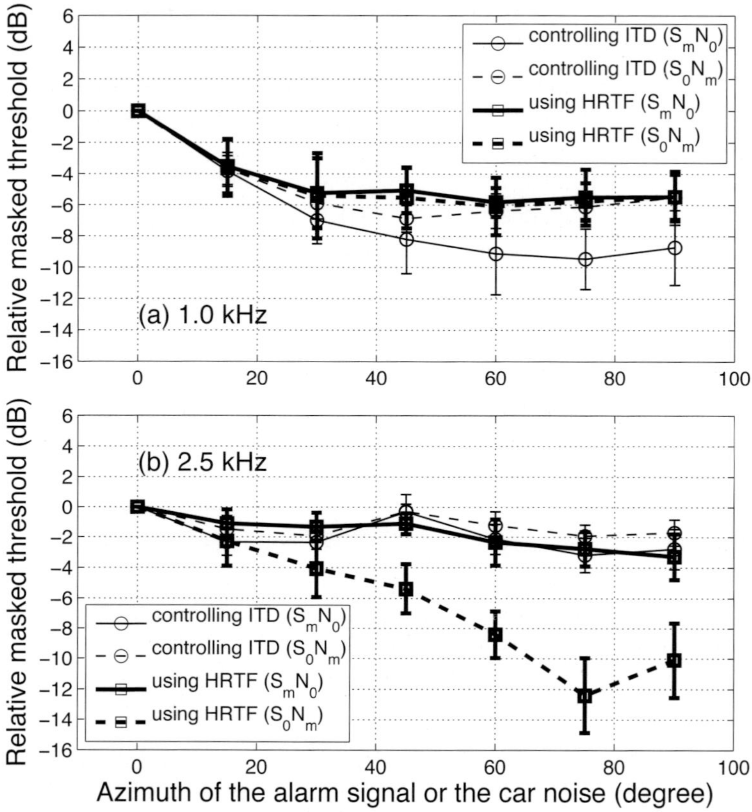

Fig. 2. Mean masked thresholds for perception of alarm signals under car noise: (a) 1.0 kHz and (b) 2.5 kHz.

for 2.5-kHz alarm signals.

Finally, we found that the most advantageous direction of presentation of alarm signals differs according to their different frequency components. When the frequency of the alarm signals was 1.0 kHz, listeners could mainly utilize ITD cues to detect them. However, when the frequency of the alarm signals was 2.5 kHz, the listeners could significantly benefit from ILD and ITD cues, although ILD cues seemed to be most important.

5. Conclusion

The aim of this study was to investigate how spatial cues influenced the detectability of alarm signals in noisy environments. We therefore measured the masked thresholds for listeners to detect alarm signals in the presence of car noise in virtual acoustical environments regenerated by using HRTFs. In summary, we obtained two main findings.

(1) When the frequency of the alarm signals was 1.0 kHz, the detectability of alarm signals was improved by using the ITD and IPD. This means listeners used ITD and IPD as significant spatial cues to perceive the alarm signals in car noise.

(2) When the frequency of the alarm signals was 2.5 kHz, ITD and IPD were used in the S_mN_0, and ITD, IPD, and ILD were used in the S_0N_m as significant spatial cues to perceive the alarm signals in car noise.

Since it is well-known that both ITD and IPD are relevant below 1.5 kHz and ILD is relevant above 1.5 kHz for the detectability of signals,[9] we found that ILD plays an important role in improving alarm-signal detection when a signal with a higher frequency component (above 1.5 kHz) was fixed in front of listeners.

References

1. JIS S 0013, Guidelines for the elderly and people with disabilities – Auditory signals on consumer products (2002).
2. T. Mizunami, K. Kurakata, H. Shimosako, and K. Matsushita, "Further examination of ON/OFF temporal patterns of auditory signals (completion signal and attention signal) recommended in JIS S 0013," *J. J. Ergonomics*, vol. 40, no. 5, pp. 264–271 (2004).
3. M. Ebata, T. Sone, and T. Nimura, "Improvement of hearing ability by directional information," *J. Acoust. Soc. Am.*, vol. 43, no. 2, pp. 289–297 (1968).
4. K. Saberi, L. Dostal, T. Sadralodabei, V. Bull, and R. D. Perrot, "Free-field release from masking," *J. Acoust. Soc. Am.*, vol. 90, no. 3, pp. 1355–1370 (1991).
5. C. Lane, N. Kopco, B. Delgutte, B. G. Shinn-Cunningham, and H. S. Colburn, "A cat's cocktail party: Psychophysical, neurophysiological, and computational studies of spatial release from masking," *Proc. ISH 2003*, pp. 341–347 (2003).
6. J. Nakanishi, M. Unoki, and M. Akagi, "Effect of ITD and Component Frequencies on Perception of Alarm Signals in Noisy Environments," *J. Signal Processing*, vol. 10, no. 4, pp. 231–234 (2006).
7. H. Uchiyama, M. Unoki, and M. Akagi, "Improvement in detectability of alarm signal in noisy environments by utilizing spatial cues," *Proc. WASPAA2007*, pp. 74–77, New Paltz, NY (2007).
8. Y. Iwaya, "Individualization of head-related transfer functions with tournament-style listening test: Listening with other's ears," *Acoust. Sci. & Tech.*, vol. 27, no 6, pp. 340–343 (2006).
9. J. Blauert, Spatial Hearing: The Psychophysics of Human Sound Localization, Revised Edition, MIT Press (1997).

Binaural Technique used for Active Noise Control Assessment

Y. WATANABE* and H. HAMADA

School of Information Environment, Tokyo Denki University,
2-1200, MuzaiGakuendai, Inzai, Chiba 270-1382, Japan
** E-mail: yuko@acl.sie.dendai.ac.jp*
www.sound.sie.dendai.ac.jp

Binaural techniques have widely various applications for noise control engineering because they represent subjective spatial impressions of sound, facilitate sound quality evaluations, and support analyses with free sound field measurement. This paper introduces examples of Active Noise Control (ANC) assessment using a binaural ear simulator, which aims to simulate sound pressure levels at the eardrum in the condition of an earphone inserted into the ear canal. The efficiency of binaural measurement for ANC evaluation will also be presented.

Keywords: Active Noise Control; binaural signals; ear simulator; spatial distribution of noise source

1. Introduction

This paper describes usage of binaural techniques for Active Noise Control (ANC) assessment using a binaural ear simulator, which aims to simulate the sound pressure level at the eardrum in the condition where there is an earphone inserted into the ear canal.

An ANC system produces an opposing signal with the same amplitude as a target noise, but with anti-phase, to reduce noise actively. Especially, it is used for low-frequency noise sources. Usually, ANC is evaluated using physical parameters such as the sound pressure level in [dB] and the noise level in [dBA] in a free sound field.

When the ANC is intended to reduce noise in a large area, referring to global ANC, the noise reduction level at the position of an error sensor microphone is used as a parameter to design ANC systems. However, when ANC solely targets noise reduction at specific points in space, referring to local ANC, such as inside a headphone and at the ears of a listener, a controlled area is affected by obstacles located in the controlled area. For example, Honda et al. introduced an ANC application around a human's head.[1]

Furthermore, recent widespread use of digital audio players introduces many commercial earphones with ANC systems. In these cases, although the error sensor should be placed ideally either at the position of eardrums or at the entrance of the ear canals, in practice, it is installed at the outside of the earphones. Therefore, signals at both listeners' ears should be evaluated to examine whether ANC earphones efficiently reduce noise.

Additionally, noise reduction seems to change depending on the installation condition of earphones into an ear canal and a pinna shape. Therefore, for ANC earphone examinations, findings in binaural technology, such as the external ear characteristics and the pinna shape, are apparently useful.

Furthermore, ANC earphones were designed originally to control specific points in space, usually at the ears of a listener with restricted head movement. Therefore, head movement strongly influences noise reduction. In other words, relative positions and spatial information of target noise sources to listeners' head (ears) are apparently also important parameters.

As presented in this paper, we first describe experiments that were undertaken to observe noise reduction performance of ANC earphones using the ear simulator. The behaviors of ANC earphones when the noise source moves in a horizontal plane are also examined using binaural techniques of virtual sound reproduction with head-related transfer functions.

2. Construction of the ear simulator

2.1. *Introduction*

In this section, we describe construction of the ear simulator using a C-coupler for assessment of ANC earphones.

First, Fig. 1 presents a conceptual diagram of ANC systems in different categories, such as global and local ANC. To evaluate ANC systems of all types inclusively, a conceptual chart is transferred into the systematic block diagram shown in Fig. 2, which supports our understanding of the evaluation system. This evaluation system (Fig. 2) enables evaluation of ANC systems of different types using similar parameters, once a set of transfer functions are understood, by simply using digital filters: one is convolution with spatial characteristics, another is convolution with the inverse filter of the ear canal.

To realize assessments of ANC earphones given above, we introduce a simplified ear simulator, which has a c-coupler and pinna. We also describe

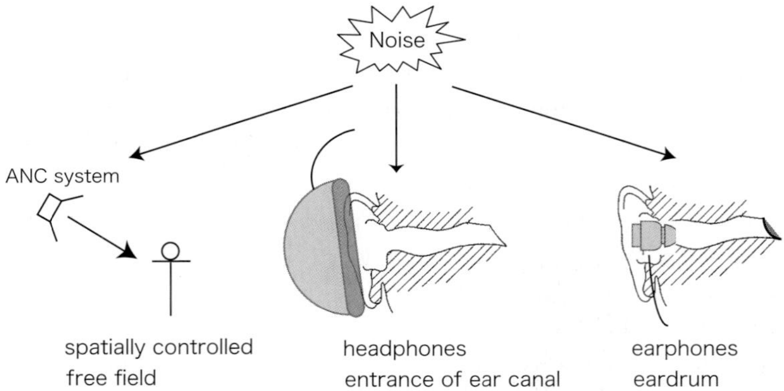

Fig. 1: Conceptual diagram of ANC evaluation process.

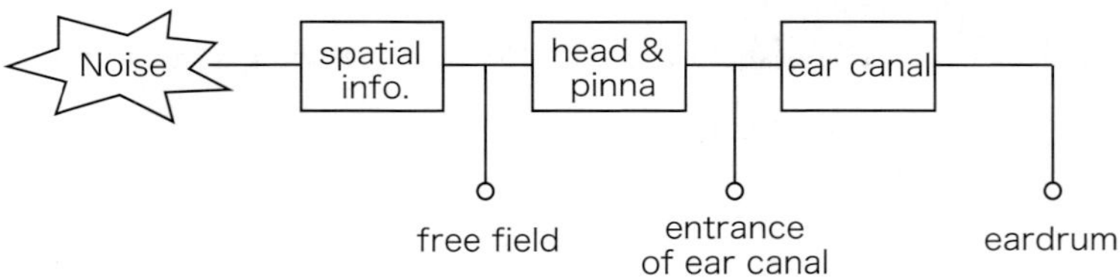

Fig. 2: Block diagram of an earphone evaluation system.

the principle of our measurement system, which enables us to measure and observe the sound pressure level at the eardrum with the occluded ear canal.

2.2. *C - coupler*

Many reports in the literature related to hearing aids and spatial hearing describe signals at the listener's eardrums.[2] For example, for spatial sound localization, signals at the eardrum vary dramatically corresponding to the incident angles of the incoming sound source, although signals observed at the position of the center of the head in free sound field condition do not change. A dummy head microphone is often used to measure signals at the eardrum of a listener using its artificial ear canal.

Okabe et al.[3] developed a simulated in-situ measurement system for hearing aids. In their reports, they described the case in which an earphone was inserted into the human's ear. In other words, an ear canal was occluded by earphones. The insertion gain was an important parameter for use in evaluation of hearing aid performance on a wearer.

To simulate the average of human eardrum impedance under a condition with an inserted earphone, they simplified the structure of an ear simulator and used it in combination with electrical compensations. They introduced an ear simulator with a C-coupler, which had equivalent acoustic impedance to that at the eardrum of a wearer of a hearing aid.

Fig. 3 and Fig. 4 show an overview and the structure of the C-coupler.

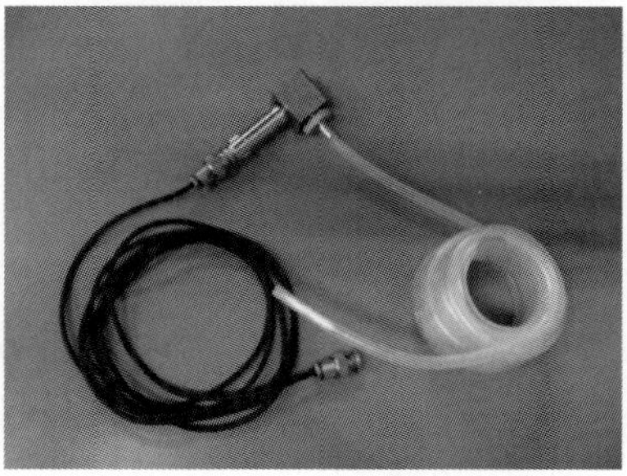

Fig. 3: C-coupler and tube.

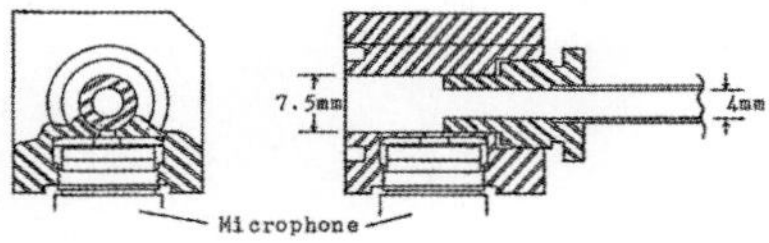

Fig. 4: Structure of C-coupler.

2.3. *Ear simulator*

Fundamental concepts and specifications of the ear simulator are given below.

(1) It can measure responses (sound pressure level) at the eardrum.
(2) It has a pinna and ear canal to fit earphones.
(3) The two ears are separated by a reasonable distance, which is necessary to attach headphones to expose noise signals to ANC systems.

To satisfy the requirements listed above, we designed the simple measurement equipment presented in Fig. 5. Replica pinnas were made using a silicon material attached to C-coupler. Those were mounted to a special frame with a variable adjuster.

2.4. *Noise source exposure*

To investigate the noise reduction level of ANC earphone system, we must provide a noise source to the system. In this investigation, noise sources were generated by the circum-aural headphone (see Fig. 6). The interval between the two pinnas was adjusted to fit the headphone adequately. To confirm the frequency spectrum of noise sources, the frequency response was measured. Fig. 7 shows a reasonably flat frequency response within a wide frequency range.

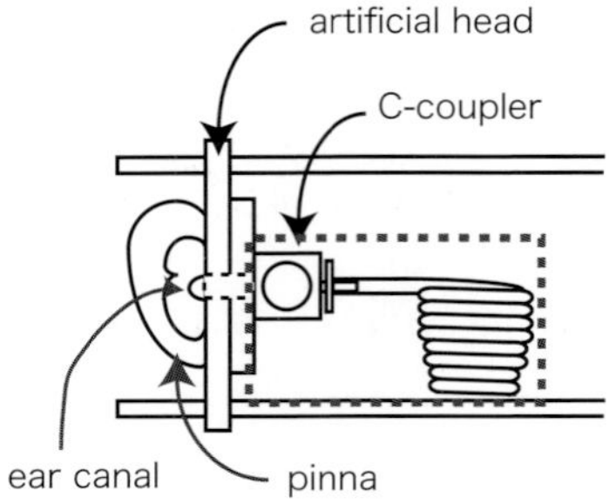

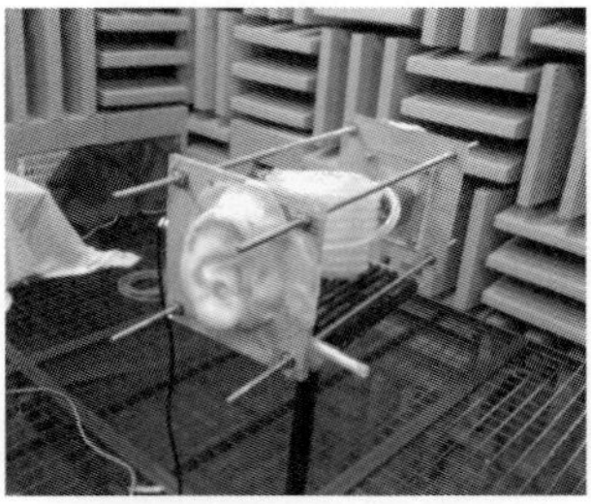

Fig. 5: Sketch of the equipment and overview.

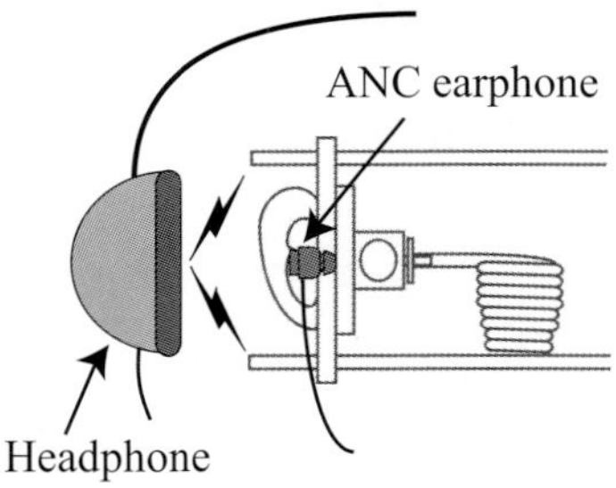

Fig. 6: Noise Exposure system.

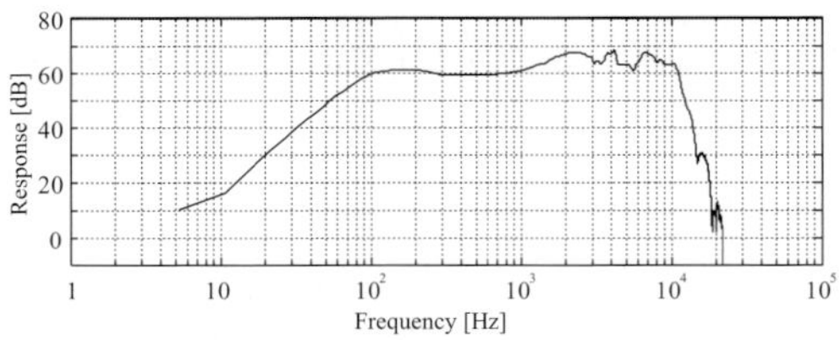

Fig. 7: Frequency spectrum of noise exposure.

2.5. *Experimental procedures*

White noise of 20 Hz to 20 kHz was used in this experiment as a noise signal. It was generated through the external headphone. Sound pressure levels of noise exposure were 84 dB at the eardrum. Five samples of ANC earphones were used for this experiment.

First, an insertion loss of each sample was measured, which was a condition when the noise signal was given by the headphone, and ANC earphone was just inserted into the ear canal (in this case, ANC system switch was off). Secondly, noise reduction was measured. Measurements of both the insertion loss and noise reduction level were repeated eight times for each earphone sample. A schematic diagram of the experiment is shown in Fig. 8.

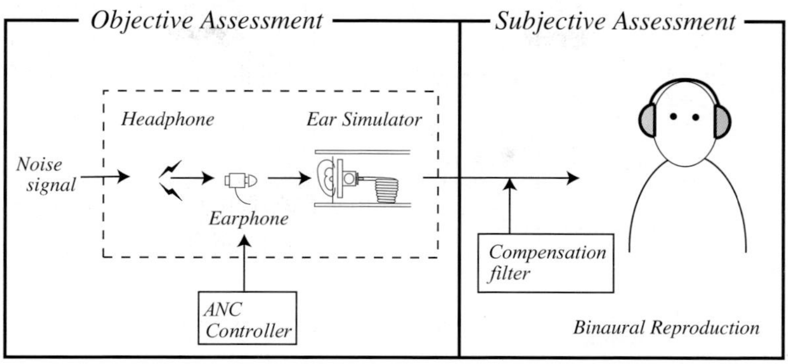

Fig. 8: Schematic diagram of ANC assessment.

3. Experiment 1: Evaluation of noise reductions of ANC earphones using an ear simulator

3.1. *Introduction*

In this section, an evaluation of noise reduction performance of ANC earphones was conducted using the binaural ear simulator, as explained in section 2.

Five different ANC earphones were used, and the SPL at the eardrums of both ears were measured under three different conditions described below.

(1) SPL under noise exposure via headphone without ANC earphones
(2) insulation loss when ANC earphones were inserted into the ear canal
(3) active noise reduction level with ANC system activated

All measured noise signals were also evaluated subjectively in terms of loudness and annoyance grades through the binaural reproduction system. Results of objective and subjective experiments are described as follows.

3.2. *Results: Objective assessment of ANC earphones*

The insertion loss level of several ANC earphones measured in this assessment is compared with the noise reduction level. The data shown in the figure with a solid line are the frequency spectrum of given noise source. The dotted line represents the insertion loss level and the dark dot line represents the noise reduction level. Fig. 9 shows examples of the results.

From the data presented in Fig. 9(a), ANC earphones give an effective insertion loss more than 10 dB whole frequency range from 20 Hz to 20

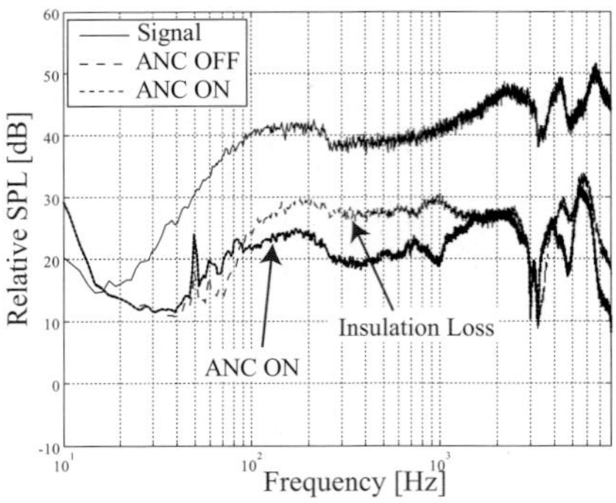

(a) Sample A

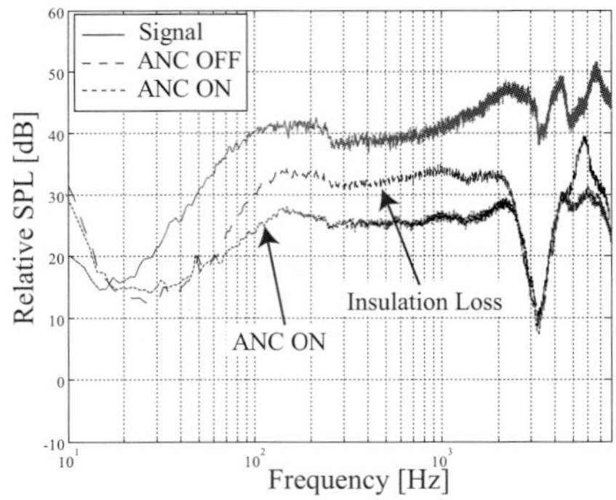

(b) Sample B

Fig. 9: Insertion loss and reduction level.

kHz; the effect of ANC noise can also be observed in the frequency range of 100 Hz and 1 kHz and in some upper frequency range. Fig. 9(b) shows a similar result of insertion loss, although the ANC system tends to increase the noise level in the frequency range greater than 3 kHz. To conclude, ANC systems sometimes increase the noise level in the upper frequency range. In this range, ANC systems do not normally treat the external noise signal as a control target.

3.3. *Results: Subjective assessment of ANC earphones*

In this section, subjective effects of ANC earphones were evaluated using the binaural sound reproduction method. Measured signals, with 5 ANC earphones and 3 noise conditions, were post-filtered to compensate the frequency characteristics of external ear canal, and presented to listeners via headphones, which had a flat frequency response over a wide frequency range. Test signals were evaluated using the paired comparison method. Listeners were asked to mark subjective impressions of test signals in loudness and annoyance. Fig. 10 presents the relation between physical quantities of noise reduction level and subjective responses. Test signals are shown in order of SPL [dB] in Fig. 10(a); Fig. 10(b) is in loudness level [sone].

With reference to Fig. 10(a), subjective judgment scores of loudness and annoyance increased with SPL in decibels, although differences of judgment occurred in cases where their noise reduction levels were equal in SPL (see sample 1 and sample 2, and sample 4 and sample 5). Good agreement of loudness levels and subjective judgments observed in Fig. 10(b).

In general, the amount of the decrease of SPL and the amount of at-

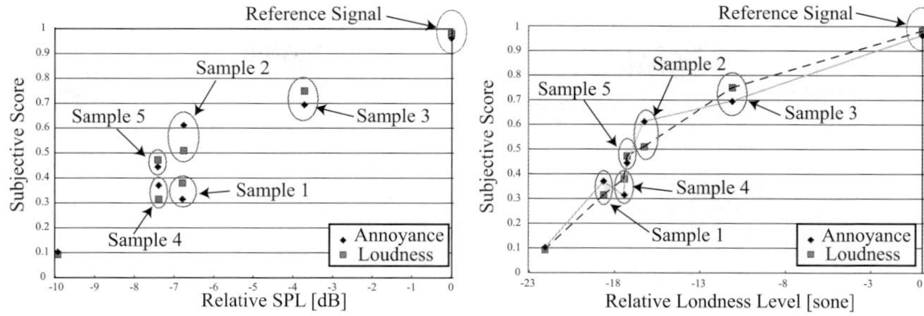

(a) Relative SPL

(b) Relative Loudness Level

Fig. 10: Relation between objective and subjective responses.

tenuation of the overalls value of SPL are used as a control parameter in the design process of ANC system. However, results shown in this section suggested that to obtain an efficient subjective effect of ANC systems, it would be necessary to introduce a psychologically-based parameter, such as loudness level, for the ANC system design process as a control parameter.

4. Experiments 2: Relation between spatial information of noise source and ANC performance

As described in this section, we experimented on space information of the noise source and the performance of ANC earphones using HRTF. Noise source signals were the same as those explained in section 2.4 (white noise, 20 Hz – 20 kHz), and convoluted with HRTF to simulate the situation when the noise source moves clockwise around the listener's head in the horizontal plane. The HRTFs measured using KEMAR[4] are used at a 10 interval degree in azimuth for the convolution. Azimuths were defined from 0 degree (front) to 360 degree clockwise. Two ANC earphones were used in these experiments, and noise reductions of those were given in Fig. 9. Here, the tested earphones have different system configurations. Sample-A has a single-error microphone for ANC at left side of earphone driver, and sample-B has dual microphones at the left and right channels of the earphone.

Therefore, in the case of sample-A, the ANC system generates anti-noise using the signal received at the left channel microphone. It feeds this signal into both left and right channels irrespective of the noise exposed at the right side. The ANC system of sample-B generates the secondary noise sources at left and right channels, independently.

Fig. 11 and Fig. 12 presents examples of noise reduction when noise

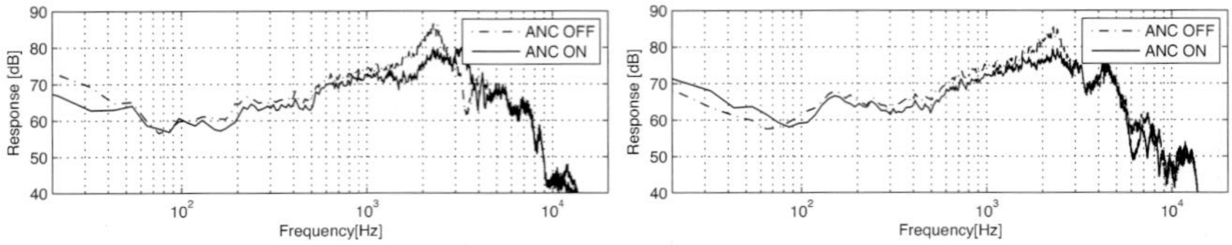

(a) Sample A (b) Sample B

Fig. 11: Noise Reduction Level at the Left channel.

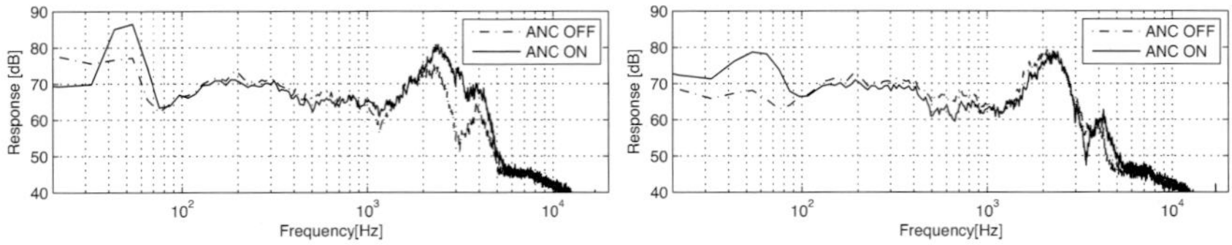

(a) Sample A (b) Sample B

Fig. 12: Noise Reduction Level at the Right channel.

source was simulated at 270 degree azimuth (right side). Solid lines represent the SPL with ANC ON, and dotted lines represent the SPL without ANC (with insulation loss).

Results of sample-A (Fig. 11(a) and Fig. 12(a)) show that ANC provides noise reduction at L-ch (error microphone installed), although SPL increases at the R-ch because of an irrelevant secondary signal that is also generated.

An excellent control sample-B is visible irrespective of the direction of the noise source at both L and R channels.

The SPL at both L and R channel in each noise source azimuth were added based on the theory of binaural summation of loudness,[7] and binaural noise reduction attributable to ANC systems were calculated. Results were presented in Fig. 13.

According to Fig. 13(b), effects of ANC are visible in sample-B earphones, though, binaural SPL remarkably increases attributable to ANC system when noise source directions were from 0 to 100 [deg] in azimuth (see Fig. 13(a)).

5. Discussion

It can be understood easily that ANC earphones with a single error microphone present problems of noise reduction when the spatial characterization in the noise source is considered. Moreover, the noise source direction will strongly influence the system performance. Therefore, it would be necessary

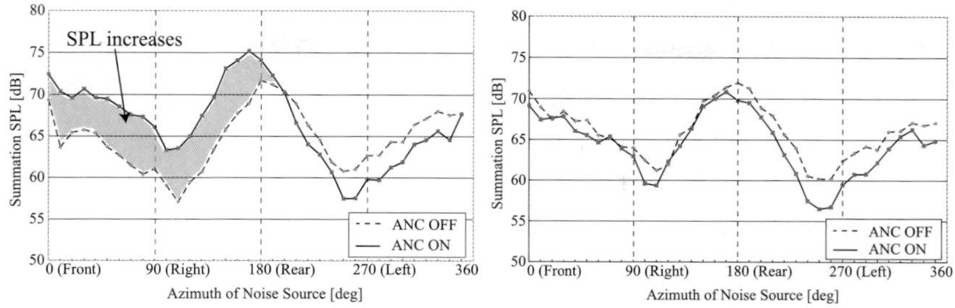

(a) Sample A (b) Sample B

Fig. 13: Binaural Summation of Noise Reduction Level.

to apply new control parameters that consider binaural summation of SPL when several noise sources with a different characteristic exist at a separate position in space. According to the results presented in section 3.3, binaural loudness theory[8–10] might also be an effective for ANC design.

6. Summary

This study conducted ANC assessments using binaural techniques. We applied a c-coupler to simulate the sound pressure level at the listener's eardrum, and introduced equipment to observe insertion loss and noise reduction level of ANC earphones so that we would be able to understand the effects of ANC earphone systems. Results will be useful to elucidate the relation between physical parameters and psychological effects of ANC earphone systems.

References

1. S. Honda and H. Hamada, *Adaptive methods in active control* , chapter 10 (2002).
2. E.A.G. Shaw, *J. Acoust. Soc. Am.* **5(6)**, p1848-1861 (1974).
3. K. Okabe, H. Hamada and T. Miura, *J. Acoust. Soc. Jpn. (E)* **5(2)**, p95-103 (1984).
4. MIT, *HRTF Measurements of a KEMAR Dummy-Head Microphone*, http://sound.media.mit.edu/resources/KEMAR.html, (1994).
5. D. Gaufer, *J. Acoust. Soc. Am.* **120(5)**, p3160 (2006).
6. H. Moller et al., *J. Audio Eng.* **43(4)**, p203-217 (1995).
7. B. Scharf and D. Fishken, *J. Exptl. Psychol.,* **86**, p374 (1970).
8. B.C.J. Moore and B.R. Glasberg, *J. Acoust. Soc. Am.* **121(3)**, p1604-1621 (2007).
9. V.P. Sivonen, *J. Acoust. Soc. Am.* **121(5)** p2852-2861 (2007).
10. V.P. Sivonen and W. Ellermeier, *J. Audio Eng. Soc.* **56(6)** p452-461 (2008).